# MULTICOMPONENT RANDOM SYSTEMS

ADVANCES IN PROBABILITY and Related Topics

Editor: Peter Ney

*Department of Mathematics*
*University of Wisconsin-Madison*
*Madison, Wisconsin*

**Vols. 1 and 2** Advances in Probability and Related Topics, edited by Peter Ney

**Vol. 3** Advances in Probability and Related Topics, edited by Peter Ney and Sidney Port

**Vol. 4** Probability on Banach Spaces, edited by James Kuelbs

**Vol. 5** Branching Processes, edited by Anatole Joffe and Peter Ney

**Vol. 6** Multicomponent Random Systems, edited by R. L. Dobrushin and Ya. G. Sinai

**Other volumes in preparation**

# MULTICOMPONENT RANDOM SYSTEMS

edited by
**R.L. Dobrushin and Ya. G. Sinai**
Institute for Information Transmission Problems
and Landau Institute of Theoretical Physics
Academy of Sciences
Moscow, USSR

contributing editor
**D. Griffeath**

series editor
**Peter Ney**

MARCEL DEKKER, Inc. New York and Basel

Library of Congress Cataloging in Publication Data
Main entry under title:

Multicomponent random systems.

(Advances in probability and related topics ; vol. 6)
Originally published in Russian.
1. Stochastic processes. 2. Stochastic systems. I. Dobrushin, R. L., II. Sinai, IAkov Grigor'evich, III. Griffeath, David. IV. Series.
QA273.A1A4 vol. 6 [QA274] 519.2s [519.2]
ISBN 0-8247-6831-0 80-17688



MARCEL DEKKER, INC.

270 Madison Avenue, New York, New York 10016

Current printing (last digit):

10 9 8 7 6 5 4 3 2 1

PRINTED IN THE UNITED STATES OF AMERICA

# INTRODUCTION TO THE SERIES

Advances in Probability and Related Topics was founded in 1970 to provide a flexible medium for publication of work in probability theory and related areas. The idea was to encourage lucid exposition of topics of current interest, while giving authors greater flexibility on the inclusion of original and expository material and on the lengths of articles than might be available in standard journals. There has not been any particular focus of subject matter, other than its relevance to probability theory.

During the past decade, a variety of series have evolved which now offer the opportunity for this kind of publication. We have therefore decided to modify our format to focus the attention of each volume on a single unified subarea of probability theory, while retaining flexibility in the format of individual articles.

To this end, Volume 4 focused on Probability on Banach Spaces, Volume 5 on Branching Processes, while the present volume is on Multicomponent Random Systems.

We intend to maintain flexible editorial arrangements. The editors for the various volumes will in general be experts in a special area of probability theory and will be responsible for the contents of their particular volume. People who might be interested in editing such a volume are invited to contact the Series Editor.

Peter Ney

# NOTE FROM THE EDITORS

The contributions to this volume are the work of mathematicians connected with the Institute for Information Transmission Problems in Moscow. The original versions were in Russian. Preliminary translations into English were provided by the authors, but in many cases there remained an extensive job of completing the translations. This was done entirely on a volunteer basis by colleagues at American universities who felt that this project was of value. Our heartiest thanks go out to the following people for their assistance in this connection: R. Arratia, M. Bramson, J. Chover, D. Dawson, S. Goldstein, L. Gray, R. Holley, H. Kesten, J. Lebowitz, S. Sawyer, and F. Spitzer. The original suggestion and impetus for the publication of this volume came from Frank Spitzer.

A casual perusal of the pages in the volume will convince the reader that the formulas and notation in some papers are unusually complex. This, together with the considerable separation in space time of the Soviet authors, volunteer translators, local editors, and the typist, made the logistics of producing this volume unusually complicated. The difficult typing task, done from copy of variable quality, was superbly carried out by Mrs. Grace Krewson. We extend our sincere thanks to her, and to Mrs. Emmy Alford for a variety of editorial assistance.

D. Griffeath

Peter Ney

# PREFACE

Over the last decade there has taken shape and, we can say, ripened a new interdisciplinary scientific field--a theory of multicomponent random systems. The main object of study here is multidimensional systems comprising a large number of homogeneous locally interacting components. These components may be of different real nature depending on the field of their application. In physical applications these are atoms in points of crystal lattices; in cybernetic applications, interacting finite-state automata, logic-informational elements, and queue systems; in biological applications, cells, neurons, etc.. It is believed that the range of the real phenomena, the mathematical description of which naturally leads to the models of the discussed type, is extremely wide.

As a branch of applied mathmatics, the theory of multicomponent random systems came into being on the intersection of the theory of probability, statistical physics, information theory, mathematical biology. Some of its basic notions, e.g., the notion of the Markov process with interaction was created in parallel by representatives of all these fields of science. But the most important proved to be the influence of statistical physics, apparently because the scientific experience accumulated by this field is incomparably deeper than that of younger sciences. The decisive factor was the development of mathematically rigourous statistical physics. At first its evolution was motivated by the desire to substantiate logically the fundamental physical notions, however, the gradual transition to a formal level, irrelevant to a direct use of physical intuition, showed that the basic ideas of classical statistical physics are connected with the fact that statistical physics studies one of the classes of multicomponent random systems with local interactions. Then the approach based on the Gibbs distribution became natural for the description of random fields of a general type independent of their nature. Singularities of the phase transition type may be described as jump discontinuities of system (macroscopic—in terms of physics) characteristics under the continuous variation of element (microscopic) characteristics and are equally relevant both to the systems of physical and informational-cybernetic nature. The methods of the non-equilibrium statistical physics are applicable to the description of the dynamics of informational-cybernetic and biological systems etc.

In its mathematical methods the theory of multicomponent random systems has borrowed much from the theory of probability and particularly from the theory of random processes. In its turn, it has lent to the theory of probability the new idea that the transition from one-dimensional random processes to multidimensional random fields provides qualitatively new opportunities. The influence of cybernetic and biological applications essentially widened the range of situations subject to study.

The present book contains a set of original scientific researches devoted to various problems of the theory of dynamic systems and multicomponent random systems, both theoretical and motivated by concrete applications. We believe that all the basic ideas and methods of this theory are reflected here and therefore the reader will get a due notion of its content.

R. L. Dobrushin
Ya. G. Sinai

# CONTENTS

# MULTICOMPONENT RANDOM SYSTEMS

chapter 1

# CLUSTER ESTIMATES FOR GIBBS RANDOM FIELDS AND SOME APPLICATIONS

F. H. Abdulla-Zadeh, R. A. Minlos, and S. K. Pogosian

Many problems of statistical physics are connected with estimating the decay of the correlations of the corresponding Gibbsian random field or with estimating the semi-invariants of a properly chosen system of random variables. Various estimates of this kind arising in connection with different problems of statistical physics and quantum field theory may be found in works by Lebowitz and Penrose [1], Lebowitz [2], Dobrushin [3], Del Grosso [4], Zurbenko [5, 6], Abraham, Gallavotti and Martin-Löf [7], in the book by Ruelle [8], in work by Denau, Souillard and Jagolnitzer [9, 10, 11, 12], Minlos and Pogosian [13], Glimm, Jaffe, Spencer [14, 15, 16], Spencer [17], Spencer, Zirilli [18], Eckman, Magnen and Seneor [19], and finally in the paper by Malyshev in this collection of works.

In this paper we consider two problems which are technically based on the most accurate estimates of semi-invariants, called <u>strong cluster estimates</u> in [12]. The first task is to establish a certain special structure of the transfer matrix in the plane Ising model which we shall call <u>cluster structure</u>. More complicated assumptions concerning the cluster structure of the transfer matrix were introduced and used in works by Minlos and Sinai [20, 21], and in the paper of Fisher and Camp [22] in investigating the spectrum of the transfer-matrix. The hypothesis is plausible that any operator with cluster structure has that special "corpuscular" type of spectrum and corresponding multichannal scattering structure as it has always been supposed for the operator $\exp\{\mathcal{H}\}$ where $\mathcal{H}$ is the Hamiltonian of a multi-particle non-relativistic system with potential interaction (see the works by Faddeev [23], Segal [24, Hepp [25]) or the Hamiltonian of quantum field theory (see the works [16, 17] mentioned above).

The second problem discussed in this work is the study of the asymptotic expansion for the logarithm of the partition function when the volume of region increases. In particular, for regions $\Lambda$ with sufficiently smooth (or piecewise smooth) boundary, the term of the asymptotic expansion following the principal one (of the order of $|\Lambda|$) turns out to be proportional to the surface area of the boundary $\Gamma(\Lambda)$ of the region $\Lambda$ (for lattice systems this result was obtained by Dobrushin [26]). The next terms depend on some geometrical characteristics of $\Gamma(\Lambda)$. This decomposition is of special interest and might be useful in solving some problems of statistical physics (for example, for obtaining an asymptotic expansion of the particle number distribution in a large region $\Lambda$).

In §1 we define some functions characterizing the decay of correlations, which we call cluster functions (semi-invariants, group (cluster) functions and Ursell functions), determine relations between them (and the formulas connecting these functions with moment functions of the field) and give the necessary estimates of cluster functions for the case of Gibbsian point field (on the lattice $Z^\nu$ or in the space $R^\nu$, $\nu = 1, 2, 3, \ldots$).

In § 2 we formulate the general notion of a cluster operator and prove the theorem that the transfer matrix in the plane Ising model for sufficiently high temperature is a cluster operator. This result in its idea goes back to [20]. Here we use some constructions from [20] and also prove assertions formulated in [20] without proof.

In § 3 we formulate a general theorem about asymptotic decomposition for the logarithm of the partition function and give the sketch of a proof. A detailed proof which is rather bulky, may be found in the work by one of the authors [27].

It should be also noted that the present paper is a detailed version of the report, made by two of the authors (by Minlos and Pogosian) at the IV International Symposium of information theory (June 1976).

## I. MOMENTS, CLUSTER FUNCTIONS AND THEIR ESTIMATES

Here, in the case of point fields (discrete and continuous), we give a definition of moment (correlation) functions, semi-invariants, group functions (also called truncated correlation functions) and in the case of Gibbs fields we define also Ursell functions. It is natural that all these functions, measuring the dependence between separate groups (clusters) of variables, should be called <u>cluster functions</u>. Finally, in the case of the Gibbs fields we give some useful estimates of cluster functions.

### 1. <u>The case of a lattice field</u>

We will denote by $(C(Z^\nu), \mathfrak{C}(Z^\nu))$ the measurable space of all subsets of the integer-valued $\nu$-dimensional lattice $Z^\nu$, $\nu = 1,2,3\ldots$ with the $\sigma$-field $\mathfrak{C}(Z^\nu)$ generated by cylinder subsets in $C(Z^\nu)$ (see [28] for details). Then by $\mathcal{K}(Z^\nu)$ we denote the collection of all non-negative integer-valued finite functions on $Z^\nu$, and by $\mathcal{L}(Z^\nu)$

we denote the collection of all subsets of $Z^\nu$. Taking into account the canonical inclusion

$$\mathcal{L}(Z^\nu) \to \mathcal{K}(Z^\nu) : s \to m_s \in \mathcal{K}(Z^\nu) \tag{1.1}$$

where $s \in \mathcal{L}(Z^\nu)$ and $m_s$ is the characteristic function of the set s, we will consider that $\mathcal{L}(Z^\nu) \subset \mathcal{K}(Z^\nu)$. By $\Lambda(Z^\nu)$ we denote the space of all (complex valued) finite functions on $Z^\nu$.

Let $\eta = \{\eta_t : t \in Z^\nu\}$ be the family of measurable bounded functions on the space $\{C(Z^\nu), \mathbb{C}(Z^\nu)\}$ and P be some generalized normed measure on the same space, i. e. a complex-valued countably additive set function of bounded variation such that

$$P[C(Z^\nu)] = 1 \tag{1.2}$$

Assume that for all sufficiently small functions $\lambda \in \Lambda(Z^\nu)$:

$$\max_{t \in Z^\nu} |\lambda(t)| < \epsilon \tag{1.3}$$

where $\epsilon > 0$ is some constant, the integral

$$Z(\lambda) = \int_{C(Z^\nu)} \exp\{ \sum_{t \in Z^\nu} \lambda(t)\eta_t(c)\} \, dP(c) \tag{1.4}$$

is defined. The expansion (convergent in the domain defined by (1.3))

$$Z(\lambda) = \sum_{m \in K(Z^\nu)} \frac{\rho(m)}{m!} \prod_{t \in Z^\nu} (\lambda(t))^{m(t)} \tag{1.5}$$

where $m! = \prod_{t \in Z^\nu} m(t)!$, defines some function $\rho(m)$ on the set $\mathcal{K}(Z^\nu)$ which is called <u>moment function</u> of the family $\eta$. The restriction of this function to the space $\mathcal{L}(Z^\nu) \subset \mathcal{K}(Z^\nu)$:

$$\rho(s) \equiv \rho(m_s) \tag{1.6}$$

is sometimes called the correlation function of the family $\eta$.

Note that if the integral (1. 4) is defined in the domain (1. 3), then the integral

$$\Xi(\lambda) = \int_{C(Z^\nu)} \Pi (1 + \lambda (t)\eta_t(c))dP(c) \tag{1. 7}$$

is also convergent and has the following expansion

$$\Xi(\lambda) = \sum_{s\in\mathcal{L}(Z^\nu)} \rho(s) \prod_{t\in s} \lambda (t) \tag{1. 8}$$

Assume also that the functionals $\ln Z(\lambda)$ and $\ln \Xi(\lambda)$ are analytic for small $\lambda$ and have the expansions

$$\ln Z(\lambda) = \sum_{m\in\mathcal{K}(Z^\nu)} \frac{\omega(m)}{m!} \prod_{t\in Z^\nu} (\lambda (t))^{m(t)} \tag{1. 9}$$

and

$$\ln \Xi(\lambda) = \sum_{m\in\mathcal{K}(Z^\nu)} \frac{\chi(m)}{m!} \prod_{t\in Z^\nu} (\lambda (t))^{m(t)} \tag{1.10}$$

The functions $\omega(m)$ and $\chi(m)$ on $\mathcal{K}(Z^\nu)$ are respectively called the semi-invariant and group (cluster) functions.

Between the functions $\rho(m)$, $\omega(m)$ and $\chi(m)$ we have the following relations:

$$\rho(\overline{m}) = \overline{m}! \sum_{g:\sum_m g(m)\cdot m = \overline{m}} \frac{1}{g!} \prod_{m\in\mathcal{K}(Z^\nu)} \left(\frac{\omega(m)}{m!}\right)^{g(m)} \tag{1.11}$$

where the sum runs over non-negative integer-valued finite functions $g$ on the space $\mathcal{K}(Z^\nu)$ satisfying the condition $\sum_m g(m)\cdot m = \overline{m}$. Then

$$\omega(\overline{m}) = \overline{m}! \sum_{g:\sum_m g(m)\cdot m=\overline{m}} \frac{(-1)^{|g|}}{g!}(|g| - 1) \left(\frac{\rho(m)}{m!}\right)^{g(m)} \tag{1.12}$$

where $|g| = \sum_m g(m)$.

Finally

$$\chi(\bar{m}) = \bar{m}! \sum_{g:\sum_s g(s)\cdot m_s = \bar{m}} \frac{1}{g!} \prod_{s\in\mathcal{L}(Z^\nu)} (\rho(s))^{g(s)} \tag{1.13}$$

where in (1.13) by $g$ we denote non-negative integer-valued finite function on the space $\mathcal{L}(Z^\nu)$.

Let $\chi(s) = \chi(m_s)$ and $\omega(s) = \omega(m_s)$ denote the restrictions of functions $\chi$ and $\omega$ to the space $\mathcal{L}(Z^\nu)$ Then by virtue of (1.11), (1.12) and (1.13) we obtain

$$\rho(s) = \sum_{\substack{(s_1,\dots,s_k:\ \cup_1^k s_i = s,\\ s_i\cap s_j = \emptyset,\ i\neq j,\ k=1,2,\dots}} \prod_{i=1}^{k} \omega(s_i), \quad s \neq \emptyset$$

$$\rho(\emptyset) = 1 \tag{1.14}$$

where the sum runs over all non-trivial partitions of the set $s\in\mathcal{L}(Z^\nu)$. Similarly

$$\omega(s) = \sum_{\substack{(s_1,\dots,s_k):\ \cup_1^k s_i = s,\\ s_i\cap s_j = \emptyset,\ i\neq j,\ k=1,2,\dots}} (-1)^k (k-1)! \prod_{i=1}^{k} \rho(s_i), \quad s \neq \emptyset$$

$$\omega(\emptyset) = 0 \tag{1.15}$$

Finally

$$\chi(s) = \omega(s), \qquad s \in \mathcal{L}(Z^\nu) \tag{1.16}$$

Consider then the family of functions on $C(Z^\nu)$ of the form

$$\eta^0 = \{\eta_t^0, \ t \in Z^\nu\}, \qquad \eta_t^0(c) = \begin{cases} 1, & t \in C \\ 0, & t \notin C \end{cases} \quad ; \ c \in C(Z^\nu) \tag{1.17}$$

By $\rho^0, \omega^0$ and $\chi^0$ we denote the moment, semi-invariant and group functions of the family $\eta^0$. Let $U_{(s)}$, $s \in L(Z^\nu)$ be some (possibly complex) function (interaction potential), such that $U(\emptyset) = 0$. Then for any set $s \in \mathfrak{L}(Z^\nu)$, the quantity

$$H(s) = \sum_{s' \subseteq s} U(s'), \qquad s \in \mathfrak{L}(Z^\nu) \tag{1.18}$$

where the sum runs over all subsets $s' \subset s$, is called the energy of the set $s$. Consider for any $\lambda \in \Lambda(Z^\nu)$ the sum

$$Q_U(\lambda) = \sum_{s \in \mathfrak{L}(Z^\nu)} \exp\{-H(s)\} \prod_{t \in s} \lambda(t) \tag{1.19}$$

which is called the partition function. Let us assume the convergence (for sufficiently small $\lambda \in \Lambda(Z^\nu)$) of the expansion

$$\ln Q_U(\lambda) = \sum_{m \in K(Z^\nu)} \frac{\varphi_U(m)}{m!} \prod_{t \in Z^\nu} (\lambda(t))^{m(t)} \tag{1.20}$$

Here the function $\varphi_U(m)$ on the space $\mathfrak{K}(Z^\nu)$ is called the Ursell function (for the potential $U$).

Next let $P_U$ denote the Gibbs measure (maybe complex) on the space $\{C(Z^\nu), \mathfrak{C}(Z^\nu)\}$, defined by the potential $U$ (see [29]; we assume the existence and the uniqueness of such a measure)* and

---

* Usually Gibbs measures are considered in the case of real potential $U(s)$, when they are probability measures. In some cases, however (one of them will be dealt with below) it is possible to define a Gibbs measure (may be non $\sigma$-additive) also for a complex potential $U(s)$.

let $\overset{0}{\rho}_U$, $\overset{0}{\omega}_U$ and $\overset{0}{\chi}_U$ denote the moment, semi-invariant and group functions of the family (1.17) with respect to this measure. It is easy to check, that the function $\overset{0}{\chi}_U$ is connected with the Ursell function by the equation

$$\overset{0}{\chi}_U(m) = \sum_{n \in K(Z^\nu)} \frac{\varphi_U(m+n)}{n!} \tag{1.21}$$

Let the translation invariant finite pair potential

$$U(s) = \begin{cases} \beta\mu & |s| = 1 \\ \gamma V(t_1 - t_2), & |s| = 2,\ s = (t_1, t_2), t_1 \neq t_2,\ t_1, t_2 \in Z^\nu \\ 0 & \text{otherwise} \end{cases} \tag{1.22}$$

be given, where $V(\tau)$, $\tau \in Z^\nu$, $\tau \neq 0$ is a finite even function on $Z^\nu \setminus \{0\}$ and the parameters $\beta$ and $\mu$ satisfy the conditions

$$0 < \beta < \beta_0, \qquad -\infty < \mu < \infty \tag{1.23}$$

where $\beta_0 = \beta_0(V) > 0$. Then, for the semi-invariant function $\overset{0}{\omega}_U(m)$ the estimate

$$|\overset{0}{\omega}_U(m)| < K \cdot m! B^{|m|} (C\beta)^{\operatorname{diam}\{s(m)\}} \tag{1.24}$$

is valid, where K, V, C are some constants depending on the function V, $s(m) = \operatorname{supp} m \in \mathcal{L}(Z^\nu)$ (see [11, 12]). The analogous estimate is valid for the functions $\overset{0}{\chi}_U$ and $\overset{0}{\varphi}_U$.

Note, that all estimates and constructions remain true also in the case of an arbitrary subset $I \subseteq Z^\nu$.

## 2. The case of a continuous point field

Let $\{C(R^\nu), \mathfrak{C}(R^\nu)\}$ denote the space of locally finite subsets of the Euclidean space $R^\nu$, $\nu = 1, 2, 3, \ldots$ with the $\sigma$-field $\mathfrak{C}(R^\nu)$

generated by the cylinder sets (see [28]). By $\mathcal{L}(R^\nu)$ we denote the collection of all finite subsets $s \subset R^\nu$. In $\mathcal{L}(R^\nu)$ one can naturally introduce the $\sigma$-field of the Borel sets and the measure of Lebesgue ds (see [28]). Then by $\Lambda(R^\nu)$ we denote the collection of all finite measurable bounded functions on $R^\nu$.

Let $\eta = \{\eta_t, t \in R^\nu\}$ be the family of measurable bounded functions on $\{C(R^\nu), \mathbb{C}(R^\nu)\}$, such that for any $c \in C(R^\nu)$ the function $f_c(t) = \eta_t(c)$, $t \in R^\nu$ has a locally finite support.

Let, then, P be some measure (maybe complex) on the space $\{C(R^\nu), \mathbb{C}(R^\nu)\}$ such that

$$P[C(R^\nu)] = 1 \tag{1. 25}$$

Assume that for all sufficiently small $\lambda \in \Lambda(R^\nu)$: $\sup_{t \in R^\nu} |\lambda(t)| < \epsilon$, where $\epsilon > 0$ is some number, the integral

$$\Xi(\lambda) = \int_{C(R^\nu)} \prod_{t \in R^\nu} (1 + \lambda(t)\eta_t(c)) dP(c) \tag{1. 26}$$

is defined and has the following representation

$$\Xi(\lambda) = \int_{\mathcal{L}(R^\nu)} \rho(s) \left( \prod_{t \in s} \lambda(t) \right) ds \tag{1. 27}$$

where $\rho$ is some measurable function on $\mathcal{L}(R^\nu)$ which is called correlation function (of the family $\eta$). Let then for sufficiently small $\lambda$, the representation

$$\ln \Xi(\lambda) = \int_{\mathcal{L}(R^\nu)} \chi(s) \left( \prod_{t \in s} \lambda(t) \right) ds \tag{1. 28}$$

be valid, where $\chi(s)$ is some measurable function on the space $\mathcal{L}(R^\nu)$ and is called the <u>group</u> (cluster) <u>function</u> of the family $\eta$. The relations between correlation and group functions exactly coincide with the formulas (1. 14) and (1. 15). In the case of the family $\eta^0 = \{\eta^0_t, t \in R^\nu\}$, where

$$\eta^0_t(c) = \begin{cases} 1, & t \in c \\ 0, & t \notin c \end{cases} \tag{1. 29}$$

we denote the correlation and group functions by $\rho^0$ and $\chi^0$.

Consider a measurable function $U(s)$, $s \in \mathcal{L}(R^\nu)$ (interaction potential) on the space $\mathcal{L}(R^\nu)$. Similar to (1. 18) we introduce the energy $H(s)$ of a set $s \in \mathcal{L}(R^\nu)$ and the integral

$$Q_U(\lambda) = \int_{L(R^\nu)} \exp\{-H(s)\}(\prod_{t \in s} \lambda(t))ds \tag{1. 30}$$

where $\lambda \in \Lambda(R^\nu)$. $Q_U(\lambda)$ is called the partition function. Let the following representation

$$\ln Q_U(\lambda) = \int_{\mathcal{L}(R^\nu)} \varphi_U(s)(\prod_{t \in s} \lambda(t))ds \tag{1. 31}$$

be valid (for sufficiently small $\lambda$), where $\varphi_U(s)$, $s \in \mathcal{L}(R^\nu)$ is some measurable function on the space $L(R^\nu)$ which is called the <u>Ursell function</u> (for the potential $U$).

Now let $P_U$ be the Gibbs measure on the space $\{C(R^\nu), \mathfrak{C}(R^\nu)\}$, defined by the potential $U$, and let $\chi^0_U$ be the group function of the family $\eta^0$ with respect to the measure $P_U$ Then it is easy to check that we have the following equality

$$\chi^0_U(s) = \int_{\mathcal{L}(R^\nu)} \varphi_U(s \cup \tilde{s})\, d\tilde{s} \tag{1. 32}$$

Finally, consider a translation invariant pair potential $U(s)$

$$U(s) = \begin{cases} \beta\mu, & |s| = 1 \\ \beta V(t_1 - t_2), & |s| = 2, \quad s = (t_1, t_2), t_1 \neq t_2, t_1, t_2 \in R^\nu \\ 0 & |s| > 2 \end{cases} \tag{1.33}$$

where $V(\tau)$, $\tau \in R^\nu$ is a function, sufficiently rapidly decreasing at infinity, such as

$$|V(\tau)| < \frac{c}{|\tau|^\gamma}, \quad |\tau| > a > 0, \quad \tau \in R^\nu \tag{1.34}$$

where $\gamma$ is sufficiently large. Then, for sufficiently large negative $\mu$:

$$\beta\mu < -C(\beta), \quad C(\beta) > 0 \tag{1.35}$$

the estimate

$$|\overset{0}{\chi}_U(s)| \leq KB^{|s|} \sum_\Gamma \prod_{(t_1, t_2) \in \Gamma} \tau(t_1 - t_2), \quad s \in \mathcal{L}(R^\nu) \tag{1.36}$$

is valid for the group function (see [12]), where $|s|$ is the number of the points of $s$, $K$ and $B$ are constants, $\tau(t)$, $t \in R^\nu$ is some non-negative function rapidly decreasing at infinity; the sum $\Sigma_\Gamma$ runs over all trees (i. e. connected graphs without closed loops) joining the points $t_1, t_2 \in s$. An estimate similar to (1.36) is valid also for the function $\varphi_U$.

It should be noted that the structure of a smooth manifold may be naturally introduced into space $\mathcal{L}(R^\nu)$. In the case when the function $V(\tau)$ in (1.33) is sufficiently smooth (namely when the derivatives are rapidly decreasing at infinity and have not more than a power singularity at zero) the derivatives $(D^m\varphi_U)(s)$ and $(D^m\overset{0}{\chi}_U)(s)$, $s \in \mathcal{L}(R^\nu)$ are defined for all non-negative integer-valued functions $m = \{m_\alpha(t),\ t \in s,\ \alpha = 1, 2, \ldots,\}$ defined on a set $s \times [1, \ldots, \nu]$ and such that $|m| = \sum_{\alpha, t} m_\alpha(t) \leq M$, where $M = M(V)$ is some constant depending on the degree of smoothness of the potential $U$. These derivatives satisfy for sufficiently large negative $\mu$ estimates similar to (1. 36) (see [13])

$$|(D^m \varphi_U^0(s)| \leq \tilde{K}\tilde{B}^{|s|} \sum_{\Gamma} \prod_{(t_1, t_2) \in \Gamma} \tilde{\tau}(t_1 - t_2) \tag{1. 37}$$

where the constants $\tilde{K}$, $\tilde{B}$ and the function $\tilde{\tau}$ depend on the order of the derivative $|m|$. Similar estimates are also valid for the derivatives $D^m \chi_U^0$ of the group function. As in the lattice case all the constructions and estimates mentioned above remain true if instead of the space $R^\nu$ we consider any region $G \subseteq R^\nu$.

## II. CLUSTER STRUCTURE OF THE TRANSFER MATRIX IN THE ISING MODEL

### 1. The operator with cluster structure

Let $L_2(\mathcal{L}(Z^\nu))$ be the Hilbert space of functions $\psi(s)$ on $\mathcal{L}(Z^\nu)$ such that

$$\|\psi\|_{L_2} = \sum_{s \in \mathcal{L}(Z^\nu)} |\psi(s)|^2)^{1/2} < \infty \tag{2. 1}$$

On the space $L_2(\mathcal{L}(Z^\nu))$ there acts the unitary representation $(T_t, t \in Z^\nu)$ of the group $Z^\nu$ by the formula

$$(T_t\psi)(s) = \psi(s - t), \quad s \in \mathcal{L}(Z^\nu), \quad t \in Z^\nu \tag{2. 2}$$

where $s - t$ is the set $s$ translated by the vector $-t$. It is evident that for any operator $A$ acting on $L_2(\mathcal{L}(Z^\nu))$

$$(A\varphi)(s) = \sum_{s' \in \mathcal{L}(Z^\nu)} a(s, s')\varphi(s') \tag{2. 3}$$

and commuting with the operators $(T_t, \ t \in Z^\nu)$, we have the equality

$$a(s+t, \ s'+t) = a(s, s'), \quad s, s' \in \mathcal{L}(Z^\nu), \quad t \in Z^\nu$$

An operator A in $L_2(\mathcal{L}(Z^\nu))$ commuting with the operators $(T_t, t \in Z^\nu)$ will be called a cluster operator, if its kernel $a(s, s')$, $s, s' \in \mathcal{L}(Z^\nu)$ has the form

$$a(s, s') = \sum_{\substack{\{s_i, s_i'\}_1^k : \bigcup_1^k s_i = s, \bigcup_1^k s_i' = s', \\ s_i \cup s_j = s_j' = \emptyset, \ i \neq j = 1, \ldots, k; \ k = 1, 2, \ldots}} \prod_{i=1}^{k} b(s_i, s_i') \tag{2. 4}$$

where the sum runs over all non-trivial partitions of the pair $(s, s')$ into pairs of sets $(s_i, s_i')$, $i = 1, \ldots, k$, $k = 1, 2, \ldots$ and the function $b(x, x')$ on $\mathcal{L}(Z^\nu \times \mathcal{L}(Z^\nu)$ is translation invariant

$$b(s, s') = b(s+t, s'+t), \quad s, s' \in \mathcal{L}(Z^\nu), \quad t \in Z^\nu \tag{2. 5}$$

and satisfies the cluster estimate

$$|b(s, s')| < AC^{|s| + |s'|} \exp\{-\alpha L(s \cup s')\} \tag{2. 6}$$

where $A, C, \alpha > 0$ are constant and $L(s \cup s')$ is the minimal length of a tree joining the points of $s \cup s'$. The quantities $b(s, s')$, $s, s' \in \mathcal{L}(Z^\nu)$ are called clusters of the cluster operator A. Evidently, the structure of the kernel of the operator A which we have described implies, as was mentioned in the introduction, some notable properties of its spectrum and scattering character, but we will not dwell on it here. Some remarks, regarding this, may be found in [20] and [22].

Let then $\mathcal{H}$ be the Hilbert space, on which acts the unitary representation $(\hat{T}_t, \ t \in Z^\nu)$ of the group $Z^\nu$. We will say that the operator A on the space $\mathcal{H}$, commuting with operators $\hat{T}_t$, has a cluster structure, if there exists a unitary mapping of $\mathcal{H}$ onto $L_2(\mathcal{L}(Z^\nu))$ which maps the operators $(\hat{T}_t,\ t \in Z^\nu)$ into operators (2. 2) and the operator A into a cluster operator.

2. Now we recall the definition of the limit transfer matrix in the Ising model, i. e. the model with pair potential $U_{Is}$ of the form (1. 22), where

$$V(\tau) = \begin{cases} \epsilon, & |\tau| = 1 \\ 0, & |\tau| > 1 \end{cases} \tag{2. 7}$$

and $P^{Is}_{\beta,\mu}$ is the probability distribution of the Gibbs point field on the lattice $Z^\nu$ with this potential. Let $\pi_{\beta,\mu}$ denote the restriction of this field to the sublattice $(Z^{\nu-1}_0 \subset Z^\nu$, where

$$(Z^{\nu-1}_0 = \{(n_1, \ldots, n_\nu) \in Z^\nu : n_\nu = 0\} \tag{2. 8}$$

Then, considering the lattice $Z^\nu$ composed of parallel horizontal sublattices $(Z^{\nu-1})_k$, $k = 0, \pm 1, \pm 2, \ldots$ where

$$(Z^{\nu-1})_k = \{(n_1, \ldots, n_k) \in Z^\nu : n_\nu = k\} \tag{2. 9}$$

we will define the conditional distribution $P^{Is}_{\beta,\mu}(\cdot/\tilde{\sigma})$, $\tilde{\sigma} \subset (Z^{\nu-1})_{-1}$ for field values on the lattice $(Z^{\nu-1})_0$ under the condition that a configuration $\tilde{\sigma}$ on the lattice $(Z^{\nu-1})_{-1}$ is given. The operator of conditional mathematical expectation in $L_2(C(Z^\nu), \pi_{\beta,\mu})$

$$(P^{Is}_{\beta,\mu} f)(\tilde{\sigma}) = \int_{C(Z^{\nu-1})} f(\sigma) dP^{Is}_{\beta,\mu}(\sigma/\tilde{\sigma}) \tag{2. 10}$$

is a self-adjoint bounded operator. This operator is called a <u>limit</u>

transfer matrix. It is evident that in the case of a translation invariant Gibbs distribution Pu operators of translation $(T_t, t \in Z^\nu)$

$$(T_t f)(\tilde{\sigma}) = f(\sigma - t), \qquad \sigma \in C(Z^{\nu-1}) \tag{2. 11}$$

are unitary operators in the space $L_2(C(Z^{\nu-1}), \pi_{\beta,\mu})$ and that the operator $P^{Is}_{\beta,\mu}$ commutes with all operators $T_t$, $t \in Z^{\nu-1}$. Later on, without mentioning in particularly, we will suppose that parameters $\beta$ and $\mu$ belong to a domain

$$W_{\beta_0, c_0} = \{0 < \beta < \beta_0; \ |\mu| < c_0 \beta^{-1}\} \tag{2. 12}$$

where $\beta_0, c_0$ are sufficiently small constants such that all estimates given below will be uniform in this domain.

Theorem 2. 1. Let $\nu = 2$ and $\beta_0, c_0$ be sufficiently small. Then for any $\beta, \mu \in W_{\beta_0, c_0}$ the operator $\hat{P}^{Is}_{\beta,\mu}$ has a cluster structure.

Proof: Let $(Z^1)_- \subset Z^1 \equiv (Z^1)_0 \subset Z^2$ denote a negative half-lattice of $Z^1$. Consider for arbitrary subset $\sigma_- \subset (Z^1)_-$ the probability distributions

$$\begin{aligned} P_1(\sigma_-) &= \pi_{\beta,\mu}(\{\sigma_+ \subset Z^1 \setminus (Z^1)_- : 0 \in \sigma_+\}/\sigma_- \\ P_0(\sigma_-) &= \pi_{\beta,\mu}(\{\sigma_+ \subset Z^1 \setminus (Z^1)_- : 0 \notin \sigma_+\}/\sigma_- \end{aligned} \tag{2. 13}$$

where $\pi_{\beta,\mu}(\cdot/\sigma_-)$ is the conditional probability distribution on the space $C(Z^1 \setminus (Z^1)_-)$, under the condition that $\sigma_- \subset (Z^1)_-$ is given, generated by the distribution $\pi_{\beta,\mu}$ on $C(Z^1)$. Consider a function on $C(Z^1)$

$$U_0(\sigma) = (-1)^{\eta_0(\sigma)} \left[\frac{p_1(\sigma_-)}{p_0(\sigma_-)}\right]^{\frac{1-2\eta_0(\sigma)}{2}} \tag{2. 14}$$

where $\sigma_- = (Z^1)_- \cap \sigma$ and the function $\eta_0(\sigma)$, $\sigma \subset Z^1$ is defined by the formula (1.17).

Lemma 2.2. The function $U_0(\sigma)$ is representable as

$$U_0(\sigma) = C_{\emptyset} + \sum_{I \neq \emptyset, I \subset (Z^1)_- \cup \{0\}} C_I \eta_I(\sigma) \tag{2.15}$$

where $\eta_I(\sigma) = \prod_{t \in I} \eta_t(\sigma)$ and constants $C_I$ satisfying the estimate

$$|C_I| < K(C \cdot \beta)^{|I \quad \{0\}|} (D \cdot \beta^{\mathrm{diam}\, \{I \cup \{0\}\}} \tag{2.16}$$

and

$$C_{\{0\}} > d > 0 \tag{2.17}$$

where K, C, D and d are constants.

Proof: Consider a square $Q_N \subset Z^2$

$$Q_N = \{(\eta_1, \eta_2) \in Z^2 : |\eta_i| \quad N, \ i = 1, 2\} \tag{2.18}$$

Let $P^{Is}_{\beta,\mu,N}$ denote the Gibbs distribution (with Ising potential) for the point field in the square $Q_N$ (with empty boundary conditions) and let $P_0^N(\sigma^N)$ and $P_1^N(\sigma^N)$, where $\sigma^N = \sigma_- \cap Q_N$, $\sigma_- \subset ((Z^1)_0)_-$, denote the conditional probabilities generated by the distribution $P^{Is}_{\beta,\mu,N}$ and defined similar to the probabilities $P_0, P_1$ (see (2.13)).

For an arbitrary set $\tilde{\sigma} \subset (Z^1)_- \cup \{0\}$ denote by $w(\tilde{\sigma}) \subset Z^2 \setminus [((Z^1)_0)_- \cup \{0\}]$ a collection of points $t \in Z^2 \setminus [((Z^1)_0)_- \cup \{0\}]$ neighboring with one of the points of $\tilde{\sigma}$ and consider the function

$$\zeta_{\tilde{\sigma}}(t) = \begin{cases} 1, & t \in w(\tilde{\sigma}) \\ 0, & t \in w(\tilde{\sigma}) \end{cases} \tag{2.19}$$

It is easy to see that

$$p_0^N(\sigma_-^N) = \Xi^{-1} \sum_{c \subset Q_N [((Z^1)_0)_- \cup \{0\}} e^{\beta(\mu N(c) - H(c))} \prod_{t \in c} \exp\{-\beta \zeta_{\sigma^N}(t)\}$$

$$p_1^N(\sigma_-^N) = \Xi^{-1} e^{\beta\mu} \sum_{c \subset Q_N [((Z^1)_0)_- \cup \{0\}} e^{\beta(\mu N(c) - H(c))} \prod_{t \in c} \exp\{\beta \zeta_{\sigma^N \cup \{0\}}(t)\}$$

$$(2.20)$$

Here $H(c)$, $c \subset Q_N [((Z^1)_0)_- \cup \{0\}]$ is the energy of the configuration $c$ and $\Xi$ is the partition function. It is immediate from (2.20) that

$$\frac{p_1^N(\sigma_-^N)}{p_0^N(\sigma_-^N)} = e^{\beta\mu} \exp\{ \sum_{\substack{m \in K(Z^2): \\ 0 \in \operatorname{supp} n \subset w(\sigma_-^N \cup \{0\})}} \frac{\omega^N(m)}{m!} (-\beta)^{|m|} \} \tag{2.21}$$

where $\omega^N(m)$, $m \in K(Z^2)$ are semi-invariants of quantities $\{\eta_t^0, t \in Q_N [((Z^1)_0)_- \cup \{0\}]\}$ computed with respect to the Gibbs field in $Q_N [((Z^1)_0) \cup \{0\}]$, and the sum in (2.21) runs over all functions $m \in K(Z^2)$ such that $\operatorname{supp} m \subset w(\sigma_-^N \cup \{0\})$ and $\operatorname{supp} m \cup w(\{0\}) \neq \emptyset$. It follows from the result of [8, 30] that for sufficiently small $\beta$ and $N \to \infty$

$$p_j^N(\sigma_-^N) \to p_j(\sigma_-), \quad \sigma_- \subset (Z^1)_-, \quad j = 1, 2,$$

$$\omega^N(m) \to \omega(m), \quad m \in K(Z^2) \tag{2.22}$$

where $\omega$ are semi-invariants of the quantities $\{\eta_t^0, t \in Z^2 \setminus [((Z^1)_0)_- \cup \{0\}]$ computed with respect to the Gibbs distribution in $Z^2 \setminus [((Z^1)_0)_- \cup \{0\}]$. Then according to (1.24) we have

$$\sum_{\substack{m \in K(Z^2): \\ 0 \in \operatorname{supp} m \subset w(\sigma_-^N \cup \{0\}}} \frac{\omega^N(m)}{m!} (-\beta)^{|m|} \to \sum_{\substack{m \in K(Z^2): \\ 0 \in \operatorname{supp} m \subset w(\sigma_-^N \cup \{0\}}} \frac{\omega(m)}{m!} (-\beta)^{|m|} \tag{2.23}$$

Thus using (2. 21), (2. 22) and (2. 23) we obtain that

$$\frac{p_1(\sigma_-)}{p_0(\sigma_-)} = e^{\beta\mu} \exp\{ \sum_{\substack{m \in K(Z^2): \\ 0 \in \operatorname{supp} m \subset w(\sigma_- \cup \{0\}}} \frac{\omega(m)}{m!} (-\beta)^{|m|} \} \tag{2.24}$$

For each function $m \in K(w((Z^1)_- \cup \{0\}))$ we denote by $I(m) \subseteq (Z^1)_- \cup \{0\}$ a minimal set such that

$$\operatorname{supp} m \subset w(I(m)) \tag{2.25}$$

and by $\tau_I$, $I \subset (Z^1)_-$ we denote the quantity

$$\tau_I = \sum_{m:\, I(m) \subset I \cup \{0\}} \frac{\omega(m)}{m!} (-\beta)^{|m|} \tag{2.26}$$

From the estimate (1. 24) it is easy to obtain that for sufficiently small $\beta_0$

$$|\tau_I| \leq K_1 \beta (C_1 \cdot \beta)^{\operatorname{diam}\{I \cup \{0\}\}} \tag{2.27}$$

where $K_1, C_1$ are positive constants. Hence

$$\frac{p_1(\sigma_-)}{p_0(\sigma_-)} = e^{\beta\mu} \exp\{ \sum_{I \in \mathcal{L}(Z^1)_-} \tau_I \eta_I(\sigma) \} \tag{2.28}$$

Consider then a functional

$$\Gamma(\lambda) = \exp\{ \frac{1}{2} \sum_{I \in \mathcal{L}(Z^1)_-} \tau_I (\prod_{t \in I} \lambda(t) \} \tag{2.29}$$

where $\lambda$ is an arbitrary complex bounded function defined on $(Z^1)_-$ and such that

$$\sup_{t \in (Z^1)_-} |\lambda(t)| < \frac{1}{2C_1\beta} \tag{2.30}$$

It is easy to compute that

$$\Big| \sum_{I \subset (Z^1)_-} \tau_I\Big( \prod_{t \in I} \lambda(t)\Big) \Big| < C_2 \tag{2. 31}$$

where $C_2$ is some constant. Then

$$\frac{\delta \Gamma(\lambda)}{\delta \lambda(t_0)} = \frac{1}{2} \exp\Big\{\frac{1}{2} \sum \tau_I\Big(\prod_{t \in I} \lambda(t)\Big\} \cdot \sum_{I: t_0 \in I} \tau_I\Big(\prod_{t \in I \setminus \{t_0\}} \lambda(t)\Big) \tag{2. 32}$$

and for any $\lambda$ satisfying (2. 30) the estimate

$$\Big| \frac{\delta \Gamma(\lambda)}{\delta \lambda(t_0)} \Big| < K_2 \beta (C_3 \cdot \beta)^{|t_0|} \tag{2. 33}$$

is valid, where $K_2$ and $C_3$ are absolute constants. Since $\Gamma(\lambda)$ is analytic with respect to $\lambda$ in the domain (2. 30) we have

$$\beta(\lambda) = \gamma_0 + \sum_{m \in K((Z^1)_-)} \frac{\gamma(m)}{m!} \prod_{t \in (Z^1)_-} (\lambda(t))^{m(t)} \tag{2. 34}$$

where

$$\gamma_0 = e^{1/2 \tau_\emptyset}$$

$$\gamma_m = \frac{\delta^{|m|} \Gamma(\lambda)}{\prod_{t \in (Z^1)_-} \delta^{m(t)} \lambda(t)} \Bigg|_{\lambda \equiv 0} = \frac{\delta^{|m'|}}{\prod_{t \in (Z^1)_-} \delta^{m'(t)} \lambda(t)} \cdot \frac{\delta \Gamma(\lambda)}{\delta \lambda(t_{ex})} \Bigg|_{\lambda \equiv 0} \tag{2. 35}$$

Here $t_{ex}$ is a point of the set $\operatorname{supp} m \subset (Z^1)_-$ with the smallest abscissa, $m' \in K((Z^1)_-)$ is such that $m'(t) = m(t)$, $t \in Z^1$, $t \neq t_{ex}$ and $m'(t_{ex}) = m(t_{ex}) - 1$. Then using Cauchy's formula we have

$$\gamma(m) = \frac{m'!}{(2\pi i)^{|\operatorname{supp} m'|}} \int_{\substack{|\xi(t)| = \frac{\gamma}{\beta} \\ t \in \operatorname{supp} m'}} \cdots \int \frac{\left.\frac{\delta \Gamma(\lambda)}{\delta \lambda(t_{ex})}\right|_{\lambda = \xi}}{\prod_{t \in (Z^1)_-} \xi^{m'(t)+1}(t)} \prod_{t \in \operatorname{supp} m'} d\xi(t) \tag{2. 36}$$

where $\xi$ is a function such that $\xi(t) = 0$ if $t \notin \mathrm{supp}\ m'$; $|\xi(t) = \frac{\kappa}{\beta}$ if $t \in \mathrm{supp}\ m'$, $\kappa = \mathrm{const}$ and the integration extends over a polycircle (of complex dimension $|\mathrm{supp}\ m'|$) of such functions. It follows from (2. 35) and (2. 32) that

$$|\gamma(m)| < K_3\, \beta m'!\ (C_4 \beta)^{|m'|} (C_3\, \beta)^{|t_{ex}|} \tag{2. 37}$$

Consider now the quantities

$$g^0_\emptyset = \gamma_0, \qquad g^0_I = \sum_{m\,:\,\mathrm{supp}\, m\, =\, I} \frac{\gamma(m)}{m!} \tag{2. 38}$$

From the estimate (2. 37) it is easy to obtain

$$|g^0_I| < K_4 (C_5 \cdot \beta)^{|I|} (C_6 \cdot \beta)^{\mathrm{diam}\,\{I \cup \{0\}\}} \tag{2. 39}$$

where $K_4, C_5, C_6$ are constants. By using the decomposition (2. 34) we find that

$$\left[\frac{p_1(\sigma_-)}{p_0(\sigma_-)}\right]^{1/2} = e^{-1/2\beta\mu} \sum_{I \subset (Z^1)_-} g^1_I\, \eta_I(\sigma) \tag{2.41}$$

where the coefficients $g^1_I$ satisfy an estimate similar to (2. 39) and $g^1_\emptyset = [g^0_\emptyset]^{-1}$. Thus according to (2. 40), (2. 41) and (2. 14) we obtain (2. 15) where

$$C_\emptyset = e^{1/2\beta\mu}\, g^0_\emptyset$$

$$C_I = \begin{cases} e^{1/2\,\beta\mu}\, g^0_I & \text{if} \quad 0 \notin I \\ \\ -g^1_{I\setminus\{0\}}\, e^{-\beta\mu/2} + g^0_{I\setminus\{0\}}\, e^{\beta\mu/2} & \text{if} \quad 0 \in I \end{cases} \tag{2.42}$$

The estimates (2.16) and (2.17) follow from (2.39) and the equality

$$C_{\{0\}} = -(g^1_\emptyset e^{-\beta\mu/2} + g^0_\emptyset e^{\beta\mu/2}) \tag{2.43}$$

The lemma is proved.

Let for any $k \in Z^1$

$$U_k(\sigma) = U_0(\sigma - k) \tag{2.44}$$

where $\sigma - k \subset Z^1$ is the set $\sigma \subset Z^1$ translated by $-k$. Put

$$U_I(\sigma) = \prod_{k \in I} U_k(\sigma), \qquad I \in \mathcal{L}(Z^1) \tag{2.45}$$

<u>Lemma 2.3.</u> <u>The functions</u> $\{U_I, I \in \mathcal{L}(Z^1)\}$ <u>form an orthonormal basis in the space</u> $L_2(C(Z^1), \pi_{\beta,\mu})$.

<u>Proof:</u> It is easy to check that $U_I$ form an orthonormal system in $L_2(C(Z^1), \pi_{\beta,\mu})$. We will show that the linear manifold determined by $U_I$ is everywhere dense in $L(C(Z^1), \pi_{\beta,\mu})$. It suffices to show that any function $\eta_I$, $I \in \mathcal{L}(Z^1)$ belongs to the subspace $\mathcal{U} \subset L_2(C(Z^1), \pi_{\beta,\mu})$ spanned by the vectors $\{U_I, I \in \mathcal{L}(Z^1)\}$. Indeed by repeating the arguments used in the proof of the previous lemma we deduce that for any $I \in \mathcal{L}(Z^1)$ the function

$$V_I = \prod_{k \in I} \frac{U_k - C_\emptyset}{C_{\{0\}}} \tag{2.46}$$

has an expansion

$$V_I = \eta_I + \sum_{I' < I} D_I^{I'} \eta_{I'} \tag{2.47}$$

where the summation extends over all $I' \in \mathcal{L}(Z^1)$ such that $k' < k$ for any $k' \in I'$, $k \in I$ and coefficients $D_I^{I'}$ satisfy the estimate

$$|D_I^{I'}| \le K_4 B^{|I|} (C_7 \cdot \beta)^{|I'|} (C_6 \cdot \beta)^{d(I', I)} \tag{2.48}$$

where $K_4, B, C_6, C_7$ are constants and

$$d(I', I) = \max_{y \in I'} \min_{x \in I : x > y} (x - y) \tag{2.49}$$

Then let $\mathfrak{E}_\xi$ be a Banach space of functions on $\mathcal{L}(Z^1)$ with the norm

$$\|\Phi\|_\xi = \sup_I \frac{|\Phi(I)|}{\xi^{\operatorname{diam} I}} \tag{2.50}$$

where $\xi$ is some constant. By using the estimate (2.48) we obtain that for sufficiently small $\beta$ the operator $D$ in $\mathfrak{E}_\xi$

$$(D\Phi)(I) = \sum_{I' < I} D_{I'}^{I} \Phi(I') \tag{2.51}$$

has norm $<1$ and hence, there exists an inverse $(E + D)^{-1} = E + G$. Thus

$$\eta_I = V_I + \sum G_{I'}^{I} V_{I'} \tag{2.52}$$

where $G_{I'}^{I}$ is a matrix of the operator $G$. Hence

$$\|V_I\|_{L_2(C(Z^1), \pi_{\beta,\mu})} \le K^{|I|} \tag{2.53}$$

where $K$ is some constant, the series $\sum G_{I'}^{I} V_{I'}$ (for abritrary $I$) converges with respect to the norm of $L_2(C(Z^1), \pi_{\beta,\mu})$ for properly chosen $\xi$. Finally, using the expansion

$$V_I = (C_{\{0\}})^{-|I|} \sum_{I' \le I} U_{I'} (-C_\emptyset)^{|I \setminus I'|} \tag{2.54}$$

we obtain that for any $I \in \mathcal{L}(Z^1)$

$$\eta_I = \sum_{I'} F_{I'}^{I} U_{I'} \tag{2.55}$$

and the series is convergent in $L_2(C(Z^1), \pi_{\beta,\mu})$. Thus $\eta_I \in \mathfrak{U}$ for any $I$ and hence $\mathfrak{U} = L_2(C(Z^1), \pi_{\beta,\mu})$.

Lemma 2. 4. The matrix elements $a(I, I')$ of the operator $P^{Is}_{\beta,\mu}$ in the basis $\{U_I, I \in \mathcal{L}(Z^1)\}$ have the form (2. 4).

Proof: Consider the following family of functions on the space $C(Z^2)$.

$$\hat{U}_t(s) = U_{t_1}(s \cap (Z^1)_{t_2}) \tag{2.56}$$

where $t = (t_1, t_2) \in Z^2$, $s \in C(Z^2)$. It follows from the definition of $\hat{P}^{Is}_{\beta,\mu}$ that

$$a(I, I') = (\hat{P}^{Is}_{\beta,\mu} U_I, U_{I'})_{L_2(C(Z^1), \pi_{\beta,\mu})}$$

$$= \int_{C(Z^2)} \prod_{k \in I} \hat{U}_{(k,0)}(x) \prod_{k' \in I'} U_{(k',1)}(s) dP^{Is}_{\beta,\mu}(s) = \rho(\tilde{I}) \tag{2.57}$$

where $\{\rho(I), I \in \mathcal{L}(Z^2)\}$ is the correlation function of the family (2. 56) and $\tilde{I} \subset Z^2$ is a set of points either of the form

$$\{(k, 0), k \in Z^1\} \quad \text{or} \quad \{(k', 1), k \in Z^1\} \tag{2.58}$$

Lemma 2. 5. The semi-invariants $\omega(I)$, $I \subset Z^2$ of the family of functions $\{U_t, t \in Z^2\}$ with respect to the Gibbs measure $P^{Is}_{\beta,\mu}$ satisfy the estimate

$$|\omega(I)| < K_7 \overline{B}^{|I|} |r(\beta)|^{\text{diam } I} \tag{2.59}$$

where $K_7, \overline{B}$ are constants and $r(\beta)$ is some function such that $r(\beta) \to 0$ when $\beta \to 0$.

Before proving this lemma, we shall show how it implies the Lemma 2. 4.

Indeed, note that for some $k \in Z^1$ and for any $I \subset (Z^2)_k$ $\rho(I) = 0$

and hence, as we may see from (1. 15), $\omega(I) = 0 \quad I \subset (Z^2)_k$ for some $k \in Z^1$. Thus, using (1. 14) and (2. 57) we find that

$$a(I, I') = \sum_{\substack{\tilde{I} = \cup \tilde{I}_i,\ \tilde{I}_i \cap (Z^2)_0 \neq \emptyset,\ \tilde{I}_i \cap (Z^2)_{-1} \neq \emptyset \\ i = 1, \dots, k;\ k = 1, 2, \dots}} \prod_{i=1}^{k} \omega(\tilde{I}_i) \tag{2. 60}$$

where the sum runs over all partitions of the set $\tilde{I}$ into sets $\tilde{I}_i$ such that $\tilde{I}_i \cap (Z^2)_0 \equiv I_i \neq \emptyset$, $\tilde{I}_i \cap (Z^2)_{-1} \equiv I'_i \neq \emptyset$. Put

$$b(I, I') = \omega(\tilde{I}) \tag{2. 61}$$

where $\tilde{I}$ is defined by the equality (2. 58); then, by using the translation invariance of the measure $P^{Is}_{\beta, \mu}$ and also (2. 59), we obtain the assertion of Lemma 2. 3.

<u>Proof of Lemma 2. 5:</u> Let $\lambda \in \Lambda(Z^2)$ and

$$H_\lambda(s) = \sum_{t \in Z^2} \lambda(t) \hat{U}_t(s) = C_\emptyset \sum_{t \in Z^2} \lambda(t)$$

$$+ \sum_{I \in \mathcal{L}(Z^2)\,:\, I \in L((Z^2)_k) \text{ for some } k} D_I(\lambda)\, \eta_I \tag{2. 62}$$

where the sum runs over all "linear horizontal" subsets $I \in \mathcal{L}(Z^2)$ and

$$D_I(\lambda) = \sum_{\substack{t \in Z^2\,:\, \{t\} \cup I \subset (Z^2)_s, \\ t < I,\ s = 0, \pm 1, \pm 2, \dots}} \lambda(t)\, C_{T^{-t}I} \tag{2. 63}$$

where the sum runs over all points, lying to the right of I, on the horizontal line, containing I. From the estimate (2. 16) we find that for

$$|\lambda(t)| \le d, \quad t \in Z^2 \tag{2. 64}$$

where $d > 0$ is some constant, and

$$|D_I(\lambda)| < K_8 \, d(C_7 \cdot \beta)^{\operatorname{diam} I} \tag{2.65}$$

where $K_8$ and $C_7$ are constants.

Now consider for any $\lambda \in \Lambda(Z^2)$ the functional

$$Z(\lambda) = \int_{C(Z^2)} \exp\{ \sum_{t \in Z^2} \hat{U}_t(s)\lambda(t) \} \, dP^{Is}_{\beta,\mu} \tag{2.66}$$

From the formula (1.9), it follows that

$$\omega(I) = \left. \frac{\delta^{|I|} \ln Z(\lambda)}{\prod_{t \in I} \delta\lambda(t)} \right|_{\lambda \equiv 0}$$

$$= \frac{1}{(2\pi i)^{|I|-2}} \int \cdots \int_{|\xi(t)|=d} \frac{1}{\prod_{t \in I \setminus \{t_1,t_2\}} \xi^2(t)} \cdot \left. \frac{\delta^2 \ln Z(\lambda)}{\delta\lambda(t_1) \cdot \delta\lambda(t_2)} \right|_{\lambda=\xi} \prod d\xi(t) \tag{2.67}$$

(we have used here Cauchy's formula). Next

$$\frac{\delta^2 \ln Z(\lambda)}{\delta\lambda(t_1)\delta\lambda(t_2)} = [Z(\lambda)]^{-1} \int \hat{U}_{t_1}(s) \cdot \hat{U}_{t_2}(s) \exp\{H_\lambda(s)\} dP^{Is}_{\beta,\mu}$$

$$- [Z(\lambda)]^{-2} \int \hat{U}_{t_1}(s) \exp\{H_\lambda(s)\} dP^{Is}_{\beta,\mu} \cdot \int \hat{U}_{t_2}(s) \exp\{H_\lambda(s)\} dP^{Is}_{\beta,\mu} \tag{2.68}$$

Let us introduce a (complex) measure $\tilde{P}_\lambda$ on $C(Z^2)$ defined by the density

$$\frac{d\tilde{P}_\lambda}{dP^{Is}_{\beta,\mu}}(s) = [Z(\lambda)]^{-1} \exp\{H_\lambda(s)\} \tag{2.69}$$

with respect to $P^{Is}_{\beta,\mu}$. It is evident that the measure $\tilde{P}_\lambda$ is a Gibbs measure, with potential (non=translation invariant)

$$\tilde{U}(s) = U^{Is}(s) + D_s(\lambda) \tag{2.70}$$

If we denote by $\rho_\lambda^0$ the correlation function of this measure, then, by using the results of [4] and [30], it can be shown that for sufficiently small d and $\beta$ we have an exponential decrease of correlations

$$|\rho_\lambda^0(I_1 \cup I_2) - \rho_\lambda^0(I_1) \cdot \rho_\lambda^0(I_2)| < K_9 \hat{B}^{|I_1|+|I_2|} [\tilde{C}(\beta)]^{d(I_1, I_2)} \tag{2.71}$$

where $d(I_1, I_2)$ is the distance between sets $I_1, I_2$; $K_9$ $\hat{B}$ are constants and $0 < \tilde{C}(\beta) < 1$. By using estimates (2. 71), (2. 16), the equality (2. 68), and the expansion (2. 15) we get that for all $\lambda$ satisfying (2. 64)

$$\left| \frac{\delta^2 \ln Z(\lambda)}{\delta\lambda(t_1)\delta\lambda(t_2)} \right| \leq D|r(\beta)|^{|t_1 - t_2|} \tag{2.72}$$

where D is a constant and $r(\beta) < 1$ is a function such that $r(\beta) \to 0$ when $\beta \to 0$. Now according to the estimate (2. 72) and the representation (2. 67), where points $t_1, t_2$ are chosen so that $|t_1 - t_2| =$ diam I, we obtain (2. 59). For any function $\Phi \in L_2(C(Z^1), \pi_{\beta,\mu})$ we have by Lemma 2.3

$$\Phi = \sum_I C_I U_I \tag{2.73}$$

Using this, we define a unitary mapping W of the space $L_2(C(Z^1), \pi_{\beta,\mu})$ onto $L_2(L(Z^1))$:

$$(W\Phi)(I) = C_I, \quad I \in L(Z^1) \tag{2.74}$$

It is evident that W maps the operator $P_{\beta,\mu}^{Is}$ into an operator with the kernal $a(I, I')$ and that is maps the representation $(T_t, t \in Z^1)$ on $L_2(C(Z^1), \pi_{\beta,\mu})$ into the operators (2. 2).

## III. THE ASYMPTOTIC EXPANSION OF ln Z

As is well known, under rather general assumptions concerning the interaction potential of a physical system and concerning the sequence of regions $\Lambda_1 \subset \Lambda_2 \subset \ldots$ one has for the quantities $\ln Z(\Lambda_k)$, $k = 1, 2, \ldots$ the asymptotic expansion (Lee-Yang theorem or Van-Hove's theorem)

$$\ln Z(\Lambda_k) = c_0(U)|\Lambda_k| + 0(|\Lambda_k|) \tag{3. 1}$$

where $|\Lambda_k|$ denotes the volume of $\Lambda_k$, and $c_0(U)$ is a quantity depending only on the interaction potential $U$ (Gibbs free energy). Here we find the subsequent term of the expansion (3. 1) (whose value is of order $|\Lambda|^{1/\nu}$). To be more exact, we develop a method permitting one to find in principle all the terms of the asymptotic expansion (3. 1) which are increasing with a growing region $\Lambda$ and in the case of a continuous system we write out the first three terms (in dimension $\nu \geq 2$). We carry out our constructions at low activity (they seem true however in the whole monophase domain). It should be noted that for the derivation of the further terms of the asymptotic expansion (3. 1) one has to assume ever increasing smoothness of the potential and regularity of the boundaries $\Gamma(\Lambda_k)$ of the regions $\Lambda_k$, $k = 1, 2, \ldots$ .

Suppose that we have a pair potential which is Euclidean invariant, sufficiently smooth outside neighborhood of zero, rapidly decreasing at infinity with its derivatives and increasing at zero so that it and its derivatives satisfy the stability condition (for the exact formulation of these conditions see [13]). Let then $\Lambda_k$, $k = 1, 2, \ldots$ be such a sequence of convex regions with smooth (or even piecewise smooth) boundaries $\Gamma(\Lambda_k)$, that the regions $\tilde{\Lambda}_k$, $k = 1, 2, \ldots$ being obtained by the contraction of each $\Lambda_k$ region $|\Lambda_k|^{1/\nu}$ times, have piecewise smooth boundaries $\Gamma(\tilde{\Lambda}_k)$ with uniformly bounded coefficients of the 2-nd and the 3-d differential forms of the surface $\Gamma(\Lambda_k)$ in their smooth points*.

Theorem 3. 1. *Under conditions mentioned above the logarithm of the partition function has the asymptotic expansion*

$$\ln Z(\Lambda_k) = c_0(U)|\Lambda_k| + c_1(U)S(\Gamma(\Lambda_k)) + 0(|\Lambda_k|^{1-1/} \tag{3. 2}$$

*when* $k \to \infty$, *where* $S(\Gamma(\Lambda_k))$ *is the area of the boundary* $\Gamma(\Lambda_k)$ *and* $c_0(U)$, $c_1(U)$ *are constants depending on the potential* $U$ *which will be explicitly written out below.*

The subsequent terms of the asymptotic expansion of $\ln Z$, are more sensitive to the smoothness properties of the boundary $\Gamma(\Lambda)$ and even have a different form for the regions with smooth and piece-wise smooth boundaries. Here we write out the asymptotic expansion for the case of a sequence of regions which have the smoothness of 4-th order uniformly bounded in the same sense, as was pointed out above.

Theorem 3. 2. *For a sequence of regions with smooth boundaries we have the following asymptotic expansion for* $\ln Z(\Lambda_k)$ *in the case* $\nu > 2$

$$\ln Z(\Lambda_k) = c_0(U)|\Lambda_k| + c_1(U)S(\Gamma(\Lambda_k))$$

$$+ c_2(U)M_1(\Gamma(\Lambda_k)) + c_3(U)M_2(\Gamma(\Lambda_k) + 0(|\Lambda_k|^{1-2/\nu}) \tag{3. 3}$$

*and in the case* $\nu = 2$

$$\ln Z(\Lambda_k) = c_0(U), \ |\Lambda_k| + c_1(U)S(\Gamma(\Lambda_k))$$

$$+ c_2(U)M_1(\Gamma(\Lambda_k)) \ln |\Lambda_k| + 0(1) \tag{3. 4}$$

---

* Apparently this condition may be weakened.

where $c_2(U)$ and $c_3(U)$ are constants depending on the potential U (see below) and the quantities $M_1, M_2$ are equal to

$$M_1(\Gamma(\Lambda_k)) = \int_{\Gamma(\Lambda_k)} (r(x), n(x)) \Big( \sum_{i=1}^{\nu-1} \kappa_i^2(x) \Big) d\sigma(x) \tag{3. 5}$$

$$M_2(\Gamma(\Lambda_k)) = \int_{\Gamma(\Lambda_k)} (r(x), n(x)) \Big( \sum_{1 \le i \le j \le \nu-1} \kappa_i(x) \kappa_j(x) \Big) d\sigma(x) \tag{3. 6}$$

where $r(x)$ is the radius vector of the point $x \in \Gamma(\Lambda_k)$, $n(x)$ is a unit exterior normal at the point $x$, $\kappa_i(x)$, $i = 1, \ldots, \nu-1$ are the principal curvatures of the surface $\Gamma(\Lambda_k)$ at the point $x$ and $d\sigma$ is an element of the surface area.

Here the appearance of the logarithmic term is unexpected. Note that in the case $\nu = 3$ a similar term appears in the fourth term of the expansion. In the case of a sequence of polygons (or polyhedra) the logarithmic term is absent. Here we sketch the proofs of Theorems 3. 1 and 3. 2 (a detailed proof will be given in a paper published by one of the authors in the near future [27]. The proof is based on a representation of $\ln Z(\Lambda)$ which follows from the formula (1. 31)

$$\ln Z(\Lambda) = \int_{C(\Lambda)} \varphi_U(c) dc \tag{3. 7}$$

where $\varphi_U(c)$, $c \in \mathcal{L}(R^\nu)$ is the Ursell function and $c(\Lambda)$ is the collection of all finite subsets of $\Lambda$.

Let then $\Lambda_L$ be the one-parameter family of regions, being obtained by the enlargement $L > 1$ times of the region $\Lambda_1$ with the center of similitude at the origin of coordinates. It is easy to obtain from the formula (3. 7) that

$$\frac{d}{dL} \ln Z(\Lambda_L) = \frac{1}{L} \int_{(\Lambda_L)} (r(x), n(x)) \psi_L(x) d\sigma(x) \tag{3. 8}$$

where

$$\psi_L(x) = \int_{C(\Lambda_L)} \varphi_U(x \cup c)dc, \qquad x \in R^\nu \tag{3. 9}$$

The asymptotic expansions (3. 2), (3. 3) and (3. 4) are obtained from the following asymptotic expansion of the function $\psi_L(x)$, $x \in \Gamma(\Lambda_L)$ when $L \to \infty$.

Consider a function

$$\omega_1(c) = \int_{C(R_1^\nu)} \varphi_U(c \cup \bar{c})\, d\bar{c} \tag{3. 10}$$

where $R_1^\nu \subset R^\nu$ is a half-space

$$R_1^\nu = \{\xi = (\xi^{(1)}, \dots, \xi^{(\nu)}) : \xi^{(\nu)} > 0\} \tag{3. 11}$$

where $\xi^{(1)}, \dots, \xi^{(\nu)}$ is an orthogonal coordinate system in $R^\nu$ and by $R^{\nu-1}$ we will denote the hyperplane $\xi^{(\nu)} = 0$.

Lemma 3. 3. *Assume the hypotheses of Theorem 3. 2. Then following asymptotic expansion*

$$\psi_L(x) = c_0 + c_1 \sum_{i=1}^{\nu-1} \kappa_i(x) + c_2 \sum_{i=1}^{\nu-1} \kappa_i^2(x)$$

$$+ c_3 \sum_{1 \le i < j \le \nu-1} \kappa_i(x)\kappa_j(x) + 0\left(\frac{1}{L^2}\right) \tag{3. 12}$$

*is valid uniformly with respect to all* $x \in \Gamma(\Lambda_L)$, *where the coefficients* $c_i$, $i = 1, 2, 3$ *are equal to*

$$c_0 = \omega_1(\{0\}) \tag{3. 13}$$

$$c_1 = -\frac{1}{2} \int_{R^{\nu-1}} (\xi^{(1)})^2 \Omega_1(\{0, \xi\}) d^{\nu-1}\xi$$

$$c_2 = \frac{1}{8} \int_{R^{\nu-1}} \int_{R^{\nu-1}} (\xi_1^{(1)})^2 (\xi_2^{(1)})^2 \omega_1(\{0, \xi_1, \xi_2\}) d^{\nu-1}\xi_1 d^{\nu-1}\xi_2$$

$$c_3 = \frac{1}{4} \int_{R^{\nu-1}} \int_{R^{\nu-1}} (\xi_1^{(1)})^2 (\xi_2^{(2)})^2 \omega_1(\{0, \xi_1, \xi_2\}) d^{\nu-1}\xi_1 d^{\nu-1}\xi_2 \qquad (3.14)$$

<u>Remark.</u> From the assumptions of Theorem 3. 1 one can only conclude the weaker asymptotic expansion of $\psi_L(x)$

$$\psi_L(x) = c_0 + c_1 \sum_{i=1}^{\nu-1} \kappa_i(x) + 0\left(\frac{1}{L}\right) \qquad (3.15)$$

uniformly with respect to all $x \in \Gamma(\Lambda_L)$. This expansion implies (3. 2).

The proof of the lemma is based on the representation

$$\psi_L(x) = \int_{L((R_x^\nu)_1 \setminus \Lambda_L)} (-1)^{N(c)} \omega_1(x \cup c) dc \qquad (3.16)$$

where $(R_x^\nu)_1 \subset R^\nu$ is a half-plane bounded by a tangent plane to the surface $\Gamma(\Lambda_L)$ at the point $x$ and containing the region $\Lambda_L$.

Then by chosing a ball $S_{L^\alpha}(x)$ of center $x$ and radius $L^\alpha$ where $\alpha$ is some number: $0 < \alpha < 1$, we have from (3. 16)

$$\int_{\mathcal{L}((R_x^\nu)_1 \setminus \Lambda_L)} (-1)^{N(c)} \omega_1(x \cup c) dc$$

$$= \int_{\mathcal{L}[((R_x^\nu)_1 \setminus \Lambda_L) \cap S_{L^\alpha}(x)]} (-1)^{N(c)} \omega_1(x \cup c) dc$$

$$+ \int_{c \not\subset ((R_x^\nu)_1 \setminus \Lambda_L) \cap S_{L^\alpha}(x)} (-1)^{N(c)} \omega_1(x \cup c) dc \qquad (3.17)$$

By using the estimate for $\omega_1$ which is obtained from the estimate for the Ursell function one can show that for some $\alpha$ the second integral in (3.17) has the order $0(\frac{1}{L^\gamma})$, where $\gamma$ is sufficiently large. Then one can also show that

$$\int_{c \in \mathfrak{L}[((R^\nu_x)_1 \setminus \Lambda_L) \cap S_{L^\alpha}(x)]} \omega_1(x \cup c)dc = 0\left(\frac{1}{L^2}\right) \tag{3.18}$$

Thus

$$\psi_L(x) = \omega_1(x) + \int_{((R^\nu_x)_1 \setminus \Lambda_L) \cap S_{L^\alpha}(x)} \omega_1(x, y)dy$$

$$+ \frac{1}{2}\int_{y_1, y_2 \in ((R^\nu_x)_1 \setminus \Lambda_L) \cap S_{L^\alpha}(x)} \omega_1(x, y_1, y_2)dy_1dy_2 + 0\left(\frac{1}{L^2}\right) \tag{3.19}$$

Let $(\xi^{(1)}, \ldots, \xi^{(\nu-1)}, \eta)$ be a coordinate system in a neighborhood of the point $x \in \Gamma(\Lambda_L)$ such that the $\eta$ axis has the direction of interior normal to $\Gamma(\Lambda_L)$ at the point $x$ and the $\xi^{(1)}, \ldots, \xi^{(\nu-1)}$ axes lie in the tangent hyperplane $R^{\nu-1}_x$ and have the directions of principal curvatures at the point $x$. Therefore, we can write the equation for $\Gamma(\Lambda_L)$ in a neighborhood of $x$ in the form

$$\eta = \sum_1^{\nu-1} \kappa_i(x)(\xi^{(i)})^2 + 0(|\xi|)^2 \tag{3.20}$$

We have in this neighborhood that

$$\omega_1(x, y) = \omega_1(x, \xi) + 0(\eta^2) \tag{3.21}$$

and

$$\omega_1(x, y_1, y_2) = \omega_1(x, \xi_1, \xi_2) + 0(|\eta_1| + |\eta_2|) \tag{3.22}$$

where $(\xi_i^{(1)}, \ldots, \xi_i^{(\nu-1)}, \eta) \equiv (\xi_i, \eta)$. Here the estimates $0(\eta^2)$ and

$0(|\eta_1| + |\eta_2|)$ follow from the estimates of the derivatives of $\omega_1$ being obtained from similar estimates (1. 37) of the derivatives of $\varphi_U$ and also from the equality

$$\left.\frac{\partial\omega_1(x, y)}{\partial\eta}\right|_{\eta = 0} = 0 \qquad (3.23)$$

which follows from the Euclidean invariance of the potential. Combining (3. 21), (3. 22) and (3. 19) and taking into accout that

$$\int_{|\xi| > L^{\alpha}} \omega_1(x, \xi)|\xi|^2 d^{\nu-1}\xi = 0(\frac{1}{L^s})$$

$$\int_{|\xi_1|+|\xi_2| > L^{\alpha}} \int \omega_1(x, \xi_1, \xi_2) d^{\nu-1}\xi_1 d^{\nu-1}\xi_2 = 0(\frac{1}{L^s}) \qquad (3.24)$$

where s in large enough, we get (3. 12). Similar to this, one can obtain the subsequent terms of the asymptotic expansion of $\psi_L(x)$ only it is necessary to make use of more distant terms of Taylor expansion in (3. 20), (3. 21) and (3. 22) and also of the subsequent integrals in (3. 19).

## REFERENCES

1. J. L. Lebowitz, O. Penrose, Analytic and clustering properties of thermodynamic functions and distributions functions for classical lattice and continuous systems, Comm. Math. Phys., 11, 99 (1968) and Decay of correlations, Comm. Math. Phys. , 39, 165, (1974).

2. J. L. Lebowitz, Bounds of the correlations and analyticity properties of ferromagnetic Ising spin systems, Comm. Math. Phys., 28, 313-321 (1972).

3. R. L. Dobrushin, The determination of systems of random values by conditional distributions. Teorija verojatnosteiy i ee primenenija V: XV., 3, 469-497 (1970). (In Russian).

4. G. Del Grosso, On the local central limit theorem for Gibbs. Comm. Math. Phys., 37, 141-160 (1974).

5. I. G. Jurbenko, About strong estimates of mixed semiinvariants of random processes. Sibirskiy matematitsheskiy journal, 13, 2 (1972), 293-308. (In Russian).

6. __________, The estimates highest spectral density of stationary processes with Cramer's condition and mixing by Rosenblatt. Litowskiy matemat. sbornik, XV, No. 1, 111-124 (1975). (In Russian).

7. D. Abraham, G. Gallavotti, A. Martin-Löf. Surface tension in two-dimensional Ising model, Physica, v. 65, 73-88 (1973).

8. D. Ruelle, Statistical mechanics (rigorous results), W. A. Benjamin Inc., N. Y. -Amsterdam, 1969.

9. M. Denau, D. Jagolnitzer, B. Souillard, Decrease properties of truncated correlation functions and analyticity properties for classical lattice and continuous systems. Comm. Math. Phys., 31, 191 (1973).

10. M. Denau, B. Souillard, D. Jagolnitzer. Analyticity and strong cluster properties for calssical cases with finite range interaction. Comm. Math. Phys., 35, 307 (1974).

11. __________, Decay of correlations for infinite range interactions. J. Math. Phys., v. 16, No. 8, 1662-1666 (1975).

12. M. Denau, D. Souillard, Cluster properties of lattice and continuous systems. Comm. Math. Phys., v. 47, 155-166 (1976).

13. R. A. Minlos, S. K. Pogosian. The estimates of Ursell's functions clusters functions and their derivatives. Teoretitscheskaj i matem. physika, v. 31, No. 2, 199-213, 1977. (In Russian).

14. J. Glimm, A. Jaffe, T. Spencer, The particle structure of the weakly coupled $P(\varphi)_2$ model and other applications of high temperature expansions. Part II. The cluster expansion. Constructive quantum field theory. Lecture notes in physics, Vol. 25, Springer Verlag, Berlin (1973).

15. __________, A convergent expansion about mean field theory. Part I, The expansion, Part II, A convergent of the expansion, Ann. of Phys., 101, I. 610-631, II. 631-669 (1976).

16. J. Glimm, A Jaffe, Two and three body equations in quantum field models. Comm. Math. Phys., 44, 293-320 (1975).

17. T. Spencer, The decay of the Bethe-Salpeter kernel in $P(\varphi)_2$ quantum field model. Comm. Math. Phys., 44, 143-164 (1975).

18. T. Spencer, F. Zirilli, Scattering states and bound states in $\lambda P$. Comm. Math. Phys., 49. 1-16 (1976).

19. J. P. Eckmann, J. Magnen, R. Seneor, Decay properties and Borel summability for the Schwinger functions if $P(\varphi)_2$ theories, Comm. Math. Phys., 39, 251-271 (1975).

20. R. A. Minlos, Y. G. Sinai, Investigation of spectrums stochasticals operators from models of lattice, gaz. Teoretitscheskaja i matem. Physica, v. 2, No. 2, 230-243 (1970). (In Russian).

21. __________, Remark about decay correlations in a thermodynamical system. Preprint 22, MGU, 1971. (In Russian).

22 M. Fisher, W. Camp, Behaviour of two-point correlation functions at high temperatures. Phys. Rev. Letters, 26, 73-77 (1971).

23 L. D. Faddeev, The mathematical questions of quantum theory scattering for three-body systems. Moscow-Leningrad, AN SSSR-Press, (1963).

24. I. M. Segal, The asymptotical completeness of many-body systems. DAN SSSR, 204, No. 4, 795-798 (1972). (In Russian).

25. K. Hepp. On the connection between Wightman and L. S. Z. quantum field theory. Axiomatic field theory. Gordon and Breach Science Publ. Inc., N. Y., London, Paris, 1965.

26. R. L. Dobrushin, The asymptotical behaviour of the gibbsian distributions for lattice systems independence of shape volume. Teoretitscheskaja i matem. physica. T. 12, No. 1, 115-134 (1972). (In Russian).

chapter 2

# STATIONARITY AND ERGODICITY OF MARKOV INTERACTING PROCESSES

V. Ya. Basis

Abstract. There are proposed certain sufficient conditions for existence and uniqueness of the stationary distributions of the multicomponent local interaction Markov processes. These conditions are free of the requirement of the phase space compactness.

The present article deals with the problem of existence and uniqueness of the stationary distributions of the multicomponent local interaction Markov processes in the infinite box. Such processes were introduced in [1]. In contrast to the similar papers [2 - 7], the proof of existence of the limiting process given in [1] did not suppose compactness of the phase spaces. Similarly, the results of this article provide an extension of the corresponding theorems proved for the finite phase spaces in [8] (discrete time) and [2, 7, 9] (continuous time) to the noncompact phase spaces. It should be noted that the conditions of uniqueness of the stationary distribution for a multicomponent process (theorem 3) are close to those of ergodicity of this process, if ergodicity of the Markov process $\xi(t)$ means that the measures

$$P(t, A|x) = \Pr\{\xi(t) \in A | \xi(0) = x\}$$

weakly converge for $t \to \infty$ and the limiting measure does not depend on $x$.

The wording of the main results are given in § I, and for the reader's convenience there is repeated the wording of the theorem of existence of the process in the infinite box given in [1].

In §II the theorem of existence and uniqueness of the stationary distributions of the multicomponent processes is proved.

§III deals with the "Gaussian" multicomponent processes whose interaction intensity decreases exponentially with the increased distance between components. A "minimum Gaussian" process is singled out, for which the stationarity conditions given in this paper coincide with the ergodicity conditions. The components of the constructed "Gaussian" processes take values in $\mathbb{R}^1$, and this shows that rejection of compactness in [1] allows us to consider an essentially larger class of the multicomponent processes.

I. For every $b$ from a countable set $T$ let there be given the standard measurable space $(X^n, \mathcal{B}^b)$, that is, a measurable space isomorphic to the Borel subset of a complete separable metric space, the elements of which will be denoted by $x^b$, $y^b$ etc. Let also the metric $\rho_b$ be given on $X^b$ and let this metric be measurable as a function of two variables and such that either the metric space $(X^b, \rho_b)$ is complete separable and its Borel $\sigma$-algebra coincides with $\mathcal{B}^b$ or $\rho_b(x, y) \geq 1$ for $x \neq y$.

Let $(X_T, \mathcal{B}_T)$ be the product of spaces $(X^b, \mathcal{B}^b)$, $b \in T$, and $\theta = (\theta^b, b \in T)$ be a fixed element from $X_T$.

Let us assume that on $T$ there is given a nonnegative function $k_b$, $b \in T$, such that the series of numbers $k_b$, $b \in T$, converges. We may consider without loss of generality that

$$\sum_{b\in T} k_b \leq 1$$

Let

$$p(x) = \sum_{b\in T} k_b \rho_b (x^b, \theta^b) \leq +\infty, \quad x = (x^b,\ b \in T) \in X_T ,$$

and $X_0$ be a domain of finiteness of the function $p(x)$, that is,

$$X_0 = \{x : x \in X_T,\ p(x) < \infty\}$$

The $\sigma$-algebra $\mathcal{B}_0$ is induced on $X_0$ by the $\sigma$-algebra $\mathcal{B}_T$.

For $x, y \in X_T$, $\alpha \subseteq T$ let $\rho_\alpha(x, y) = \sum_{b\in\alpha} \rho_b(x^b, y^b)$ and $X_\alpha = \{x : x \in X_T,\ \rho_{T\setminus\alpha}(x, \theta) = 0\}$. The $\sigma$-algebra $\mathcal{B}_\alpha$ is induced on $X_\alpha$ by the $\sigma$-algebra $\mathcal{B}_T$.

It is clear that if $\alpha \subseteq \beta \subseteq T$ then $X_\alpha \subseteq X_\beta \subseteq X_0$ and $\mathcal{B}_\alpha \subseteq \mathcal{B}_\beta \subseteq \mathcal{B}_0$ .

For $x \in X_T$, $\alpha \in T$ define $x_\alpha \in X_\alpha$ from the relation

$$\rho_\alpha(x_\alpha, x) + \rho_{T\setminus\alpha}(x_\alpha, \theta) = 0$$

Let us assume that there is fixed a sequence of finite subsets $\alpha_n \subset T$, $n = 1, 2, \ldots$, such that $\alpha_m \subseteq \alpha_n$ for $m \leq n$ and $\bigcup_{n=1}^{\infty} \alpha_n = T$. For every $\alpha_n$ from this sequence let there be given a family

$$Q_n = \{q_n^\alpha(x),\ \Pi_n^\alpha(\Gamma|x),\ \alpha \subseteq \alpha_n,\ \Gamma \in \mathcal{B}_{\alpha_n},\ x \in X_T\}$$

consisting of the nonnegative values called "the local characteristics in the box $\alpha_n$" and having the following properties:

1. When $\alpha$ and $\Gamma$ are fixed, the functions $q_n^\alpha(x)$ and $\Pi_n^\alpha(\Gamma|x)$ of the variable $x$ depend only on $x_{\alpha_n}$ and are measurable with respect to $\mathcal{B}_{\alpha_n}$ .

2. When $\alpha$ and $\Gamma$ are fixed, the value $\Pi_n^{\alpha}(\Gamma|x)$ is a probability measure on $\mathcal{B}_{\alpha_n}$ with respect to $\Gamma$, concentrated on the set $\{y : y^a = x^a_{\alpha_n},\ a \in T\setminus\alpha,\ y^b \neq x^b,\ b \in \alpha\}$.

3. $q_n^{\phi}(x) \equiv 0$.

We shall handle the regular families of the local characteristics only, that is, such families which meet the following regularity conditions:

there exist the functions

$$\varphi_n(x) \geq \sum_{\alpha \subseteq \alpha_n} q_n^{\alpha}(x), \qquad x \in X_{\alpha_n},$$

$$\rho_n(x) \geq \rho_{\alpha_n}(x, \theta), \qquad x \in X_{\alpha_n},$$

and the constants $A_n \geq 0$, $0 \leq a_n \leq 1$ such that

$$\int_{X_{\alpha_n}} \varphi_n(y)\Pi_n(dy|x) \leq A_n(1 - a_n) + a_n\varphi_n(x), \qquad x \in X_{\alpha_n}$$

$$\int_{X_{\alpha_n}} \rho_n(y)\Pi_n(dy|x) \leq A_n(1 - a_n) + \rho_n(x), \qquad x \in X_{\alpha_n}$$

$$\int_{X_{\alpha_n}} \rho_n(y)\varphi_n(y)\Pi_n(dy|x) \leq A_n(1 - a_n) + \rho_n(x)\varphi_n(x), \qquad x \in X_{\alpha_n}$$

where

$$\Pi_n(\Gamma|x) = \sum_{\alpha \subseteq \alpha_n} (q_n^{\alpha}(x)/q_n(x))\Pi_n^{\alpha}(\Gamma|x), \qquad x \in X_{\alpha_n},\quad \Gamma \in \mathcal{B}_{\alpha_n}$$

and

$$q_n(x) = \sum_{\alpha \subseteq \alpha_n} q_n^{\alpha}(x), \qquad x \in X_{\alpha_n}$$

Definition. The nonnegative function $f(x)$ given on $X_{\alpha_n}$ will be called $Q_n$-regular if for all $x \in X_{\alpha_n}$

$$\int_{X_{\alpha_n}} f(y)\Pi_n(dy|x) \leq A_n(1 - a_n) + f(x)$$

$$\int_{X_{\alpha_n}} f(y)\varphi_n(y)\Pi_n(dy|x) \leq A_n(1 - a_n) + f(x)\varphi_n(x)$$

Let us consider the time-homogeneous jump Markov process $\xi_n(t)$ in $X_{\alpha_n}$ the transition function $P_n(t, \Gamma|x)$, $t \geq 0$, $\Gamma \in \mathcal{B}_{\alpha_n}$, $x \in X_{\alpha_n}$, of which is the minimum nonnegative solution of the Kolmogorov equation

$$\frac{d}{dt} P_n(t, \Gamma|x) = \sum_{\alpha \subseteq \alpha_n} q_n^{\alpha}(x) \int_{X_{\alpha_n}} (P_n(t, \Gamma|y) - P_n(t, \Gamma|x)) \Pi_n^{\alpha}(dy|x) \qquad (1.1)$$

under the initial condition

$$P_n(0, \Gamma|x) = \chi_{\Gamma}(x)$$

where $\chi_{\Gamma}(\cdot)$ is the indicator of the set $\Gamma$.

As shown in [1], if the family of the local characteristics $Q_n$ is regular, then the process $\xi_n(t)$ is also regular, that is, the process $\xi_n(t)$ makes the finite number of jumps over any finite time interval. The latter is equivalent to the fact that the solution of the equation (1.1) is unique in the class of the nonnegative functions answering the given initial condition.

Let $\lambda \subseteq \alpha \subseteq T$ and $P$ be the measure on $(X_{\alpha}, \mathcal{B}_{\alpha})$. The measure $P'$ defined on $(X_{\lambda}, \mathcal{B}_{\lambda})$ is said to be a projection of the measure $P$ if

$$P'(\Gamma) = P(\{x : x \in X_{\alpha}, x_{\lambda} \in \Gamma\}), \quad \Gamma \in \mathcal{B}_{\lambda}$$

If $P_1$ and $P_2$ are the probability measures on the spaces

$(X_\alpha, \mathcal{B}_\alpha)$ and $(X_\beta, \mathcal{B}_\beta)$ respectively and $\lambda \subseteq \alpha \cap \beta$ let

$$R_\lambda(P_1, P_2) = \inf \int_{X_\lambda}\int_{X_\lambda} \rho_\lambda(x, y) P(dx, dy)$$

where the greatest lower bound is computed over all the measures P on $(X_\lambda, \mathcal{B}_\lambda) \times (X_\lambda, \mathcal{B}_\lambda)$, such that $P(\Gamma \times X_\lambda) = P_1'(\Gamma)$, $P(X_\lambda \times \Gamma) = P_2'(\Gamma)$, $\Gamma \in \mathcal{B}_\lambda$, where $P_1'$ and $P_2'$ are the projections of the measures $P_1$ and $P_2$ onto $(X_\lambda, \mathcal{B}_\lambda)$. In other words, $R_\lambda$ is the Kantorovich-Rubinstein-Waserstein distance (the KRW-distance) between the probability measures on $(X_\lambda, \mathcal{B}_\lambda)$ computed with respect to the metric $\rho_\lambda$ (see [8, 10, 11]).

We shall constantly suppose further on that m and n are the natural numbers, $m \leq n$.

For $\alpha \subseteq \alpha_m$, let

$$q^\alpha_{n|m}(x) = \sum_{\beta:\beta\subseteq\alpha_n,\ \beta\cap\alpha_m = \alpha} q^\beta_n(x);$$

$$\Pi^\alpha_{n|m}(\cdot\,|x) = \begin{cases} \displaystyle\sum_{\beta:\beta\subseteq\alpha_n,\ \beta\cap\alpha_m=\alpha} (q^\beta_n(x)/q^\alpha_{n|m}(x))\Pi^\beta_n(\cdot\,|x), & \text{if } q^\alpha_{n|m}(x) \neq 0 \\ \Pi^\alpha_m(\cdot\,|x) & \text{otherwise} \end{cases}$$

$$q^\alpha_{m,n}(x_1, x_2) = \min\{q^\alpha_m(x_1), q^\alpha_{n|m}(x_2)\};$$

$$J^\alpha_{m,n}(x_1, x_2) = \begin{cases} \displaystyle\int_{X_{\alpha_m}} \rho_\alpha(y, x_2)\Pi^\alpha_m(dy|x_1) & \text{if } q^\alpha_m(x_1) \geq q^\alpha_{n|m}(x_2) \\ \displaystyle\int_{X_{\alpha_n}} \rho_\alpha(y, x_1)\Pi^\alpha_{n|m}(dy|x_2) & \text{otherwise} \end{cases}$$

$$D^b_{m,n}(x_1, x_2) = \sum_{\alpha : b \in \alpha \subseteq \alpha_m} \{q_{m,n}(x_1, x_2)[R_\alpha(\Pi^\alpha_m(\cdot\ |x_1), \Pi^\alpha_{n|m}(\cdot\ |x_2))$$
$$- \rho_b(x_1, x_2)] + |q^\alpha_m(x_1) - q^\alpha_{n|m}(x_2)|(J^\alpha_{m,n}(x_1, x_2) - \rho_b(x_1, x_2))\}$$

Now the main result of [1] can be formulated as

<u>Theorem 1.</u> <u>For existence of the Markov process with the values in $X_0$ and the transition function</u> $P(t, \cdot\ |x)$ <u>such that for every finite</u> $\lambda \subset T$

$$R_\lambda(P(t, \cdot\ |x), P_n(t, \cdot\ |x)) \underset{n}{\to} 0, \quad x \in X_0, \quad t \geq 0 \tag{1. 2}$$

<u>the following conditions are sufficient:</u>

1. <u>There exists a constant</u> $G < \infty$, <u>such that</u>

$$q_n(x) \int_{X_{\alpha_n}} (p(y) - p(x)) \Pi_n(dy\,|x) \leq G(1 + p(x)), \quad x \in X_0;$$

2. <u>There exist the numbers</u> $g^{c,b}$, $b, c \in T$, <u>which are nonnegative if</u> $b \neq c$ <u>and the nonnegative numbers</u> $g^b_{m,n}$, $b \in \alpha_m$, <u>such that for</u> $t \geq 0$, $b \in T$

$$\lim_{m,n \to \infty} (g^b_{m,n} + \sum_{r=1}^{\infty} \frac{t^r}{r!} \sum_{c_1, \ldots, c_r \in \alpha_m} |g^{c_1}_{m,n} g^{c_1,c_2} \ldots g^{c_r,b}|) = 0^{*} \tag{1. 3}$$

<u>and for all</u> $m \leq n$, $b \in \alpha_m$, $x_1, x_2 \in X_0$ <u>the inequalities hold true</u>

$$D^b_{m,n}(x_1, x_2) \leq g^b_{m,n}(1 + p(x_1) + p(x_2)) + \sum_{c \in \alpha_m} g^{c,b} \rho_c(x_1, x_2) \tag{1.4}$$

---

* The following condition, for example, is sufficient for this equality to take place:

$g^b_{m,n} \to 0$ as $m, n \to \infty$ and $\sup_{m,n,b \in \alpha_m} g^b_{m,n} + \sup_{b \in T} \sum_{c \in T} |g^{c,b}| < \infty$.

Here the convergence in (1. 2) is uniform in $t$ from any finite interval and in $x \in X_0^N = \{x : p(x) \leq N\}$ for any finite $N$, while for every fixed $t$ the function $P(t, \cdot\,|x)$ continuously maps the space $X_0$ with the topology of the p-bounded coordinatewise convergence ** into the space of the probability measures on $(X_0, \mathcal{B}_0)$ with the topology of the KRW-convergence of all finite-dimensional distributions.

Let us discuss the condition of Theorem 1.

In view of (1. 4) it appears natural to construe the number $g^{c,b}$ as the intensity with which the component $c$ affects the component $b$ and the number $g^b_{m,n}$ as the intensity with which the layer $\alpha_n, \alpha_m$ and the boundary of the box $\alpha_m$ affect the component $b \in \alpha_m$.

If we enumerate quite arbitrarily the elements of $T$ and define the distance on $T$ as $|j - i|$, where $i$ and $j$ are the numbers of the corresponding elements, then we can say that (1. 3) is a condition of decrease in the component interaction intensity with the increased distance between components.

Further on in this article we shall assume that the conditions of theorem 1 are satisfied. The Markov process taking the values from $X_0$ with the transition function $P(t, \cdot\,|x)$ will be called $Q_n$-limiting local interaction Markov process.

Theorem 2. Given $b \in T$ let $h_b(x^b)$, $x^b \in X^b$, be the nonnegative bicompact function, that is, such a function that for any $d$, $0 \leq d < \infty$, the set $\{x^b : h_b(x^b) \leq d\}$ is bicompact (see [II]) and let also the following conditions be satisfied:

1. For some $\rho \geq 0$

$$\rho_b(x^b, \theta^b) \leq \rho + h_b(x^b), \qquad b \in T;$$

---

** The sequence $x_n \in X_0$ converges to $x \in X_0$ in the above-mentioned topology, if

$$\sup_n p(x_n) < \infty \quad \text{and} \quad \rho_b(x_n, x) \underset{n}{\to} 0, \quad b \in T.$$

2. For all $n = 1, 2, \ldots$ the function $h_n(x) = \sum_{b\in\alpha_n} h_b(x^b)$, $x \in X_{\alpha_n}$ is $Q_n$-regular.

3. There exist such constants $0 \leq H < \infty$, $\eta > 0$ and $h^{c,b}$, $b, c \in T$, nonnegative for $b \neq c$, that for all $n = 1, 2, \ldots$

$$\sum_{\alpha : b\in\alpha\subseteq\alpha_n} \int_{X_{\alpha_n}} (h_b(y^b) - h_b(x^b)) q_n^{\alpha}(dy\,|\,x) \leq H + \sum_{c\in\alpha_n} h^{c,b} h_c(x^c) \qquad (1.5)$$

and

$$\sum_{c\in T} h^{c,b} \leq -\eta < 0, \qquad \sum_{c\in T} h^{b,c} \leq -\eta < 0 \qquad (1.6)$$

Then

(A) the $Q_n$-limiting local interaction Markov process has the stationary distribution $\pi$ for which

$$\int_{X_0} h_b(x^b)\pi(dx) \leq H/\eta, \qquad b \in T \qquad (1.7)$$

(B) all of the processes $\xi_n(t)$, $n = 1, 2, \ldots,$ have stationary distributions and for any stationary distribution $\pi_n$ of the process $\xi_n(t)$

$$\int_{X_{\alpha_n}} h_b(x^b)\pi_n(dx) \leq H/\eta, \qquad b \in \alpha_n \qquad (1.8)$$

(C) the stationary distribution of the $Q_n$-limiting process can be obtained as a weak limit of a certain subsequence of the $\pi_n$'s.

Theorem 3. Let for all $n = 1, 2, \ldots$

$$D^b_{n,n}(x, y) \leq \sum_{c\in\alpha_n} \kappa^{c,b} \rho_c(x, y) + (1 + p(x) + p(y))\kappa^b_n$$

where the coefficients $\kappa^{c,b} \geq 0$ for $b \neq c$ $(b, c \in T)$ and for some $\kappa > 0$, $K > 0$

$$\sum_{c \in T} \kappa^{c,b} \leq -\kappa < 0, \qquad b \in T \tag{1.9}$$

$$\sum_{c \in T} |\kappa^{c,b}| \leq K < \infty, \qquad b \in T \tag{1.10}$$

Then

(A) the $Q_n$-limiting process has at most one stationary distribution $\pi$ for which

$$\sup_{b \in T} \int_{X_0} \rho_b(x, \theta)\, \pi(dx) < \infty \tag{1.11}$$

If $\pi$ is the distribution mentioned above then for all $x \in X_0$ such that

$$\sup_{b \in T} \rho_b(x, \theta) < \infty$$

and for any finite $\lambda \subset T$

$$R_\lambda(P(t, \cdot\, |x),\ \pi(\cdot)) \leq K(\lambda, x) e^{-\kappa t}$$

where the coefficient $K(\lambda, x)$ does not depend on $t$;

(B) if $\kappa_n^b = 0$, $b \in \alpha_n$, then the multicompomponent process $\xi_n(t)$ has at most one stationary distribution $\pi_n$ for which

$$\int_{X_{\alpha_n}} \rho_b(x, \theta) \pi_n(dx) < \infty, \qquad b \in \alpha_n$$

If there exists such a distribution $\pi_n$, then $\xi_n(t)$ is ergodic and

$$R_{\alpha_n}(P_n(t, \cdot\, |x),\ \pi_n(\cdot)) \leq K_n(x) e^{-\kappa t}, \qquad x \in X_0$$

where the coefficient $K_n(x)$ does not depend on $t$.

II. When proving theorem 2 we shall make use of the following proposition:

Proposition 1. Let in the phase space $(X, \mathcal{B})$ there be given the Markov chain with such a transition kernel $P(A|x)$, $A \in \mathcal{B}$, $x \in X$, that for some nonnegative measurable function $h(x)$, $x \in X$,

$$\int_X h(y) P(dy|x) \le C + c\,h(x) \tag{2. 1}$$

where $C < \infty$, $0 \le c < 1$. Then for any stationary distribution $\pi$ of this chain

$$\int_X h(x)\pi(dx) \le C/(1 - c)$$

Proof. Let $P^n(A|x)$ be the n-fold convolution of the kernel with itself. Then by iterating (2. 1) we obtain

$$\int_X h(y) P^{(n)}(dy|x) \le C(1 + c + \ldots + c^{n-1}) + c^n h(x) \tag{2. 2}$$

Given $N < \infty$, let

$$h^{(N)}(x) = \min(h(x), N)$$

It is easy to see that

$$\int_X h^{(N)}(x)\pi(dx) \le C/(1 - c) \tag{2. 3}$$

Really

$$\int_X h^{(N)}(x)\pi(dx) = \int_X \pi(dx) \int_X h^{(N)}(y) P^{(n)}(dy|x)$$

$$\le \int_X \pi(dx) \limsup_n \int_X h^{(N)}(y) P^{(n)}(dy|x)$$

In view of (2. 2) the last integral is less than $C/(1-c)$ and this fact proves the inequality (2. 3) and, therefore proposition 1.

Proof of Theorem 2. In compliance with proposition 2 from [I],

$$\frac{d}{dt}\int_{X_{\alpha_n}} h_b(y^b)P_n(t,dy\,|\,x) \le H + \sum_{c\in\alpha_n} h^{c,b}\int_{X_{\alpha_n}} h_c(y^c)P_n(t,dy\,|\,x), \quad b\in\alpha_n,$$

hence

$$\int_{X_{\alpha_n}} h_b(y^b)P_n(t,dy\,|\,x) \le e^{tH}-1+\sum_{c\in\alpha_n}\eta_n^{c,b}(t)h_c(x^c), \quad b\in\alpha_n \tag{2.4}$$

where

$$\eta_n^{c,b}(t) = \delta^{c,b} + th^{c,b} + \sum_{r=2}^{\infty}\frac{t^r}{r!}\sum_{c_1,\dots,c_r\in\alpha_n} h^{c,c_1}\dots h^{c_r,b}, \quad b,c\in\alpha_n$$

($\delta^{c,b}$ - is the Kronecker delta).

Let

$$E_n^b(t) = \sum_{c\in\alpha_n}\eta_n^{c,b}(t), \quad b\in\alpha_n$$

then

$$\frac{d}{dt}E_n^b(t) = \sum_{c\in\alpha_n} h^{c,b}E_n^c(t), \quad b\in\alpha_n$$

$$E_n^b(0) = 1, \quad b\in\alpha_n,$$

hence in view of (1.6)

$$\sum_{c\in\alpha_n}\eta_n^{c,b}(t) \le e^{-\eta t}, \quad b\in\alpha_n \tag{2.5}$$

Similarly,

$$\sum_{c\in\alpha_n}\eta_n^{b,c}(t) \le e^{-\eta t}, \quad b\in\alpha_n \tag{2.6}$$

Summing (2.4) over all $b\in\alpha_n$, we obtain from (2.6) that

$$\int_{X_{\alpha_n}} h_n(y)P_n(t,dy\,|\,x) \le |\alpha_n|(e^{tH}-1)+e^{\eta t}h_n(x), \quad x\in X_{\alpha_n} \tag{2.7}$$

Fix $t > 0$ and consider the Markov chain with the values from $X_{\alpha_n}$, which is obtained by observing the process $\xi_n(\cdot)$ at the moments $kt$, $k = 0, 1, \ldots$ .

Let $\mathfrak{P}_n^{(t)}$ be the set of all stationary distributions of the chain under consideration. This it follows from (2. 7), the theorem of existence of the random field with the given conditional distributions (the theorem proved by Dobrushin in [II ] ) and proposition 1 proved above that, firstly, $\mathfrak{P}_n^{(t)} \neq \phi$ for all $n = 1, 2, \ldots$ and $t > 0$, secondly, in view of (2. 4) for any $\pi_n^{(t)} \in \mathfrak{P}_n^{(t)}$

$$\int_{X_{\alpha_n}} h_b(x^b)\pi_n^{(t)}(dx) \leq (e^{tH} - 1)/(1 - e^{-t\eta}), \quad b \in \alpha_n \tag{2. 8}$$

It is clear that the families $\mathfrak{P}_n^m = \mathfrak{P}_n^{(2^{-m})}$, $m = 1, 2, \ldots$, are compact in the topology of weak convergence of finite-dimensional distributions (see [ ii] , lemma 2) and $\mathfrak{P}_n^{m_1} \supseteq \mathfrak{P}_n^{m_2}$ for $m_1 \leq m_2$. Therefore $\bigcap_{m=1}^{\infty} \mathfrak{P}_n^m \neq \phi$. Let $\pi_n \in \bigcap_{m=1}^{\infty} \mathfrak{P}_n^m$. It is easy to see that $\pi_n$ is the stationary distribution of the process $\xi_n(t)$. Indeed, let $t \geq 0$ and $t_k \xrightarrow[k]{} t$, where $t_k$, $k = 1, 2, \ldots$, are the binary fractions. Then for any bounded measurable function $f(x)$, $x \in X_{\alpha_n}$,

$$\int_{X_{\alpha_n}} \pi_n(dx) \int_{X_{\alpha_n}} f(y) P_n(t, dy \mid x) = \lim_{t_k \to t} \int_{X_{\alpha_n}} \pi_n(dx) \int_{X_{\alpha_n}} f(y) P_n(t_k, dy \mid x)$$

$$= \int_{X_{\alpha_n}} f(x)\pi_n(dx)$$

hence the stationarity of the measure $\pi_n$ follows.

Passing to the limit when $t \to 0$ we obtain (1. 8) from (2. 8). It follows from (1. 8) and lemma 2 mentioned above that the family of measures $\pi_n$, $n = 1, 2, \ldots$, is compact in the topology of weak convergence of all finite-dimensional distributions. Let $\pi$ be the limit point of this family, that is, $\pi$ is the limit of the subsequence

$\pi_{n_k}$, $k = 1, 2, \ldots$, in the aforementioned topology. It is clear that for $\pi$ the inequalities (1. 7) hold true. Thus, to complete proving theorem 2 it remains to show that $\pi$ is the stationary distribution of the $Q_n$ - limiting Markov process. For this reason we shall show that for any finite $\lambda \subset T$ and any bounded function $f(x)$, $x \in X_0$ which depends on $x_\lambda$ only and meets the Lipschitz condition with respect to metric $\rho_\lambda$

$$\int_{X_0} f(x)\pi(dx) = \int_{X_0} \pi(dx) \int_{X_0} f(y) P(t, dy \mid x) \tag{2. 9}$$

Let "$A \overset{(\epsilon)}{=} B$" denote that $|A - B| \leq \epsilon$. Then for every $\epsilon > 0$ and sufficiently large integers $k, m, N$ the following $\epsilon$ -equalities are true:

$$\int_{X_0} \pi(dx) \int_{X_0} f(y) P(t, dy \mid x) \overset{(\epsilon)}{=} \int_{X_0} \pi(dx) \int_{X_{\alpha_m}} f(y) P_m(t, dy \mid x_{\alpha_m})$$
$$\overset{(\epsilon)}{=} \int_{X_0} \pi_{n_k}(dx) \int_{X_{\alpha_m}} f(y) P_m(t, dy \mid x_{\alpha_m}) \overset{(\epsilon)}{=} \int_{X_0^N} \pi_{n_k}(dx) \int_{X_{\alpha_{n_k}}} f(y) P_{n_k}(t, dy \mid x_{\alpha_{n_k}})$$
$$\overset{(\epsilon)}{=} \int_{X_0} f(x)\pi_{n_k}(dx) \overset{(\epsilon)}{=} \int_{X_0} f(x)\pi(dx)$$

Equation (2. 9) is an immediate consequence from these $\epsilon$-equalities.

We shall not prove in detail all the equalities, as far as it is quite a standard procedure, but note instead that the proof of the thirs $\epsilon$ -equality makes use of the fact that for every $\epsilon > 0$ there exists $N < \infty$, such that

$$\pi_n \{x : p(x) \leq N\} > 1 - \epsilon$$

for all $n = 1, 2, \ldots$ simultaneously (it follows from condition 1 and inequality (1. 8)).

Theorem 2 is completely proved.

<u>Proof of Theorem 3.</u> Using Lemma 1 and proposition 2 proved in [1], it is easy to derive from the conditions of the theorem that

$$R_b(P(t,\cdot\,|x),\ P(t,\cdot\,|y)) \le \limsup_n R_b(P_n(t,\cdot\,|x),\ P_n(t,\cdot\,|y))$$

$$\le \limsup \sum_{c\in\alpha_n} r_n^{c,b}(t)\rho_c(x,y), \quad b\in T,\quad x,y\in X_0$$

where

$$r_n^{c,b}(t) = \delta^{c,b} + t\kappa^{c,b} + \sum_{r=2}^{\infty}\frac{t^r}{r!}\sum_{c_1,\dots,c_{r-1}\in\alpha_n}\kappa^{c,c_1}\kappa^{c_1,c_2}\dots\kappa^{c_{r-1},b}$$

($\delta^{c,b}$ is the Kronecker delta).

In view of (1.10) it is clear that the coefficients $r_n^{c,b}(t)$ converge to

$$r^{c,b}(t) = \delta^{c,b} + t\kappa^{c,b} + \sum_{r=2}^{\infty}\frac{t^r}{r!}\sum_{c_1,\dots,c_{r-1}\in T}\kappa^{c,c_1}\kappa^{c_1,c_2}\dots\kappa^{c_{r-1},b}$$

for $n\to\infty$.

It is also clear that

$$R_b(P(t,\cdot\,|x),\ P(t,\cdot\,|y)) \le \sum_{c\in T} r^{c,b}(t)\rho_c(x,y) \qquad (2.10)$$

and

$$r^{c,b}(t)\ge 0, \qquad b,c\in T$$

The following inequalities may be derived from (1.9) in the same way as we derived inequalities (2.5) when proving theorem 2:

$$\sum_{c\in\alpha_n} r_n^{c,b}(t) \le e^{-\kappa t}, \qquad b\in\alpha_n$$

and

$$\sum_{c\in\alpha_n} r^{c,b}(t) \le e^{-\kappa t}, \qquad b\in T \qquad (2.11)$$

as a limit.

Now it is easy to obtain proposition A of theorem 3. Indeed, let $\pi$ be the stationary distribution of the $Q_n$ - limiting process for which (1. 11) is true and let also for $x \in X_0$

$$\sup_{b \in T} \rho_b(x, \theta) < \infty$$

Then from convexity of metric $R_\lambda(\cdot, \cdot)$ in each of its arguments and from (2. 10) and (2. 11) it follows that for any finite $\lambda \subset T$

$$R_\lambda(P(t, \cdot \mid x), \pi(\cdot)) \leq \int_{X_0} \pi(dy) R_\lambda(P(t, \cdot \mid x), \; P(t, \cdot \mid y))$$

$$\leq \sum_{b \in \lambda} \sum_{c \in \lambda} r^{c,b}(t)[\rho_c(x, \theta) + \int_{X_0} \rho_c(y, \theta)\pi(dy)]$$

$$\leq K(\lambda, x) e^{-\kappa t}$$

These inequalities prove that the stationary distribution satisfying (1. 11) is unique. Thus, proposition A is proved.

Proposition B of theorem 3 may be proved in the same way as proposition A.

III. Let $T = \mathbb{Z}^\nu$ be the integer lattice in the $\nu$-dimensional Euclidean space, $X^b = \mathbb{R}^1$, $\rho_b(x^b, y^b) = |x^b - y^b|$, $b \in T$, $\theta$-zero vector in $\mathbb{R}^T$, that is, $\theta^b = 0$, $b \in T$.

For an arbitrary sequence of the finite subsets $\alpha_n$, $n = 1, 2, \ldots$, that converges monotonically to $T$ we define a family of the local characteristics as follows:

$$q_n^\alpha(x) \equiv 0, \qquad \text{if} \qquad |\alpha| > 1$$

$$q_n^b(x)\Pi_n^b(\cdot \mid x), \quad b \in \alpha_n, \qquad \text{is the probability}$$

measure on $(X_{\alpha_n}, \mathcal{B}_{\alpha_n})$ the projection of which onto $(X^b, \mathcal{B}^b)$ is the Gaussian measure with the dispersion $1/f^{b,b}$ and the mean $-\sum_{c\in\alpha_n\setminus b}(f^{c,b}/f^{b,b})x^c$, where the coefficients $f^{c,b}$, $b, c \in T$, satisfy the following conditions:

1. $f^{c,b} = f(|b-c|) \geq 0, \quad b, c \in T;$ (3.1)

2. There exist such constants $\gamma > 0$, $F < \infty$, that

$$\sum_{c\in T} f^{c,b} e^{\gamma|b-c|} \leq F, \quad b \in T; \tag{3.2}$$

3. $f^{b,b} > 0, \quad \sum_{c\in T\setminus b} f^{c,b} < f^{b,b}$

(in view of (3.1) these values do not depend on b);

4. For all n the matrix $\|f^{c,b}\|_{b,c\in\alpha_n}$ is positive definite.

It is easy to see that the aforecited conditions are satisfied, for example, by the values $f^{b,c} = e^{-\alpha|b-c|}$, $b, c \in T$, for $\alpha > 0$.

Assuming

$$p(x) = k \sum_{c\in T} e^{-\tilde{\gamma}|c|}|x^c|$$

where k is the normalization factor, $0 < \tilde{\gamma} < \gamma$, it is easy to see that the aforesaid family of the local characteristics $Q_n$, $n = 1,2,\ldots$, are regular and satisfy the conditions of theorem 1. Therefore, in every subspace

$$X^{\tilde{\gamma}} = \{x : \sum_{c\in T} e^{-\tilde{\gamma}|c|}|x^c| < \infty\} \subseteq X_T, \quad 0 < \tilde{\gamma} < \gamma,$$

there exists the $Q_n$-limiting Markov process $\xi^{\tilde{\gamma}}(t)$, which we shall call "Gaussian". Since the transition functions $P^{\tilde{\gamma}}(t,\cdot|x)$ of the processes $\xi^{\tilde{\gamma}}(t)$, $0 < \tilde{\gamma} < \gamma$, are obtained from the transition functions of the up-to-limit processes by means of the passage to the limit, then for $0 < \tilde{\gamma}_1 < \tilde{\gamma}_2 < \gamma$ the transition function of the process $\xi^{\tilde{\gamma}_1}(t)$ is evidently a restriction of the transition function of the process $\xi^{\tilde{\gamma}_2}(t)$ from the subspace $X^{\tilde{\gamma}_2}$ to the subspace $X^{\tilde{\gamma}_1}$. It is evident as well that the restriction $P^0(t, \cdot \mid x)$ of all transition functions $P^{\tilde{\gamma}}(t, \cdot \mid x)$, $0 < \tilde{\gamma} < \gamma$, to the subspace $Y_0 = \bigcap_{0 < \tilde{\gamma} < \gamma} X^{\tilde{\gamma}}$ is the conservative transition function on $Y_0$, that is, $P^0(t, Y_0 \mid x) = 1$, $x \in Y_0$, $t \geq 0$. The Markov process in the phase space $Y_0$ corresponding to the transition function $P^0(t, \cdot \mid x)$ will be called the minimum $Q_n$-limiting Gaussian process.

It follows from the results of S. S. Vallander [12] that if $N(m_1, \sigma_1)$ and $N(m_2, \sigma_2)$ are the Gaussian distributions in $\mathbb{R}^1$ with means $m_1$ and $m_2$ and dispersions $\sigma_1$ and $\sigma_2$ respectively, then the KRW-distance admits the following estimation:

$$R(N(m_1, \sigma_1), N(m_2, \sigma_2)) \leq |m_1 - m_2| + \sigma_1 - \sigma_2|$$

Using this estimation and taking into account those conditions which were imposed on the coefficients $f^{c,b}$, $b, c \in T$, one can easily make sure that the conditions of theorems 2 and 3 are satisfied and, therefore, each of the processes $\xi^{\tilde{\gamma}}(t)$, $0 < \tilde{\gamma} < \gamma$, has the unique stationary distribution.

The aforesaid implies that the stationary distributions of all processes coincide, are concentrated on $Y_0$ and make up the unique stationary distribution of the minimum Gaussian process.

Proposition 2. The mimimum $Q_n$-limiting Gaussian processes $\xi^0(t)$ is ergodic.

Proof: Introduce the following values:

$$\varphi^{c,b} = (-1)^{\delta^{c,b}} f^{c,b}/f^{b,b}, \quad b, c \in T$$

$$\Phi^{c,b}(t) = \delta^{c,b} + t\varphi^{c,b} + \sum_{r=2}^{\infty} \frac{t^r}{r!} \sum_{c_1, \ldots, c_{r-1} \in T} \varphi^{c,c_1} \varphi^{c_1,c_2} \ldots \varphi^{c_{r-1},b}, b, c \in T \tag{3.3}$$

(cf. $r^{c,b}(t)$ introduced in the previous section). Then

$$R_\lambda(P^0(t,\cdot\,|x),\ P^0(t,\cdot\,|y)) \le \sum_{b\in\lambda}\ \sum_{c\in T} \Phi^{c,b}(t)\,|x^c - y^c|,\quad x,y\in Y_0 \qquad (3.4)$$

where $P^0(t,\cdot\,|x)$ is the transition function of the process $\xi^0(t)$. It is easy to see that

$$\Phi^{c,b}(t+s) = \sum_{a\in T} \Phi^{c,a}(t)\,\Phi^{a,b}(s), \qquad b,c\in T \qquad (3.5)$$

and

$$\sum_{c\in T} \Phi^{c,b}(t) = \sum_{c\in T} \Phi^{b,c}(t) \le e^{-\varphi t}, \qquad b\in T \qquad (3.6)$$

where

$$\varphi = 1 - \sum_{c\in T\setminus b} f^{c,b}/f^{b,b} > 0$$

(in view of (3. 1) this expression does not depend on b). It is not difficult to show by induction that if

$$\sum_{c\in T} |\varphi^{c,b}|\, e^{\gamma|b-c|} \le \Phi < \infty$$

then

$$\sum_{c_1,\ldots,c_r\in T} |\varphi^{c,c_1}\varphi^{c_1,c_2}\cdots\varphi^{c_r,b}| \le \Phi^{r+1} e^{-\gamma|b-c|}$$

Therefore, we receive from (3. 2) and (3. 3) that

$$\Phi^{c,b}(t) \le e^{t\Phi} e^{-\gamma|b-c|}, \qquad b,c\in T \qquad (3.7)$$

where

$$\Phi = F/f^{b,b}$$

Fix $t_0 > 0$. Then it follows from (3. 6) and (3. 7) that there exists such a number $d$, $0 < d < 1$, that for all $\delta$ less than some $\delta_0 > 0$

$$\sum_{c \in T} \Phi^{c,b}(t_0) e^{\delta|b-c|} \leq d \tag{3.8}$$

Let $\pi^0$ be the stationary distribution of the process $\xi^0(t)$, then for $t = (n+1)t_0 + \bar{t}$, where $0 \leq \bar{t} < t$ it follows from (3.4) - (3.8) that for all $x \in Y_0$ and all finite $\lambda \subset T$

$$R_\lambda(P^0(t, \cdot\,|x), \pi^0(\cdot)) \leq \int_{Y_0} \pi^0(dy) R_\lambda(P^0(t, \cdot\,|x), P^0(t, \cdot\,|y))$$

$$\leq \sum_{b \in \lambda} \sum_{c \in T} \Phi^{c,b}(t)|x^c| + O(e^{-\varphi t})$$

$$\leq \sum_{b \in \lambda} \sum_{c \in T} \sum_{c_1, \ldots, c_{n+1} \in T} |x^c| \Phi^{c,c_1}(t_0) \ldots \Phi^{c_n, c_{n+1}}(t_0) \Phi^{c_{n+1}, b}(\bar{t})$$

$$+ O(e^{-\varphi t}) \leq d^{n+1} e^{t_0 \Phi} \sum_{b \in \lambda} \sum_{c \in T} e^{-\delta|b-c|} |x^c| + O(e^{-\varphi t})$$

The last equality entails ergodicity of the process $\xi^0(t)$:

$$\lim_{t \to \infty} R_\lambda(P^0(t, \cdot\,|x), \pi^0(\cdot)) = 0, \quad x \in Y_0$$

Proposition 2 is proved.

It is not hard as well to find an explicit form of this ergodic distribution. This is the distribution of the Gaussian random field given by the system of the conditional distributions

$$\{\pi^b(\cdot\,|x^{T \setminus b}), \ b \in T, \ x^{T \setminus b} \in \mathbb{R}^{T \setminus b}\}$$

where $\pi^b(\cdot\,|x^{T \setminus b})$ is the Gaussian measure on $\mathbb{R}^1$ with the dispersion $1/f^{b,b}$ and the mean

$$m_b(x) = \begin{cases} -\sum_{c \in T \setminus b} (f^{c,b}/f^{b,b})\, x^c, & \text{if the series converges} \\ 0, & \text{otherwise} \end{cases}$$

Indeed, by arguments similar to those in [2], one can demonstrate that each of the processes $\xi^{\tilde{\gamma}}(t)$, $0 < \tilde{\gamma} < \gamma$, as well as the minimum process $\xi^0(t)$ are reversible; the distribution of the aforementioned Gaussian field serves as the stationary distribution of these reversible processes and due to the uniqueness of the stationary distribution coincides with $\pi^0$.

## REFERENCES


1. V. Ya. Basis, Infinite-dimensional Markov processes with almost local interaction of components, Theory Probability Appl. 21, 727-740, 1976. (In Russian).
2. R. L. Dobrushin, Markov processes with a large number of locally interacting components, Probl. Peredaci Inform. 7, 70-87, 1971. (In Russian).
3. R. Holley, A class of interactions in an infinite particle system, Advances in Math. 5, 291-309, 1970.
4. R. Holley, Markov interaction processes with finite range interactions, Annals of Math. Stat. 43, 1961-1967, 1972.
5. T. Harris, Nearest-neighbor Markov interaction processes on multidimensional lattices, Advances in Math. 9, 66-89, 1972.
6. T. Liggett, Existence theorems for infinite particle system, Trans. Amer. Math. Soc. 165, 471-481, 1972.
7. W. Sullivan, A unified existence and ergodic theorem for Markov evolution of random fields, Z. Wahrcheinlichkeitstheorie und Verw. Gebiete, 31, 47-56, 1974.

8. L. N. Wasserstein, Markov processes over denumerable products of spaces describing large systems of automata, Problemy Pekedaci Informacii 5, 64-73, 1969. (In Russian).

9. D. Griffeath, Ergodic theorems for graph interactions, Adv. App. Prob. 7, 179-194, 1975.

10. R. V. Kantorovich and G. S. Rubinstein, On the functional space and some extremal problems, Dokl. Akad. Nauk SSSR, 115, 1058-1061, 1957. (In Russian).

11. R. L. Dobrushin, Definition of random variables by conditional distributions, Theory Probl. Appl. 15, 469-497, 1970. (In Russian).

12. S. S. Vallander, Calculation of the Wasserstein distance between probability distributions on the line, Theory Prob. Appl. 18, 824-827, 1973. (In Russian).

chapter 3

# ϵ-EXPANSION FOR AUTOMODEL RANDOM FIELDS

P. M. Bleher

## INTRODUCTION

The renormalization group (RG) method has made it possible to deal with many important problems of statistical physics. It has in particular explained the scaling theory of critical behavior which was suggested earlier in several physical works (see works by Kadanoff et al. [1], Patashinsky and Pokrovsky [2], Widom [3], Stanley [4]). The RG method was proposed by Wilson (see [5, 6]) and is based on the Gell-Mann-Low transformation in quantum field theory and an analogy between quantum field theory and statistical mechanics. This analogy had been established earlier by different authors (see, for example, papers by Gribov, Migdal [8], Polyakov [9]). It seems worth pointing out that Kadanoff's idea of "block-spin interactions" (see [10]) was also very important for Wilson (see [6]). The contribution of other scientists to the development of the RG method is described in the works by Wilson and Kogut [6] and Patashinsky and Pokrovsky [30].

According to Wilson the behavior of statistical mechanics systems at the critical point is determined by a universal effective hamiltonian. The latter is constructed as a fixed point of some nonlinear transformation (the RG transformation) and the critical exponents of the system can be expressed via the eigenvalues of this transformation differential for the effective hamiltonian.

There exist various physical applications of the RG method (most of them are mentioned in [2, 6, 11--13]). But the RG method also poses some mathematical problems the importance of which was pointed out by different authors (see, for example, [6, 11]). These problems are concerned both with a rigorous approach to the RG method and the solution of the equations arising from its application. Bleher and Sinai have investigated second order phase transitions in so-called asymptotically-hierarchical models which are a slight generalization of Dyson's well-known hierarchical models (see [14--17] and [18, 19]). The main nonlinear equation obtained and investigated in [14--17] can be considered as the equation of the RG transformation for asymptotically-hierarchical models (see [20]). Thus the applicability of the RG method to these models is actually proved in the above mentioned works. The papers by Gallavotti, Knops [20], Jona-Lasinio [21] and Baker [22] contain some non-rigorous results concerning the $\epsilon$-expansion for hierarchical models. It should be emphasized that there exists a deep connection between the block structure of Dyson's hierarchical model hamiltonians and the RG transformation. But at the same time hierarchical models have the serious defect of not being translation-invariant.

A mathematical approach to the RG method for translation-invariant random fields was suggested by Sinai in connection with second order phase transition in ferromagnet models with long-range potentials (see [23-- 25]). In these papers a rigorous definition was given of the RG transformation for discrete random fields (in Kadanoff's style) and Gaussian automodel fields were constructed.

An $\epsilon$-expansion construction algorithm for non-Gaussian automodel fields was suggested which is similar to the Wilson algorithm (see [6]). In an interesting paper by Dobrushin a natural RG for continuous random fields was introduced and its connection with the Sinai RG was demonstrated (see [28]).

The present work may be considered as a continuation of Sinai's investigations. Our main results enable us to make the RG transformation differential spectrum more precise and to find non-Gaussian automodel fields in second order perturbation theory. It is demonstrated tha the RG fixed point equation in second order perturbation theory has a unique solution in some class of (formal) random fields (the definition of formal random fields is given below). The random fields of this class are restricted by two important conditions which have natural physical interpretations.

§ 1 gives the main definitions and notation and contains the derivation of the linear equation for the automodel field spectral density. Solving this equation we obtain again the formula for the automodel field spectral density which was found before in [23]. In § 2 similar equations are obtained for the eigenfunctions of the RG transformation differential for Gaussian automodel fields. The solution of these equations makes it possible to investigate the differential spectrum. It is shown further, that for a random field which is a linear transformation of a given one the differential is written as an integral Gauss transformation (for the definition and properties of the latter, see [26]). As a consequence it is proved that the spaces of Hermite polynomials are differential-invariant.

The central point of this paper is § 3, where the equations of the perturbation theory for non-Gaussian automodel fields are analyzed. The solvability of the equations of the perturbation theory is investigated and it is shown that the second order equation has a unique solution under some natural regularity and non-degeneration conditions. It should be emphasized that this equation is solved without any iteration procedure as was suggested by Wilson, Kogut [6] and Sinai [23].

The main idea here is to represent the quadratic part of the RG transformation for a Gaussian automodel field in a special way. For this purpose the theory of generalized functions is used. In particular, the solvability of the second order equation of perturbation theory is connected with the non-homogeneity of the generalized function $f(x) = \text{reg } |x|^{-d}$ which is defined by the equation

$$(\text{reg}\,|x|^{-d}, \varphi(x)) \overset{\text{def}}{=} \int_{-\infty}^{\infty} |x|^{-d} (\varphi(x) - \varphi(0)\chi_{\{|x|<1\}}(x)) d^d x, \; \varphi(x) \in C_0^{\infty}$$

(see [29]). Namely it is easy to see that

$$f(\lambda x) = \lambda^{-d}(f(x) + S_{d-1} \ln \lambda\, \delta(x))$$

for $\lambda > 0$, where $S_{d-1}$ is the volume of the $(d-1)$-dimensional sphere. The formula for the solution of the second order equation of the perturbation theory contain several regularized functions of the above type.

We express our gratitude to Prof. Ja. G. Sinai for his valuable help in preparing this paper for publication.

## I. MAIN DEFINITIONS AND GENERAL RESULTS

I. 1. Definition of the RG for random fields. Let $\Sigma$ be a set of real-valued functions $\xi = \xi(x)$, $x \in \mathbb{Z}^d$, on the cubic lattice $\mathbb{Z}^d$ ($\Sigma$ is the configuration space of random fields) and $\alpha > 0$, $n \geq 1$ respectively real and integral numbers, otherwise arbitrary.

Definition I. 1 (see [23]). The renormalization group (RG) transformation is the mapping of the configuration space $\Sigma$ into itself which is defined by the formula

$$R_n(\alpha) : \xi = \xi(x) \to \xi_n = \xi_n(x) \equiv \sum_{r \in \mathbb{Z}_n^d} \xi(nx + r)/n^{\frac{\alpha}{2}} \qquad (1.1)$$

where $\mathbb{Z}_n^d = \underbrace{\mathbb{Z}_n \oplus \ldots \oplus \mathbb{Z}_n}_{d} = \{r = r_1, \ldots, r_d) | 0 \le r_i \le n-1, r_i \in \mathbb{Z}, i = 1, \ldots, d\}$

It is clear, that $R_n(\alpha) R_k(\alpha) = R_{nk}(\alpha)$.

Definition 1. 2 (see [23]). The renormalization group is the commutative semigroup of the transformations $\{R_n(\alpha),\ n = 1, 2, \ldots\}$.

Definition 1. 3 (see [23]). The conjugate RG $R^*(\alpha)$ is the commutative semigroup of the dual mapping of random fields

$$R_n^*(\alpha) : P(\xi) \to P((R_n(\alpha))^{-1} \xi)$$

In what follows the parameter $\alpha$ is fixed and therefore the dependence of $\alpha$ is as a rule not indicated. It is noteworthy, that the RG is not a group because the inverse operation is not defined. For the field of independent random values the RG transformation reduces to sums of independent random variables.

Definition 1. 4 (see [23]). A random field P is automodel if $R_n^* P = P$ for all $n = 1, 2, \ldots$. The hamiltonian defining the automodel field is called the effective hamiltonian.

The main problem of RG theory is to find automodel fields and to investigate their stability properties with respect to RG transformations. We should like to point out that the RG defined above fits Kadanoff's picture (see [1], [10]) of critical phenomena. In Wilson's papers a slightly different RG based on the impulse representation of random fields was considered.

I. 2. Correlation functions of automodel fields. It follows

directly from definition 1. 3 that the correlation functions of the transformed field $P_n = R_n^* P$ satisfy the following equation

$$\rho_n(x^{(1)},\dots x^{(N)}) \equiv \langle \xi_n(x^{(1)})\dots \xi_n(x^{(N)})\rangle_{P_n}$$

$$= \langle (n^{-\frac{\alpha}{2}} \sum_{r^{(1)}\in \mathbb{Z}_n^d} \xi(nx^{(1)}+r^{(1)}))\dots (n^{-\frac{\alpha}{2}} \sum_{r^{(N)}\in \mathbb{Z}_n^d} \xi(nx^{(N)}+r^{(N)}))\rangle_P$$

$$= n^{-\frac{\alpha N}{2}} \sum_{r^{(1)},\dots,r^{(N)}\in \mathbb{Z}_n^d} \langle \xi(nx^{(1)}+r^{(1)})\dots \xi(nx^{(N)}+r^{(N)})\rangle_P$$

$$= n^{-\frac{\alpha}{2}N} \sum_{y^{(1)}\in V_n(x^{(1)}),\dots,y^{(N)}\in V_n(x^{(N)})} \rho(y^{(1)},\dots,y^{(N)}) \qquad (1.\ 2)$$

where $V_n(x) = \{y \in \mathbb{Z}^d \,|\, y = nx + r,\ r\in \mathbb{Z}_n^d\}$, $\rho(y^{(1)},\dots,y^{(N)}) = \langle \xi(y^{(1)})\dots \xi(y^{(N)})\rangle_P$, $y^{(i)} = (y_1^{(i)},\dots,y_d^{(i)})$. This equation describes the RG transformation for correlation functions. The same equation is valid for cumulants.

For an automodel field $\rho_n(x^{(1)},\dots,x^{(N)}) = \rho(x^{(1)},\dots,x^{(N)})$ and so for all $n = 1, 2, \dots$

$$\rho(x^{(1)},\dots,x^{(N)}) = n^{-\frac{\alpha N}{2}} \sum_{y^{(1)}\in V_n(x^{(1)}),\dots,y^{(N)}\in V_n(x^{(N)})} \rho(y^{(1)},\dots,y^{(N)}) \qquad (1.\ 3)$$

To construct solutions of this equation let us consider an arbitrary homogeneous function $q(x^{(1)},\dots,x^{(N)})$,

$$q(\lambda x^{(1)},\dots,\lambda x^{(N)}) = \lambda^{-\gamma} q(x^{(1)},\dots,x^{(N)}), \qquad \lambda > 0,$$

and find the limit

$$q_\infty(x^{(1)},\dots,x^{(N)}) = \lim_{n\to\infty} n^{-\frac{\alpha N}{2}} \sum_{y^{(1)}\in V_n(x^{(1)}),\dots,y^{(N)}\in V_n(x^{(N)})} q(y^{(1)},\dots,y^{(N)})$$

We have

$$q_n(x^{(1)},\dots,x^{(N)}) \equiv n^{-\frac{\alpha N}{2}} \sum_{y^{(1)}\in V_n(x^{(1)}),\dots,y^{(N)}\in V_n(x^{(N)})} q(y^{(1)},\dots,y^{(N)})$$

$$= n^{-\frac{\alpha N}{2}} \sum_{r^{(1)},\dots r^{(N)}\in \mathbb{Z}_n^d} q(nx^{(1)}+r^{(1)},\dots,nx^{(N)}+r^{(N)})$$

$$= n^{-\frac{\alpha N}{2}-\gamma+Nd} \sum_{r^{(1)},\dots,r^{(N)}\in \mathbb{Z}_n^d} q(x^{(1)}+n^{-1}r^{(1)},\dots,x^{(N)}+n^{-1}r^{(N)})n^{-Nd}$$

but $\sum_{r^{(1)},\dots,r^{(N)}\in \mathbb{Z}_n^d} q(x^{(1)}+n^{-1}r^{(1)},\dots,x^{(N)}+n^{-1}r^{(N)})n^{-Nd}$ is the integral sum for the integral

$$\int_{\Lambda(x^{(1)})}\cdots\int_{\Lambda(x^{(N)})} q(y^{(1)},\dots,y^{(N)})dy^{(1)}\dots dy^{(N)},$$

where $\Lambda(x) = \{y\in \mathbb{R}^d \mid y = x+t,\ t = (t_1,\dots,t_d), 0\le t_1,\dots,t_d\le 1\}$. So the nontrivial limit $q_\infty(x^{(1)},\dots,x^{(N)}) = \lim_{n\to\infty} q_n(x^{(1)},\dots,x^{(N)})$ exists only if the homogeneity order $(-\gamma)$ equals $-(Nd - \frac{N\alpha}{2})$ and in this case

$$q_\infty(x^{(1)},\dots,x^{(N)}) = \int_{\Lambda(x^{(1)})}\cdots\int_{\Lambda(x^{(N)})} q(y^{(1)},\dots,y^{(N)})dy^{(1)}\dots dy^{(N)}$$

under the condition that all the right side integrals exist. This equation gives the general solution of the fixed point equations (1.3) for the class of functions regular at infinity $\rho(x^{(1)},\dots,x^{(N)})$. So for an automodel field we have

$$\rho(x^{(1)},\dots,x^{(N)}) = \int_{\Lambda(x^{(1)})} \cdots \int_{\Lambda(x^{(N)})} q(y^{(1)},\dots,y^{(N)})dy^{(1)}\dots dy^{(N)} \qquad (1.\ 4)$$

where $q(y^{(1)},\dots,y^{(N)})$ is a homogeneous function of order

$$-\gamma_N = -Nd + \frac{N\alpha}{2} \qquad (1.\ 5)$$

If $\xi_t$, $t \in \mathbb{R}^d$, is a random field in $\mathbb{R}^d$ and $\xi(x) = \int_{\Lambda(x)} \xi_t dt$, $x \in \mathbb{Z}^d$, its discretization, then the correlation functions of the fields $\xi(x)$ and $\xi_t$ are connected by the same equation (1. 4). So there exists a correspondence between automodel random fields in $\mathbb{Z}^d$ and homogeneous random fields in $\mathbb{R}^d$ (for this correspondence see the paper by Dobrushin [28]).

In what follows we restrict ourselves for the sake of definiteness to stationary isotropic random fields.

Definition 1. 5. A stationary random field $P(\xi)$, $\xi = \xi(x)$, is called isotropic if and only if its cumulants $\rho^T(x^{(1)},\dots,x^{(N)})$ are asymptotically isotropic, that is

$$\lim_{\max|x^{(i)}-x^{(j)}|\to\infty} \frac{\rho^T(Ux^{(1)},\dots,Ux^{(N)})}{\rho^T(x^{(1)},\dots,x^{(N)})} = 1$$

for any isometric transformation $U$ of $\mathbb{R}^d$.

Remark. The cumulants $\rho^T(x^{(1)},\dots,x^{(N)})$ are defined inductively by the following equations

$$\rho(X) = \sum_{X_1 \cup \dots \cup X_r = X} \rho^T(X_1)\dots\rho^T(X_r)$$

(see, for instance, [30]).

For a stationary isotropic automodel field the equation (1. 4) gives

$$\rho(x^{(1)}, x^{(2)}) = \int_{\Lambda(x^{(1)})} \int_{\Lambda(x^{(2)})} q(y^{(1)}y^{(2)})dy^{(1)}dy^{(2)}$$

where $q(y)$ is isotropic, so $q(y) = C|y|^{\alpha-2d}$ and

$$\rho(x^{(1)}, x^{(2)}) = C \int_{\Lambda(x^{(1)})} \int_{\Lambda(x^{(2)})} |y^{(1)}y^{(2)}|^{\alpha-2d} dy^{(1)}dy^{(2)} \tag{1. 6}$$

The higher correlation functions of a stationary isotropic automodel field are not defined uniquely by the equation (1. 4).

I. 3. <u>The RG differential.</u> Let $P(\xi)$ be a random field on $\mathbb{Z}^d$ and $P_n(\xi) = R_n^* P(\xi)$. The RG differential for $P(\xi)$ is an operator from the space of random variables of $P(\xi)$ to the one of $P_n(\xi)$. Namely let us consider the conditional random fields $P(\xi \mid R_n\xi = \xi') = P_{\xi'}(\xi)$ induced by $P(\xi)$ for almost all $\xi'$. The corresponding conditional mathematical expectation is denoted by $M(\cdot \mid R_n\xi = \xi')$.

<u>Definition 1. 6.</u> <u>The conditional mathematical expectation operator</u>

$$D_n = D_n(\alpha, P) : H(\xi) \to H'(\xi') = M(H(\xi) \mid R_n\xi = \xi')$$

<u>is called the RG transformation differential for the random field</u> $P(\xi)$.

Now we would like to make some remarks elucidating this definition.

Let us assume that $P(\xi)$ and $P_n(\xi) = R_n^* P(\xi)$ are Gibbsian random fields with hamiltonians $H$ and $H_n$ respectively. In physical literature it is just the mapping $H \to H_n$ that is referred to as the RG transformation (see, for instance [6]). Generally speaking this definition is not correct, because in the multiphase region the Gibbsian field $P$ corresponding a given hamiltonian $H$ is not unique and the hamiltonian $H_n$ actually depends on P and not only on H. But this definition of the RG transformation is correct in the one-phase region up to the critical point and very useful for calculations. The above defined differential is the formal linearization of the non-linear transformation $r_n : H \to H_n$. To show this let us consider an arbitrary random field P. The probabilistic measure on the configuration space which determines the random field P will be denoted by $\mu(d\xi)$.

Proposition 1.1. Let $H(\xi)$ be a bounded random variable and $\frac{\mu'(d\xi)}{\mu(d\xi)} = 1 + H(\xi)$. Then $\frac{R_n^* \mu'(d\xi)}{R_n^* \mu(d\xi)} = 1 + D_n H(\xi)$.

Proof: Let $H'(\xi)$ be a random variable. We have (in obvious notation)

$$\int H'(\xi')R_n^*\mu'(d\xi) \equiv \int H'(R_n\xi)\mu'(d\xi) = \int H'(R_n\xi)(1 + H(\xi))\mu(d\xi)$$

$$= \int H'(R_n\xi)(1 + H(\xi))\mu(d\xi)R_n \,|R_n\xi = \xi')R_n^*\mu(d\xi')$$

$$= \int H'(\xi')D_n(1 + H(\xi'))R_n^*\mu(d\xi') = \int H'(\xi')(1 + D_nH(\xi'))R_n^*\mu(d\xi')$$

Since $H'(\xi)$ is arbitrary the proposition is proved.

Proposition 1.2. Let $H(\xi)$ be a bounded random variable, $\epsilon \in \mathbb{R}^1$ and $\frac{\mu'(d\xi)}{\mu(d\xi)} = \exp(\epsilon H(\xi))$. Then $\frac{R_n^*\mu'(d\xi)}{R_n^*\mu(d\xi)} = \exp(\epsilon D_n H(\xi) + O(\epsilon^2))$, where $O(\epsilon^2)$ is in the uniform norm.

Proof: Using the previous proposition we have

$$\frac{R_n^*\mu'(d\xi)}{R_n^*\mu(d\xi)} = 1 + D_n(e^{\epsilon H(\xi)} - 1) = 1 + D_n(\epsilon H(\xi) + O(\epsilon^2))$$

$$= 1 + \epsilon D_n H(\xi) + O(\epsilon^2) = \exp(\epsilon D_n H(\xi) + O(\epsilon^2)) .$$

Thus the proposition is proved.

This result justifies the above definition of the RG transformation differential. For further calculations we need an extension of the differential action.

Let $H(\xi)$ be a hamiltonian, that is,

$$H(\xi) = \sum_{n=1}^{\infty} \sum_{x^{(1)},\dots,x^{(n)} \in \mathbb{Z}^d} a_n(x^{(1)},\dots,x^{(n)})\xi(x^{(1)}\dots\xi(x^{(n)})$$

a formal monomial sum which satisfies the translation invariance condition $a_n(x^{(1)}+j,\dots,x^{(n)}+j) = a_n(x^{(1)},\dots,x^{(n)})$ for any $j \in \mathbb{Z}^d$, n. Let P be a random field all the moments of which are finite. Then $\xi(x^{(1)})\dots\xi(x^{(N)})$ is a random variable for P. Let us assume now that the random variable $D_n(\xi(x^{(1)})\dots\xi(x^{(N)}))$ can be written in the form

$$D_n(\xi(x^{(1)})\dots\xi(x^{(N)})) = \sum_{m=0}^{\infty} \sum_{y^{(1)},\dots,y^{(m)}} b_m^{(N)}(y^{(1)},\dots,y^{(m)};$$

$$x^{(1)},\dots,x^{(N)})\xi(y^{(1)}\dots\xi(y^{(m)}) \tag{1.7}$$

for any $x^{(1)},\dots,x^{(N)} \in \mathbb{Z}^d$, $n = 1, 2, \dots$ . Below we shall calculate the coefficients $b_m^{(N)}$ exactly in the case of a Gaussian automodel random field $P(\xi)$. The operator $D_n = D_n(\alpha, P)$ is a linear one so it is natural to define for

$$H(\alpha) = \sum_{N=1}^{\infty} \sum_{x^{(1)},\dots,x^{(N)} \in \mathbb{Z}^d} a_n(x^{(1)},\dots,x^{(N)})\xi(x^{(1)}\dots\xi(x^{(N)})$$

that

$$D_n H(\xi) = \sum_{N=1}^{\infty} \sum_{x^{(1)},\dots,x^{(N)} \in \mathbb{Z}^d} a_n(x^{(1)},\dots,x^{(N)})D_n(\xi(x^{(1)}\dots\xi(x)^{(N)}))$$

$$= \sum_{m=0}^{\infty} \sum_{y^{(1)},\dots,y^{(m)} \in \mathbb{Z}^d} c_m(y^{(1)},\dots,y^{(m)})\xi(y^{(1)})\dots\xi(y^{(m)}) \tag{1.8}$$

where

$$c_m(y^{(1)},\dots,y^{(m)}) = \sum_{N=1}^{\infty} \sum_{x^{(1)},\dots,x^{(N)}\in\mathbb{Z}^d} b_m^{(N)}(y^{(1)},\dots,y^{(m)};$$

$$x^{(1)},\dots,x^{(N)})a(x^{(1)},\dots,x^{(N)}) \qquad (1.9)$$

provided that these series converge.

Definition 1.7. The RG transformation differential in the space of hamiltonians is defined by the equations (1.7)--(1.9).

I.4. The RG transformation for formal hamiltonians. The above definition of the RG transformation was framed in terms of random fields, while for the $\epsilon$-expansion calculations it is actually necessary to know the transformation of hamiltons. The introduction of formal power series enables one to define the RG transformation for hamiltonians. Let $P_0$ be a Gaussian automodel isotropic random field and $H_0$ its hamiltonian (see section I.2).

Definition 1.8. The formal power series $H = H_0 + \sum_{k=1}^{\infty} \epsilon^k H_k$ with arbitrary hamiltonian $H_k$ is called a formal hamiltonian.

Definition 1.9. The RG transformation $r_n = r_n(\alpha, P_0)$ in the space of formal hamiltonians is the composition of the following three transformations:

1. Exponential expansion

$$\exp(-\sum_{k=1}^{\infty} \epsilon^k H_k) = 1 - \frac{1}{1!}(\sum_{k=1}^{\infty} \epsilon^k H_1) + \frac{1}{2!}(\sum_{k=1}^{\infty} \epsilon^k H_k) - \dots = 1 - \sum_{k=1}^{\infty} \epsilon^k G_k$$

2. Differential action

$$\sum_{k=1}^{\infty} \epsilon^k G_k \to \sum_{k=1}^{\infty} \epsilon^k D_n G_k$$

3. Logarithm application

$$\ln(1-\sum_{k=1}^{\infty}\epsilon^k D_n G_k) = -\sum_{k=1}^{\infty}\epsilon^k D_n G_k - \frac{1}{2}(\sum_{k=1}^{\infty}\epsilon^k D_n G_k)^2 - \dots = -\sum_{k=1}^{\infty}\epsilon^k Q_k$$

that is $\quad r_n : H_0 + \sum_{k=1}^{\infty}\epsilon^k H_k \to H_0 + \sum_{k=1}^{\infty}\epsilon^k Q_k .$

It is easy to verify that

$$Q' = D_n H' - \frac{1}{2!}(D_n H'^2 - (D_n H')^2) + \frac{1}{3!}(D_n H'^3 - 3D_n H' D_n H'^2 + 2(D_n H')^3) + \dots \tag{1.10}$$

where

$$H' = \sum_{k=1}^{\infty}\epsilon^k H_k, \quad Q' = \sum_{k=1}^{\infty}\epsilon^k Q_k, \quad D_n = D_n(\alpha, P_0)$$

and the sum runs over the expressions of cumulant type. Substituting the $\epsilon$-expansion of $H'$ in the last equation we have

$$Q_1 = D_n H_1, \quad Q_2 = D_n H_2 - \frac{1}{2!}(D_n H_1^2 - (D_n H_1)^2) \tag{1.11}$$

and so on.

The reason for such a definition of the formal hamiltonian RG transformation is illustrated by the following proposition.

Proposition 1.3. Let $H_k(\xi)$, $k = 1, 2, \dots$, be random variables uniformly bounded in $k$ and $\xi$, and

$$\frac{\mu_\epsilon(d\xi)}{\mu_0(d\xi)} = e^{-\sum_{k=1}^{\infty}\epsilon^k H_k},$$

where $\mu_0(d\xi)$ is the measure generating random field $P_0$ and $|\epsilon| < 1$. Then

$$\frac{R_n^* \mu_\epsilon(d\xi)}{R_n^* \mu_0(d\xi)} = e^{-\sum_{k=1}^{\infty}\epsilon^k Q_k},$$

where $Q_k$ are constructed by the formulae of definition 9.

Proof: By the proposition 1. 1

$$\frac{R_n^* \mu_\epsilon (d\xi)}{R_n^* \mu_0 (d\xi)} = 1 + D_n \left(e^{-\sum_{k=1}^{\infty} \epsilon^k H_k} - 1\right)$$

$$= 1 - \sum_{k=1}^{\infty} \epsilon^k D_n G_k = e^{-\sum_{k=1}^{\infty} \epsilon^k Q_k}$$

Q. E. D.

In reality the $\epsilon$-expansion arises in situations when $\epsilon$ is not a free parameter but is connected with the RG parameter $\alpha$ in such a way that $\alpha = \alpha_0 + \epsilon$ where $\alpha_0$ is some fixed number. In this case the differential $D_n$ and hence the coefficients $Q_k$ are functions of $\epsilon$ and the definition 1. 9 is slightly changed. Namely a fourth step is added: the expansion of the formal series

$$\sum_{k=1}^{\infty} \epsilon^k Q_k(\epsilon) = \sum_{k=1}^{\infty} \epsilon^k \sum_{\ell=0}^{\infty} Q_{k\ell} \epsilon^\ell = \sum_{m=1}^{\infty} \Big( \sum_{k+\ell=m} Q_{k\ell} \Big) \epsilon^m$$

in the formal series whose coefficients do not depend on $\epsilon$ .

Definition 1. 9'. The RG transformation when $\alpha = \alpha_0 + \epsilon$ and $\alpha_0$ is fixed is the composition of the four above given transformations.

The main problem of formal RG theory is to calculate $\epsilon$-expansions for formal effective hamiltonians. In what follows we intend to consider the RG transformation of formal hamiltonians in the case when the main random field $P_0$ is an isotropic Gaussian automodel field. Moreover we will consider only formal hamiltonians

$$H = H_0 + \sum_{k=1}^{\infty} \epsilon^k H_k ,$$

$$H_k = \sum_{m=1}^{\infty} \sum_{x^{(1)},\ldots,x^{(m)} \in \mathbb{Z}^d} a_m^{(k)}(x^{(1)},\ldots,x^{(m)}) \xi(x^{(1)}) \ldots \xi(x^{(m)})$$

such that $a_m^{(k)}(x^{(1)},\ldots,x^{(m)}) \equiv 0$ if $m$ is greater than some $M = M_{(k)}$, i. e. $H_k$ are finite-particle hamiltonians.

I. 5. The $\epsilon$-expansion uniqueness conditions and the main theorem. The existence of non-quadratic formal effective hamiltonians is possible only at some bifurcation values $\alpha_0$ (now $\alpha = \alpha_0 + \epsilon$). These values are characterized by the condition that the spectrum of the operator $D_n = D_n(\alpha_0, P_0)$ contains one. Such values $\alpha_0$ were discussed in [23], but the procedure used there needs to be made more precise. The key lies in the correct definition of the space of hamiltonians $H_k$ which are the coefficients of an effective hamiltonian. As we shall see below the bifurcation values $\alpha_0$ occupy the whole segment if no conditions are posed on the space of hamiltonians. It turns out that the bifurcation values $\alpha_0$ pointed out in [23] are extracted by a natural condition of the hamiltonian's coefficients decrease. The following definition will be useful.

Definition I. 10. A formal hamiltonian $H = \sum_{k=1}^{\infty} \epsilon^k H_k$,

$$H_k = \sum_{m=1}^{M_k} \sum_{x^{(1)},\ldots,x^{m} \in \mathbb{Z}^d} a_m^{(k)}(x^{(1)},\ldots,x^{(m)})\xi(x^{(1)}\ldots\xi(x^{(m)})$$

has the coefficients' decrease property (CDP) if for any $a_m^{(k)}$

$$\lim_{d(X_1,X_2)\to\infty} \frac{a_m^{(k)}(x^{(1)},\ldots,x^{(m)})}{d(X_1,X_2)^{\alpha}} = 0$$

where $X_1 \cup X_2 = \{x^{(1)},\ldots,x^{(m)}\}$, $X_1 = \{x^{(i_1)},\ldots,x^{(i_p)}\}$, $X_2 = \{x^{(i_{p+1})},\ldots,x^{(i_m)}\}$ is an arbitrary division of the variables $x^{(1)},\ldots,x^{(m)}$ into two non-empty sets of variables, $d(X_1,X_2) = \inf_{k \le p < \ell} \left|x^{(i_k)} - x^{(i_\ell)}\right|$, and $\alpha$ is a RG parameter.

It is noteworthy that the coefficients $a_0(x^{(1)} - x^{(2)})$ of a quadratic isotropic effective hamiltonian decrease as const. $\left|x^{(1)}-x^{(2)}\right|^{-\alpha}$, when $|x^{(1)} - x^{(2)}| \to \infty$, so CDP means that the coefficients $a_m^{(k)}$, $k \geq 1$, decrease faster than $a_0(x^{(1)} - x^{(2)})$. To formulate our main result we introduce now the following definition.

Definition 1.11. A formal hamiltonian H is automodel in m-th order of the perturbation theory if the first m coefficients of the $\epsilon$ power series $r_n(H) - H$ are equal to zero.

Main Theorem. If the dimension d is not divisible by 4, then for bifurcation values $\alpha_0 = 2d - \frac{d}{j}$, $j = 2, 3, \ldots$, and only for them, there exist nonquadratic isotropic automodel in 2-nd order perturbation theory formal hamiltonians, which have CDP.

Remark. A procedure for the $\epsilon$-expansion construction of the effective hamiltonians was suggested earlier in [23], but this procedure was too formal. Namely the convergence of the iteration process used in that paper is not clear. We construct the effective hamiltonian directly without any iteration process.

Moreover we would like to emphasize that the dimension condition is essential. If d is divisible by 4 then an effective hamiltonian which has CDP does not exist.

I. 6. Impulse representation of the RG transformation. In the remainder of this section another representation of the RG transformation will be given and an equation for the spectral density of an automodel field will be obtained.. The same type of equation will arise in §II in the eigenvalues problem for the RG differential. A general approach to the solution of these equations is given in the present section.

It follows directly from Definition 1.1, that the RG transformation may be written in impulse representation in the following way (see also [23]):

$$\tilde{R}_n : \tilde{\xi}(k) \to \tilde{\xi}_n(k) = n^{-\frac{\alpha}{2}} p(k) T_n[\tilde{\xi}(k) p^{-1}(k)] , \tag{1.12}$$

where $\tilde{\xi}(k) = \sum_{j \in \mathbb{Z}^d} e^{ijk} \xi(j)$, $p(k) = \prod_{r=1}^{d} (e^{-ik_r} - 1)$,

$$T_n[\varphi(k)] \equiv n^{-d} \sum_{j \in \mathbb{Z}^d} \varphi\left(\frac{k+2\pi j}{n}\right) = n^{-d} \sum_{j_1,\dots,j_d=0} \varphi\left(\frac{k_1+2\pi j_1}{n},\dots,\frac{k_d+2\pi j_d}{n}\right) \tag{1.13}$$

Using substitutions

$$\theta(k) = p^{-1}(k)\xi(k) = \prod_{r=1}^{d} (e^{-ik_r} - 1)\xi(k), \qquad \theta_n(k) = p^{-1}(k)\xi_n(k) \tag{1.14}$$

we obtain that

$$\bar{R}_n : \theta(k) \to \theta_n(k) = n^{-\frac{\alpha}{2}} T_n[\theta(k)] \tag{1.15}$$

where $\bar{R}_n \theta(k) = p^{-1}(k)\tilde{R}_n(p(k)\theta(k))$, $\bar{R}_n = p^{-1}\tilde{R}_n p$ .

The operators $\tilde{R}_n$, $\bar{R}_n$ are equivalent to the RG transformation $R_n$ and will be also referred to as the RG transformations. Similarly the transformations of generalized random fields

$$\tilde{R}_n^* : \tilde{P}(\tilde{\xi}) \to \tilde{P}(\tilde{R}_n^{-1}\tilde{\xi}) \tag{1.16}$$

$$\bar{R}_n^* : \bar{P}(\theta) \to \bar{P}(\bar{R}_n^{-1}\theta) \tag{1.17}$$

are equivalent to $R_n^*$ and they will be referred to as conjugate RG transformations. It is noteworthy that $\tilde{P}(\tilde{\xi}) = P(\xi)$ is a generalized random field with respect to the test functions space

$$E = C^\infty_{per}(\mathbb{R}^d) = \{\varphi(k) \in C^\infty | \varphi(k + 2\pi j) = \varphi(k) \ \ \forall j \in \mathbb{Z}^d\}$$

of $C^\infty$ periodic functions while the test function space corresponding to $\bar{P}(\theta) = \tilde{P}(p\theta)$ is

$$E_0 = \{\varphi(k) | \varphi(k) = p(k)\psi(k), \quad \psi(k) \in E\}$$

The $\bar{P}(\theta)$ binary correlation function is written in the form

$$\langle \theta(k)\theta(k') \rangle_{\bar{P}} = G(k)\delta_p(k+k') \quad G_k \equiv \sum_{j\in\mathbb{Z}^d} \delta(k_1 + k_2 - 2\pi j)$$

where $G(k)$ is the spectral density of stationary random field $\bar{P}(F[\eta])$, $\theta = F[\eta]$ is Fourier transformation. Next we have

$$\langle \theta_n(k)G_n(k') \rangle_{\bar{R}_n^* \bar{P}} = \langle \bar{R}_n \theta(k) \bar{R}_n \theta(k') \rangle_{\bar{P}}$$

$$= n^{-\alpha} \langle T_n[\theta(k)]\, T_n[\theta(k')] \rangle_{\bar{P}} = n^{-\alpha-2d} \sum_{j,j'\in \mathbb{Z}_n^d} \langle \theta(\frac{k+2\pi j}{n})$$

$$\times\, \theta(\frac{k'+2\pi j'}{n}) \rangle_{\bar{P}} = n^{-\alpha}\, \delta_p(k+k')T_n[G(k)]$$

Thus the spectral density $G(k)$ is transformed by the rule

$$G_n(k) = n^{-\alpha} T_n[G(k)] \tag{1. 18}$$

Fixed points of the last transformation correspond to automodel random fields. We seek fixed points as $\lim_{n\to\infty} n^{-\alpha} T_n[G_0(k)]$, where $G_0(k)$ is some initial function. Let $Q(k)$ be a homogeneous function of order $(-\gamma)$ such that $\sum_{j\in\mathbb{Z}^d} |Q(k+2\pi j)| < \infty$ for any $k \neq 0$. Then

$$n^{-\alpha} T_n[Q(k)] = n^{-\alpha-d} \sum_{j=(j_1,\dots,j_d):\{-\pi \le \frac{k_r + 2\pi j_r}{n} < \pi\}_{r=1}^d} Q(\frac{k+2\pi j}{n})$$

$$= n^{-\alpha-d+\gamma} \sum Q(k+2\pi j)$$

so for $\gamma = \alpha + d$

$$n^{-\alpha} T_n[Q(k)] \xrightarrow[n\to\infty]{} \sum_{j\in\mathbb{Z}^d} Q(k+2\pi j) = G_Q(k) \tag{1. 19}$$

The limit function $G_Q(k)$ is a fixed point of the transformation (1.18). Thus there exist various fixed points of this transformation. Our aim is to find the one which corresponds to the isotropic random field P. We have

$$<\tilde{\xi}(k)\tilde{\xi}(k')>_{\tilde{P}} = p(k)p(k') <\theta(k)\theta(k')>_{\bar{P}} = |p(k)|^2 G(k)\delta_p(k+k')$$

If a random field $P(\xi)$ is isotropic in the sense of the definition 1. 5 then the function $|p(k)|^2 G(k)$ is asymptotically isotropic when $k \to 0$. Only for

$$Q(k) = \left(\prod_{r=1}^{d} k_r^2\right)^{-1} |k|^{-\alpha+d} \tag{1. 20}$$

is the function $|p(k)|^2 G_Q(k)$ asymptotically isotropic. Thus we have

Proposition 1. 4 (see also [23]). The spectral density of an automodel isotropic random field has the form

$$\rho_{iso}(k) = |p(k)|^2 \sum_{j \in \mathbb{Z}^d} Q(k + 2\pi j)$$

where $p(k) = \prod_{r=1}^{d} (e^{-ik_r} - 1)$ and $Q(k) = |k|^{-\alpha+d} \left(\prod_{r=1}^{d} k_r\right)^{-2}$.

Remark. One can obtain this result directly from the equation (1. 6).

## II. THE RG DIFFERENTIAL

II. 1. The main equation for the RG differential. In this section an equation will be given which is very useful for RG differential calculation. Let $P_0(\xi)$ be an automodel Gaussian isotropic random field and $D_n = D_n(\alpha, P_0)$ be the RG differential corresponding ot it. Let us consider now an arbitrary hamiltonian

$$H(\xi) = \sum_{m=1}^{M} \sum_{\bar{x} \in \mathbb{Z}^{md}} a_m(\bar{x}) \xi(x^{(1)}) \ldots \xi(x^{(m)}$$

where $\bar{x} = (x^{(1)}, \ldots, x^{(m)}) = (x_1^{(1)}, \ldots, x_d^{(m)})$. It is convenient for us to rewrite it using Wick ordering in the form

$$H(\xi) = \sum_{m=1}^{M} \sum_{\bar{x} \in \mathbb{Z}^{d(m)}} b_m(\bar{x}) : \xi(x^{(1)}) \ldots \xi(x^{(m)}) : \qquad (2.1)$$

where : : is taken with respect to $P_0$ (the constant term is thrown away). It will be shown that the hamiltonian $H_n(\xi) = D_n H(\xi)$ has the same structure

$$H_n(\xi) = \sum_{m=1}^{M} \sum_{\bar{x} \in \mathbb{Z}^{md}} b_m^{(n)}(\bar{x}) : \xi(x^{(1)}) \ldots \xi(x^{(m)}) :$$

and the coefficients $b_m^{(n)}$ are calculated by the equations

$$b_m^{(n)}(\bar{x}) = \sum_{\bar{y} \in \mathbb{Z}^{(md)}} b(\bar{y}) c(x^{(1)}, y^{(1)}) \ldots c(x^{(m)}, y^{(m)}) \qquad (2.2)$$

where $c(x, y)$ is an explicitly calculated function of $x, y \in \mathbb{Z}^d$, $\bar{y} = (y^{(1)}, \ldots, y^{(m)})$.

The equation (2. 2) is rather complicated for analysis. It is essentially simplified if one introduces a new random field

$$\zeta(t) = \sum_{x \in \mathbb{Z}^d} \Gamma_0(x - t)\xi(x) = \Gamma_0 \xi(t) \qquad (2.3)$$

where

$$\Gamma_0(x) = (2\pi)^{-d} \int_{-\infty}^{\infty} e^{-ikx} [G(k)|k|^{\alpha-d} \prod_{m=1}^{d} (ik_m e^{-ik_m} - 1))]^{-1} d^d k \qquad (2.4)$$

and

$$G(k) = \sum_{j \in \mathbb{Z}^d} Q(k + 2\pi j), \quad Q(k) = |k|^{-\alpha+d} \prod_{m=1}^{d} k_m^{-2}$$

(see (1. 20)). It is necessary to emphasize that the argument $t \in \mathbb{R}^d$ is a continuous parameter, so $\zeta(t)$ is a continuous (not discrete) random field. The main properties of $\Gamma_0(x)$ are

$$\int_{-\infty}^{\infty} \Gamma_0(x)\, d^d x = 1, \quad \Gamma_0(x) \in C^d(\mathbb{R}^d), \quad |\Gamma_0(x) < \frac{\text{const.}}{(1+|x|)^{\alpha+1}} \tag{2. 5}$$

If

$$H(\zeta) = \sum_{m=1}^{M} \int_{-\infty}^{\infty} h_m(\bar{t}) : \zeta(t^{(1)}) \ldots \zeta(t^{(m)}) : d^{md}\bar{t}$$

is a hamiltonian in $\zeta$-representation then one obtains a hamiltonian $H'(\xi) = H(\Gamma_0 \xi)$ by the substitution $\zeta = \Gamma_0 \xi$ :

$$H'(\xi) = H(\Gamma_0 \xi) = \sum_{m=1}^{M} \sum_{\bar{x} \in \mathbb{Z}^{md}} b_m(\bar{x}) : \xi(x^{(1)} \ldots \xi(x^{(m)}) :$$

where

$$b_m(\bar{x}) = \int_{-\infty}^{\infty} h_m(\bar{t}) \prod_{i=1}^{m} \Gamma_0(x^{(i)} - t^{(i)})\, d^{md}\bar{t} \tag{2. 6}$$

Let $\bar{x} = (x^{(1)}, \ldots, x^{(m)} \in \mathbb{R}^{md}$, $x^{(i)} \in \mathbb{R}^d$, $i = 1, \ldots, m$. Let us assume that $m$ is an even number and divide any $2p$ variables $x^{(i_1)}, \ldots, x^{(i_{2p})}$, $1 \le i_1 < \ldots < i_{2p} \le m$, into $p$ pairs $\{(x^{(i_{j_1})}, x^{(i_{j_2})}) \ldots (x^{(i_{j_{2p-1}})}, x^{(i_{j_{2p}})})\} = \pi$. The set of all the divisions for given $p$ is denoted by $\Pi_p$. To each division $\pi$ corresponds the function

$$\Phi_\pi^{(\alpha)}(\bar{x}) = \prod_{r=1}^{p} \left(1 + \left| x^{(i_{j_{2r-1}})} - x^{(i_{j_{2r}})} \right| \right)^{-\alpha+2d} \prod_{i \notin I_\pi} (1+|x|)^{\alpha+1} \tag{2. 7}$$

where $I_\pi = \{i_{1p}, \ldots, i_{2p}\}$ and $i \notin I_\pi$ means that $i \in \{1, \ldots, m\}$ $\{i_1, \ldots, i_{2p}\}$. Finally

$$\Phi^{(\alpha)}(\bar{x}) = \sum_{p=0}^{m/2-1} \sum_{\pi \in \Pi_p} \Phi_\pi^{(\alpha)}(\bar{x}) \tag{2. 8}$$

With the help of the function $\Phi^{(\alpha)}(\bar{x})$, we introduce the following Banach space

$$X_\alpha(\mathbb{R}^{md}) = \{\varphi(\bar{x}) \in C^d(\mathbb{R}^{md}) \mid \|\varphi\|_\alpha$$

$$= \sup_{\bar{x}\in\mathbb{R}^{md}} \Big( \sum_{|\beta|\le d} |D^\beta \varphi(\bar{x})|\Big)\Phi^{(\alpha)}(\bar{x}) < \infty\} \;;$$

here $\beta = (\beta_1^{(1)}, \ldots, \beta_d^{(m)})$ is a multiindex, $D^\beta$ is partial derivative, $|\beta| = \beta_1^{(1)} + \ldots + \beta_d^{(m)}$.

We will say that a hamiltonian $H(\zeta) = \sum_{m=1}^{M} \int_{-\infty}^{\infty} h_m(\bar{t}) \times$ $\times : \zeta(t^{(1)}) \ldots \zeta(t^{(m)}) : d^{md}\bar{t}$ is admissible if $h_m(\bar{t}) \in X'_\alpha(\mathbb{R}^{md})$, $m = 1, \ldots, M$, where $X'_\alpha(\mathbb{R}^{md})$ is a dual space to $X_\alpha(\mathbb{R}^{md})$. If $H(\zeta)$ is admissible then the integrals (2. 6) are defined and uniformly bounded for $\bar{x} \in \mathbb{R}^{md}$.

Now we turn to the formulation of the main result of this section.

Theorem 2. 1. If a hamiltonian $H(\zeta)$ is admissible then $D_n H(\Gamma_0 \xi)$ exists and can be written in the form $D_n H(\Gamma_0 \xi) = H_n(\Gamma_0 \xi)$, where

$$H_n(\zeta) = \sum_{m=1}^{M} n^{-\frac{m\alpha}{2}} \int_{-\infty}^{\infty} h_m(n\bar{t}) : \zeta(t^{(1)}) \ldots \zeta(t^{(m)}) : d^{md}\bar{t} .$$

The hamiltonian $H_n(\zeta)$ is also admissible.

Remarks. 1. A closely related equation was obtained in [25].

2. A slight generalization of the theorem will also be helpful for us. Namely, for any positive integers $p_1, \ldots, p_m$

$$D_n \int_{-\infty}^{\infty} h(\bar{t}) : \zeta^{p_1}(t^{(1)}) \ldots \zeta^{p_m}(t^{(m)}) : d^{md}\bar{t}$$

$$= n^\gamma \int_{-\infty}^{\infty} h(n\bar{t}) : \zeta^{p_1}(t^{(1)}) \ldots \zeta^{p_m}(t^{(m)}) : d^{md}\bar{t} \tag{2. 9}$$

where $\gamma = \frac{|p|\alpha}{2} - |p|d + md$, $|p| = p_1 + \ldots + p_m$. The last equation is derived as a particular case of the theorem when

$h_{|p|}(t^{(1)},\dots,t^{(|p|)}) \equiv h(t^{(1)}, t^{(q_1)},\dots,t^{(q_{m-1})}) \times \delta(t^{(1)}-t^{(2)})\dots\delta(t^{(1)}-t^{(q_1-1)})\delta(t^{(q_1)}-t^{(q_1+1)})\dots\delta(t^{(q_{m-1})}-t^{(|p|)})$, $q_i = p_1 +\dots+p_{i-1}+1$, $i = 1,\dots,m-1$, and $h_r \equiv 0$, $r \neq |p|$.

3. The equation for $D_n H(\Gamma_0 \xi)$ emphasizes the scaling character of the RG transformations.

4. The choice of a function $\Gamma_0(x)$ in (2. 3), (2. 4) will be clarified in the proof.

As a supplement to Theorem 2. 1 we indicate some properties of the Fourier transform of the function $\Gamma_0(x)$

$$q_0(k) = \int_{-\infty}^{\infty} e^{ikx}\Gamma_0(x)d^dx = G^{-1}(k)\rho(k)\prod_{m=1}^{d}(e^{-ik_m}-1)^{-1} \tag{2.10}$$

and of two point correlation function

$$\psi(x, x') - \langle\xi(x)\xi(x')\rangle_0$$

$$= \int_{-\infty}^{\infty}\int_{-\infty}^{\infty} e^{i(xk+x'k')}\rho(k)\rho(k')G^{-1}(k)\delta_p(k\ k')d^dkd^dk' \tag{2. 11}$$

Here $\rho(k) = |k|^{d-\alpha}\prod_{m=1}^{d}(ik_m)^{-1}$.

<u>Supplement to Theorem 2. 1.</u> <u>The function</u> $q_0(k) \in C^{\ell,\delta}(\mathbb{R}^d)$, $\ell = [\alpha - d+1]$, $\delta = \alpha - d + 1 - \ell$, <u>i. e. the</u> $\ell$-th <u>partial derivatives</u> of $q_0(k)$ <u>belong to Hölder space with exponent</u> $\delta$; $q_0(0) = 1$ <u>and</u>

$$|q_0(k)| < \text{const.}\ |k|^{d-\alpha}\prod_{m=1}^{d}(1 + |k_m|)^{-1} \tag{2. 12}$$

<u>The function</u> $\psi(x, x') \in C^d(\mathbb{R}^{2d})$ <u>and</u>

(i) $\psi(x + j,\ x' + j) = \psi(x, x')$ <u>for any</u> $j \in \mathbb{Z}^d$

(ii) <u>for</u> $|x - x'| \to \infty$

$$\psi(x, x') = C|x - x'|^{\alpha-2d} + O(|x - x'|^{-\alpha-2}) \tag{2. 13}$$

where

$$C = 2^{2d-\alpha}\pi^{d/2}\Gamma(d - \frac{\alpha}{2})\Gamma^{-1}(\frac{\alpha - d}{2}) \tag{2.14}$$

The remainder of this section is concerned with the proof of Theorem 2.1 and the supplement to it, and has a technical character. Moreover the considerations of this section will not be used below, so if the reader has become acquainted with the formulation of Theorem 2.1 he may proceed to §III.

The proof of Theorem 2.1 will be performed in several steps. It will be convenient for us to use the impulse representation of the RG transformation.

II. 2. Decomposition of Gaussian automodel fields. Let us consider now the RG transformation

$$\overline{R}_n : \theta(k) \to \theta_n(k) = n^{-\frac{\alpha}{2}} T_n[\theta(k)] = n^{-\frac{\alpha}{2}-d} \sum_{j\in\mathbb{Z}_n^d} \theta(\frac{k+2\pi j}{n})$$

and let $\mu_0(d\theta)$ be a Gaussian automodel (generalized) random field with spectral density

$$G(k) = \sum_{j\in\mathbb{Z}^d} Q(k+2\pi j), \quad Q(k) = |k|^{-\alpha+d}(\prod_{r=1}^{d} k_r^2)^{-1} \tag{2.15}$$

(see Proposition 1.4). Let us recall that such a choice of $Q(k)$ ensures the isotropy property of the random field

$$\tilde{\xi}(k) = p(k)\theta(k) = \prod_{r=1}^{d} (e^{-ik_r} - 1)\theta(k)$$

The RG transformation differential is a conditional functional integral. Now we would like to consider the differential $\overline{D}_n$ for the automodel Gaussian field $\mu_0(d\theta)$. In this case the conditional functional integral is computed with the help of an orthogonal decomposition

$$\theta(k) = a(k)\theta_n(nk) + \omega(k) \tag{2.16}$$

where $\langle\theta_n(nk)\omega(k')\rangle_0 = 0$ for any $k, k' \in \mathbb{R}^d$. Here $\langle\cdot\rangle_0$ is the mathematical expectation with respect to the measure $\mu_0(d\theta)$. The last equation defines uniquely a (deterministic) function $q(k)$. Namely we have

$$\langle\theta_n(nk')\theta(k)\rangle_0 = a(k)\overline{\langle\theta_n(nk')\theta_n(nk)\rangle_0},$$

$$n^{-\frac{\alpha}{2}-d}\sum_{j\in\mathbb{Z}_n^d}\langle\theta(k'+\frac{2\pi j}{n})\theta(k)\rangle_0 = a(k)\langle\theta(nk')\theta(nk)\rangle_0$$

(we use the fact that $\theta(k)$ is an automodel random field). Due to the equation $\langle\theta(k)\theta(k')\rangle_0 = G(k)\delta_p(k-k')$ this leads to the relation

$$n^{-\frac{\alpha}{2}-d}G(k)\sum_{j\in\mathbb{Z}_n^d}\delta_p(k'+2\pi j/n-k) = a(k)G(nk)\delta_p(nk'-nk)$$

But

$$\sum_{j\in\mathbb{Z}_n^d}\delta_p(k'+2\pi j/n-k) = n^d\delta_p(nk'-nk)$$

and therefore

$$n^{-\frac{\alpha}{2}}G(k)\delta_p(nk'-nk) = a(k)G(nk)\delta_p(nk'-nk),\quad a(k) = n^{-\frac{\alpha}{2}}\frac{G(k)}{G(nk)} \tag{2.17}$$

This is just the desired equation for $a(k)$.

The Gaussian automodel field $\theta(k)$ under consideration is generated by the quadratic hamiltonian $\overline{H}_0(\theta) = \int_{-\pi}^{\pi}\frac{|\theta(k)|^2}{2G(k)}d^dk$. It follows immediately from the orthogonal decomposition (2.16) that

$$\overline{H}_0(\theta) = \overline{H}_0(\theta_n) + \overline{H}_0(\omega) \tag{2.18}$$

Also, it follows from (2.8), (2.9), that $\omega(k)$ satisfies the equation

$$T_n[\omega(k)] = 0 \tag{2.19}$$

The two point function $\langle\omega(k)\omega(k')\rangle_0$ is easily computed using (2.16):

$$\langle\omega(k)\omega(k')\rangle_0 = \langle(\theta(k) - a(k)\theta_n(nk))(\theta(k') - a(k')\theta_n(nk'))\rangle_0$$

$$= G(k)(\delta_p(k+k') - n^{-\alpha}G(k')G^{-1}(nk')\delta_p(nk+nk')) \qquad (2.20)$$

Straightforward calculations show that the Fourier transform

$$K(x, x') = (2\pi)^{-2d}\int_{-\infty}^{\infty}\int_{-\infty}^{\infty} e^{-i(kx+k'x')}\langle\omega(k)\omega(k')\rangle d^dk d^dk'$$

of the two point function satisfies the periodicity condition

$$K(x+nj, x'+nj) = K(x, x') \qquad (2.21)$$

for any $j \in \mathbb{Z}^d$ and the estimate

$$|K(x, x')| < \text{const.}\ (1 + |x - x'|)^{-\alpha-2d} \qquad (2.22)$$

II. 3. Diagonal part of the RG differential. Now let

$$H(\theta) = \int_{-\pi}^{\pi} h(\bar{k})\theta(k^{(1)})\dots\theta(k^{(r)})d^{rd}\bar{k}, \quad \int_{-\pi}^{\pi} = \underbrace{\int_{-\pi}^{\pi}\dots\int_{-\pi}^{\pi}}_{rd}$$

be a homogeneous hamiltonian; here $\bar{k} = (k^{(1)},\dots,k^{(r)}) = (k_1^{(1)},\dots,k_d^{(r)}) \in \mathbb{R}^{rd}$. Let us substitute the orthogonal decomposition (2.16) in this hamiltonian:

$$H(\theta) = \int_{-\pi}^{\pi} h(\bar{k}) \prod_{m=1}^{r} (a(k^{(m)})\theta_n(mk^{(m)}) + \omega(k^{(m)}))\, d^{rd}\bar{k}$$

and open the brackets. Due to equation (2.18) we have $\mu_0(d\theta) = \mu_1(d\theta_n)\mu_2(d\omega)$, where $\mu_1, \mu_2$ are mutually independent Gaussian measures. So the conditional measure $\mu_0(d\theta)|n^{-\frac{\alpha}{2}}T_n[\theta(k)] = \theta_n(k))$ does not depend on its conditioning and coincides with $\mu_2(d\omega)$. Thus

$$\bar{D}_n H(\theta_n) = \int_{-\pi}^{\pi} h(\bar{k}) \sum_{J \subseteq \{1,\ldots,n\}} \prod_{m \in J} a(k^{(m)}) \theta_n (nk^{(m)}) \times$$

$$\times \int \mu_2(d\omega) \prod_{m \notin J} \omega(k^{(m)}) \, d^{rd}\bar{k} \tag{2.23}$$

By the Wick theorem, higher moments in this formula can be expressed via the $\mu_2(d\omega)$ -- two point function which is given in (2.20). As a result we obtain that $\bar{D}_n H(\theta_n)$ is a sum of homogeneous hamiltonians:

$$\bar{D}_n H(\theta_n) = \sum_{p=0}^{r} \int_{-\pi}^{\pi} g_p^{(n)}(\bar{k}) \theta_n (nk^{(1)}) \ldots \theta_n (nk^{(p)}) d^{pd}\bar{k}$$

$$= \sum_{p=0}^{r} \int_{-n\pi}^{n\pi} g_p^{(n)}(n^{-1}\bar{k}) \theta_n (k^{(1)}) \ldots \theta_n (k^{(p)}) n^{-pd} d^{pd}\bar{k}$$

$$= \sum_{p=0}^{r} \int_{-\pi}^{\pi} n^{-pd} \sum_{\bar{j} \in \mathbb{Z}^{pd}} g_p^{(n)}\left(\frac{\bar{k} + 2\pi \bar{j}}{n}\right) \theta_n (k^{(1)}) \ldots \theta_n (k^{(p)}) d^{pd}\bar{k}$$

$$= \sum_{p=0}^{r} \int_{-\pi}^{\pi} h_p^{(n)}(\bar{k}) \theta_n (k^{(1)}) \ldots \theta_n (k^{(p)}) d^{pd}\bar{k}$$

where

$$h_p^{(n)} = n^{-pd} \sum_{\bar{j} \in \mathbb{Z}_n^{pd}} g_p^{(n)}\left(\frac{\bar{k} + 2\pi \bar{j}}{n}\right) = T_n [g_p^{(n)}(\bar{k})] \tag{2.24}$$

Now one can consider the maps

$$\bar{\partial}_n^{(r,p)} : h(k^{(1)}, \ldots, k^{(r)}) \to h_p^{(n)}(k^{(1)}, \ldots, k^{(p)})$$

All the maps $\bar{\partial}_n^{(r,p)}$ are linear integral operators whose kernels are defined uniquely by the function $G(k)$.

The map $\bar{\partial}_n^{(r,r)}$ will be called the diagonal part of the RG differential. In view of the triangular character of the RG differential (that is $\bar{\partial}_n^{(r,p)} \equiv 0$ for $p > r$) the spectrum of the RG differential is determined in the Gaussian case by the diagonal part $\bar{\partial}_n^{(r,r)}$.

It follows from (2. 23), (2. 24) that

$$\bar{\partial}_n^{(r,r)} h(\bar{k}) = T_n[h(\bar{k}) \prod_{m=1}^{r} a(k^{(m)})]$$

Let us use now equation (2. 17). It states that

$$a(k) = n^{-\frac{\alpha}{2}} G(k) G^{-1}(nk)$$

so

$$\bar{\partial}_n^{(r,r)} h(\bar{k}) = T_n[h(\bar{k}) \prod_{m=1}^{r} (n^{-\frac{\alpha}{2}} G(k^{(m)}) G^{-1}(nk^{(m)}))]$$

$$= n^{-\frac{r\alpha}{2}} \prod_{m=1}^{r} G^{-1}(k^{(m)}) T_n[h(\bar{k}) \prod_{m=1}^{r} G(k^{(m)})] \qquad (2.\ 25)$$

Using the notation

$$g(\bar{k}) = h(\bar{k}) \prod_{m=1}^{r} G(k^{(m)}), \quad g_n(\bar{k}) = (\bar{\partial}_n^{(r,r)} h(\bar{k})) \prod_{m=1}^{r} G(k^{(m)}) \qquad (2.\ 26)$$

the last equation can be rewritten in a very simple form:

$$g_n(\bar{k}) = n^{-\frac{r\alpha}{2}} T_n[g(\bar{k})] \qquad (2.\ 27)$$

It is noteworthy that the equation obtained is analogous to equation (1. 18) describing the RG transformation of the spectral density. This remark will be useful in seeking eigenvectors of the operator $\bar{\partial}_n^{(r,r)}$.

Summing up our considerations we have

Proposition 2. 2. The diagonal part of the RG differential is given by the equation (2. 25).

II. 4. Eigenvectors of the diagonal part of the RG differential. Our next step is to find eigenvectors of the diagonal part. It suffices to consider eigenvectors of the transformation (2. 27) because after that one obtains eigenvectors of the diagonal part of the RG differential using the substitution (2. 26).

Our considerations are the same as in Section I. 6. As before let $Q(\bar{k})$, $\bar{k} = (k^{(1)}, \ldots, k^{(r)}) \in \mathbb{Z}^{rd}$, be a homogeneous function of order $(-\gamma)$ and

$$Q_0(\bar{k}) = \delta_p(k^{(1)} + \ldots + k^{(r)})Q(\bar{k}) \tag{2. 28}$$

when $|k_1^{(1)}|, \ldots, |k_d^{(r)}| \leq \pi$. An eigenfunction of the transformation (2. 27) arises as a leading term of the functions $T_n[Q_0(\bar{k})]$ when $n \to \infty$. We have

$$T_n[Q_0(\bar{k})] = n^{-rd} \sum_{\bar{j} \in \mathbb{Z}_n^{rd}} Q_0(\bar{k} + 2\pi\bar{j})$$

$$= n^{-rd} \sum_{\bar{j} \in \mathbb{Z}^{rd} : \{-\pi \leq \frac{k_p^{(m)} + 2\pi j_p^{(m)}}{n} < \pi\}} \delta_p\left(\sum_{m=1}^{r} \frac{k^{(m)} + 2\pi j^{(m)}}{n}\right) Q\left(\frac{\bar{k} + 2\pi\bar{j}}{n}\right)$$

but

$$n^{-d}\delta_p(n^{-1}k) \xrightarrow[n \to \infty]{} \delta(k)$$

so

$$\lim_{n \to \infty} (n^{rd-\gamma-d} T_n[Q_0(\bar{k})] = \sum_{\bar{j} \in \mathbb{Z}^{rd}} \delta\left(\sum_{m=1}^{r} (k^{(m)} + 2\pi j^{(m)})\right) Q(\bar{k} + 2\pi\bar{j}) \tag{2. 29}$$

under condition that the last series converges. Thus if the series does converge the function

$$g_Q(\bar{k}) = \sum_{\bar{j} \in \mathbb{Z}^{rd}} \delta\left(\sum_{m=1}^{r} (k^{(m)} + 2\pi j^{(m)})\right) Q(\bar{k} + 2\pi\bar{j}) \tag{2. 30}$$

is an eigenfunction of the operator (2. 27) with eigenvalue

$$\lambda_Q = n^{-\frac{r\alpha}{2} + \gamma - d(r-1)} \tag{2. 31}$$

It is easy to verify that the function

$$g_0(\bar{k}) = \delta_p(k^{(1)} + \ldots + k^{(r)}) \tag{2.32}$$

is also an eigenfunction of the operator (2.27) with eigenvalue

$$\lambda_0 = n^{-\frac{r\alpha}{2}} \tag{2.33}$$

This eigenfunction arises as a limit of $T_n[Q_0(\bar{k})]$ in the case when the function $Q_0(\bar{k})$ has integrable singularity at the origin.

Now using the equation (2.26) one obtains eigenfunctions of the diagonal part of the differential of the RG transformation $\bar{R}_n$. It is more convenient for us to reformulate this result in terms of the diagonal part of the differential of the RG transformation $\tilde{R}_n$. The latter is equivalent to $\bar{R}_n$ by equation (1.14). Analogous equivalence takes place between the differentials $\tilde{D}_n$ and $\bar{D}_n$ and hence between their diagonal part operators $\tilde{\partial}_n^{(r,r)}$ and $\bar{\partial}_n^{(r,r)}$. Thus eigenfunctions of the operator $\tilde{\partial}_n^{(r,r)}$ are connected with those of the operator $\bar{\partial}_n^{(r,r)}$ by the additional multiplier $\prod_{m=1}^{r} p^{-1}(k^{(m)}) = \prod_{m=1}^{r} \prod_{p=1}^{d} (e^{-ik_p^{(m)}} - 1)^{-1}$. Thus we have

<u>Theorem 2.3.</u> <u>Let</u>

$$q(k) = (G(k) \prod_{p=1}^{d} (e^{-ik_p} - 1))^{-1}, \quad q^{(r)}(\bar{k}) = \prod_{m=1}^{r} q(k^{(m)}) \tag{2.34}$$

<u>Then there are the following eigenfunctions of the diagonal part $\tilde{\partial}_n^{(r,r)}$ of the differential of the RG transformation $\tilde{R}_n$:</u>

(i) $\tilde{h}_0(\bar{k}) = q^{(r)}(\bar{k})\delta_p(k^{(1)} + \ldots + k^{(r)})$, <u>eigenvalue</u> $\lambda_0 = n^{-\frac{r\alpha}{2}}$;

(ii) $\tilde{h}_Q(\bar{k}) = q^{(r)}(\bar{k}) \sum_{\bar{j} \in \mathbb{Z}^{rd}} \delta(\sum_{m=1}^{r} (k^{(m)} + 2\pi j^{(m)})) Q(\bar{k} + 2\pi\bar{j})$

(2.35)

where $Q(\lambda\bar{k}) = \lambda^{-\gamma}Q(\bar{k})$, $\lambda > 0$, is a homogeneous function such that the series defining the function $\tilde{h}_Q(\bar{k})$ converges; eigenvalue $\lambda_Q = n^{-\frac{r\alpha}{2} + \gamma - d(r-1)}$.

Remarks. 1. The necessary condition for series convergence is the inequality $\gamma > d(r-1)$, so we always have $\lambda_Q > \lambda_0$.

2. Generalized homogeneous functions $Q(\bar{k})$ are admissible in (ii) in principle. For example one can use $Q(\bar{k}) = \frac{\delta(k_1^{(1)})}{|\bar{k}|^{rd}}$ and so on.

3. Presumably all the eigenfunctions of the diagonal part of the differential $\tilde{D}_n$ have the above described structure.

II. 5. Nondegenerate eigenfunctions. As we have seen in the previous section there exist many eigenfunctions of the diagonal part of the RG differential. Some of them were found earlier by Sinai [23]. It turns out that Sinai's eigenfunctions are extracted by a non-degeneracy condition. Namely let us consider an arbitrary eigenfunction $\tilde{h}_Q(\bar{k})$ (see Theorem 2. 3) such that the homogeneous function $Q(\bar{k})$ is smooth away from the origin. Then the singularity of $\tilde{h}_Q(\bar{k})$ at the origin is

$$\delta(k^{(1)}+\ldots+k^{(r)})Q(\bar{k}) \prod_{m=1}^{r} (G(k^{(m)}) \prod_{p=1}^{d} (e^{-ik_p^{(m)}} - 1))^{-1} \sim$$

$$\sim \delta(k^{(1)}+\ldots+k^{(r)})Q(\bar{k}) \prod_{m=1}^{r} (|k^{(m)}|^{\alpha-d} \prod_{p=1}^{d} (ik_p^{(m)}))$$

Definition 2. 1. An eigenfunction $\tilde{h}_Q(\bar{k})$ is called non-degenerate if

$$\tilde{h}_Q(\bar{k}) = \delta(k^{(1)}+\ldots+k^{(p)})(1 + o(1)) \quad (2.\ 37)$$

for $\bar{k} \to 0$.

Proposition 2.4. *If*

$$Q(\overline{k}) = \prod_{m=1}^{r} \rho(k^{(m)}) \tag{2.38}$$

$\rho(k) = |k|^{-\alpha+d} \prod_{p=1}^{d} (ik_p)^{-1}$ then the eigenfunction $\tilde{h}_Q(\overline{k})$ is non-degenerate.

Proof: This follows from (2.36).

One can verify that Proposition 2.4 gives the full description of non-degenerate eigenfunctions. Thus for each $r = 1, 2, \ldots$ we have a unique non-degenerate eigenfunction $\tilde{h}^{(r)}(\overline{k}) \equiv \tilde{h}_Q(\overline{k})$ where $Q(\overline{k})$ is given by (2.38). It follows from the general formula of Theorem 2.3 that the corresponding eigenvalues are

$$\lambda^{(r)} = n^{\frac{r\alpha}{2} - rd + d} \tag{2.39}$$

Precisely the eigenfunctions $\tilde{h}^{(r)}(\overline{k})$ were found by Sinai. These eigenfunctions appear in the construction of non-Gaussian formal non-degenerate random fields.

Let us consider now formal hamiltonians

$$\tilde{H}^{(r)}(\tilde{\xi}) = \int_{-\pi}^{\pi} \tilde{h}^{(r)}(\overline{k}) \prod_{m=1}^{r} \tilde{\xi}(k^{(m)}) d^{rd}\overline{k}, \quad \int_{-\pi}^{\pi} = \underbrace{\int_{-\pi}^{\pi} \cdots \int_{-\pi}^{\pi}}_{rd}$$

Due to the special character of the homogeneous function $Q(\overline{k})$ in (2.38) the $\tilde{H}^{(r)}(\tilde{\xi})$ can be written in a very compact form. We have

$$\delta(k) = (2\pi)^{-d} \int_{-\infty}^{\infty} e^{iyk} d^d y$$

so

$$\tilde{h}^{(r)}(\overline{k}) = \prod_{m=1}^{r} q(k^{(m)}) \sum_{\overline{j} \in \mathbb{Z}^{rd}} \delta\Big(\sum_{m=1}^{r} (k^{(m)} + 2\pi j^{(m)})\Big) Q(\overline{k} + 2\pi \overline{j})$$

$$= \prod_{m=1}^{r} q(k^{(m)}) (2\pi)^{-d} \int_{-\infty}^{\infty} d^d y \sum_{\overline{j} \in \mathbb{Z}^{rd}} \prod_{m=1}^{r} \Big(e^{iy(k^{(m)} + 2\pi j^{(m)})} \rho(k^{(m)} + 2\pi j^{(m)})\Big) =$$

$$= (2\pi)^{-d} \int_{-\infty}^{\infty} d^d y \prod_{m=1}^{r} (q(k^{(m)}) \sum_{j \in \mathbb{Z}^d} e^{iy(k^{(m)} + 2\pi j} \rho(k^{(m)} + 2\pi j))$$

$$= (2\pi)^{-d} \int_{-\infty}^{\infty} d^d y \prod_{m=1}^{r} S(k^{(m)}, y)$$

where

$$S(k, y) = q(k) \sum_{j \in \mathbb{Z}^d} e^{iy(k + 2\pi j)} \rho(k + 2\pi j) \tag{2. 40}$$

Thus

$$\tilde{H}^{(r)}(\tilde{\xi}) = \int_{-\pi}^{\pi} d^{rd}\bar{k}(2\pi)^{-d} \int_{-\infty}^{\infty} d^d y \prod_{m=1}^{r} (S(k^{(m)}, y)\tilde{\xi}(k^{(m)}))$$

$$= (2\pi)^{-d} \int_{-\infty}^{\infty} d^d y \prod_{m=1}^{r} (\int_{-\pi}^{\pi} S(k^{(m)}, y)\tilde{\xi}(k^{(m)}) d^d k^{(m)})$$

$$= (2\pi)^{-d} \int_{-\infty}^{\infty} \zeta^r(y) d^d y \tag{2. 41}$$

where

$$\zeta(y) = \int_{-\pi}^{\pi} S(k, y)\tilde{\xi}(k) d^d k$$

$$= \int_{-\pi}^{\pi} \sum_{j \in \mathbb{Z}^d} q(k + 2\pi j) e^{i(k+2\pi j)y} \rho(k + 2\pi j)\tilde{\xi}(k + 2\pi j) d^d k$$

$$= \int_{-\infty}^{\infty} e^{iky} q(k)\rho(k)\tilde{\xi}(k) d^d k \tag{2. 42}$$

The interchange of integrals in (2. 41) is formal, but these non-rigorous considerations are the intuitive foundation for the introduction of the random field $\zeta(y)$.

II. 6. Differentials as Gauss integral operators. The proof of Theorem 2. 1 will be obtained as a consequence of the following result which is interesting by itself.

Theorem 2.5. Let

$$H(\zeta) = \sum_{r=1}^{N} \int_{-\infty}^{\infty} f_r(\bar{y})\zeta(y^{(1)})\dots\zeta(y^{(r)})d^{rd}\bar{y}$$

where $f_r(\bar{y}) \in X^1_\alpha(\mathbb{R}^{rd})$, $r = 1, \dots, N$. Then

$$D_n H(\Gamma_0 \xi) = H_n(\Gamma_0 \xi)$$

where

$$H_n(\zeta) = \sum_{r=1}^{N} \int_{-\infty}^{\infty} \frac{n^{rd} f_r(n\bar{y})}{\sqrt{(2\pi)^{rd} \det M^{(r)}(\bar{y})}} \int_{-\infty}^{\infty} \prod_{m=1}^{r} (n^{-d+\frac{\alpha}{2}} \zeta(y^{(m)}) + \kappa^{(m)}) \times$$

$$\times \exp(-\frac{1}{2}(M^{(r)}(\bar{y}))^{-1}\bar{\kappa}, \bar{\kappa}))d^r\bar{\kappa}\, d^{rd}\bar{y} \tag{2.43}$$

and where $\bar{y} = (y^{(1)}, \dots, y^{(r)}) = (y_1^{(1)}, \dots, y_d^{(r)}) \in \mathbb{R}^{rd}$, $\bar{\kappa} = (\kappa^{(1)}, \dots, \kappa^{(r)} \in \mathbb{R}^r$,

$$M^{(r)}(\bar{y}) = \|M(y^{(i)}, y^{(j)})\|^r_{i,j=1}$$

is an $r \times r$ matrix whose elements are defined by the function

$$M(y, y') = \psi(ny, ny') - n^{\alpha-2d}\psi(y, y') \tag{2.44}$$

and where

$$\psi(y, y') = \langle\zeta(y)\zeta(y')\rangle_0 = (-1)^d \int_{-\infty}^{\infty}\int_{-\infty}^{\infty} e^{i(yk+y'k')}\delta_p(k+k') \times$$

$$\times (|k|\,|k'|)^{-\alpha+d} G^{-1}(k) \prod_{m=1}^{d} (k_m k'_m)^{-1} \tag{2.45}$$

Remark. Gaussian integration in the equation (2.43) can be considered as an integral Gauss transformation [26], applied to the function $\prod_{m=1}^{r} \kappa^{(m)}$. It is interesting that the linearization of the main nonlinear equation of the heirarchical models theory leads also to integral Gauss transformations [14]--[17].

Proof: It is sufficient to consider the homogeneous hamiltonian

$$H(\zeta) = \int_{-\infty}^{\infty} f(\bar{y})\zeta(y^{(1)})\ldots \zeta(y^{(r)})d^{rd}\bar{y}$$

because the differential is a linear operator. We have from (2. 32)

$$\zeta(y) = \int_{-\infty}^{\infty} e^{iyk}\rho(k)q(k)\xi(k)d^dk = \int_{-\infty}^{\infty} e^{iyk}\rho(k)q(k)p(k)\theta(k)\, d^dk$$

where

$$\rho(k) = |k|^{-\alpha+d}\prod_{m=1}^{d}(ik_m)^{-1}, \quad p(k) = \prod_{m=1}^{d}(e^{-ik_m}-1),$$

$$q(k) = (G(k)\prod_{m=1}^{d}(e^{-ik_m}-1))^{-1}$$

so

$$\zeta(y) = \int_{-\infty}^{\infty} e^{iyk}q_1(k)\theta(k)\, d^d k$$

where

$$q_1(k) = \rho(k)G^{-1}(k) = |k|^{-\alpha+d}\prod_{m=1}^{d}(ik_m)^{-1}G^{-1}(k)$$

The orthogonal decomposition (2. 16) of the generalized random field $\theta(k)$ induces an analogous decomposition of $\zeta(y)$:

$$\begin{aligned}\zeta(y) &= \int_{-\infty}^{\infty} e^{iyk}q_1(k)(a(k)\theta_n(nk) + \omega(k))d^dk \\ &= \int_{-\infty}^{\infty} e^{iyk}\rho(k)n^{-\frac{\alpha}{2}}G^{-1}(nk)\theta_n(nk)d^dk + \int_{-\infty}^{\infty} e^{iyk}\rho(k)G^{-1}(k)\omega(k)d^dk \\ &= \int_{-\infty}^{\infty} e^{in^{-1}yk_n}{}^{\frac{\alpha}{2}-d}\rho(k)G^{-1}(k)\theta_n(k)d^dk + \kappa(y) \\ &= n^{\frac{\alpha}{2}-d}\zeta_n(n^{-1}y) + \kappa(y)\end{aligned}$$

Random variables

$$\kappa(y) = \int_{-\infty}^{\infty} e^{iyk}\rho(k)G^{-1}(k)\omega(k)\, d^dk \tag{2. 46}$$

and $\zeta_n(y')$ are linear combinations of $\omega(k)$ and $\theta_n(nk)$ resp. so $\kappa(y)$ and $\zeta_n(y')$ are mutually orthogonal for any $y, y'$. Therefore the application of the differential to the hamiltonian

$$H(\zeta(y)) = H(n^{\frac{\alpha}{2}-d}\zeta_n(n^{-1}y) + \kappa(y))$$

reduces to the calculation of a Gaussian integral with respect to $\kappa(y)$. Thus

$$D_n H(\zeta_n) = \int H(n^{\frac{\alpha}{2}-d}\zeta_n(n^{-1}y) + \kappa(y))\mu(d\kappa(y))$$

where $\mu(d\kappa(y))$ is the distribution of the Gaussian (generalized random field $\kappa(y)$. Denoting $n^{-1}y$ by $y$ we come to the desired equation

$$D_n H(\zeta) = \int_{-\infty}^{\infty} n^{rd}\int f(n\bar{y}) \prod_{m=1}^{r} (n^{\frac{\alpha}{2}-d}\zeta(y^{(m)}) + \kappa(ny^{(m)}))\mu(d\kappa)d^{rd}\bar{y}$$

Now we turn to the calculation of the two point function $<\kappa(ny)\kappa(ny')>_{\mu(d\kappa)}$ for the measure $\mu(d\kappa)$. From (2. 20) and (2. 46)

$$<\kappa(ny)\kappa(ny')>_{\mu(d\kappa)} = \int_{-\infty}^{\infty}\int_{-\infty}^{\infty} e^{i(yk+y'k')n}\rho(k)G^{-1}(k)\rho(k')G^{-1}(k') \times$$

$$\times <\omega(k)\omega(k')>_{\mu_2(d\omega)} d^dk d^dk' = \int_{-\infty}^{\infty}\int_{-\infty}^{\infty} e^{in(yk+y'k')}\rho(k)G^{-1}\rho(k')G^{-1}(k') \times$$

$$\times (G(k)\delta_p(k+k') - n^{-\alpha}G(k)G(k')G^{-1}(nk')\delta_p(nk+nk'))d^dk d^dk' =$$

$$= S_1 - S_2$$

where

$$S_1 = \int_{-\infty}^{\infty}\int_{-\infty}^{\infty} e^{in(yk+y'k')}\rho(k)\rho(k')G^{-1}(k')\delta_p(k+k')d^dk d^dk' = \psi(ny, ny')$$

Similarly

$$S_2 = \int_{-\infty}^{\infty}\int_{-\infty}^{\infty} e^{in(yk+y'k')}\rho(k)\rho(k')n^{-\alpha}G^{-1}(nk')\delta_p(nk+nk')d^dk\,d^dk'$$

$$= n^{\alpha-2d}\int_{-\infty}^{\infty}\int_{-\infty}^{\infty} e^{i(yk+y'k')}\rho(k)\rho(k')G^{-1}(k')\delta_p(k+k')d^dk\,d^dk'$$

$$= n^{\alpha-2d}\psi(y, y')$$

Thus equation (2. 44) is proved.

The given proof of equation (2. 43) is rather formal. A rigorous proof may be done in the same way with help of the fact that $f_r(\overline{y}) \in X'_\alpha(\mathbb{R}^{rd})$. We omit the details.

II. 7. Proof of Theorem 2. 1. Let $A > B > 0$ be $r \times r$ matrices and $c > 0$ a real number. Then it is well known that Gauss integral operator

$$\mathcal{G}f(x) = ((2\pi)^r \det B)^{-1/2}\int_{-\infty}^{\infty} f\left(\frac{x}{\sqrt{c}} + u\right)\exp\left(-\frac{1}{2}B^{-1}u, u\right)d^r u$$

maps the Hermite polynomial $:x_1 \dots x_r:_A$, corresponding to the Gaussian measure with correlation matrix $A$, to the Hermite polynomial $c^{-r/2} :x_1 \dots x_r:_{\overline{A}}$ with correlation matrix

$$\overline{A} = c(A - B) \tag{2. 47}$$

One can consider the integral with respect to $\overline{\kappa}$ in equation (2. 43) as a Gauss operator with

$$A = \|\psi(ny^{(i)}, ny^{(j)})\|^r_{i,j=1}, \quad B = \|M(y^{(i)}, y^{(j)})\|^r_{i,j=1}, \quad c = n^{2d-\alpha}$$

Therefore

$$D_n\int_{-\infty}^{\infty} h(\overline{y}) : \zeta(y^{(1)})\dots\zeta(y^{(r)}) : d^{rd}\overline{y}$$

$$= n^{\frac{r\alpha}{2}}\int_{-\infty}^{\infty} h(n\overline{y}) : \zeta(y^{(1)}\dots\zeta(y^{(r)}) :_{\overline{A}} d^{rd}\,\overline{y}$$

where

$$\bar{A} = n^{2d-\alpha}(A-B) = \|n^{2d-\alpha}(\psi(ny^{(i)}, ny^{(j)}) - M(y^{(i)}, y^{(j)})\|_{i,j=1}^{r}$$

$$= \psi(y^{(i)}, y^{(j)})\|_{i,j=1}^{r} = A$$

by equation (2. 44). Theorem 2. 1 is proved.

Now we turn to proof of the Supplement to Theorem 2. 1. We have

$$q_0(k) = G^{-1}(k)|k|^{d-\alpha} \prod_{m=1}^{d} (ik_m(e^{-ik_m} - 1))^{-1}$$

Singularities of $q_0(k)$ are located at points $k = 2\pi j$, $j \in \mathbb{Z}^d$. Let us consider $j = 0$. In this case

$$G(k) = |k|^{d-\alpha} \prod_{m=1}^{d} k_m^{-2} + G_0(k)$$

where $G_0(k)$ is analytic in a neighborhood of the origin. So

$$q_0(k) = (|k|^{d-\alpha} \prod_{m=1}^{d} k_m^{-2} + G_0(k))^{-1} |k|^{d-\alpha} \prod_{m=1}^{d} (ik_m(e^{-ik_m} 1))^{-1}$$

$$= (1 + |k|^{\alpha-d} \prod_{m=1}^{d} k_m^2 G_0(k))^{-1} \prod_{m=1}^{d} (\frac{-ik_m}{e^{-ik_m} 1}) = 1 + A_1(k) + A_2(k)$$

where $A_1(k)$ is analytic at the origin and $A_1(0) = 0$, $A_2(k) = O(|k|^{\alpha+d})$ when $k \to 0$.

If now $j \neq 0$, $\tau = k - 2\pi j$, then $G(k) = |\tau|^{d-\alpha} \prod_{m=1}^{d} \tau_m^{-2} + G_1(k)$, where $G_1(k)$ is analytic at $k = 2\pi j$, so

$$q_0(k) = (1 + |\tau|^{\alpha-d} \prod_{m=1}^{d} \tau_m^2 G_1(k))^{-1} \frac{|\tau|^{\alpha-d}}{|k|^{\alpha-d}} \prod_{m=1}^{d} \frac{\tau_m^2}{k_m(e^{-i\tau_m} 1)} = O(|\tau|^{\alpha+d+1})$$

Thus we prove that $q_0(k) \in C^{\ell,\delta}$, $\ell = [\alpha - d+1]$, $\delta = \alpha - d + 1 - \ell$.

As regards the estimate (2. 12) we remark that the function $((\prod_{m=1}^{d} (e^{-ik_m} - 1))G(k))^{-1}$ is periodic and bounded, so

$$|q_0(k)| = \left|(G(k) \prod_{m=1}^{d} (e^{-ik_m} - 1))^{-1} |k|^{-\alpha+d} \prod_{m=1}^{d} \frac{e^{-ik_m} - 1}{k_m}\right|$$

$$\leq \text{const } |k|^{-\alpha+d} \prod_{m=1}^{d} (1 + |k_m|)^{-1}$$

Q. E. D.

Properties of the two point function $\psi(y, y')$ are established in an analogous way.

## III. $\epsilon$ - EXPANSION

III. 1. Bifurcation values of renormalizing exponents. In this section we analyze $\epsilon$-expansion equations for an effective hamiltonian $H = H_0 + \sum_{r=1}^{\infty} \epsilon^r H_r$. According to (1. 11) the first equation is

$$D_n H_1 = H_1 \tag{3. 1}$$

i. e. $H_1$ is an eigenvector of the differential $D_n$ with eigenvalue $\lambda = 1$. As we have seen in the previous section finite-particle hamiltonian spaces are differential-invariant, so we seek $H_1$ in finite-particle form

$$H_1 = \sum_{m=1}^{M} \sum_{\bar{x} \in \mathbb{Z}^{md}} a_m(\bar{x}) : \xi(x^{(1)}) \ldots \xi(x^{(m)}) :$$

where $\bar{x} = (x^{(1)}, \ldots, x^{(m)})$, $a_m(x^{(1)}+j, \ldots, x^{(m)}+j) = a_m(\bar{x})$ for any $j \in \mathbb{Z}^d$. We restrict the choice of $H_1$ by two conditions. First, $H_1$ must have CDP in the sense of Definition 1. 10, i. e. its coefficients must decrease stronger than those of $H_0$. Secondly, we shall assume that $H_1$ is positive in the sense that

$$\sum_{x^{(2)}, \ldots, x^{(M)} \in \mathbb{Z}^d} a_M(0, x^{(2)}, \ldots, x^{(M)}) > 0 \tag{3. 2}$$

As a result we come to the problem of finding differential eigenvectors satisfying these two conditions. Such eigenvectors were considered in Section II. 4. They are

$$H^{(m)}(\zeta) = \int_{-\infty}^{\infty} :\zeta^m(y): d^d y \tag{3. 3}$$

$m = 1, 2, \ldots$ .

We should like also to explain this result in view of Theorem 2. 1. Namely, it states that Hermite polynomials spaces are differential-invariant and

$$D_n \int_{-\infty}^{\infty} h(\bar{y}) : \zeta(y^{(1)}) \ldots \zeta(y^{(m)}) : d^{md}\bar{y} = n^{\frac{r\alpha}{2}} \int_{-\infty}^{\infty} h(n\bar{y}) : \zeta(y^{(1)}) \ldots \zeta(y^{(m)}) : d^{md}\bar{y}$$

so solutions of the scaling equation

$$h(n\bar{y}) = n^{\gamma} h(\bar{y})$$

corresponds to $D_n$ eigenvectors. If we demand that $h(\bar{y})$ be translation-invariant, strongly decreasing off the diagonal $y^{(1)} = \ldots = y^{(m)}$ and positive, i. e. $\int_{-\infty}^{\infty} h(0, y^{(2)}, \ldots, y^{(m)}) d^d y^{(2)} \ldots d^d y^{(m)} > 0$ , then solutions of the scaling equation are

$$h(\bar{y}) = \delta(y^{(1)} - y^{(2)}) \ldots \delta(y^{(1)} - y^{(m)})$$

i. e. the $D_n$ eigenvectors are (3. 3).

It follows from (2. 9) that

$$D_n H^{(m)} = n^{\frac{m\alpha}{2} - md + d} H^{(m)} \tag{3. 4}$$

so (3. 1) leads to the relation $n^{\frac{m\alpha}{2} - md + d} = 1$ or $\frac{m\alpha}{2} - md + d = 0$, i. e.

$$\alpha = \frac{2d(m-1)}{m} \tag{3. 5}$$

This equation defines the bifurcation values of the renormalizing exponent $\alpha$.

In what follows we restrict ourselves to even m. The case m = 2 is trivial, because it is connected with normed factor uncertainty in the quadratic effective hamiltonian. The next case m = 4 is the most interesting one from the statistical physics point of view. This interest is caused by the fact that there arises for m = 4 a class of non-Gaussian automodel random fields which is thermodinamically stable for d = 1, 2, 3 [6, 14--17]. That means that there is exactly one relevant, i. e. exceeding one in modulus, eigenvalue in the RG differential spectrum for these branch random fields. Thermodinamically stable RG fixed points are of great interest for ferromagnet critical behavior description. For m = 4

$$D_n(H^{(4)}) = n^{2\alpha - 3d} H^{(4)} \tag{3.6}$$

$$\alpha_{bif} = \frac{3}{2} d \tag{3.7}$$

III. 2. The fixed point equation in second order perturbation theory. Let

$$\epsilon = \alpha - \alpha_{bif} = \alpha - \frac{3}{2} d$$

In $\epsilon$ -expansion analysis it is convenient to consider formal hamiltonians, whose coefficients are functions of $\epsilon$, or more precisely formal hamiltonians themselves (the coefficients being $\epsilon$ -independent). This permits us to avoid additional $\epsilon$ -independent decomposition in Definition 1. 9' and solve perturbation theory equations not exactly but with $O(\epsilon)$ - error. This error can be taken into account in the next equations. In such an approach we define

$$H_1 = a_1 H^{(4)} = a_1 \int_{-\infty}^{\infty} :\zeta^4(y): d^d y \tag{3.8}$$

the constant $a_1$ is defined by a second order equation. Thus $H = H_0 + a_1 \epsilon \int_{-\infty}^{\infty} :\zeta^4(y): d^d y + O(\epsilon^2)$ so up to $O(\epsilon^2)$, H is a so-called Landau-Ginzburg type hamiltonian (see [11]). As $n^{2\alpha - 3d} = n^{2\epsilon} = 1 + 2\epsilon \ln n + \ldots$ there arises a $O(\epsilon)$-error in equation (3. 1). Taking it into account we obtain the second order equation in the form

$$-(D_n - 1)H_2 = a_1^2 D_n^{(2)}(H^{(4)}) + \frac{n^{2\epsilon} - 1}{\epsilon} a_1 H^{(4)} \tag{3. 9}$$

where

$$D_n^{(2)}(H) = -\frac{1}{2}[D_n H^2 - (D_n H)^2] \tag{3. 10}$$

is the RG transformation's quadratic part.

Let us compute $D_n^{(2)}(H^{(4)})$. As $D_n H^{(4)} = n^{2\epsilon} H^{(4)}$,

$$D_n^{(2)}(H^{(4)}) = -\frac{1}{2}(D_n - n^{4\epsilon}) \int_{-\infty}^{\infty}\int_{-\infty}^{\infty} :\zeta^{(4)}(y)::\zeta^{(4)}(y'): d^d y d^d y'$$

Next

$$:\zeta^4(y)::\zeta^4(y'): = :\zeta^4(y)\zeta^4(y'): + 16\psi(y,y'):\zeta^3(y)\zeta^3(y'):$$
$$+ 72\psi^2(y,y'):\zeta^2(y)\zeta^2(y'): + 96\psi^3(y,y'):\zeta(y)\zeta(y'): + 24\psi^4(y,y')$$

and by (2. 9)

$$(D_n - n^{4\epsilon}) \int_{-\infty}^{\infty}\int_{-\infty}^{\infty} :\zeta^4(y)\zeta^4(y'): d^d y d^d y' = 0$$

So

$$D_n^{(2)}(H^{(4)}) = -\frac{1}{2}(D_n - n^{4\epsilon}) \int_{-\infty}^{\infty}\int_{-\infty}^{\infty} (16\psi(y,y'):\zeta^3(y)\zeta^3(y'):$$
$$+ 72\psi^2(y,y'):\zeta^2(y)\zeta^2(y'): + 96\psi^3(y,y'):\zeta(y)\zeta(y'):) d^d y d^d y' \tag{3. 11}$$

(the constant term is thrown away). According to the Supplement to Theorem 2. 1

$$\psi(y,y') = C|y-y'|^{-\frac{d}{2}+\epsilon} + \psi_1(y,y') \tag{3.12}$$

$$\psi^2(y,y') = C^2|y-y'|^{-d+2\epsilon} + \psi_2(y,y') \tag{3.13}$$

$$\psi^3(y,y') = C^3|y-y'|^{-\frac{3}{2}d+3\epsilon} + \psi_3(y,y') \tag{3.14}$$

where for $|y-y'| \to \infty$, $i = 1, 2, 3$

$$|\psi_i(y,y')| < \text{const}\ |y-y'|^{-\frac{3}{2}d-2} \tag{3.15}$$

Let us consider the hamiltonian

$$H(\zeta) = \int_{-\infty}^{\infty}\int_{-\infty}^{\infty} (16\psi(y,y') :\zeta^3(y)\zeta^3(y'):$$

$$+ 72\psi^2(y,y') :\zeta^2(y)\zeta^2(y'): + 96\psi^3(y,y') :\zeta(y)\zeta(y'):)\, d^dy\, d^dy'$$

so that

$$D_n^{(2)}(H^{(4)}) = \frac{1}{2}(D_n - n^{4\epsilon})H \tag{3.16}$$

and substitute $\zeta = \Gamma_0 \xi$ in it (see (2.3)). Due to the fact that $|\Gamma_0(x)| < \text{const}\ (1 + |x|)^{-\alpha-1}$ the hamiltonian $H(\Gamma_0\xi)$ has the same coefficient decreasing properties as $H(\zeta)$ itself. As a consequence $H(\Gamma_0\xi)$ lacks CDP (see Definition 1.10), because the functions $\psi^i(y,y')$, $i = 1, 2, 3$ in (3.12)--(3.14) decrease more slowly than $|y-y'|^{-\alpha} = |y-y'|^{-\frac{3}{2}d-\epsilon}$. We get rid of this lack by some regularization of $H(\zeta)$, which slightly modifies equation (3.16). Final formulae are given by equations (3.29), (3.30) below.

Let us consider the hamiltonian

$$H'(\zeta) = \int_{-\infty}^{\infty}\int_{-\infty}^{\infty} (16\psi_1(y,y') :\zeta^3(y)\zeta^3(y'): + 72\psi_2(y,y') :\zeta^2(y)\zeta^2(y'):$$

$$+ 96\psi_3(y,y') :\zeta(y)\zeta(y'): d^dy\, d^dy' \tag{3.17}$$

which is obtained from the $H(\zeta)$ by subtracting the leading terms from the functions $\psi^i(y, y')$, $i = 1, 2, 3$. In view of estimates (3. 15) we may expect that $H'(\Gamma_0 \xi)$ has CDP. Moreover it follows formally from (2. 9) that

$$(D_n - n^{4\epsilon} \int_{-\infty}^{\infty}\int_{-\infty}^{\infty} |y - y'|^{i(-\frac{d}{2}+\epsilon)} : \zeta^i(y)\zeta^i(y') : d^d y\, d^d y' = 0$$

for $i = 1, 2, 3$, so

$$(D_n - n^{4\epsilon})(H - H') = 0 \tag{3. 18}$$

and

$$D_n^{(2)}(H^{(4)}) = -\frac{1}{2}(D_n - n^{4\epsilon})H'$$

But it is necessary to note that the hamiltonian $H'(\Gamma_0 \xi)$ is not defined because of non-integrable singularities of the functions $\psi_2(y, y')$ and $\psi_3(y, y')$ at the diagonal $y = y'$. More precisely the $\psi_2$ singularity $C^2 |y - y'|^{-d+2\epsilon}$ becomes non-integrable for $\epsilon = 0$ and the $\psi_3$ one $C^3 |y - y'|^{-\frac{3}{2}d+3\epsilon}$ is non-integrable for all small $\epsilon$. As a result, corresponding integrals in (2. 6) diverge. To avoid these divergences we regularize the functions $\psi_2$ and $\psi_3$ at the distance $y = y'$ using generalized function theory (see, for instance [29]). First we consider $\psi_2$. Let us define

$$[\psi_2(y, y')]_{reg} = \psi^2(y, y') - C^2 \operatorname{reg} |y - y'|^{-d+2\epsilon} \tag{3. 19}$$

(cf. (3. 13)) where

$$(\operatorname{reg} |y - y'|^{-d+2\epsilon}, \varphi(y, y')) \overset{def}{=} \int_{-\infty}^{\infty}\int_{-\infty}^{\infty} |y - y'|^{-d+2\epsilon}$$

$$\times (\varphi(y, y') - \varphi(\tfrac{y+y'}{2}, \tfrac{y+y'}{2})\chi(y - y'))d^d y\, d^d y' \tag{3. 20}$$

for any $C_0^\infty$ - function $\varphi(y, y')$ : $\chi(y)$ is the characteristic function of the sphere $\{|y| \leq 1\}$. It is essential for us that

$[\psi_2(y,y')]_{reg}\,\delta(y-z)\delta(y'-z')$ is an element of $X'_\alpha(\mathbb{R}^{4d})$ (see Section II. 1) and so the hamiltonian

$$\int_{-\infty}^{\infty}\int_{-\infty}^{\infty}[\psi_2(y,y')]_{reg} : (\Gamma_0\xi)^2(y)(\Gamma_0\xi)^2(y') : d^dy\,d^dy'$$

is defined.

Let us now compute

$$D_n\int_{-\infty}^{\infty}\int_{-\infty}^{\infty}[\psi_2(y,y')]_{reg} : \zeta^2(y)\zeta^2(y') : d^dy\,d^dy'$$

$$= D_n\int_{-\infty}^{\infty}\int_{-\infty}^{\infty}\psi^2(y,y') : \zeta^2(y)\zeta^2(y') : d^dy\,d^dy'$$

$$- D_n\int_{-\infty}^{\infty}\int_{-\infty}^{\infty}C^2 \text{ reg } |y-y'|^{-d+2\epsilon} : \zeta^2(y)\zeta^2(y') : d^dy\,d^dy'$$

By Theorem 2. 1

$$D_n\int_{-\infty}^{\infty}\int_{-\infty}^{\infty}C^2 \text{ reg } |y-y'|^{-d+2\epsilon} : \zeta^2(y)\zeta^2(y') : d^dy\,d^dy'$$

$$= n^{d+2\epsilon}\int_{-\infty}^{\infty}\int_{-\infty}^{\infty}C^2 \text{ reg } |ny-ny'|^{-d+2\epsilon} : \zeta^2(y)\zeta^2(y') : d^dy\,d^dy'$$

Next for $\varphi(y,y')\in C_0^\infty(\mathbb{R}^{2d})$

$$(\text{reg } |ny-ny'|^{-d+2\epsilon},\ \varphi(y,y')) \overset{\text{def}}{=} \int_{-\infty}^{\infty}\int_{-\infty}^{\infty}|y-y'|^{-d+2\epsilon}$$

$$\times\left(\varphi\left(\frac{y}{n},\frac{y'}{n}\right)-\chi(y-y')\varphi\left(\frac{y+y'}{2n},\frac{y+y'}{2n}\right)\right)n^{-2d}\,d^dy\,d^dy'$$

$$= \int_{-\infty}^{\infty}\int_{-\infty}^{\infty}|ny-ny'|^{-d+2\epsilon}(\varphi(y,y')-\chi(ny-ny')$$

$$\times\varphi\left(\frac{y+y'}{2},\frac{y+y'}{2}\right))d^dy\,d^dy' = n^{-d+2\epsilon}((\text{reg } |y-y'|^{-d+2\epsilon},\ \varphi(y,y'))$$

$$+\int_{-\infty}^{\infty}\int_{-\infty}^{\infty}|y-y'|^{-d+2\epsilon}(\chi(y-y')-\chi(ny-ny'))\varphi\left(\frac{y+y'}{2},\frac{y+y'}{2}\right)d^dy\,d^dy')$$

$$= n^{-d+2\epsilon}((\text{reg } |y-y'|^{-d+2\epsilon},\ \varphi(y,y'))+\int_{n^{-1}<|t|<1}^{\infty}|t|^{-d+2\epsilon}d^dt\int_{-\infty}^{\infty}\varphi(z,z)d^dz)$$

so

$$\operatorname{reg} |ny - ny'|^{-d+2\epsilon} = n^{-d+2\epsilon} (\operatorname{reg} |y - y'|^{-d+2\epsilon} + C_0 \delta(y-y')) \quad (3.21)$$

where

$$C_0 = \int_{n^{-1}<|t|<1} |t|^{-d+2\epsilon} d^d t = \frac{1-n^{-2\epsilon}}{2\epsilon}$$

Thus

$$(D_n - n^{4\epsilon}) \int_{-\infty}^{\infty} \int_{-\infty}^{\infty} C^2 \operatorname{reg} |y - y'|^{-d+2\epsilon} : \zeta^2(y) \zeta^2(y') : d^d y \, d^d y'$$

$$= n^{4\epsilon} \int_{-\infty}^{\infty} \int_{-\infty}^{\infty} C^2 C_0 \delta(y-y') : \zeta^2(y) \zeta^2(y') : d^d y d^d y' = n^{4\epsilon} C_1 \int : \zeta^4(y) : d^d y$$

where $C_1 = C^2 C_0$, so

$$(D_n - n^{4\epsilon}) \int_{-\infty}^{\infty} \int_{-\infty}^{\infty} [\psi_2(y, y')]_{\text{reg}} : \zeta^2(y) \zeta^2(y') : d^d y \, d^d y'$$

$$= (D_n - n^{4\epsilon}) \int_{-\infty}^{\infty} \int_{-\infty}^{\infty} \psi^2(y, y') : \zeta^2(y) \zeta^2(y') : d^d y \, d^d y'$$

$$- n^{4\epsilon} C_1 \int_{-\infty}^{\infty} : \zeta^4(y) : d^d y \quad (3.22)$$

The obtained equation is the first step in the $(D_n - n^{4\epsilon}) H'$ calculation. The second one is connected with regularization of the function $\psi_3(y, y')$.

III. 3. <u>Quadratic part regularization in the second order</u> perturbation theory. We need to regularize the function $\psi_3(y, y') = \psi^3(y, y') - C^3 |y-y'|^{-\frac{3}{2}d+3\epsilon}$ at the diagonal $y = y'$. By analogy with equation (3. 19) we write

$$\psi_3(y, y') = \psi^3(y, y') = C^3 \operatorname{reg} |y - y'|^{-\frac{3}{2}d + 3\epsilon} \quad (3.23)$$

where $\operatorname{reg} |y - y'|^{-\frac{3}{2}d+3\epsilon}$ is a generalized function coinciding with $|y - y'|^{-\frac{3}{2}d+3\epsilon}$ off the diagonal $y = y'$. There exist various regularizations of the power function $|y - y'|^{-\frac{3}{2}d+3\epsilon}$. To simplify the

$(D_n - n^{4\epsilon})H'$ calculation we choose the so-called canonical regularization c. r. $|y - y'|^{-\frac{3}{2}d+3\epsilon}$ (see [29]), defining a homogeneous generalized function, i. e.

$$\text{c. r. } |ny - ny'|^{-\frac{3}{2}d+3\epsilon} = n^{-\frac{3}{2}d+3\epsilon} \text{ c. r. } |y-y'|^{-\frac{3}{2}d+3\epsilon} \tag{3. 24}$$

To investigate the $\epsilon$ dependence of c. r. $|y - y'|^{-\frac{3}{2}d+3\epsilon}$ let us consider its Fourier transform. One can calculate that

$$\int_{-\infty}^{\infty}\int_{-\infty}^{\infty} e^{i(yk+y'k')} \text{ c. r. } |y-y'|^{-\frac{3}{2}d+3\epsilon} d^d y d^d y' = C_0 |k+k'|^{\frac{d}{2}-3\epsilon} \tag{3. 25}$$

where

$$C_0 = 2^{-\frac{d}{2}+3\epsilon} \pi^{\frac{d}{2}} \Gamma(-\frac{d}{4}+\frac{3}{2}\epsilon)\Gamma^{-1}(\frac{3d}{4}-\frac{3\epsilon}{2}) \tag{3. 26}$$

(cf. [29]).

It is very important that $C_0 = C_0(\epsilon)$ is analytic at $\epsilon = 0$ if the dimension d is not divisible by 4. Due to this fact the function

$$[\psi_3(y, y')]_{\text{c. r.}} = \psi^3(y, y') - C^3 \text{ c. r. } |y - y'|^{-\frac{3}{2}d+3\epsilon} \tag{3. 27}$$

defines $\epsilon$ on power series whose coefficients belong to $X'_\alpha(\mathbb{R}^{2d})$. However, if d is divisible by 4, then $C_0(\epsilon)$ has a simple pole at $\epsilon = 0$ (thanks to $\Gamma(-\frac{d}{4}+\frac{3}{2}\epsilon)$), so $[\psi_3(y, y')]_{\text{c. r.}}$ depends on $\epsilon$ meromorphically and such a regularization is not correct.

Using Theorem 2. 1 we have now

$$\begin{aligned} D_n \int_{-\infty}^{\infty}\int_{-\infty}^{\infty} C^3 \text{ c. r. } |y-y'|^{-\frac{3}{2}d+3\epsilon} &: \zeta(y)\zeta(y') : d^d y\, d^d y' \\ = n^{\frac{3}{2}d+\epsilon} \int_{-\infty}^{\infty}\int_{-\infty}^{\infty} C^3 \text{ c. r. } |ny - ny'|^{-\frac{3}{d}+3\epsilon} &: \zeta(y)\zeta(y') : d^d y\, d^d y' \\ = n^{4\epsilon} \int_{-\infty}^{\infty}\int_{-\infty}^{\infty} C^3 \text{ c. r. } |y - y'|^{-\frac{3}{2}d+3\epsilon} &: \zeta(y)\zeta(y') : d^d y\, d^d y' \end{aligned}$$

so

$$(D_n - n^{4\epsilon}) \int_{-\infty}^{\infty} \int_{-\infty}^{\infty} C^3 \text{ c. r. } |y - y'|^{-\frac{3}{2}d+3\epsilon} : \zeta(y)\zeta(y') : d^d y \, d^d y' = 0 \quad (3.28)$$

Let us finally consider the hamiltonian

$$H''(\zeta) = \int_{-\infty}^{\infty} \int_{-\infty}^{\infty} (16\psi_1(y, y') : \zeta^3(y)\zeta^3(y') : + 72\,[\psi_2(y, y')]_{\text{reg}}$$

$$\times :\zeta^2(y)\zeta^2(y'): + 96[\psi_3(y,y')]_{\text{c. r.}} :\zeta(y)\zeta(y'): d^d y \, d^d y' \quad (3.29)$$

which is a $H'(\zeta)$ regularization (see (3. 17), (3. 19), (3. 27)). It follows from (3. 22) and (3. 28) that

$$(D_n - n^{4\epsilon}) H''(\zeta) = (D_n - n^{4\epsilon}) \int_{-\infty}^{\infty} \int_{-\infty}^{\infty} (16\psi(y, y') : \zeta^3(y)\zeta^3(y') :$$

$$+ 72\psi^2(y, y') : \zeta^2(y)\zeta^2(y') : + 96\psi^3(y,y') : \zeta(y)\zeta(y') :) d^d y \, d^d y'$$

$$- 72\, n^{4\epsilon} C_1 \int_{-\infty}^{\infty} : \zeta^4(y) : d^d y = (D_n - n^{4\epsilon}) H(\zeta) - C_2 \int_{-\infty}^{\infty} : \zeta^4(y) : d^d y$$

where $C_2 = 72\, n^{4\epsilon} C_1$. Substituting this equation in (3. 16) we obtain the desired representation

$$D_n^{(2)}(H^{(4)}) = -\frac{1}{2}(D_n - n^{4\epsilon}) H'' - \frac{C_2}{2} H^{(4)} \quad (3.30)$$

It is noteworthy that $H'(\Gamma_0 \xi)$ has CDP.

III. 4. <u>Second order equation resolution.</u> Let us return to the equation (3. 9). To solve it we use (3. 30). Namely combining (3. 9) and (3. 30) we have

$$(D_n - 1)(H_2 - \frac{a_1^2}{2} H'') = (\frac{C_2}{2} a_1^2 - \frac{n^{2\epsilon} - 1}{\epsilon} a_1) H^{(4)} - \frac{a_1^2}{2}(n^{4\epsilon} - 1) H'' \quad (3.31)$$

In view of $n^{4\epsilon} - 1 = O(\epsilon)$ the last term may be neglected. Moreover because $H^{(4)} = \int_{-\infty}^{\infty} : \zeta^4(y) : d^d y$ belongs to the $(D_n - 1)$ kernel (with

$O(\epsilon)$ precision) the left and right hand sides of this equation are independent. So

$$\frac{C_2}{2} a_1^2 - \frac{n^{2\epsilon} - 1}{\epsilon} a_1 = O(\epsilon)$$

or

$$a_1 = 4 \ln n \, C_2^{-1} + O(\epsilon) = \frac{1}{18} C^{-2} + O(\epsilon) = \frac{(2\pi)^{-d}}{18} + O(\epsilon) \qquad (3.32)$$

(the last term $O(\epsilon)$ may be thrown away) and

$$(D_n - 1)(H_2 - \frac{a_1^2}{2} H'') = O(\epsilon) \qquad (3.33)$$

The solution of the last equation is

$$H_2 = \frac{a_1^2}{2} H'' + a_2 H^{(4)} + O(\epsilon) \qquad (3.34)$$

where $a_2$ is a constant, which can be obtained from the third equation of perturbation theory.

There is no other "good" solution of the equation (3.31). To show this we introduce a functional space containing possible solutions and prove an uniqueness theorem.

Definition 3.1. A hamiltonian $H(\zeta) = \sum_{r=2}^{2N} \int_{-\infty}^{\infty} h_r(\bar{y}) \times :\zeta(y^{(1)}) \ldots \zeta(y^{(r)}): d^{rd}\bar{y}$ belongs to the space $Y_\alpha$ if and only if it is an admissible even (i. e. $h_{2p+1}(\bar{y}) \equiv 0$, $p = 1, 2, \ldots$) hamiltonian such that the functions $h_r(\bar{y})$, $r = 2, 4, \ldots$, are even ones, satisfying the periodicity condition

$$h_r(y^{(1)} + j, \ldots, y^{(r)} + j) = h_r(y^{(1)}, \ldots, y^{(r)}) \qquad (3.35)$$

for any $j \in \mathbb{Z}^d$, and their Fourier transforms $\tilde{h}_r(\bar{k})$ have the form

$$\tilde{h}_r(\bar{k}) = \sum_{j \in \mathbb{Z}^d} \delta(k^{(1)} + \ldots + k^{(r)} - 2\pi j)\, g_r^{(j)}(\bar{k}) \qquad (3.36)$$

Here $g_r^{(j)}(\bar{k})$ are assumed to be smooth functions of $\bar{k} \in \mathbb{R}^{rd}$ such that

$$g_r^{(j)}(k^{(1)} + a, \ldots, k^{(r)} + a) = g_r^{(j)}(k^{(1)}, \ldots, k^{(r)})$$

for any $a \in \mathbb{R}^d$ and

$$\|g_r^{(j)}(\bar{k})\|_{a,r}^{(j)} \equiv \sup_{|\beta| \leq \ell} \sup_{\bar{k} \in \mathbb{R}^{rd}} (1+|j|)^{-\ell+|\beta|} \Psi_{r,\beta}(\bar{k}) |D^\beta g_r^{(j)}(\bar{k})| < C \quad (3.37)$$

where

$$\ell = [\alpha - d] + 1, \quad \Psi_{r,\beta}(\bar{k}) = \prod_{m=2}^{r} \Psi_\beta(k^{(m)} - k^{(m-1)}), \quad \Psi_\beta(k) = (1+|k|)^{-\ell+|\beta|}$$

and $C$ does not depend on $j$.

For $H(\zeta) \in Y_\alpha$ we will use the notation

$$\|H(\zeta)\|_\alpha = \sum_{r=2}^{2N} \sup_{j \in \mathbb{Z}^d} \|g_r^{(j)}(\bar{k})\|_{\alpha, r}^{(j)} \quad (3.38)$$

Now we would like to introduce some differential-invariant subspaces of the $Y_\alpha$ and their projections. First of all Hermite polynomials spaces are differential-invariant. Let $Q_{2m}$,

$$Q_{2m} \sum_{r=2}^{2N} \int_{-\infty}^{\infty} h_r(\bar{y}) : \zeta(y^{(1)}) \ldots \zeta(y^{(r)}) : d^{rd}\bar{y}$$

$$= \int_{-\infty}^{\infty} h_{2m}(\bar{y}) : \zeta(y^{(1)}) \ldots \zeta(y^{(2m)}) : d^{2md}\bar{y}$$

be a projection on the 2m-order Hermite polynomials space.

Next let

$$P \sum_{r=2}^{2N} \int_{-\infty}^{\infty} h_r(\bar{y}) : \zeta(y^{(1)}) \ldots \zeta(y^{(r)}) : d^{rd}\bar{y}$$

$$= \sum_{r=2}^{2N} \int_{-\infty}^{\infty} h_r'(y) : \zeta(y^{(1)}) \ldots \zeta(y^{(r)}) : d^{rd}\bar{y}$$

where $\tilde{h}_r'(\bar{k}) = \delta(k^{(1)} + \ldots + k^{(r)}) g_r^{(0)}(\bar{k})$. On other words the projection $P$ remains in the right hand side of (3. 36) the only term corresponding to $j = 0$. One can derive from Theorem 2.1 that P and $D_n$ commute in the space $Y_\alpha$ (see also considerations below), so the subspace $P(Y_\alpha)$ is differential-invariant.

At last let us consider the hamiltonians

$$H_\beta(\zeta) = \int_{-\infty}^{\infty}\int_{-\infty}^{\infty} D^\beta\delta(y'-y) : \zeta(y)\zeta(y') : d^d y d^d y' = \int_{-\infty}^{\infty} : \zeta(y) D^\beta \zeta(y) : d^d y \quad (3.39)$$

where $\beta$ are multiindices such that $|\beta| \le [\alpha - d]$, $|\beta|$ even. It follows from Theorem 2.1 that

$$D_n H_\beta = n^{\frac{d}{2} + \epsilon - |\beta|} H_\beta$$

i.e. $H_\beta$ are eigenvectors of the differential $D_n$.

A uniqueness theorem for equation (3.32) is a consequence of the following statement.

<u>Proposition 3.1.</u> <u>If</u> $H \in Y_\alpha$ <u>then</u> $D_n H \in Y_\alpha$ <u>and</u>

(i) $\|D_n(1 - P)H\|_\alpha \le 2^{d/2+1} n^{-1/3} \|(1-P)H\|_\alpha$

(ii) $\|D_n(1 - Q_2 - Q_4)PH\|_\alpha \le n^{-d/2 + 1/3} \|(1-Q_2-Q_4)PH\|_\alpha$

(iii) <u>There exist real numbers</u> $a$, $\{a_\beta\}_{|\beta| < \ell}$, $\ell = [\alpha - d] + 1$, $\alpha_\beta = 0$ <u>if</u> $|\beta|$ <u>is odd, such that for</u>

$$\hat{H} = (Q_2 + Q_4)PH - a\int_{-\infty}^{\infty} : \zeta^4(y) : d^d y - \sum_{|\beta| < \ell} a_\beta H_\beta$$

<u>the estimate</u>

$$\|D_n \hat{H}\|_\alpha \le C n^{-1/6}$$

<u>is valid the constant</u> C <u>does not depend on</u> <u>n</u>, <u>but generally speaking depends on</u> $\hat{H}$.

<u>Remark.</u> We use in the proof that $0 > \epsilon < 0.1$.

<u>Proof:</u> Let $H = \sum_{r=2}^{2N} \int_{-\infty}^{\infty} h_r(\bar{y}) : \zeta(y^{(1)}) \dots \zeta(y^{(r)}) : d^{rd}\bar{y} \in Y_\alpha$ and

$$H_n = D_n H = \sum_{r=2}^{2N} \int_{-\infty}^{\infty} h_{r,n}(\bar{y}) : \zeta^{(1)} \dots \zeta(y^{(r)}) : d^d y$$

By Theorem 2.1

$$h_{r,n}(\bar{y}) = n^{\frac{r\alpha}{2}} h_r(n\bar{y})$$

so

$$\tilde{h}_{r,n}(\bar{k}) = n^{\frac{r\alpha}{2}-rd} \tilde{h}_r(n^{-1}\bar{k} = n^{\frac{r\alpha}{2}-rd} \sum_{j \in \mathbb{Z}^d} g_r^{(j)}(n^{-1}\bar{k}) \times$$

$$\times + (\frac{k^{(1)}+\dots+k^{(r)}}{n} - 2\pi j) = n^{\frac{r\alpha}{2}-rd+d} \sum_{j \in \mathbb{Z}^d} g_r^{(j)} (n^{-1}\bar{k}) + (k^{(1)}+\dots+k^{(r)} - 2\pi n j)$$

so

$$\tilde{h}_{r,n}(\bar{k}) = \sum_{j \in \mathbb{Z}^d} g_{r,n}^{(j)}(\bar{k}) \delta(k^{(1)}+\dots+k^{(r)} - 2\pi j)$$

where $g_{r,n}^{(j)}(\bar{k}) \equiv 0$ if $j$ is not divisible by $n$ and for $j$ divisible by $n$, i. e. for $j = nj'$, $j' \in \mathbb{Z}^d$,

$$g_{r,n}^{(j)}(\bar{k}) = n^{\frac{r\alpha}{2}-rd+d} g_r^{(j')}(n^{-1}\bar{k}) \tag{3.40}$$

Now we consider (i), (ii), (iii) in succession.

(i) We have

$$\|D_n(1-P)H\|_\alpha = \sum_{r=2}^{2N} \sup_{j \neq 0} \|g_{r,n}^{(j)}(\bar{k})\|_{\alpha,r}^{(j)}$$
$$= \sum_{r=2}^{2N} n^{\frac{r\alpha}{2}-rd+d} \sup_{j' \neq 0} \|g_r^{(j')}(n^{-1}\bar{k})\|_{\alpha,r}^{(j)}$$

Next

$$\|g_r^{(j')}(n^{-1}\bar{k}\|_{\alpha,r}^{(j)} = \sup_{|\beta| \leq \ell} \sup_{\bar{k}} (1+|j|)^{-\ell+|\beta|} \Psi_{r,\beta}(\bar{k}) |D^\beta(g_r^{(j')}(n^{-1}\bar{k}))|$$

$$= \sup_{|\beta| \leq \ell} \sup_{\bar{k}'} (1+n|j'|)^{-\ell+|\beta|} \Psi_{r,\beta}(n\bar{k}') n^{-|\beta|} |D^\beta g_r^{(j')}(\bar{k}')|$$

Here we use $\bar{k}' = n^{-1}\bar{k}$. For $j' \neq 0$

$$(1+n|j'|)^{-\ell+|\beta|} \le C_\beta n^{-\ell+|\beta|}(1+|j'|)^{-\ell+|\beta|}$$

$C_\beta = 2^{\ell-|\beta|} < 2^\ell = C$, and for all $\bar{k}' \in \mathbb{R}^{rd}$

$$\Psi_{r,}(n\bar{k}') \le \Psi_{r,\beta}(\bar{k}')$$

so

$$\|g_r^{(j')}(n^{-1}\bar{k})\|_{\alpha,r}^{(j)} \le \sup_{|\ |\le \ell}\ \sup_{\bar{k}'} Cn^{-\ell}(1+|j'|)^{-\ell+|\beta|}$$

$$\times\ \Psi_{r,\beta}(\bar{k}')|D^\beta g_r^{(j')}(\bar{k}')| = Cn^{-\ell}\|g_r^{(j')}(\bar{k})\|_{\alpha,r}^{(j')} \tag{3.41}$$

Thus

$$\|D_n(1-P)H\|_\alpha \le \sum_{r=2}^{2N} Cn^{\frac{r\alpha}{2}-rd+d-\ell} \sup_{j\neq 0}\|g_r^{(j)}(\bar{k})\|_{\alpha,r}^{(j)}$$

but $\frac{r\alpha}{2}-rd+d-\ell = \frac{r(d-2\epsilon)}{4}+d-[\frac{d}{2}+\epsilon]-1 < -\frac{1}{3}$, if $\epsilon < 0.1$ and $r \ge 2$,

so

$$\|D_n(1-P)H\|_\alpha \le \sum_{r=2}^{2N} C_n^{-1/3} \sup_{j\neq 0}\|g_r^{(j)}(\bar{k})\|_{\alpha,r} = Cn^{-1/3}\|(1-P)H\|_\alpha$$

Q. E. D.

(ii) We have

$$\|D_n(1-Q_2-Q_4)PH\|_\alpha = \sum_{r=6}^{2N}\|g_{r,n}^{(0)}(\bar{k})\|_{\alpha,r}^{(0)}$$

$$= \sum_{r=6}^{2N} n^{\frac{r\alpha}{2}-rd+d}\|g_r^{(0)}(n^{-1}\bar{k})|_{\alpha,r}^{(0)} \le \sum_{r=6}^{2N} n^{\frac{r\alpha}{2}-rd+d}\|g_r^{(0)}(\bar{k})\|_{\alpha,r}^{(0)}$$

because

$$\|g_r^{(0)}(n^{-1}\bar{k})\|_{\alpha,r}^{(0)} \le g_r^{(0)}(\bar{k})\|_{\alpha,r}^{(0)}$$

The last relation is established in the same way as (3.41). Besides

$$n^{\frac{r\alpha}{2}-rd+d} \leq n^{3\alpha-5d} = n^{-\frac{d}{2}+3\epsilon} < n^{-\frac{d}{2}+\frac{1}{3}}$$

if $r \geq 6$. Therefore

$$\|D_n(1-Q_2-Q_4)PH\|_\alpha \leq n^{-d/2+1/3} \sum_{r=6}^{2N} \|g_r^{(0)}(\bar{k})\|_{\alpha,r}^{(0)}$$

$$= n^{-d/2+1/3} \|(1-Q_2-Q_4)PH\|_\alpha$$

Q. E. D.

(iii) We have

$$Q_2PH = \int_{-\infty}^{\infty} h(y^{(1)}, y^{(2)}) : \zeta(y^{(1)})\zeta(y^{(2)}) : d^{2d}\bar{y}$$

where

$$\tilde{h}(k^{(1)}, k^{(2)}) = \delta(k^{(1)}+k^{(2)})\, g_2^{(0)}(k^{(1)}, k^{(2)}) = \delta(k^{(1)}+k^{(2)}) g_2^{(0)}(k^{(1)}-k^{(2)}, 0)$$

Let us denote $a_\beta = (\beta!)^{-1} D_k^\beta g_2^{(0)}(k,0)|_{k=0}$, i. e. $a_\beta$ are Taylor coefficients of the function $g_2^{(0)}(k,0)$ at the origin. Let

$$\hat{g}_2^{(0)}(k,0) = g_2^{(0)}(k,0) - \sum_{|\beta|<\ell} a_\beta k^\beta \tag{3.42}$$

$$\tilde{\hat{h}}_2(k^{(1)}, k^{(2)}) = \delta(k^{(1)}+k^{(2)})\, \hat{g}_2^{(0)}(k^{(1)}-k^{(2)}, 0)$$

and

$$\hat{H}_2(\zeta) = \int_{-\infty}^{\infty} \hat{h}_2(y^{(1)}, y^{(2)}) : \zeta(y^{(1)})\zeta(y^{(2)}) : d^{2d}\bar{y}$$

Now we estimate $\|D_n\hat{H}_2\|_\alpha$. In view of (3. 42) and the fact that $\|g_2^{(0)}(k^{(1)}-k^{(2)}, 0)\|_{\alpha,2}^{0} < \infty$

$$|D^\beta \hat{g}_2^{(0)}(k,0)| \leq C|k|^{\ell-|\beta|}$$

for $|\beta| \leq \ell$. Next in obvious notation

$$\hat{g}_{2,n}^{(0)}(k,0) = n^{\alpha-d}\hat{g}_2^{(0)}(n^{-1}k, 0)$$

by (3. 40). So

$$\|D_n \hat{H}_2\|_\alpha = \sup_{|\beta|\le \ell} \sup_k n^{\alpha-d} |D_k^\beta \hat{g}_2^{(0)}(n^{-1}k, 0)| (1+|k|)^{-\ell+|\beta|}$$

$$= \sup_{|\beta|\le \ell} \sup_{k'} n^{\alpha-d-|\beta|} |D^\beta \hat{g}_2^{(0)}(k', 0)| (1+n|k'|)^{-\ell+|\beta|}$$

$$\le C \sup_{|\beta|\le \ell} \sup_k n^{\alpha-d-|\beta|} |k'|^{\ell-|\beta|} (1+n|k'|)^{-\ell+|\beta|}$$

$$= C \sup_{|\beta|\le \ell} \sup_k n^{\alpha-d-\ell} (|k|/(1+|k|))^{\ell-|\beta|} = Cn^{\alpha-d-\ell} \le Cn^{-1/6}$$

In the same way $\hat{H}_4$ is defined and estimated. The proposition is proved.

## III. 5. Uniqueness theorem.

Theorem 3. 1. Let $H(\zeta) = \sum_{r=2}^{2N} \int_{-\infty}^{\infty} h_r(\bar{y}) : \zeta(y^{(1)}) \ldots \zeta(y^{(r)}) : d^{rd}\bar{y} \in Y_\alpha$ not depend on $n$ and be a solution of the equation (3. 31) in the sense that

$$\|(D_n - 1)(H - \frac{a_1^2}{1} H'') + (\frac{n^{2\epsilon} - 1}{\epsilon} a_1 - \frac{C_2}{2} a_1^2) H^{(4)}\|_\alpha = O(\epsilon)$$

for any fixed $n = 2, 3, \ldots$ ($O(\epsilon)$ is not assumed to be uniform with respect to $n$). Then for some real $a_2$

$$\|H - \frac{a_1^2}{2} H'' - a_2 H^{(4)}\|_\alpha = O(\epsilon)$$

where $a_1$, $H''$ are given resp. by (3. 32), (3. 33).

Proof: Let

$$\hat{H} = H - \frac{a_1^2}{2} H''$$

so that

$$\| (D_n - 1)\hat{H} \|_\alpha = O(\epsilon)$$

by (3. 31) -- (3. 33). Let us decompose $\hat{H}$ according to Proposition 3. 1

$$\hat{H} = aH^{(4)} + \sum_{|\beta| < \ell} a_\beta H_\beta + \hat{H}'$$

By this proposition

$$\| D_n \hat{H}' \|_\alpha < C n^{-1/6}$$

so

$$\| - \hat{H}' + \sum_{|\beta| < \ell} a_\beta (n^{\alpha - d - |\beta|} - 1) H_\beta \|_\alpha = O(\epsilon) + O(n^{-1/6})$$

The sum runs over even $|\beta|$ so if the dimension $d$ is not divisible by 4, $\alpha - d|\beta| > \frac{1}{3}$ for all the $\beta$. Therefore we can estimate subsequently from the last relation all the $a_\beta$ and $\hat{H}'$. The theorem is proved.

III. 6. Concluding remarks.

1. We would like to present the obtained formula for the effective hamiltonian:

$$H = H_0 + \epsilon a_1 \int_{-\infty}^{\infty} :\zeta^4(y): d^d y + \epsilon^2 \left( \frac{a_1^2}{2} \int_{-\infty}^{\infty} \int_{-\infty}^{\infty} (16\psi_1(y, y') \right.$$

$$\times :\zeta^3(y)\zeta^3(y'): + 72[\psi_2(y, y')]_{\text{reg}} :\zeta^2(y)\zeta^2(y'):$$

$$\left. + 96[\psi_3(y, y')]_{\text{c. r.}} :\zeta(y)\zeta(y'): d^d y d^d y' + a_2 \int_{-\infty}^{\infty} :\zeta^4(y): d^d y \right) + O(\epsilon^3)$$

where

$$a_1 = \frac{(2\pi)^{-d}}{18}$$

and $a_2$ is defined from the third order equation of perturbation theory.

2. Canonical regularization of the function $\psi_2(y, y')$ at the diagonal $y = y'$ has as a function of $\epsilon$ a simple pole at $\epsilon = 0$, so it is not correct. This phenomenon has the same nature as the above discussed pole of $\psi_3(y, y')$ canonical regularization in the case when d divides by 4.

3. Higher order equations of perturbation theory for an effective hamiltonian can be investigated in principle in the same manner. But an existence and uniqueness theorem for them has not yet been proved.

4. In recent work of Baker and Krinsky [31] (see also seom references therein) a correlation functions description of the RG transformation is given and analyzed. This analysis is somewhat close to ours in Sections I. 2, I. 6 and II. 4.

## REFERENCES

1. L. P. Kadanoff, W. Götze, D. Hamblen, R. Hecht, E. A. S. Lewis, V. V. Palciauskas, M. Rayl, J. Swift, D. Aspens, J. Kane, Static Phenomena Near Critical Points: Theory and Experiment, Rev. Mod. Phys., 39, 2, 395 (1967).

2. A. Z. Patashinsky, V. L. Pokrovsky, Second Order Phase Transition in Bose-Liquid, JETP, 64, 3, 994 (1964).
V. L. Pokrovsky, Scaling Hypothesis in Phase Transition Theory, Usp. Fis. Nauk, 94, 1, 127 (1968). (In Russian).

3. B. Widom, Surface Tension and Molecular Correlations near the Critical Point, J. Chem. Phys., 43, 3892 (1965).

4. H. E. Stanley, Introduction to Phase Transitions and Critical Phenomena, Clarendom Press. Oxford, 1974.

5. K. G. Wilson, J. Kogut, The Renormalization Group and the $\epsilon$-expansion, Phys. Rep., 12C, 2, 75 (1974).

6. K. G. Wilson, J. Kogut, The Renormalization Group and the $\epsilon$-expansion, Phys. Rep., 12C, 2, 75 (1974).
7. M. Gell-Mann, F. Low, Quantum Electrodynamics at Small Distances, Phys. Rev. 95, 1300 (1954).
8. V. N. Gribov, A. A. Migdal, Strong Coupling in the Problem of Pomeranchuk Pole, JETP, 55, 4, 1498 (1968).
9. A. M. Polyakov, Microscopic Description of Critical Phenomena, JETP, 55, 3, 1026 (1968).
10. L. P. Kadanoff, Physics, 2, 263, (1966).
11. S.-K. Ma, Introduction to the Renormalization Group, Rev. Mod. Phys., 45, 4, 589 (1973).
12. M. E. Fisher, The Renormalization Group in the Theory of Critical Behaviour, Rev. Mod. Phys., 46, 4, 597 (1974).
13. L. P. Kadanoff, The Application of Renormalization Group Techniques to Quarks and Strings, preprint, Brown Univ., Providence, Rhode Island, 1976.
14. P. M. Bleher, Ya. G. Sinai, Investigation of the Critical Point in Models of the Type of Dyson's Hierarchical Models, Comm. Math. Phys., 33, 23 (1973).
15. P. M. Bleher, Second Order Phase Transition in Some Ferromagnet Models, Tr. Mosc. Mat. Obsch., 33, 155 (1975). (In Russian).
16. P. M. Bleher, Ya. G. Sinai, Critical Indices for Dyson's Asymptotically-Hierarchical Models, Comm. Math. Phys., 45, 3, 247 (1975).
17. P. M. Bleher, Critical Exponents of Long-Range Models (Numerical Results), preprint Inst. Prikl. Mat. AN USSR 3, 1975. (In Russian).
18. F. J. Dyson, Existence of a Phase-Transition in a One-Dimensional Ising Ferromagnet, Comm. Math. Phys. 12, 2, 91 (1969).

19. F. J. Dyson, An Ising Ferromagnet with Discontinuous Long-Range order, Comm. Math. Phys., 21, 4, 269 (1971).

20. G. Gallovotti, H. Knops, The Hierarchical Model and the Renormalization Group, preprint, Inst. Theor. Fys. Univ. Nijmegen, Netherlands, 1974.

21. G. Jona-Lasinio, The Renormalization Group: a Probabilistic View, preprint, Inst. Fis. Univ. Padova, 1974.

22. J. Baker, Ising Model with a Scaling Interaction, Phys. Rev. B5, 7, 2633 (1972).

23. Ya. G. Sinai, Automodel Probabilistic Distributions, Teor. Ver. Prim., 21, 1, 63 (1976). (In Russian).

24. Ya. G. Sinai, Some Rigorous Results in the Theory of Phase Transitions, Proc. Int. Conf. Stat. Phys., Budapest, Publ. House Hung. Ac. Sci., 1975.

25. Ya. G. Sinai, Some Mathematical Probelsm in the Theory of Phase Transitions, Publ. House Hung. Ac. Sci., 1977.

26. H. Bateman, A. Erdelyi, Higher Transcendental Functions, v. 2, New York, Toronto, London, McGraw Hill Book Co, Inc., 1953.

27. M. E. Fisher, S.-K. Ma, B. G. Nickel, Critical Exponents for Long-Range Interactions, Phys. Rev. Lett., 28, 14, 917 (1972).

28. R. L. Dobrushin, Automodelity and Renorm-Group of Generalized Random Fields, see this volume.

29. I. M. Gelfand, G. E. Shilov, Generalized Functions and Actions upon Them, v. 1, Fiz. Mat. Giz., Moscow, 1958.

30. D. Ruelle, Statistical Mechanics, Rigorous Results, W. A Benjamin, Inc., New York, Amsterdam, 1969.

31. G. A. Baker, Jr., S. Krinsky, Renormalization Group Structure for Translationally Invariant Ferromagnets, J. Math. Phys., 18, 4, 590 (1977).

chapter 4

# GAUSSIAN RANDOM FIELDS–GIBBSIAN POINT OF VIEW

R. L. Dobrushin

## I. INTRODUCTION

In recent years there has been a great interest in the theory of Gibbsian random field with continuous spaces of values. The non-compactness of the space of values brings additional difficulties, and it is difficult to obtain here the complete description of the class of the fields having a given potential. Thus it is interesting to consider a partial case of linear-quadratic potentials leading to Gaussian fields. Here it is possible to use traditional methods of the linear theory of random processes, the situation can be investigated more completely and the inferences illustrate what can happen in the general case.

The main result of the paper states that any stationary Gaussian random field on the $\nu$-dimensional lattice $\mathbb{Z}^{\nu}$ having the spectral density $f(k)$ and mean value $h$ can be interpretated as a Gibbsian random field with the pair quadratic potential

$$U(x_t, x_s) = U(t - s)x_s x_t, \quad x_s, x_t \in \mathbb{R}^1, \qquad s, t \in \mathbb{Z}^{\nu} \tag{1.1}$$

where $U(t)$ are Fourier coefficients of the function $(2\pi)^{-\nu}[f(k)]^{-1}$

and with the linear-quadratic one particle potential

$$U(x_t) = \frac{1}{2} U(0)(x_t)^2 + h(\sum_{t \in \mathbb{Z}^\nu} U(t))x_t, \quad x_t \in \mathbb{R}^1, \; t \in \mathbb{Z}^\nu \tag{1. 2}$$

The set of all Gibbsian fields having such a potential coincides with the set of all convolutions of the stationary Gaussian field with spectral density $f(k)$ and mean value $0$ and random fields for which all their realizations $(a_t, \; t \in \mathbb{Z}^\nu)$ satisfy functional equations

$$\sum_{t \in \mathbb{Z}^\nu} a_t U(t - s) = - h, \quad z \in \mathbb{Z}^\nu \tag{1. 3}$$

Earlier some of the questions considered in this paper were investigated in Rosanov's paper [1], and Chay's paper [2] developed the previous one from other methodological positions which are traditional for the theory of stationary random processes. A brief discussion of the question is contained in the review paper [3].

## II. RANDOM FIELDS

Let $\mathbb{Z}^\nu$ be the $\nu$-dimensional integer lattice and let $V^\nu$ be the set of all finite subsets of the set $\mathbb{Z}^\nu$. We give $V^\nu$ the structure of a direct system in which the ordering is defined by the subset structure and we shall interpret the limit of any function $q(V)$, $V \in V^\nu$ when $V \to \infty$ as the limit with respect to this structure. Sums of the series of the type $\sum_{t \in \mathbb{Z}^\nu} q_t$ shall be interpreted as the limit of sums $\sum_{t \in V} q_t$ when $V \to \infty$. If $q \in \mathbb{R}^1$ are real numbers then the convergence of the series $\sum_{t \in \mathbb{Z}^\nu} q_t$ is equivalent to the condition $\sum_{t \in \mathbb{Z}^\nu} |q|_t < \infty$.

Let $X = (\mathbb{R}^1)^{\mathbb{Z}^\nu}$ be a space of functions $x = (x_t, \; t \in \mathbb{Z}^\nu)$ on $\mathbb{Z}^\nu$ with the values on the real line $\mathbb{R}^1$. For all $x \in X$ and $V \in \mathbb{Z}^\nu$ we let $x_V = (x_t, \; t \in V) \in (\mathbb{R}^1)^V$. Let $\Phi \subset X$ be a set of finite-range functions of $t \in \mathbb{Z}^\nu$. We let

$$\xi_\varphi(x) = \sum_{t \in \mathbb{Z}^\nu} \varphi(t)x_t, \quad x \in X, \varphi \in \Phi$$

$$\xi_t(x) = x_t, \quad x \in X, \ t \in \mathbb{Z}^\nu \tag{2.1}$$

Let $\mathcal{B}_V$ where $V \subset \mathbb{Z}^\nu$ denotes the smallest $\sigma$-subalgebra of the space X with respect to which all the functions $\xi_t$, $t \in V$ are measurable. *) Finally the $\sigma$-algebra $\mathcal{B}_{\mathbb{Z}^\nu}$ will be denoted simply $\mathcal{B}$.

We shall call probability measures on the space $(X, \mathcal{B})$ <u>probability distributions of the random fields.</u> We shall call the linear functional

$$A^P(\varphi) = \int_X \xi_\varphi(x)P(dx) = \sum_{t \in \mathbb{Z}^\nu} A_t^P \varphi(t), \quad \varphi \in \Phi \tag{2.2}$$

where

$$A_t^P = \int_X \xi_t(x)P(dx), \qquad t \in \mathbb{Z}^\nu \tag{2.3}$$

as the <u>mean value</u> of the field having the probability distribution P.

If the mean value exists we let

$$\eta_\varphi^P(x) = \xi_\varphi(x) - A^P(\varphi), \qquad \varphi \in \Phi$$

$$\eta_t^P(x) = \xi_t(x) - A_t^P, \qquad t \in \mathbb{Z}^\nu \tag{2.4}$$

Finally we introduce the <u>covariance functional</u>

$$B^P(\varphi_1, \varphi_2) = \int_X \eta_{\varphi_1}^P(x)\eta_{\varphi_2}^P(x)P(dx)$$

$$= \sum_{s,t \in \mathbb{Z}^\nu} \varphi_1(s)\varphi_2(t)B^P(s,t), \quad \varphi_1, \varphi_2 \in \Phi \tag{2.5}$$

where

---

* For finite V we shall always identify $\mathcal{B}_V$ with the Borel $\sigma$-algebra of the corresponding Euclidean space.

$$B^P(s,t) = \int_X \eta_t^P(x)\eta_s^P(x)P(dx), \qquad s,t \in \mathbb{Z}^\nu \tag{2.6}$$

We shall say that a random field having the probability distribution $P$ is covariance-stationary if it has a finite covariance functional and

$$B^P(\varphi_1,\varphi_2) = B^P(\varphi_1^s,\varphi_2^s), \qquad \varphi_1,\varphi_2 \in \Phi, \quad s \in \mathbb{Z}^\nu \tag{2.7}$$

where $\varphi_i^s(t) = \varphi_i(t-s)$, $s,t \in \mathbb{Z}^\nu$, $i = 1,2$. It is known (see [4], §4.2)* that in the case of covariance-stationary field

$$B^P(\varphi_1,\varphi_2) = \int_{(-\pi,\pi]} \tilde{\varphi}_1(k)\tilde{\varphi}_2(k)F^P(dk), \qquad \varphi_1,\varphi_2 \in \Phi \tag{2.8}$$

where

$$(-\pi,\pi]^\nu = \{k = (k_1,\ldots,k_\nu) \in \mathbb{R}^\nu : -\pi < k_i \le \pi,\ i = 1,\ldots,\nu\} \tag{2.9}$$

and where

$$\tilde{\varphi}_i(k) = \sum_{t \in \mathbb{Z}^\nu} \varphi_i(k)e^{ikt}, \qquad k \in (-\pi,\pi]^\nu, \quad i = 1,2 \tag{2.10}$$

are Fourier series with coefficients $\varphi_i(t)$ and where $F^P$ is a measure on the $\sigma$-algebra $\mathcal{B}((-\pi,\pi]^\nu)$ of Borel subsets of the set $(-\pi,\pi]^\nu$ called the spectral measure of the field having the probability distribution $P$. If the spectral measure

$$F(D) = \int_D f(k)dk, \qquad D \in \mathcal{B}((-\pi,\pi]^\nu)^{**} \tag{2.11}$$

then the function $f(k) \ge 0$ is called the spectral density of the field. The field is called Gaussian if (see [4], §3.1) its characteristic

---

* The fields with discrete arguments are considered in this book only in the case $\nu = 1$, but the necessary generalizations are direct.

** Here and in the following we shall denote integrals with respect to the Lebesgue measure in this way.

functional

$$\chi^P(\varphi) = \int_X \exp\{i\xi_\varphi(x)\}P(dx)$$
$$= \exp\{-\frac{1}{2}B^P(\varphi,\varphi) + iA^P(\varphi)\}, \qquad \varphi \in \Phi \qquad (2.12)$$

where $A^P$ and $B^P$ are mean value and correlation functional of the field. A Gaussian field is called stationary with parameters $(a, F)$ if for some $a \in \mathbb{R}^1$

$$A^P(\varphi) = a \sum_{t \in \mathbb{Z}^\nu} \varphi(t), \qquad \varphi \in \Phi \qquad (2.13)$$

and it is a covariance-stationary field with the spectral measure $F$. If the measure $F$ has the density $f$ we shall call the field as a stationary Gaussian field with parameters $(a, f)$.

For any probability distribution $P$ having finite covariance functional we shall denote by $L_2(P)$ the Hilbert space of real functions which are square-integrable with respect to the measure $P$, and by $L_2^1(P)$ the subspace of the space $L_2(P)$ which is the closure in $L_2(P)$ of the set of functions $\eta_\varphi^P$, $\varphi \in \Phi$. Let $P$ now be the probability distribution of a covariance-stationary random field with a spectral measure $F$. Let $L_2(F)$ be a real Hilbert space of even complex functions square integrated with respect to the measure $F$. The relations (2.6) and (2.8) imply easily the existence of the unique isomorphism: $L_2(F) \longleftrightarrow L_2^1(P) : \varphi \to Z_\varphi$ such that

$$Z_{\tilde{\varphi}} = \eta_\varphi^P, \qquad \varphi \in \Phi^* \qquad (2.14)$$

---

* The function $Z(D) = Z_{\chi_D}$ $D \in \mathcal{B}((-\pi,\pi]^\nu)$ where $\chi_D$ is the indicator of the set $D$ is called a random spectral measure in the theory of random fields. In such cases $Z_{\tilde{\varphi}}$ is denoted by $\int \tilde{\varphi}(k)Z(dk)$ (see for example [4], §4.5).

Let $P$ be a probability distribution of some random field and $V \in \mathbf{V}^{\nu}$. We shall denote by $P_V(A|x)$, $A \in \mathcal{B}_V$, $x \in X$ the restriction of the $\sigma$-algebra $\mathcal{B}_V$ of the conditional probability distribution with respect to the $\sigma$-algebra $\mathcal{B}_{\mathbb{Z}^{\nu} \setminus V}$ generated by the state $P$, i. e. (see for example [4], §1. 3) a function of $x \in X$ which is defined almost everywhere with respect to the measure $P$ and is measurable with respect to the $\sigma$-algebra $\mathcal{B}_V$ and such that its values are probability measures on the $\sigma$-algebra $\mathcal{B}_V$ and finally

$$P(\{x \in X : x_V \in A\} \cap B) = \int_B P_V(A|x) P(dx), \quad A \in \mathcal{B}_V, \ B \in \mathcal{B}_{\mathbb{Z}^{\nu} \setminus V} \tag{2. 15}$$

If the measures $P_V(\cdot|x)$ are absolutely continuous with respect to the Lebesgue measure on $(\mathbb{R}^1)^V$ for almost all $x \in X$ the densities of the measures $P_V(\cdot|x)$ with respect to the Lebesgue measure shall be called the conditional densities and denoted $P_V(\hat{x}_V|x)$, $\hat{x}_V \in (\mathbb{R}^1)^V$, $x \in X$.

We shall note now some results of the correlation theory of random fields which will be useful in the following.

Let $P$ be a probability distribution of a random field having a finite covariance functional. Let $H_V \subseteq L_2^1(P)$, $V \subset \mathbb{Z}^{\nu}$ be the closed subspace generated by the vectors $\eta_t^P$, $t \in V$. Let $d_{t,V}^P \in L_2^1(P)$, $t \in V$ be the projection of the vector $\eta_t^P$ on the subspace $H_{\mathbb{Z}^{\nu} \setminus V}$. Also let

$$B_V^P(s, t) = \int_X (\eta_t^P(x) - d_{t,V}^P(x))(\eta_s^P(x) - d_{s,V}^P(x)) P(dx), \quad s, t \in V \tag{2. 16}$$

Proposition 2. 1. (See [5], §§ 2. 3, 4. 3, 7. 12). *Let P be a probability distribution of a Gaussian field. Then for any $V \in \mathbf{V}^{\nu}$ the conditional distribution $P_V(\cdot|x)$ has the following characteristical* functional,

$$\chi(\lambda_t, t \in V|x) = \int_{(\mathbb{R}^1)} \exp\{i \sum_{t\in V} \lambda_t \xi_t(\hat{x}_V)\} P_V(d\hat{x}_V|x)$$

$$= \exp\{-\frac{1}{2} \sum_{s,t\in V} B_V^P(s,t)\lambda_s\lambda_t$$

$$+ i \sum_{s\in V} \lambda_s (d_{s,V}^P(x) + A_s^P)\}, \quad x \in X \tag{2.17}$$

i. e., it is Gaussian and has the covariance matrix $B_V^P = \{B_V^P(s,t), s,t \in V\}$ and the mean values vector $d_V^P = \{d_{t,V}^P + A_t^P, t \in V\}$.

The random field is called minimal if $\eta_t^P \notin H_{\mathbb{Z}^\nu \setminus \{t\}}$, for all $t \in \mathbb{Z}^\nu$. By the application of the proposition 2.1 we find that a Gaussian field is minimal if and only if the conditional distributions $P_{\{t\}}(\cdot|x)$ are Gaussian with the nonvanishing variances $(\sigma_t)^2 = B_{\{t\}}^P(t,t)$ and the mean values $d_{t,\{t\}}^P + A_t^P$.

<u>Proposition 2. 2</u> (See [2]). <u>A covariance-stationary field having a spectral density</u> f <u>is minimal if and only if there exist the spectral density</u> f <u>such that</u>

$$\int_{(-\pi,\pi)^\nu} [f(k)]^{-1} dk < \infty \tag{2.18}$$

If the field is minimal then (the "interpolation error")

$$\sigma^2 = (\sigma_t)^2 = (2\pi)^{2\nu} [\int_{(-\pi,\pi]^\nu} [f(k)]^{-1} dk]^{-1}, \quad t \in \mathbb{Z}^\nu \tag{2.19}$$

and (the "best prediction")

$$d_t = Z_{\ell^{itk}} [1 - (2\pi)^{-\nu} \sigma^2 (f(k))^{-1}] \tag{2.20}$$

A random field P is called linearly-regular if the intersection of the subspaces $H_{\mathbb{Z}^\nu \setminus V}$ or all $V \in \mathcal{V}^\nu$ is the zero vector. We shall note that if a field is linearly-regular then the definition (2. 16) implies that

$$B^P(s, t) = \lim_{V \to \infty} B_V^P(s, t), \quad s, t \in \mathbb{Z}^\nu \tag{2. 21}$$

Proposition 2. 3. (See [1]). If the condition (2. 18) is true the field having the probability distribution $P$ is linearly-regular.

We shall say that the set of vectors $\{\eta_t^P, t \in \mathbb{Z}^\nu\}$ is a weak basis of the space $L_2^1(P)$ if any vector $\zeta \in L_2^1(P)$ can be represented uniquely as

$$\zeta = \sum_{t \in \mathbb{Z}^\nu} c_t \eta_t^P \tag{2. 22}$$

where $c_t$, $t \in \mathbb{Z}^\nu$ are real numbers and the series (2. 22) converges in the sense of the weak convergence in the Hilbert space $L_2^1(P)$.

Proposition 2. 4. If the condition (2. 18) is true then the vector system $\{\eta_t^P, t \in \mathbb{Z}^\nu\}$ is a weak basis of the space $L_2^1(P)$. The space $L_2(F) \subseteq L_1(\mathbb{R}^1)$ where $L_1(\mathbb{R}^\nu)$ is the space of the functions integrated with respect to the Lebesgue measure and if $\zeta = Z_\varphi$, $\varphi \in L_2(F)$ the coefficients $c_t$, $t \in \mathbb{Z}^\nu$ in (2. 22) are described as the Fourier coefficients

$$c_t = \frac{1}{(2\pi)^\nu} \int_{(-\pi, \pi]^\nu} \varphi(k) e^{-itk} dk, \quad t \in \mathbb{Z}^\nu \tag{2. 23}$$

This fact was proven for the case $\nu = 1$ in the Rosanov book ([6], §2.11). It can be proved in the general case in the same way if the propositions 2. 2 and 2. 3 are used.

## III. GIBBSIAN FIELDS WITH LINEAR-QUADRATIC POTENTIAL

A pair $(h, U)$ where $h$ is a real number and $U(t)$, $t \in \mathbb{Z}^\nu$ is a real even function such that

$$\sum_{t \in \mathbb{Z}^\nu} |U(t)| < \infty \tag{3.1}$$

will be called a <u>parameter of a linear-quadratic potential</u>. Let $Y(U) \subseteq X$ be the set of functions $x = (x_t, t \in \mathbb{Z}^\nu) \in X$ such that

$$\sum_{t \in \mathbb{Z}^\nu} |U(t-s)x_t)| < \infty, \quad s \in \mathbb{Z}^\nu \tag{3.2}$$

We shall introduce the <u>interaction energy</u>

$$H_V^{(h, U)}(\hat{x}_V | x) = \sum_{\{s,t\} \subset V} U(t-s)\hat{x}_s \hat{x}_t + \frac{1}{2} \sum_{s \in V} U(0)(\hat{x}_s)^2$$

$$+ \sum_{s \in V} \hat{x}_s + \sum_{s \in V} \hat{x}_s ( \sum_{t \in \mathbb{Z}^\nu \setminus V} U(t-s)x_t),$$

$$\hat{x}_V = (\hat{x}_t, t \in V) \in (\mathbb{R}^1)^V, \quad x = (x_t, t \in \mathbb{Z}^\nu) \in Y(U), \quad V \in V^\nu \tag{3.3}$$

and <u>Gibbs density</u>

$$P_V^{(h, U)}(\hat{x}_V | x) = \frac{\exp \{- H_V^{(h, U)}(x_V | x)\}}{Z_V(x)} \tag{3.4}$$

where the <u>partition function</u>

$$Z_V(x) = \int_{(\mathbb{R}^1)^V} \exp \{- H_V^{(h, U)}(\hat{x}_V | x)\} d\hat{x}_V \tag{3.5}$$

Of course the previous definition makes sense only if the integral (3.5) converges. So we shall formulate the following simple result.

<u>Proposition 3.1.</u> <u>The partition function</u> $Z_V(x) < \infty$ <u>for all</u> $V \in V^\nu$ <u>and all (or for some)</u> $x \in Y(U)$ <u>if and only if for all</u> $V \in V^\nu$ <u>the following condition of positive definiteness is true:</u>

$$\sum_{s,t \in V} z_s \bar{z}_t U(t-s) > 0, \quad z_s \in \mathbb{C}, \ s \in V, \quad z_s \neq 0, \ s \in V \tag{3.6}$$

The condition (3. 5) is equivalent to the following condition: for almost all $k \in (-\pi, \pi]^{\nu}$ the function

$$g(k) = \sum_{t \in \mathbb{Z}^{\nu}} U(t) e^{itk} \geq 0 \tag{3. 7}$$

and this function does not vanish identically.

Proof: The condition (3. 6) is evident. It can be rewritten in spectral form as

$$\int_{(-\pi, \pi]^{\nu}} |h(k)|^2 g(k) dk > 0 \tag{3. 8}$$

for any non zero trigonometric polynomial

$$h(k) = \sum_{t \in V} z_t e^{ikt}, \quad z_t \in \mathbb{C}, \quad t \in V, \quad V \in V^{\nu} \tag{3. 9}$$

The sufficiency of the condition of nonnegativeness of the function $g(k)$ follows from the fact that a trigionmetric polynomial can vanish only on a set of zero measure which is easy to check by using for example induction on $\nu$ and the fact that a fixation of a value of $\nu$-th variable in trigonometric polynomial $h(k)$ defines a trigonometric polynomial of $\nu - 1$ variables. The necessity follows from the fact that the trigonometric polynomials $h(k)$ are dense in the Hilbert space of functions of $k \in (-\pi, \pi]^{\nu}$ having the scalar product

$$\int_{(-\pi, \pi]^{\nu}} \varphi(k) \psi(k) |g(k)| dk \tag{3. 10}$$

and so it is possible to choose trigionmetric polynomials approximating arbitrarily well the function

$$g^1(k) = \begin{cases} 1 & \text{if} \quad g(k) < 0 \\ 0 & \text{if} \quad g(k) \geq 0 \end{cases} \tag{3. 11}$$

in this space.

In the following we shall consider only the potentials for which the condition (3. 6) is true. We shall note that the definition (3. 4) can be rewritten as

$$p_V^{(h,U)}(\hat{x}_V|x)=[(2\pi)^{\text{Card } V}\det U_V]^{-\frac{1}{2}}\exp\{-\frac{1}{2}\sum_{s,t\in V} U(t-s)(\hat{x}_s-a_s^V(x))(\hat{x}_t-a_t^V(x))\} \tag{3. 12}$$

where $a_t^V(x)$, $t \in V$ are defined by the system of linear equations

$$\sum_{t\in V} U(t-s)a_t^V(x) + \sum_{t\in \mathbb{Z}^\nu \setminus V} U(t-s)x_t = -h, \quad s\in V \tag{3. 13}$$

and $\det U_V$ is the determinant of the matrix $U_V$ of this system which is positive because of the relation (3. 6). It means that $P_V^{(h,U)}$ is the density of the Gaussian probability distribution with the covariance matrix $(U_V)^{-1}$ and the mean values vector $(a_t^V(x),\ t \in V)$ (see (2. 17)). For the case when $U(t) = 0$ if $|t| > 1$, Spitzer [7] has noted an interesting possibility to find these mean values and covariances in terms of a random walk on the $\nu$-dimensional lattice.

By using the usual scheme we shall call the random field having the probability distribution $P$ a <u>Gibbsian field with a</u> $(h, U)$-<u>linear-quadratic potential</u> (or briefly $(h, U)$-<u>Gibbsian</u>) if

$$P(Y(U)) = 1 \tag{3. 15}$$

for any $V \in V^\nu$ there exists the conditional density $p_V$ (see §2) and for $P$-- almost all $x \in X$ and $P_V(\cdot\,|x)$ -- almost all $\hat{x}_V \in (\mathbb{R}^1)^V$

$$p_V(x_V|x) = p_V^{(h,U)}(x_V|x) \tag{3. 16}$$

<u>Proposition 3. 2.</u> <u>A probability distribution</u> $P$ <u>is</u> $(h, U)$-<u>Gaussian iff the condition</u> (3. 15) <u>and the conditions</u> (3. 16) <u>for sets</u> $V = \{t\}$ <u>containing one point</u> $t \in \mathbb{Z}^\nu$ <u>are fulfilled.</u>

Proof: The necessity is evident. To prove the sufficiency we must at first note that $\alpha$ and $\beta$ are two arbitrary random vectors such that there exist positive continuous conditional densities $p_{\alpha|\beta}$ and $p_{\beta|\alpha}$ of the vector $\alpha$ for fixed $\beta$ and the vector $\beta$ for fixed $\alpha$; then there exists also the mutual density $p_{\alpha|\beta}$ of these vectors which is defined in a unique way by the densities $p_{\alpha|\beta}$ and $p_{\beta|\alpha}$. This fact follows from the elementary formula

$$p_{\alpha|\beta}(a, b) = \frac{p_{\alpha|\beta}(a|b)}{\int p_{\alpha|\beta}(\hat{a}|b)[p_{\beta|\alpha}(b|\hat{a})]^{-1} d\hat{a}} \tag{3. 17}$$

Then we must note that the relations (3. 4) for $p_V^{(h, U)}$ are compatible for different V and so if we find by using the relation (3. 4) the conditional densities $p_{\{t\}}(\hat{x}_t|x)$ and $p_{V^1}(\hat{x}_{V^1}|x)$ where $t \in V$ and $V^1 = V \setminus \{t\}$ these densities are defined by the same relation. By using the last note and the previous note for $\alpha = \hat{x}_t$, $\beta = \hat{x}_{V^1}$ we see that the relation (2. 16) can be obtained with the help of induction with respect to the number of the points in V.

## IV. THE MAIN THEOREM

We shall introduce now some definitions which are needed for the formulation of the main result of the paper; the description of all (h, U)-Gibbsian fields. We shall denote by $A(h, U) \subseteq Y(U)$ the set of all functions $a = (a_t, t \in \mathbb{Z}^\nu) \in X$ such that

$$\sum_{t \in \mathbb{Z}^\nu} U(t - s)a_t = h, \quad s \in \mathbb{Z}^\nu \tag{4. 1}$$

(and of course the sums in (4. 1) exist). The convolution

$$P = P_1 * P_2 \tag{4. 2}$$

of two probability distributions $P_1$ and $P_2$ of random fields is by the definition the probability distribution

$$P(A) = \int_X P_1^a(A) P_2(da), \quad A \in \mathcal{B} \tag{4. 3}$$

where $P_1^a$, $a \in X$ is the shift of the probability distribution $P$ on $a$, i. e. , the probability distribution

$$P_1^a(A) = P_1(A - a), \quad A \in \mathcal{B} \tag{4. 4}$$

Let (see (4. 7))

$$f(k) = (2\pi)^{-\nu}[g(k)]^{-1}, \quad k \in (-\pi, \pi]^{\nu} \tag{4. 5}$$

If

$$\int_{(-\pi, \pi]^{\nu}} [g(k)]^{-1} dk < \infty \tag{4. 6}$$

there exists a stationary Gaussian field with parameters $(0, f)$. Its probability distribution shall be denoted $P^f$.

*Theorem 4. 1. The set of* $(h, U)$-*Gibbsian fields is non-empty iff the condition* (1. 6) *is fulfilled and the set* $A(h, U)$ *is non-empty. If the condition* (4. 6) *is fulfilled a probability distribution* $P$ *is* $(h, U)$-*Gibbsian iff*

$$P = P^f * P^l \tag{4. 7}$$

*where the probability distribution* $P^l$ *is such that*

$$P^l(A(h, U)) = 1 \tag{4. 8}$$

*Proof:* First we must check that the condition (4. 6) implies that the probability distribution $P^f$ is $(0, U)$-Gibbsian. The condition (3. 1) implies that the function $g(k)$ is continuous and so propositions 2. 1 and 2. 2 imply that the conditional density $P^f_{\{t\}}(\hat{x}_t | x)$ induced by the probability distribution $P^f$ is defined by the relation

$$P^f_{\{t\}}(\hat{x}_t|x) = \frac{1}{\sqrt{2\pi\sigma}} \exp\{-\frac{1}{2\sigma^2}(x_\sigma - d(x))^2\} \tag{4.9}$$

Propositions 2. 2 and 2. 4 show that

$$\sigma^2 = (U(0))^{-1} \tag{4.10}$$

and

$$d_t(x) = -\sigma^2 \sum_{s \neq t} U(t-s)\xi_s(x) \tag{4.11}$$

where the series converges in the weak sense in the space $L^1_2(P^f)$. If the condition (3. 1) is fulfilled the series $\sum_{s \neq t} U(t-s)e^{iks}$ converges in $L_2(F)$ and so the series (4. 11) converges in $L^1_2(P^f)$ in the strong sense. If the condition (3. 1) is fulfilled the mean value with respect to the measure $P^f$

$$E_f(\sum_{t \in \mathbb{Z}^\nu} |U(t-s)||\xi_t|) \leq E_f(|\xi_0|) \sum_{t \in \mathbb{Z}^\nu} |U(t-s)| < \infty \tag{4.12}$$

and so the condition (3. 15) and the relation (4. 11) are fulfilled in the sense of the convergence for almost all with respect to the measure $P^f$ values $x \in X$. So the right hand part of (4. 9) coincides almost everywhere with $p^{(0,U)}_{\{t\}}(\hat{x}_t|x)$ and so proposition 3. 2 implies that $P^f$ is a $(0, U)$-Gibbsian probability distribution.

Further we note that in the case (4. 6) the shift $P^{f,a}$, $a \in X$ (see (4. 4)) of the probability distribution $P^f$ has the conditional densities $p^{f,a}_V$ which are connected with conditional densities $p^f_V$ of the probability distribution $P^f$ by the relation

$$p^{f,a}_V(\hat{x}_V|x) = p^f_V(\hat{x}_V - a_V|x - a), \quad \hat{x}_V \in (\mathbb{R}^1)^V, \quad x \in X, \quad V \in \mathcal{V}^\nu \tag{4.13}$$

The relation (3. 3) shows that if (4. 1) is true the difference $H^{(h,U)}_V(\hat{x}_V|x) - H^{(h,U)}_V(\hat{x}_V - a_V|x - a)$ does not depend on $\hat{x}_V$ and

$$p^{(h,U)}_V(\hat{x}_V|x) = p^{(h,U)}_V(\hat{x}_V - a_V|x - a), \quad \hat{x}_V \in (\mathbb{R}^1)^V, \quad x \in X \tag{4.14}$$

The fulfillment of the condition (3. 15) for the probability distribution $P^{f,a}$ follows from its fulfillment for the probability distribution $P^f$ and the absolute convergence of the series (4. 1). Now the comparison of the relations (4. 13) and (4. 14) shows that in the case when the condition (4. 1) is true the field having the probability distribution $P^{f,a}$ is (h, U) - Gibbsian. The usual considerations (see for example [8], §2) show that any weighing integral mean of (h, U)-Gibbsian probability distributions is again a (h, U)-Gibbsian probability distribution. It implies that the probability distributions of the type (4. 7) are probability distributions of (h, U)-Gibbsian fields and so the sufficiency statement of the theorem has been proved.

In proving that the condition (4. 6) is necessary for the existence of (h, U)-Gibbsian fields we shall use the following.

Lemma 4. 2. Let $P^i$, $i = 1, 2$ be probability distributions of a covariance-stationary random field having spectral densities $f^i$ such that

$$f^1(k) \geq f^2(k), \qquad k \in (-\pi, \pi]^\nu \tag{4. 15}$$

Let the variances $B_V^i(t, t) = B_V^{P^i}(t, t)$, $i = 1, 2$ be defined by the relation (2. 16). Then

$$B_V^1(t, t) \geq B_V^2(t, t), \qquad t \in V, \ V \in \mathcal{V}^\nu \tag{4. 16}$$

Proof: We define $f^3(k) = f^1(k) - f^2(k)$ and consider the Hilbert spaces $L_2^1(P^i)$, $i = 1, 2, 3$ corresponding to the probability distributions of Gaussian fields with the parameters $(0, f^i)$. We also consider the direct sum $\overline{L}_2 = L_2^1(P^2) \oplus L_2^1(P^3)$ of Hilbert spaces $L_2^1(P^i)$, $i = 2, 3$. It is clear that if $\xi_t^1 = \xi_t^2 + \xi_t^3$, $t \in \mathbb{Z}^\nu$ where $\xi_t^i \in \overline{L}_2$, $i = 2, 3$ are the images of the corresponding elements $\xi_t \in L_2^1(P^i)$ in the natural isometric imbedding $L_2^1(P^i) \subseteq \overline{L}_2$, $i = 2, 3$ the transformation $\xi_t \in L_2^1(P^1) \rightarrow \xi_t^1 \in \overline{L}_2$, $t \in \mathbb{Z}^\nu$ is continued to the isometric imbedding $L_2^1(P^1) \in \overline{L}_2$. Let $H_V^i$, $i = 1, 2, 3$, $V \subseteq \mathbb{Z}^\nu$ be the closed subspaces of the space $\overline{L}_2$ generated by the vectors $\xi_t^i$, $t \in V$. Let $d_{t,V}^i$, $V \in \mathcal{V}^\nu$, $t \in V$, $i = 1, 2, 3$ be projections of the vectors $\xi_t^i$ on the subspaces $H_{\overline{V}}^i$ where $\overline{V} = \mathbb{Z}^\nu$ V. The isometric property of the imbeddings and the relation (2. 16) imply that

$$B_V^i(t, t) = \|\xi_t^i - d_{t,V}^i\|^2, \quad t \in V, \ V \in \mathcal{V}^\nu, \quad i = 1, 2, 3 \qquad (4.\ 17)$$

where $\|\cdot\|$ is the norm in the Hilbert space $\overline{L}_2$. It is clear that the space $H_{\overline{V}}^1 \subseteq H_{\overline{V}}^2 \oplus H_{\overline{V}}^3$ and that the projection of the vector $\xi_t^1$ on the subspace $H_{\overline{V}}^2 \oplus H_{\overline{V}}^3$ is equal to $d_{t,V}^2 + d_{t,V}^3$. So

$$B_V^1(t, t) = \|\xi_t^1 - d_{t,V}^2 - d_{t,V}^3\|^2 = \|\xi_t^2 - d_{t,V}^2\|^2 + \|\xi_t^3 - d_{t,V}^3\|^2 \geq B_V^2(t, t) \qquad (4.\ 18)$$

and this inequality proves the lemma.

We suppose now that the potential is such that the intregral (4. 6) diverges. Let $b_V(t)$, $t \in V$ be the diagonal elements of the matrix which is inverse to the matrix $\{U(t - s), \ s, t \in V\}$, i. e. , be the variations of random variables $\hat{x}_t$ described by the densities (3. 12). Let also

$$g_n(k) = g(k) + \frac{1}{n}, \quad k \in (-\pi, \pi]^\nu, \quad n = 1, 2, \ldots \qquad (4.\ 19)$$

Now we denote by $p_V^{(h, U_n)}$ the density of the probability distribution described by the relation (3. 12) in which the coefficients $U(t - s)$ are changed on $U_n(t - s)$ where (comp. (3. 7)) $U_n(t)$, $t \in \mathbb{Z}^\nu$ are the Fourier coefficients of the function $g_n$ and we denote $b_V^n(t)$ the variances of the random variables $\hat{x}_t$, $t \in V$ described by the densities $p_V^{(h, U_n)}$. It is clear that

$$\lim_{n \to \infty} b_V^n(t) = b_V(t), \quad t \in V, \ V \in \mathcal{V}^\nu \qquad (4.\ 20)$$

By using the sufficiency statement of the theorem which was proved above, to the field $P^{f_n}$ where $f_n = [g_n]^{-1}$ we find that

$$b_V^n(t) = B_V^{P^{f_n}}(t, t) \tag{4. 21}$$

(comp. (2. 16)). By using the lemma 4. 2 and the definition (4. 19) we find that the sequence $b_V^n(t)$ is monotone increased on n and so the relation (4. 20) implies that

$$b_V(t) \geq b_V^n(t), \quad t \in V, \quad V \in \mathcal{V}^\nu \tag{4. 22}$$

The continuity of the function $g_n$ and the proposition 2. 3 imply that the field having the probability distribution $P^{f_n}$ is linearly regular and (see (2. 21) and (2. 8))

$$\lim_{V \to \infty} b_V^n(t) = B^{P^{f_n}}(t, t) = \int_{(-\pi, \pi]^\nu} f_n(k)dk \tag{4. 23}$$

The divergence of the integral (4. 6), the relation (4. 20) and the inequality (4. 22) imply now that

$$\lim_{V \to \infty} b_V(t) = \infty, \quad t \in \mathbb{Z}^\nu \tag{4. 24}$$

Now let P be a (h, U)-Gibbsian probability distribution. The relations (3. 16) and (3. 12) imply that for any $V \in \mathcal{V}^\nu$, $t \in V$ the random variable

$$\xi_t(x) = a_t^V(x) + \zeta_t^V(x) \tag{4. 25}$$

where $\zeta_t^V(x) = \xi_t(x) - a_t^V(x)$ has the Gaussian probability distribution with the mean value 0 and the variance $b_V(t)$ and such that the summands in (4. 25) are independent. It is clear that

$$P(|\xi_t| > (b_V(t))^{1/2}) \geq P(a_t^V > 0)P(\zeta_t^V > (b_V(t))^{1/2}$$
$$+ P(a_t^V \leq 0)P(\zeta_t^V < -(b_V(t))^{1/2}) \geq P(\zeta_t^V > (b_V(t))^{1/2} \tag{4. 26}$$

and the left hand side of this inequality is a constant independent of $V$. So the inequality (4. 24) contradicts the relation $P(|\xi_t| > C) \to 0$ when $C \to \infty$. This contradiction proves the necessity of the condition (4. 6).

We suppose now that the condition (4. 6) is fulfilled and begin to prove the statement about the necessity of the condition (4. 7). We must remember that a random field having the probability distribution $P$ is called <u>regular</u> if $P(B)$ has only values 0 or 1 on the sets $B \in \mathcal{B}_\infty$ where, $\mathcal{B}_\infty$ is the $\sigma$-algebra which is the intersection of all $\sigma$-algebras $\mathcal{B}_{\mathbb{Z}^\nu \setminus V}$, $V \in \mathcal{V}^\nu$. The well-known general facts (see for example [8], §2) of the theory of Gibbsian fields imply that any $(h, U)$-Gibbsian field is a weighted integral mean of regular $(h, U)$-Gibbsian fields. Because the operation of such weighing transforms the fields of the type (4. 7) to the fields of the same type it is enough to check the relation (4. 7) only for regular $(h, U)$-Gibbsian fields. So the statement of the theorem will be an implication of the following proposition.

<u>Proposition 4. 3.</u> <u>Any regular</u> $(h, U)$-<u>Gibbsian random field (having the probability distribution</u> $P$) <u>is a Gaussian covariance-stationary field having the spectral density</u> $f$ <u>and mean values</u> $a_t^P$ (see (2. 3)) <u>such that</u> $(a_t^P, t \in \mathbb{Z}^\nu) \in A(h, U)$.

<u>Proof:</u> Let $\chi_P(\varphi)$ be (comp. (2. 12)) the characteristic functional of the field. We shall consider for any $V \in \mathcal{V}^\nu$ and any $x \in X$ the characteristic functionals

$$\chi_{V,x}(\varphi) = \int_{(\mathbb{R}^1)^V} \exp\{i \sum_{t \in V} \varphi(t)\hat{x}_t\} p_V^{(h, U)}(\hat{x}_V | x) d\hat{x}_V$$

$$= \exp\{-\frac{1}{2}(\sum_{s,t \in V} b_V(s,t)\varphi(s)\varphi(t)) + i \sum_{t \in V} \varphi(t) a_t^V(x)\}, \quad \varphi \in \Phi, \; x \in X \tag{4. 27}$$

where $B_V = (b_V(s,t),\ s,t \in V)$ is a matrix inverse to the matrix $U_V = \{U(t-s),\ s,t \in V\}$ and $\{a_t^V(x),\ t \in V\}$ is a solution of the equations (3.13). The condition (3.16) and the well known properties of conditional expectations (see for example [5], Theorem 7.4 4) imply that

$$\lim_{V \to \infty} \chi_{V,x}(\varphi) = \chi_P(\varphi) \tag{4.28}$$

for any $\varphi \in \Phi$ and almost all with respect to the measure $P$ values $x \in X$. By using the statement of the theorem which has been proved, proposition 2.3 and the relation 3.2 we find similarly to the relation (4.23) that

$$\lim_{V \to \infty} b_V(s,t) = B^{P^f}(s,t), \quad s,t \in \mathbb{Z}^\nu \tag{4.29}$$

Now, the relations (4.27) and (4.28) imply, for almost all with respect to the measure $P$, values $x \in X$ the existence of the limits

$$\lim_{V \to \infty} a_t^V(x) = a_t^P, \qquad t \in \mathbb{Z}^\nu \tag{4.30}$$

where $a_t$ are some constants and the equality

$$\chi_P(\varphi) = \exp\{-\frac{1}{2} B^{P^f}(\varphi,\varphi) + \sum_{t \in \mathbb{Z}^\nu} a_t^P \varphi(t)\} \tag{4.31}$$

So we have shown that the field having the probability distribution $P$ is Gaussian and has the spectral density $f$. It remains only to prove that $(a_t^P,\ t \in \mathbb{Z}^\nu) \in A(h, U)$. To prove it we note that the condition (3.15) shows that the series (3.2) converges with probability 1 with respect to the measure $P$ and because the convergence in probability of some sequence of Gaussian random variables implies their convergence in the mean, the series (3.2) converge also in the mean with respect to the measure $P$. So the relation (2.3) and the convergence in the mean of the series (3.2) imply the convergence of the series in the left hand side of the relation (4. 1). Further the convergence of the series (3. 2), the condition (3. 1) and the covariance stationarity of the field having the probability distribution P imply that

$$\sum_{t\in\mathbb{Z}^\nu} |U(s-t)| \int_X |\xi_t(x)| P(dx) \le \sum_{t\in\mathbb{Z}^\nu} |a_t| |U(s-t)|$$

$$+ \sum_{t\in\mathbb{Z}^\nu} |U(s-t)| \int_X |\eta_t(x)| P(dx) < \infty \tag{4. 32}$$

We shall note now that the condition (3. 16) and the relation (3. 12) means that for almost all $x \in X$ with respect to the measure P

$$a_t^V(x) = E_P(\xi_t | \mathcal{B}_{\mathbb{Z}^\nu \setminus V})(x), \quad t \in V, \ V \in \mathcal{V}^\nu \tag{4. 33}$$

where the conditional mathematical expectation with respect to the $\sigma$-algebra $\mathcal{B}_{\mathbb{Z}^\nu \setminus V}$ induced by the probability distribution P is used in the right hand side of the equality. By using a well known general inequality for conditional mathematical expectations (see for example [9], sect. 2. 48) we see that

$$\int_X |E_P(\xi_t | \mathcal{B}_{\mathbb{Z}^\nu \setminus V})(x)| P(dx) \le \int_X |\xi_t(x)| P(dx), \quad t \in V, \ V \in \mathcal{V}^\nu \tag{4. 34}$$

So the relation (4. 32) implies that each of the series $\sum_{t\in\mathbb{Z}^\nu} U(s-t)\hat{a}_t^V(x)$, $s \in \mathbb{Z}^\nu$ where $\hat{a}_t^V = a_t^V$, $t \in V$, $\hat{a}_t^V = x_t$, $t \in \mathbb{Z}^\nu \setminus V$ (comp. (3. 13)) converges in the mean with respect to the measure P uniformly with respect to $V \in \mathcal{V}^\nu$. This fact together with the relation (4. 30) justifies the possibility of the limit approach in the sense of the convergence in measure P in the equalities (3. 13) when $V \to \infty$, and we obtain in the limit the relation (4. 1) sought for. The proof of the proposition and also the proof of the theorem is finished.

## V. COMMENTS ON THE MAIN THEOREM AND ITS IMPLICATIONS

Note 1. The statement that the stationary Gaussian field having parameters $(0, f)$ is a $(0, U)$ - Gibbsian field is a reformulation of the known facts about the interpolation of stationary processes ($\nu = 1$) and stationary fields. For the case $\nu = 1$ these facts are achieved as the result of the series of the papers of Kolmogorov [10], Jaglom [11], Rosanov [12] and are written in the book [6], § 1. 10). For the case $\nu > 1$ they are obtained by Rosanov [1] and Chay [2]. The proposition 3. 2 gives the possibility of reducing the general interpolation problem to the problem of the interpolation in one point.

Note 2. The field having the probability distribution P is called Markov if the conditional probabilities $P_V(A|x)$ are measurable as a function of x with respect to the $\sigma$-algebra $\mathcal{B}_{V^1}$ where the set $V^1 = V^1(V) \in \mathcal{V}^\nu$. It is clear that $(h, U)$-Gibbsian field is Markov iff the function U is finite-range. The stationary Markov Gaussian fields were introduced by Levy in the year 1948 (see [13], §28). Theorem 4. 1 gives the description of such fields due to Rosanov [1]. We note also that a more general class of Gaussian Markov fields which have not conditional densities p was considered in Rosanov's paper [1] and in Chay's paper [2] and Pitt's paper [14], which developed and made more precise, Rosanov's paper.

Note 3. A Gaussian stationary field having the probability distribution P is $(h, U)$ - Gibbsian field if and only if its spectral density is described by the relation (4. 5) and its mean value is such that equations (4. 1) have the constant solution $\{a_t \equiv a,\ t \in \mathbb{Z}^\nu\}$. It is clear that this mean value is defined by the relation

$$a = -h\Big(\sum_{t\in \mathbb{Z}^\nu} U(t)\Big)^{-1} = -hf(0) \qquad (5.1)$$

if we exclude the special case $g(0) = 0$ or what is the same

$$\sum_{t \in \mathbb{Z}^\nu} U(t) = 0 \qquad (5.2)$$

In this special case any $a$ defines a constant solution of the equations (4. 1) if $h = 0$. If $h \neq 0$ there are no such solutions.

Note 4. The condition (5. 2) is equivalent to the following statement: if $\hat{g}_V = (g_t \equiv g,\ t \in V) \in (\mathbb{R}^1)^V$, $\hat{g} = (g_t \equiv g,\ t \in \mathbb{Z}^\nu) \in X$ where $g$ is any real number

$$p_V^{(0,U)}(\hat{x}_V + \hat{g}_V | x + g) = p_V^{(0,U)}(\hat{x}_V | x), \quad \hat{x}_V \in (\mathbb{R}^1)^V, \quad x \in Y(U) \qquad (5.3)$$

and so it is the case when the transition probabilities are shift invariant. The nonuniqueness of stationary (0, U)-Gibbsian fields is connected, as it usual for phase transition, with the breakdown of symmetry. (It is symmetry with respect to the shifts in $\mathbb{R}^1$).

The concrete examples of such nonuniqueness appear when the conditions (3. 1), (4. 6) and (5. 2) are true simultaneously. If the potential $U$ is finite-range the function $g(k)$ is differentiable and has a minimum in $k = 0$ and the integral (4. 6) diverges for $\nu = 1, 2$ and can converge only if $\nu \geq 3$. So the breakdown of symmetry can exist for the considered Markov fields only if the dimension $\nu \geq 3$. The first such breakdown of symmetry of Gaussian fields was considered by Kac and Berlin [15]. The examples of the breakdown of the symmetry in the smaller dimensions $\nu = 1, 2$ are given by functions $g(k)$ which are smooth when $k \neq 0$ and have an asymptotic $g(k) \sim c|k|^\beta$ when $k \to 0$ where $\beta < \nu$. The usual methods of the theory of Fourier series show that such functions correspond to functions $U(t)$ such that the rate of their decrease when $|t| \to \infty$ is $a|t|^{-\gamma}$ where $\gamma < 2\nu$. (For the case $\nu = 1$ see [16], chap. 5, problem 11). These notes agree with the general results about the absence of the breakdown of the continuous symmetry in dimensions $\nu = 1, 2$ for enough quickly decreasing potentials (see [17] and illustrate the bounds of the action of such results.

Note 5. The realtion (4. 7) implies that the (0, U)-Gibbsian field having the probability distribution P is a covariance-stationary field with the mean value $A_t^P \equiv 0,\ t \in \mathbb{Z}^\nu$ iff the field having the probability distribution $P^l$ is such a field. The conditions (4. 8), (4. 1) mean in the spectral form that (comp. (2. 8) and (3. 7))

$$\int_{(-\pi,\pi]^\nu} |e^{isk} g(k)|^2 F^{P^l}(dk) \equiv 0, \qquad s \in \mathbb{Z}^\nu \tag{5. 4}$$

It is clear that this is possible iff the spectral measure $F^{P^l}$ is concentrated on the set

$$H_U = \{k \in (-\pi, \pi]^\nu : g(k) = 0\} \tag{5. 5}$$

It implies in particular that a stationary Gaussian field with mean value 0 is (0, U)-Gibbsian iff its spectral measure is the sum of the measure described by the relation (2. 11) and a measure concentrated on the set $H_U$. This fact is equivalent to a result of Rosanov [1]. All previous results are also true for (h, U)-Gibbsian fields but here the mean value is defined by the relation (5. 1). All observations made in note 4 about the case of the function g(k) vanishing at k = 0 can be extended to the case when g(k) vanishes at another point $k \in (\pi, \pi]^\nu$.

Note 6. Let $\overline{H}_U$ be a set of all $\nu$-dimensional complex $z = (z_1, \ldots, z_\nu)$ such that $z_i \neq 0$, $i = 1, \ldots, \nu$,

$$\sum_{t=(t_1,\ldots,t_\nu)\in\mathbb{Z}^\nu} U(t)(z_1)^{t_1} \ldots (z_\nu)^{t_\nu} = 0 \tag{5. 6}$$

and the corresponding series in (5. 6) converges absolutely. It is easy to understand that the sequences

$$a_t = c(\rho_1)^{t_1} \ldots (\rho_\nu)^{t_\nu} \cos \lambda t, \quad t = (t_1, \ldots, t_\nu) \in \mathbb{Z}^\nu \tag{5. 7}$$

where

$$z = (\rho_1 e^{i\lambda_1}, \dots, \rho_\nu e^{i\lambda_\nu}) \in \overline{H}_U, \quad \lambda = (\lambda_1, \dots, \lambda_\nu) \in \mathbb{R}^2,$$

$$\rho_1, \dots, \rho_\nu \in (0, \infty), \quad c \in \mathbb{R}^1$$

are in $A(0, U)$. The elements $k \in H_U$ (see (5. 5)) correspond to bounded solutions of the type (5. 7) having $\rho_1 = \dots = \rho_\nu = 1$. In the case of a finite $U$ the set $\overline{H}_U$ is always nonempty and if $\nu > 1$ it is even a continuum but if the function $g$ is positive all the solutions of the type (5. 7) have exponential increase. The following constructions give a noncomplete answer to the following question: can nonzero nonemponentially increasing elements of $A(0, U)$ exist if $g$ is positive.

Let $\hat{g}(k)$ be the function of $k \in \mathbb{R}^\nu$ which is the periodical continuation of the function $g(k)$ (i. e. , the function defined by the relation (3. 7) for all $k \in \mathbb{R}^\nu$). Let us suppose that the function $\hat{g}(k)$ is infinitely differentiable. (It is true if $\sum_{t \in \mathbb{Z}^\nu} |U(t)|\,|t|^m < \infty$ for all $m < \infty$). Let

$$G = \{a = (a_t,\ t \in \mathbb{Z}^\nu) : |a_t| \le c(|t|^m + 1),\ t \in \mathbb{Z}^\nu$$

$$\text{for some } m = m(a) < \infty, \quad c = c(a) < \infty\} \tag{5. 8}$$

and so the generalized function

$$A(k) = \sum_{t \in \mathbb{Z}^\nu} a_t e^{ikt}, \quad k \in \mathbb{R}^\nu,\ a \in G \tag{5. 9}$$

is contained in Schwartz space $T^*(\mathbb{R}^\nu)$. Therefore (see for example [18], §7) the system of equations (4. 1) can be written (if $h = 0$) in the form

$$A(k)\hat{g}(k) = 0 \tag{5. 10}$$

It is clear that if the function $g(k)$ is positive the relation (5.1) implies that $A(k) \equiv 0$ and so the intersection $A(0, U) \cap G$ is empty.

This result can be extended to a wide class of more slowly increased potentials if we use the Gelfand representation of Banach algebras (see for example [19], chap. 9).* A sequence $B = \{B_t, t \in \mathbb{Z}^\nu\}$ of the numbers $B_t > 0$ will be called admitted if the following four conditions are fulfilled:

1) The sum

$$\sum_{t \in \mathbb{Z}^\nu} |B_t| < \infty \tag{5.11}$$

2) For any $c \neq 0$ and $k = 1, \dots, \nu$

$$\sum_{t = (t_1, \dots, t_\nu) \in \mathbb{Z}^\nu} |B_t| e^{ct_k} = \infty \tag{5.12}$$

3) For any $s \in \mathbb{Z}^\nu$

$$\lim_{|t| \to \infty} \frac{B_{t-s}}{B_t} = 1 \tag{5.13}$$

4) The upper limit

$$\overline{\lim_{|t| \to \infty}} \frac{\sup_{|s| \geq \frac{|t|}{2}} B_s}{B_t} < \infty \tag{5.14}$$

An example of admitted sequences are sequences $B_t$ such that $B_t \sim d|t|^{-\alpha}$, $|t| \to \infty$ where the constants $d > 0$, $\alpha > \nu$.

Proposition 5.1. Let $U = \{U(t), t \in \mathbb{Z}^\nu$ such that the function $g$ is positive and the admitted sequence $B$ is such that there exists a

---

* The authors wish to thank B. S. Mitjagin who pointed out such a possibility.

_finite limit_ $\lim_{|t|\to\infty} U(t)|B_t$. _Let_ $G(B)$ _be the set of all sequences_ $a = (a_t,\ t \in \mathbb{Z}^\nu)$ _such that_

$$\sum_{t\in\mathbb{Z}^\nu} B_t|a_t| < \infty \qquad (5.15)$$

_Then the intersection_ $A(0, U) \cap \hat{G}(B)$ _contains only the zero sequence._

_Proof:_ We shall consider the Banach space $G(B)$ of all complex sequences $x = (x_t,\ t \in \mathbb{Z}^\nu)$ such that the finite limit $x_\infty = \lim_{|t|\to\infty} x_t B_t$ exists with the norm

$$\|x\| = \sup_{t\in\mathbb{Z}^\nu} [(B_t)^{-1}|x_t|] \qquad (5.16)$$

It is easy to check (see for example [19], §4.9; the general case is reduced to the case $B_t \equiv 1$ considered in the reference by the evident change of variables) that the space $G^1(B)$ conjugated to the space $G(B)$ is the space of the sequences and $\varphi = (\varphi_\infty, \varphi_t,\ t \in \mathbb{Z}^\nu)$ where

$$\varphi(x) = \sum_{t\in\mathbb{Z}^\nu} \varphi_t x_t + \varphi_\infty x_\infty, \quad x \in G(B),\ \varphi \in G^1(B) \qquad (5.17)$$

We shall interpret the composition

$$(a * b)_t = \sum_{s\in\mathbb{Z}^\nu} a_{t-s} b_s, \quad t \in \mathbb{Z}^\nu,\ a, b \in G(B) \qquad (5.18)$$

as the multiplication operation in $G(B)$. We must note for it that the condition (5.13) implies the existence of a function $\alpha(|t|)$ such that $\alpha(|t|) \le \frac{|t|}{2}$, $\alpha(|t|) \to \infty$ and $\sup_{|s|\le\alpha(|t|)} B_{t-s} B_t \to 1$ when $|t| \to \infty$.

We note that

$$\sum_{s:\,|s|\le\alpha(|t|)} a_{t-s} b_s \sim B_t a_\infty \sum_{s\in\mathbb{Z}^\nu} b_s, \quad |t| \to \infty \qquad (5.19)$$

Further the relations (5. 16), (5. 14) and (5. 11) show that

$$\left| \sum_{s:|s|>\alpha(|t|);\ |s|\leq \frac{|t|}{2}} a_{t-s} b_s \right| \leq \left( \sup_{u:|u|\geq \frac{|t|}{2}} |a_u| \right) \sum_{s:|s|>\alpha(|t|)} |b_s|$$

$$= 0\,(B_t), \qquad |t| \to \infty \tag{5. 20}$$

By using a similar estimate for the sum in the domains $\{s: |t-s| \leq \alpha(|t|)\}$ and $\{s: |t-s| > \alpha(|t|),\ |s| > \frac{|t|}{2}\}$ we see that $a * b \in G(B)$ if $a, b \in G(B)$ and

$$(a * b)_\infty = b_\infty \sum_{s \in \mathbb{Z}^\nu} a_s + a_\infty \sum_{t \in \mathbb{Z}^\nu} b_t \tag{5. 21}$$

The similar estimates show also that the composition operation is continuous in the norm (5. 16) and therefore $G(B)$ is a commutative Banach algebra with respect to the usual component addition and the multiplication introduced above. The conditions of the proposition imply that the element $U = \{U(t),\ t \in \mathbb{Z}^\nu\} \in G(B)$.

We want to describe now the nonzero linear multiplicative functionals on the algebra $G(B)$. The condition (5. 12) implies that the restriction of any such functional on the subalgebra $G_0(B)$ consisting of the elements $x \in G(B)$ having $x_\infty = 0$ is

$$\varphi_k(x) = \sum_{t \in \mathbb{Z}^\nu} e^{i(k,t)} x_t, \qquad x \in G_0(B) \tag{5. 22}$$

where $k \in \mathbb{R}^\nu$. Its extension to $G(B)$ must be described by the relation

$$\varphi(x) = \varphi_k(\hat{x}) + \varphi_\infty x_\infty, \qquad x \in G(B) \tag{5. 23}$$

The multiplication property and the relation (5. 21) imply that

$$\varphi(a * b) = \varphi_k(\hat{a})\varphi_k(\hat{b}) + \varphi_\infty(b_\infty \sum_{s\in\mathbb{Z}^\nu} a_s + a_\infty \sum_{t\in\mathbb{Z}^\nu} b_t)$$

$$= (\varphi_k(\hat{a}) + \varphi_\infty a_\infty)(\varphi_k(\hat{b}) + \varphi_\infty b_\infty), \qquad a, b \in G(B) \tag{5.24}$$

By using $a = b \in G(B)$ such that $\varphi_k(\hat{a}) = \sum_{s\in\mathbb{Z}^\nu} a_s = 0$ but $a_\infty \neq 0$ we see that the relation (5. 24) implies that $\varphi_\infty = 0$. So the relation (5. 22) defines the general formula for the linear multiplicative continuous functional on $G(B)$.

The usual properties of Fourier series show now that the algebra $G(B)$ is semisimple and so the Gelfand isomorphism transforms the elements $a \in G(B)$ in their Fourier series (5. 9). So the positivity of the function $g$ shows that the element $U \in G(B)$ has an inverse element $U^{-1}$. Let $a \in \hat{G}(B)$. By using the natural embedding $\hat{G}(B) \subset G^1(B)$ it is possible if $h = 0$ to rewrite the equality (4. 1) as $a(U * \delta^S) = 0$, $S \in \mathbb{Z}^\nu$ where $(\delta^S)_t$, $t \neq s$ and $(\delta^S)_s = 1$. Because the linear combinations of the elements $\delta^S$ are dense in $G(B)$ it is possible to find that $a(U * b) = 0$, $b \in G(B)$ if $a \in A(0, U) \cap \hat{G}(B)$. If we let $b = U^{-1} * c$ we see that $a(c) \equiv 0$, $c \in G(B)$ and $a_t \equiv 0, t \in \mathbb{Z}^\nu$. This proves the proposition.

If $B = U$ the set $A(0, U) \subset \hat{G}(B)$. So if the sequence $U$ is admitted and the function $g$ is positive the set $A(0, U)$ contains only the zero sequence.

Note 7. It is evident that if $a \in A(h, U)$ then

$$A(h, U) = a + A(0, U) \tag{5.25}$$

So the question which is specific for the case $h \neq 0$ is only the question about the nonemptiness of the set $A(h, U)$. In the correspondence to note 3 it is necessary to investigate only the special case when the equality (5. 2) is true or (equivalently) $g(0) = 0$, and there are no constant elements $a = \{a_t, t \in \mathbb{Z}^\nu\}$, $a_t$ = const in $A(h, U)$. We want to show that usually in such situations the set $A(h, U)$ contains only the sequences which increase with power rate. Indeed in the analogy with (5.10) we can rewrite the system (4.1) in the form:

$$A(k)\hat{g}(k) = -h(2\pi)^\nu \hat{\delta}(k) \tag{5.26}$$

where $\hat{\delta}$ is the Dirac $\delta$-function on $(-\pi, \pi]^\nu$ which is continued periodically to all $\mathbb{R}^\nu$. For any multi-index $m = (m_1, \ldots, m_\nu)$ we denote by $D^m$ the operator of partial derivate $\dfrac{\partial^{|m|}}{\partial^{m_1} k_1 \ldots \partial^{m_\nu} k_\nu}$ where $|m| = m_1 + \ldots + m_\nu$. Let the multi-index $m$ be such that $|m|$ is even, $D^m \hat{g}(0) = d \neq 0$, but $D^{m^1} g(0) = 0$ if $|m^1| < |m|$. Then the function

$$A(k) = -\frac{(2\pi)^\nu h}{d} D^m \hat{\delta}(k) \tag{5.27}$$

satisfies the equation (5.26) and so the sequence

$$a_t = \frac{(-1)^{\frac{|m|}{2}+1} h}{d} (t_1)^{m_1} \ldots (t_\nu)^{m_\nu}, \quad t = (t_1, \ldots, t_\nu) \in \mathbb{Z}^\nu \tag{5.28}$$

is contained in $A(h, U)$. Because the function $g(\lambda)$ is even the corresponding multi-index $m$ exists always if the function $\hat{g}$ is analytical and in particular if the potential $U$ has a finite range. The similar considerations show that there exists no bounded sequences contained in $A(h, U)$ in the considered case.

The condition of the existence of all derivatives of the function $\hat{g}$ can be changed to the condition

$$\sum_{t = (t_1, \ldots, t_\nu) \in \mathbb{Z}^\nu} |t_1|^{m_1} \ldots |t_\nu|^{m_\nu} |U(t)| < \infty \tag{5.29}$$

if we use (compare the proof of Proposition 5. 1) the Gelfand representation of the corresponding algebras. However in the case of nonfinite range potential the situation is possible when A(h, U) is empty.

Proposition 5. 2. Let the sequence $U = \{U(t),\ t \in \mathbb{Z}^\nu\}$ be such that the function $\hat{g}$ vanishes identically in some neighborhood V of the point $0 \in \mathbb{R}^\nu$ and let U be admitted and such that for some $c > 0$, $n > 0$

$$U(t) \geq c\,|t|^{-n}, \qquad t \in \mathbb{Z}^\nu \tag{5. 30}$$

Then the set A(h, U) is empty.

Using the usual methods of the study of the asymptotical behavior of the Fourier coefficients (see [16], chap. 5) it is easy to construct explicitly examples of the sequences which satisfy the conditions of the proposition. For example this can be achieved if the domain V has a smooth boundary and the analytical singularities of the function g are reduced to the discontinuity of its derivatives on this boundary.

Proof: It is necessary to use in the case B = U the same construction as in the proof of Proposition 5. 1. The equations (4. 1) for $a \in A(h, U)$ correspond to the identity

$$a(U * b) = -\,h(2\pi)^\nu \sum_{t \in \mathbb{Z}^\nu} b_t, \qquad b \in G(U) \tag{5. 31}$$

(comp. (5. 26)). The condition (5. 30) implies the possibility of constructing an element $b \in G(U)$ such that its Fourier series vanishes outside V and does not vanish at the point $k = 0$. Then $\sum_{t \in \mathbb{Z}^\nu} b_t \neq 0$ but the Gelfand isomorphism implies that $U * b = 0$ and so the relation (5. 31) cannot be true if $h \neq 0$.

Note 8. Of course the results of the previous discussion can be applied to the problem of the existence and the uniqueness of the (h, U)-Gibbsian fields. Note 7 implies that in the case of finite-range potentials there always exist an infinite number of such fields, and uniqueness can occur only if we restrict the class of fields under consideration. There are two natural types of such restrictions. We shall say that the probability distribution P of a random field is a probability distribution of slowly increasing type if

$$P(G) = 1 \tag{5.32}$$

where the set G was defined in (5.8). We shall say that the probability distribution P is a probability distribution of the slowly increasing type in the mean if there exists constants $\alpha > 0$, $m < \infty$, $C > 0$ such that

$$\int_X |\xi_t(x)|^\alpha P(dx) \leq C|t|^m, \qquad t \in \mathbb{Z}^\nu \tag{5.33}$$

Using Chebyshev's inequality and the Borel-Cantelli lemma it is not difficult to check that the condition (5.32) implies the condition (5.33). Notes 6 and 7 imply in particular that if the function g is analytic the (h, U)-Gibbsian probability distributions contained in each of these two classes exist and, in the case of positive g, are unique. If the potential U is admitted and the function g is positive the uniqueness takes place in the class of all (h, U)-Gibbsian probability distributions. If $g(0) = 0$ and $h \neq 0$, (h, U)-Gibbsian states may not exist, and a typical situation is when there are no stationary (h, U)-Gibbsian fields even though (h, U)-Gibbsian probability distributions of slowly increasing type in the mean exist.

Note 9. The condition (3.1) restricts essentially the class of stationary Gibbsian fields having a Gibbs interpretation. The class can be extended if we modify somewhat the main definitions. Let P be a probability distribution having a finite covariance functional and the zero mean value and let a vector system $\{\xi_t, t \in \mathbb{Z}^\nu\}$ be a weak basis of the space $L_2^1(P)$. We shall say that P is a probability distribution of a (h, U)-Gibbsian in the weak sense field if for any $V \in \mathcal{V}^\nu$ and $s \in V$ the series $\sum_{t \in \mathbb{Z}^\nu \setminus V} U(t-s)$ converges in the weak sense, in the definition of the conditional density (3. 3), (3. 4), the infinite sums are interpreted as the sums in the weak sense and the equality (3. 16) is true for P -- almost all $x \in X$. The Propositions 2. 2 and 2. 4 imply that any stationary Gaussian field having the zero mean value and such that the condition (2. 18) is fulfilled is a (0, U)-Gibbsian field with potential U described by the relations (4. 5), (3. 7).

## REFERENCES


1. Yu. A. Rosanov, On Gaussian Fields with Given Conditional Distribution, Theory of Prob. and its Appl. , 7, No. 3, 1967, 433-443.
2. S. C. Chay, On Quasi-Markov Random Fields, J. of Multiv. Anal. , 2, No. 1, 1972, 14-76.
3. R. L. Dobrushin, S. A. Pirogov, Theory of Random Fields, Proc. of the 1975 JEEE-USSR Joint Workshop on Information Theory, 1976, IEEE, N. Y. , 39-49.
4. I. I. Gihman, A. V. Shorohod, Theory of Random Processes, Vol. I, "Nauka", Moscow, 1971 (in Russian).
5. J. L. Doob, Stochastic Processes, John Wiley and Sons, N. Y. , Lnd. , Champan Hall, 1953.
6. Yu. A. Rosanov, Stationary random processes, Fismatgis, Moscow, 1963 (in Russian).
7. F. L. Spitzer, Introduction aux processus de Markov parametre dans $\mathbb{Z}^\nu$, Lect. Notes, Math. , No. 390, 114-189, Springer Verlag, 1974.

8. C. Preston, Random fields, Lect. Notes, Math., No. 534, Springer Verlag, 1976.

9. P. A. Meyer, Probability and Potentials, Blaisdell Publ. Com., Waltham, Tor., Lnd., 1966.

10. A. N. Kolmogorov, Interpolation and Extrapolation of Stationary Random Sequences, Commun. (Isvestija) Sovjet. Acad. Sci. (Ser. Mathem.), 5, No. 1, 3-14, 1941.

11. A. M. Jaglom, To the Question of the Linear Interpolation of stationary random sequences and processes. Progress (Uspechi) of Math. Sci., 4, No. 9, 171-178, 1949.

12. Yu. A. Rosanov., On the interpolation of stationary processes with discrete time, Report of Sovjet Acad. Sci., 130, No. 4, 730-733, 1960.

13. P. Levy, Processus Stochastiques et movement Brownien, Gautheir-Villars, Paris, 1965.

14. D. Pitt, Deterministic Gaussian Markov Fields, J. Multiv. Anal., 5, No. 3, 312-313, 1975.

15. M. Kac, T. Berlin, The spherical model of ferromagnet, Phys. Rev., 86, No. 6, 821-835, 1952.

16. A. Zygmund, Trigonometric series, vol. I, Cambridge Univ. Press, Vol. 1.

17. R. L. Dobrushin, S. B. Shlosman, Non-existence of One-dimensional and Two-dimensional Gibbsian Fields with Noncompact Continuous Symmetry Group, this volume.

18. V. S. Vladimirov, Generalized functions in Mathematical Physics, Nauka, Moscow, 1976 (in Russian).

19. K. Yosida, Functional Analysis, Springer-Verlag, Berlin-Gottingen-Heidelberg, 1965.

chapter 5

# AUTOMODEL GENERALIZED RANDOM FIELDS AND THEIR RENORM-GROUP

R. L. Dobrushin

## I. INTRODUCTION

The well-known results (see for example [1]) of probability theory about the central limit theorem for the sums of weakly dependent random variables can be reformulated in the following way. The "large-scale" behavior of random processes which belong to a wide class of processes having sufficiently small correlations between their values for distant arguments is described by the generalized Gaussian process with uniform spectrum (white noise). In the more general situation when the second moments do not exist such limits are described by the stable generalized processes with independent values. The probabilistic papers in which the question about the possibility of other limit laws in the situation when the correlations are stronger (but nevertheless asymptotically decreasing) are rare and of special type. It is possible to note here an interesting example which has been constructed by Rosenblatt [2], the papers of Lamperti connected with the notion of semistable random processes [3], and finally the recent interesting paper of Taqqu [4].

But this complex of problems has also another history which is parallel and independent of the previous one. The investigations of essentially the same problem have been lead by the specialists of statistical physics who until recently were not even aware of the existence of the theory of probability. (Of course the probability specialists behaved in the same way.) The physicists have understood that although the situation in which the new limit laws apply is rare and refined in the case of random processes, such a situation is the usual one for second order phase transitions for the most natural Markov fields. Because the problem of limit theorems in its physical interpretation is the problem of the macroscopic description of a physical system, this problem has in recent years been at the center of interests for statistical physicists. (See, for example, the basic papers, Kadanoff [5], Wilson [6], and the review paper, Ma [7]). The deep ideas developed in this connection in the physical literature are of a form which is very far from the norm of mathematical rigor. The series of mathematical papers of recent years (see [8] - [14]) has developed the program of mathematically capturing these ideas, but the program is difficult and the results as yet are meager.

The main distinction of this paper vis-a-vis previous ones is that the main objects of consideration here are generalized random fields. This provides some essential advantages. One of the main objects of the theory, which at first, in the physical literature, was the set of scale transformations on the arguments of the random field (called the renormalization group) is here a one-parameter group of transformations of the probability distributions of the random fields. The tradition of the physical literature requires the consideration of discrete-argument fields; in such cases "the renorm-group" is a discrete semigroup. (For this case the main concept was introduced at the mathematical level by Gallavotti and John-Lasinio [14] and Sinai [9]).

The fact that the renorm-group is really a group permits us to combine the usual investigation of the large scale limit properties of the fields (the limit theorems of probability theory) with the investigation of their small scale properties (the local structure of the random fields). The class of possible limit random fields is the same in both cases. It is the class of random fields which are invariant with respect to the renorm-group. The Euclidean approach to quantum field theory developed in the past few years (see [15]) allows us to intrepret the well-known difficulties of quantum field theory as the difficulties of the local description of the Markov fields. In this way the main problem of this paper is connected with the problems of the construction of quantum field theories. The existence of such a connection is known in the physical literature, but it is described there in other terms. Using Sinai's terminology [9] we shall call the fields which are invariant with respect to the renorm-group the automodel fields. It is natural in the tradition of probability theory to call such fields stable fields (as in the paper [15]) because of this concept directly generalizes the concept of stable random processes with independent values. But it is inconvenient because the word "stable" is widely used in modern mathematics in another sense. The usual automodel random processes were introduced by Lamperti [3] who called them "semi-stable processes". We note that the transition to the consideration of generalized processes is important because there are no nontrivial usual automodel stationary processes. We also note that the class of processes with stationary increments which are invariant with respect to the similarity transformations introduced by Kolmogorov [16] and studied later by Pinsker [17] will be a class of automodel stationary processes if we use the language of the theory of generalized random processes. So apparently Kolmogorov's paper [35] published 35 years ago is the first publication on the theme.

The content of this paper, which is the first in a series, is essentially limited to the introduction of the main concepts and the discussion of their mutual relations. The advantages of the point of view expressed here can be exposed more explicitly in the case of Gaussian fields and especially in the possibility of constructing a nontrivial class of stationary automodel random fields which can be used for the description of the limit behavior of functionals of Gaussian processes (comp. [4]). The subsequent publications will be devoted to such questions.

The author is sincerely thankful to P. Blecher, R. Minlos and J. Sinai. Their influence determined to a significant extent the content of this paper.

## II. RENORM-GROUP AND AUTOMODEL RANDOM FIELDS

Let $\mathfrak{R}^\nu$ be the linear space of all functions of $\bar{x} \subseteq \mathbb{R}^\nu$ having real values. We shall introduce four groups of transformations on $\mathfrak{R}^\nu$. The group of similarity transformations in the space of values, $T = \{T_c, c \in (0,\infty)\}$, consists of the transformations

$$T_c\varphi(x) = c\varphi(x), \quad \varphi \in \mathfrak{R}^\nu, \quad x \in \mathbb{R}^\nu, \quad c \in (0,\infty) \tag{2.1}$$

The group of similarity transformations in the space of arguments, $U = \{U_\lambda, \lambda \in (0,\infty)\}$, consists of the transformations*

$$U_\lambda \varphi(x) = \lambda^{-\nu}\varphi(\lambda^{-1}x), \quad \varphi \in \mathfrak{R}^\nu, \quad x \in \mathbb{R}^\nu, \quad \lambda \in (0,\infty) \tag{2.2}$$

The group of shift transformations, $E = \{E_a, a \in \mathbb{R}^\nu\}$, consists of

$$E_a\varphi(x) = \varphi(x-a), \quad \varphi \in \mathfrak{R}^\nu, \quad x \in \mathbb{R}^\nu, \quad a \in \mathbb{R}^\nu. \tag{2.3}$$

Finally, fixing the number $\kappa \in \mathbb{R}^1$, we introduce the transformation group $S_\kappa = \{S_{\kappa,\lambda}, \lambda \in (0,\infty)\}$:

$$S_{\kappa,1}\varphi(x) = T_{\lambda^\kappa} U_\lambda \varphi(x) = U_\lambda T_{\lambda^\kappa}\varphi(x) = \lambda^{\kappa-\nu}\varphi(\lambda^{-1}x), \varphi \in \mathfrak{R}^\nu, x \in \mathbb{R}^\nu, \lambda \in (0,\infty) \tag{2.4}$$

*The inclusion of the factor $\lambda^{-\nu}$ in the definition is natural, as will be clear from formula (2.17).

Now let $\mathcal{L} \subset \mathcal{R}^{\nu}$ be a linear topological space of functions such that the restrictions of all transformations $E_a$, $a \subseteq \mathbb{R}^{\nu}$, $U_l$, $\lambda \in (0,\infty)$ to $\mathcal{L}$ are continuous transformations from $\mathcal{L}$ to $\mathcal{L}$ and that for all $\varphi \in \mathcal{L}$ the transformations $\mathbb{R}^{\nu} \to \mathcal{L} : a \to E_a\varphi$ and $\mathbb{R}^1 \to \mathcal{L} : \lambda \to U_\lambda \varphi$ are continuous. The most important examples of function spaces having these properties are Schwartz spaces $\mathcal{T}(\mathbb{R}^{\nu})$ and $\mathcal{D}(\mathbb{R}^{\nu})$ (see for example [18, 19]) which consist respectively of rapidly decreasing and finite-range infinitely differentiable functions.

A system of random variables $\{\Phi_\varphi, \varphi \in \mathcal{L}\}$, defined on the probability space $(W, \mathcal{B}, m)$ (here $\mathcal{B}$ is a $\sigma$-algebra of subsets of $W$, $m$ is a probability measure on $\mathcal{B}$ and the <u>random variables</u> are elements of the space $L_0 = L_0(W, \mathcal{B}, m)$ of equivalence classes of real measurable functions which coincide almost everywhere) such that the following two conditions are fulfilled is called a <u>random field upon $\mathcal{L}$</u>:

<u>condition</u> (1) for all $\varphi_1, \varphi_2 \in \mathcal{L}$, $\alpha_1, \alpha_2 \subseteq \mathbb{R}^1$

$$\alpha_1 \Phi_{\varphi_1} + \alpha_2 \Phi_{\varphi_2} = \Phi_{\alpha_1\varphi_1 + \alpha_2\varphi_2} \tag{2. 5}$$

<u>condition</u> (2) the transformation $\mathcal{L} \to L_0(W, \mathcal{B}, m) : \varphi \to \Phi_\varphi$ is continuous with respect to the topology of convergence in probability in $L_0(W, \mathcal{B}, m)$ (see for example [20]).

The set of one dimensional probability distributions

$$P_\varphi(B) = m(w \in W : \Phi_\varphi(w) \in B), \quad B \in \mathcal{B}_{\mathbb{R}^1} \tag{2. 6}$$

which are elements of the space $\mathcal{P}_1$ of probability measures on the $\sigma$-algebra $\mathcal{B}_{\mathbb{R}^1}$ of Borel subsets of $\mathbb{R}^1$, is called the <u>probability distribution</u> $P = \{P_\varphi, \varphi \in \mathcal{L}\}$ of the random field $\{\Phi_\varphi, \varphi \in \mathcal{L}\}$. We note that the linearity condition (2. 5) implies that the probability distribution $P$ also defines the joint finite-dimensional probability distributions $P_{\varphi_1, \dots, \varphi_n}$ of the random variables $(\Phi_{\varphi_1}, \dots, \varphi_{\varphi_n})$. The set of all probability distributions of the random fields upon $\mathcal{L}$ is denoted by $\mathcal{P}(\mathcal{L})$. The probability distribution of the random field $\{\Phi_\varphi, \varphi \in \mathcal{L}\}$ is also conveniently described with the help of the <u>characteristic functionals</u> $L^P = \{L^P(\varphi), \varphi \in \mathcal{L}\}$:

$$L^P(\varphi) = \int_{\mathbb{R}^1} \exp\{ix\} P_\varphi(dx) = \int_W \exp\{i\Phi_\varphi(w)\} m(dw) \tag{2.7}$$

It is easy to show (comp. [21], ch. 4, §4) that the transformation $P \to L^P$ is a one-to-one transformation of the space $\mathcal{P}(L)$ onto the set of all continuous nonnegative definite functionals $L^P = \{L^P(\varphi), \varphi \in \mathcal{L}\}$ with $L^P(0) = 1$.

Let $\mathcal{L}^1$ be the dual space of $\mathcal{L}$, let the functionals

$$\Phi_\varphi(f) = f(\varphi), \quad f \in \mathcal{L}^1, \quad \varphi \in \mathcal{L} \tag{2.8}$$

and let $\mathcal{B}(\mathcal{L})$ be the smallest $\sigma$-algebra of subsets of $L^1$ with respect of which all functionals $\Phi_\varphi$, $\varphi \in \mathcal{L}$ are measurable. A random field upon $\mathcal{L}$ defined by the probability space $(\mathcal{L}^1, \mathcal{B}(\mathcal{L}), \hat{P})$, where $\hat{P}$ is a probability measure upon $\mathcal{B}(\mathcal{L})$, and by the random variables (2.8) is called a state upon $\mathcal{L}$. It is well known (Minlos theorem; see [21, 22]) that in a wide class of cases, and, in particular, in the cases $\mathcal{L} = \mathfrak{T}(\mathbb{R}^\nu)$, $\mathcal{L} = \mathcal{D}(\mathbb{R}^\nu)$, any probability distribution $P \in \mathcal{P}(\mathcal{L})$ is the probability distribution of some state upon $\mathcal{L}$.

We shall introduce in $\mathcal{P}(\mathcal{L})$ <u>the topology of weak convergence of the finite-dimensional distributions</u> which is defined as the weakest on $\mathcal{P}(\mathcal{L})$ for which the transformations $\mathcal{P}(\mathcal{L}) \to \mathcal{P}_1 : P \to P_\varphi$, $\varphi \in \mathcal{L}$ are continuous if $\mathcal{P}_1$ is regarded as equipped with the usual topology of weak convergence of probability measures. This topology induces the topology of pointwise convergence in the space of characteristic functionals by the transformation $P \to L^P$. So it is understandable that convergence in this topology implies the convergence of all finite-dimensional joint probability distributions $P_{\varphi_1,\ldots,\varphi_n}$.

We introduce now the transformation groups $T^* = \{T^*_c,\ c \in (0,\infty)\}$, $U^* = \{U^*_\lambda,\ \lambda \in (0,\infty)\}$, $E^* = \{E^*_a,\ a \in \mathbb{R}^\nu\}$, $S^*_\kappa = \{S^*_{\kappa,\lambda},\ \lambda \in (0,\infty)\}$, $\in \mathbb{R}^1\}$ on the space $\mathcal{P}(\mathcal{L})$ with the help of the relations

$$(T^*_c P)_\varphi = P_{T_c\varphi}, \qquad c \subseteq (0,\infty),$$
$$(U^*_\lambda P)_\varphi = P_{U_\lambda\varphi}, \qquad \lambda \in (0,\infty),$$
$$(E^*_a P)_\varphi = P_{E_a\varphi}, \qquad a \in \mathbb{R}^\nu,$$
$$(S^*_{\kappa,\lambda} P)_\varphi = (T^*_{\lambda^\kappa} U^*_\lambda P)_\varphi = (U^{\lambda} T^*_{\lambda^\kappa} P)_\varphi =$$
$$= P_{S_{\kappa,\lambda}\varphi}, \qquad \lambda \in (0,\infty), \quad \in \mathbb{R}^1, \quad \varphi \in \mathcal{L} \tag{2.9}$$

It is evident that all transformations $T^*_c = U^*_\lambda$, $E^*_a$, $S^*_{\kappa,\lambda}$ are continuous transformations from $\mathcal{P}(\mathcal{L})$ to $\mathcal{P}(\mathcal{L})$. Condition (2) of the definition of random field and the continuity of the transformations $c \to T_c\varphi$, $\lambda \to U_\lambda\varphi$, $a \to E_a\varphi$ for any $\varphi \in \mathcal{L}$ imply that for any $P \in \mathcal{P}(\mathcal{L})$ the transformation $(0,\infty) \to \mathcal{P}(\mathcal{L}): c \to T^*_c P$, the transformation $(0,\infty) \to \mathcal{P}(\mathcal{L}): \lambda \to U^*_\lambda P$ and the transformation $\mathbb{R}^\nu \to \mathcal{P}(\mathcal{L}): a \to E^*_a P$ are continuous. The continuous transformation group $S^*_n$ is called <u>the renorm-group upon $\mathcal{L}$ with parameter</u> $\kappa$. The action of the group on the characteristic functional is described by the formula

$$L^{S^*_{\kappa,\lambda}P}(\varphi) = L^P(S_{\kappa,\lambda}\varphi), \quad \varphi \in \mathcal{L}, \quad \lambda \in (0,\infty) \tag{2.10}$$

We shall say that the probability distribution $P \in \mathcal{P}(\mathcal{L})$ (and also the random field having the distribution) is <u>stationary</u> if

$$E^*_a P = \mathcal{P}, \qquad a \in \mathbb{R}^\nu \tag{2.11}$$

and is <u>automodel with parameter</u> $\kappa \in \mathbb{R}^1$ if

$$S^*_{\kappa,\lambda} P = P, \quad \lambda \in (0,\infty) \tag{2.12}$$

The set of all automodel (similarly the set of all stationary automodel) probability distributions with parameter $\kappa$ is denoted by $\mathcal{A}_\kappa(\mathcal{L})$ (similarly $\mathcal{A}^{st}_\kappa(\mathcal{L})$). The probability distributions $P \in \mathcal{A}_\kappa(L)$ may be described, with the help of the characteristic functional, by the relation

$$L^P(\varphi) = L^P(S_{\kappa,\lambda}\varphi), \quad \varphi \in \mathcal{L}, \quad \lambda \in (0,\infty) \tag{2.13}$$

If the probability distribution $P \in \mathcal{A}_\kappa(\mathcal{L})$ has the finite n-th moment (see [21] ), sect. 3. 2. 1)

$$B^P_n(\varphi_1,\dots,\varphi_n) = (-i)^n \frac{\partial^n}{\partial\gamma_1 \dots \partial\gamma_n} L^P(\partial_1\varphi_1 + \dots + \gamma_n\varphi_n)\Big|_{\lambda_1 = \dots = \lambda_n = 0},$$
$$\varphi_1,\dots,\varphi_n \in \mathcal{L}, \quad n = 1,2,\dots \tag{2.14}$$

the linearity condition (2. 5) implies that

$$B^P_n(\varphi_1,\dots,\varphi_n) = \lambda^{nn} B^P_n(U_\lambda\varphi_1,\dots,U_\lambda\varphi_n),$$
$$\lambda \in (0,\infty), \ \varphi_1,\dots,\varphi_n \in \mathcal{L}, \ n = 1,2,\dots \tag{2.15}$$

If all moments of the probability distribution $P \in \mathcal{P}(\mathcal{L})$ are finite and the known conditions (see for example [2, 3] , § 7. 4 and 15. 4) which guarantee that the probability distributions $P_\varphi$, $\varphi \in \mathcal{L}$ may be recovered in a unique way from its moments are fulfilled, then condition (2. 15) for $\varphi_1 = \varphi_2 = \dots = \varphi_n \in L$, $n = 1, 2, \dots$ is also sufficient for $P \in \mathcal{A}_\kappa(L)$.

We note that the probability distribution $P$ can be automodel only for one value of $\kappa$. The only exception is the probability distribution of the <u>zero field</u>, i. e. , the probability distribution for which all measures $P_\varphi$, $\varphi \in \mathcal{L}$ are concentrated in zero point. Such a probability distribution is simultaneously automodel for all $\kappa \in \mathbb{R}^1$. It will be called also the zero probability distribution.

The concepts just introduced become clearer when applied to usual random fields. We shall call a usual random field upon $\mathbb{R}^\nu$ any system of random variables $\{\xi(x),\ x \in \mathbb{R}^\nu\}$. We shall say that the random field $\{\Phi_\varphi,\ \varphi \in \mathcal{L}\}$ upon $\mathcal{L}$ is generated by the usual random field $\{\xi(x),\ x \in \mathbb{R}^\nu\}$ if

$$\Phi_\varphi = \int_{\mathbb{R}^\nu} \varphi(x)\xi(x)\,dx, \qquad \varphi \in \mathcal{L} \tag{2.16}$$

where we intrepret such integrals as limits of Riemann sums in the sense of convergence in $L_0$. (Nothing will change if we use any other reasonable intrepretation of the integral).

So if $P \in \mathcal{P}(\mathcal{L})$ is the probability distribution of the random field upon $\mathcal{L}$, generated by the usual random field $\{\xi(x),\ x \in \mathbb{R}^\nu\}$, the random fields

$$\xi_c^T(x) = c\xi(x), \quad x \in \mathbb{R}^\nu, \quad c \in (0,\infty),$$

$$\xi_\lambda^U(x) = \xi(\lambda x), \quad x \in \mathbb{R}^\nu, \quad \lambda \in (0,\infty),$$

$$\xi_0^E(x) = \xi(x + a), \quad x \in \mathbb{R}^\nu, \quad a \in \mathbb{R}^\nu,$$

$$\xi_{\kappa,\lambda}(x) = \lambda^\kappa \xi(\lambda x), \quad x \in \mathbb{R}^\nu, \quad \lambda \in (0,\infty), \quad \kappa \in \mathbb{R}^1 \tag{2.17}$$

generate the random fields upon L having the probability distributions $T_c^*P$, $U_\lambda^*P$, $E_a^*P$ and $S_{\kappa,\lambda}^*P$ we shall say that the usual random field $\{\xi(x),\ x \in \mathbb{R}^\nu\}$ is <u>automodel with the parameter</u> $\kappa$ if the joint distributions of the systems of random variables $\{\xi(x_1),\ldots,\xi(x_n)\}$ and $\{\xi_{\kappa,\lambda}(x_1),\ldots,\xi_{\kappa,\lambda}(x_n)\}$ coincide for any $x_1,\ldots,x_n \in \mathbb{R}^\nu$, $n = 1,2,\ldots$, and all $\lambda \in (0,\infty)$. It is clear that if a usual field is automodel the generalized field with it generates is an automodel random field with the same parameter*. The usual random fields upon $\mathbb{R}^1$ were introduced by Lamperti [3], who called them the semistable random processes.

We shall now make more precise the statement of the introduction about the absence of interesting examples of usual stationary automodel random fields. We shall show that any automodel stationary (i. e., such that the systems of random variables $\{\xi(x_1),\ldots,\xi(x_n)\}$ and $\{\xi(x_1+a),\ldots,\xi(x_n+a)\}$ have for all $a, x_1,\ldots,x_n \in \mathbb{R}^\nu$, $n = 1,2,\ldots$ the same joint probability distributions) stochastically continuous (i. e., such that $\xi(x) \to \xi(x_0)$, in the sense of convergence in $L_0$, ) if $x \to x_0$, for all $x_0 \in \mathbb{R}^\nu$ usual random field $\{\xi(x), x \in \mathbb{R}^\nu\}$ is a random constant. Indeed, the fact that the field is automodel and stationary implies that the one dimensional distributions of the random variables $\xi(x)$ and $\lambda^\kappa \xi(\lambda x)$ coincide; if $x \neq 0$ this can occur only if $\xi(x) \equiv 0$. Furthermore if $\kappa = 0$ the fact that the field is automodel and stationary implied that the differences $\xi(x_1) - \xi(x_2)$ and $\xi(\lambda x_1) - \xi(\lambda x_2)$ have the same probability distributions. By considering limit approach $\lambda \to 0$ we see that the stochastic continuity of $\xi(x)$ at $x = 0$ implies that $\xi(x_1) - \xi(x_2) \equiv 0$. However, we note that nontrivial examples of automodel stationary fields can be constructed by differentiating usual automodel random fields with stationary increments (see §5 of the paper and Jaglom paper [24]).

If probability distribution P describes a state upon the space $\mathcal{L}$, and the elements of $\mathcal{L}^1$ are interpreted as generalized functions in the usual way, the relations (2.17) are also applicable. We wish only to remark that for any function $\xi(\cdot) \in \mathcal{L}^1$ the function $\xi(\lambda\cdot)$ is interpreted as the functional on $\mathcal{L}$ whose value on $\varphi \in \mathcal{L}$ is equal to

---

* The converse statement is also true if some natural additional conditions are fulfilled.

the value of $\xi(\cdot)$ on $U_\lambda \varphi$. $\xi(\cdot + a)$ is intrepretated in a similar way.

Sometimes it is useful to consider the Fourier transforms of the realizations of the random field. Let $F : \mathfrak{T}^1(\mathbb{R}^\nu) \to \tilde{\mathfrak{T}}^1(\mathbb{R}^\nu)$ be the transformation which transforms the function $\xi \in \mathfrak{T}^1(\mathbb{R}^\nu)$ into its Fourier transform $\tilde{\xi} \in \tilde{\mathfrak{T}}^1(\mathbb{R}^\nu)$, where $\tilde{\mathfrak{T}}^1(\mathbb{R}^\nu)$ is the space of complex valued generalized functions of the type $\varphi + i\psi$, $\varphi, \psi \in \mathfrak{T}^1(\mathbb{R}^\nu)$ (see [18, 19]). The state upon $\tilde{\mathfrak{T}}(\mathbb{R}^\nu)$ defined by the relation

$$\tilde{P}(B) = \hat{P}(F^{-1}B), \qquad B \in F(\mathcal{B}(J(\mathbb{R}^\nu))) \tag{2.18}$$

will be called the <u>Fourier transform of the state</u> $\hat{P}$ <u>upon</u> $\mathfrak{T}(\mathbb{R}^\nu)$. When the measure $\hat{P}$ describes a stationary random field, the random field described by the measure $\tilde{P}$ is usually called in probability theory the <u>random spectral measure of the stationary field.</u> (There another perhaps more artifical construction is usually used; see for example [21], §3. 4.) By using the Fourier transform in the last of the relations (2. 17) we see that if the state described by the measure $\hat{P}$ has the probability distribution P, the random field upon $\tilde{\mathfrak{T}}(\mathbb{R}^\nu)$ defined by the relation

$$\tilde{\xi}_{\kappa,\lambda}(k) = \lambda^{\kappa - r}\tilde{\xi}(k\lambda^{-1}), \qquad k \in \mathbb{R}^\nu \tag{2.19}$$

has the same probability distribution as the state $S^*_{\kappa,\lambda}P$. In physical literature (where the situation considered is usually a little different) the renorm-group defined by the last of the relations (2. 17) is called the Kadanoff renorm-group and the renorm group defined by the relation (2. 19) is called the Wilson renorm-group.

## III. LIMIT DISTRIBUTIONS

We shall say that the probability distribution $P \in \mathcal{P}(\mathcal{L})$ has a <u>large-scale limit</u> $P^\infty \in \mathcal{P}(\mathcal{L})$ <u>with normalization</u> $f(\lambda)$, where $f(\lambda) > 0$

is a measurable function of $\lambda \in (0, \infty)$, if the limit

$$\lim_{\lambda \to \infty} T^*_{f(\lambda)} U^*_\lambda P = P^\infty \tag{3.1}$$

exists. Similarly, say $P$ has a small-scale limit $P^0$ if

$$\lim_{\lambda \to 0} T^*_{f(\lambda)} U^*_\lambda P = P^0 \tag{3.2}$$

exists. To demonstrate the sense of these definitions we note that if

$$V_\lambda = \{x = (x_1, \ldots, x\ ) \in \mathbb{R}^\nu : |x_i| \le \frac{\lambda}{2}, \ i = 1, \ldots, \nu\}, \quad \lambda \in (0, \infty) \tag{3.3}$$

are cubes in $\mathbb{R}^\nu$ and $\mathcal{L}$ contains the indicators $\chi_{V_\lambda}$ of these cubes, the existence of the limit $P^\infty$ $(P^0)$ for the field of the type (2.16) implies that the probability distributions of the random variables

$$\eta_\lambda = f(\lambda) \int_{V_\lambda} \xi(x)\, dx \tag{3.4}$$

weakly converge when $\lambda \to \infty$ $(\lambda \to 0)$ to $P^\infty_{\chi_{V_1}}$ $(P^0_{\chi_{V_1}})$. Thus the notion of large-scale limit generalizes the constructions usually used in the limit theorems of probability theory, and the small-scale limit describes the local behavior of the field near the origin.

The following proposition shows that the large and small-scale limits and the corresponding normalizations are defined by the probability distribution $P$ in essentially a unique way.

<u>Proposition 3.1.</u> <u>Let a probability distribution</u> $P \in \mathcal{P}(\mathcal{L})$ <u>have non-zero large-scale (small-scale) limits</u> $P^\infty_i$ $(P^0_i)$, $i = 1, 2$ <u>with normalizations</u> $f_i(\lambda)$, $i = 1, 2$. <u>Then there exists a constant</u> $C \in (0, \infty)$ <u>such that</u>

$$P^\infty_1 = T^*_C P^\infty_2, \qquad \lim_{\lambda \to \infty} \frac{f_1(\lambda)}{f_2(\lambda)} = C \tag{3.5a}$$

$$P^C_1 = T^*_C P^0_2, \qquad \lim_{\lambda \to 0} \frac{f_1(\lambda)}{f_2(\lambda)} = C \tag{3.5b}$$

Proof: At first we shall make the following evident remark. Let $Q \in \mathcal{P}(\mathcal{L})$ be a nonzero probability distribution and let $c_1, c_2$ be positive numbers. If

$$T^*_{c_1} Q = T^*_{c_2} Q \tag{3.6}$$

then

$$c_1 = c_2 \tag{3.7}$$

This remark and the continuity of the transformation $c \to T^*_c Q$ observed in sect. 2 imply that if the probability distributions $Q^\mu_i$, $Q_i$, $\mu \in (0, \infty)$, $i = 1, 2$ are such that

$$\begin{gathered} Q^\mu_1 = T^*_h(\mu)\, Q^\mu_2, \qquad \mu \in (0, \infty) \\ \lim_{\mu \to \infty} Q^\mu_i = Q_i, \qquad i = 1, 2 \end{gathered} \tag{3.8}$$

and the probability distributions $Q_1, Q_2$ are nonzero, then

$$\lim_{\mu \to \infty} h(\mu) = \hat{h}, \qquad Q_1 = T^*_{\hat{h}} Q_2 \tag{3.9}$$

for some positive $\hat{h} > 0$. In the situation of the proposition the relations (3.8) hold for

$$\begin{gathered} Q^\mu_1 = T^*_{f_1(\mu)} U^*_\mu P, \qquad Q^\mu_2 = T^*_{f_2(\mu)} U^*_\mu P \\ Q_1 = P^\infty_1, \quad Q_2 = P^\infty_2, \quad h(\mu) = f_1(\mu)/f_2(\mu) \end{gathered} \tag{3.10}$$

So the relation (3.9) implies the relation (3.5a). The relation (3.5b) can be proved in a similar way.

We shall recall (see for example [1], Appendix 1 for the details) that a function $f(\lambda)$ of $\lambda \in (0, \infty\_$ is called slowly-changing at infinity (near zero) if for any $t \in (0, \infty)$

$$\lim_{\lambda \to \infty} \frac{f(t\lambda)}{f(\lambda)} = 1 \tag{3.11a}$$

$$\left( \lim_{\lambda \to \infty} \frac{f(t\lambda)}{f(\lambda)} = 1 \right) \tag{3.11b}$$

We shall say that the function $f(\lambda)$ of $\lambda \in (0,\infty)$ is *locally bounded at infinity* (*near zero*) if the function is bounded for all enough large $d$ in any interval $(d, d+s)$, $s > 0$ (for all enough small $d$ in any interval $(d-s, d)$, $0 < s < d$).

The following proposition indicates the importance of the class of automodel probability distributions.

*Proposition 3.2. Let the probability distribution* $P \in \mathcal{P}(\mathcal{L})$ *have a nonzero large-scale limit* $P^{\infty}$ (*small-scale limit* $P^{0}$) *with normalization* $f(\lambda)$. *Then there exists a number* $\kappa^{\infty} = \kappa^{\infty}(P)$ ($\kappa^{0} = \kappa^{0}(P)$) *such that the probability distribution* $P^{\infty}$ *is automodel with parameter* $\kappa^{\infty}$ (*the probability distribution* $P^{0}$ *is automodel with parameter* $\kappa^{0}$) *and there exists a function* $g^{\infty}(\lambda)$ *which is locally bounded and slowly-changing at infinity such that*

$$f(\lambda) = \lambda^{\kappa^{\infty}} g^{\infty}(\lambda), \qquad \lambda \in (0,\infty) \tag{3.12a}$$

(*there exists a function* $g^{0}(\lambda)$ *which is locally bounded and slowly changing near zero such that*

$$f(\lambda) = \lambda^{\kappa^{0}} g^{0}(\lambda), \qquad \lambda \in (0,\infty)) \tag{3.12b}$$

*If the probability distribution* $P$ *is stationary then the probability distributions* $P^{\infty}$ *and* $P^{0}$ *are also stationary.*

*If the probability distribution* $P \in \mathcal{A}_{\kappa}(\mathcal{L})$ *is automodel with parameter* $\kappa$ *it has large-scale and small-scale limits*

$$P^{\infty} = P^{0} = P \tag{3.13}$$

with normalization

$$f(\lambda) = \lambda^{\kappa} \tag{3.14}$$

Proof. Let

$$P^{\infty} = \lim_{\mu \to \infty} T^{*}_{f(\mu)} U^{*}_{\mu} P \tag{3.15}$$

Let $\lambda > 0$ be fixed. The relation (3.15) and the continuity of the transformation $P \to U^{*}_{\lambda} P$ (see sect. 2) imply that the relations (3.8) hold for

$$Q^{\mu}_{1} = T^{*}_{f(\lambda\mu)} U^{*}_{\lambda\mu} P, \quad Q^{\mu}_{2} = U^{*}_{\lambda} T^{*}_{f(\mu)} U^{*}_{\mu} P ,$$

$$Q_1 = P^{\infty}, \quad Q_2 = U^{*}_{\lambda} P^{\infty}, \quad h(\mu) = f(\lambda\mu)/f(\mu) \tag{3.16}$$

So the relations (3.9) imply the existence of the limit

$$\lim_{\mu \to \infty} \frac{f(\lambda\mu)}{f(\mu)} = \hat{h}(\lambda) \tag{3.17}$$

such that

$$P^{\infty} = T^{*}_{h(\lambda)} U^{*}_{\lambda} P^{\infty} \tag{3.18}$$

The application of the operator $T^{*}_{h(\gamma)} U^{*}_{\gamma}$, where $\gamma > 0$, to both parts of (3.18) implies that

$$P^{\infty} = T^{*}_{\hat{h}(\gamma)\hat{h}(\lambda)} U^{*}_{\gamma\lambda} P^{\infty} = T^{*}_{\hat{h}(\gamma\lambda)} = U^{*}_{\gamma\lambda} P^{\infty} \tag{3.19}$$

By using (3.7) for $Q = U^{*}_{\gamma\lambda} P^{\infty}$ we find that

$$\hat{h}(\gamma\lambda) = \hat{h}(\gamma)\hat{h}(\lambda), \qquad \gamma, \lambda \in (0, \infty) \tag{3.20}$$

Then the relation (3.18), the continuity of the transformation $\lambda \to T^{*}_{\lambda} P$, $\lambda \to U^{*}_{\lambda} P^{\infty}$ and the relation (3.7) imply that the function $h(\lambda)$ of $\lambda \in (0, \infty)$ is continuous. So

$$\hat{h}(\lambda) = \lambda^{\kappa^{\infty}}, \qquad \lambda \in (0, \infty) \tag{3.21}$$

for some $\kappa^{\infty} \in \mathbb{R}^1$. Now the relations (3.21), (3.15) and (3.17) easily imply the statement of the proposition about the large-scale limit. We want only to explain the demonstration of the local boundedness of $g^{\infty}(\lambda)$ or, what is the same, the local boundedness of $f(\lambda)$. To demonstrate this it is necessary to use the relation (3.15), the continuity of the transformation $\mu \to U^*_{\mu} P$ and the fact that for any $\mu_0$ there exists the neighborhoods $W_{\mu_0}$ of the probability distribution $U^*_{\mu_0} P$ and $W^{\infty}$ of the probability distribution $P^{\infty}$ such that the set of $c \in \mathbb{R}^1$ for which $P^1 \in W_{\mu_0}$ and simultaneously $T^*_c P^1 \in W^{\infty}$ for some probability distribution $P^1$ is bounded. The statement of the proposition about the small-scale limit may be demonstrated similarly. The statements (3.13) and (3.14) of the proposition are evident.

The results of proposition 3.2 extend the well known results of Levy and Khinchin (the case of independent random variables), of Ibragimov, Rosanov and Wolknosky (the case of weakly dependent random variables) about convergence to stable random variables (see [1], theorems 2.1.1 and 18.1.2) and the Lamperti result [3] about convergence to semi-stable random processes.

The usual formulations of the limit theorems of probability theory include the possibility of the subtraction of a normalizing constant. We shall introduce the corresponding definitions within the context of this paper. For any $g \in \mathbb{R}^1$ and any probability distribution $P \in \mathcal{P}(\mathcal{D}(\mathbb{R}^{\nu}))$, we shall denote by $G_g P \in \mathcal{P}(\mathcal{D}(\mathbb{R}^{\nu}))$ the probability distribution such that

$$G_g P_{\varphi}(A) = P_{\varphi}\left(A + g \int_{\mathbb{R}^{\nu}} \varphi(x)dx\right), \quad A \in \mathcal{B}_{\mathbb{R}^1}, \quad \varphi \in \mathcal{D}(\mathbb{R}^{\nu}) \tag{3.22}$$

We shall say that the probability distribution $P \in \mathcal{P}(\mathcal{D}(\mathbb{R}^\nu))$ has a large-scale limit $P^\infty$ in the wide sense with normalization $(f(\lambda), g(\lambda))$, where $f(\lambda) > 0$ and $g(\lambda)$ are measurable functions of $\lambda \in (0, \infty)$ if (comp. (3. 1)) the limit

$$\lim_{\lambda \to \infty} T^*_{f(\lambda)} G_{g(\lambda)} U^*_\lambda P = P^\infty \tag{3. 23}$$

exists*). It is possible to note in analogy with (3. 4) that the relation (3. 23) implies, for the field of the type (2. 16), the weak convergence of the sequence of random variables

$$\eta_\lambda = f(\lambda)[\int_{V_\lambda} \xi(x)\,dx - g(\lambda)] \tag{3. 24}$$

and this fact indicates the sense of the definition. We shall say that a probability distribution $P \in \mathcal{P}(\mathcal{D}(\mathbb{R}^\nu))$ is automodel in the wide sense with the parameter $\kappa \in \mathbb{R}^1$ if (comp. (2. 12)) there exists an measurable function $g(\lambda)$ of $\lambda \in (0, \infty)$ such that

$$T^*_{\lambda^\kappa} G_{g(\lambda)} U^*_\lambda P = P, \qquad \lambda \in (0, \infty) \tag{3. 25}$$

The concepts of limits and automodel fields in the wide sense can be reduced to the previous ones if we consider a more restricted space of basic functions. We shall denote by $\mathcal{D}^1(\mathbb{R}^\nu)$ the closed subspace of the space $\mathcal{D}(\mathbb{R}^\nu)$ which consists of the functions $\varphi \in \mathcal{D}(\mathbb{R}^\nu)$ such that

$$\int_{\mathbb{R}^\nu} \varphi(x)\,dx = 0 \tag{3. 26}$$

We shall introduce in $\mathcal{D}^1(\mathbb{R}^\nu)$ the topology induced by the topology in $\mathcal{D}(\mathbb{R}^\nu)$. It is evident that the conditions about the topology in $\mathcal{L}$

---

* This definition can be extended with obvious modifications to the case of the small-scale limit.

introduced in sect. 2 are fulfilled in the case $\mathcal{L} = \mathcal{D}^1(\mathbb{R}^\nu)$. The random fileds upon the space $\mathcal{D}^1(\mathbb{R}^\nu)$ were introduced by Jaglom [24]. A probability distribution $P^1 \in \mathcal{P}(\mathcal{D}^1(\mathbb{R}^\nu))$ will be called the <u>restriction of the probability distribution</u> $P \in \mathcal{P}(\mathcal{D}(\mathbb{R}^\nu))$ if

$$P^1_\varphi = P_\varphi, \quad \varphi \in \mathcal{D}^1(\mathbb{R}^\nu) \tag{3.27}$$

The consideration of the restriction of the probability distribution $P \in \mathcal{P}(\mathcal{D}(\mathbb{R}^\nu))$ means that we consider only the increments of the field and do not take into account the constant term.

The following simple fact is evident.

<u>Proposition 3.3.</u> (I) <u>If a probability distribution</u> $P \in \mathcal{P}(\mathcal{D}(\mathbb{R}^\nu))$ <u>has a large-scale limit in the wide sense</u> $P^\infty \in \mathcal{P}(\mathcal{D}(\mathbb{R}^\nu))$ <u>with normalization</u> $(f(\lambda), g(\lambda))$ <u>then the restriction</u> $P^1 \in \mathcal{P}(\mathcal{D}^1(\mathbb{R}^\nu))$ <u>of the probability distribution</u> $P$ <u>has the large-scale limit</u> $P^{\infty,1} \in \mathcal{P}(\mathcal{D}^1(\mathbb{R}^\nu))$, <u>which is the restriction of the probability distribution</u> $P^\infty$, <u>with normalization</u> $f(\lambda)$.

(II). <u>If a probability distribution</u> $P \in \mathcal{P}(\mathcal{D}(\mathbb{R}^\nu))$ <u>is automodel in the wide sense with parameter</u> $\kappa$ <u>then its restriction</u> $P^1 \in \mathcal{P}(\mathcal{D}^1(\mathbb{R}^\nu))$ <u>is automodel with the same parameter</u> $\kappa$.

The converse of statement I of the proposition is false. This is connected, roughly speaking, with the fact that the limit of the restrictions can exist even if the constant $g(\lambda)$ in (3.23) is a random variable. The following question is open. Is every automodel probability distribution $P^1 \in \mathcal{P}(\mathcal{D}^1(\mathbb{R}^\nu))$ the restriction of a probability distribution $P \in \mathcal{P}(\mathcal{D}(\mathbb{R}^\nu))$ which is automodel in the side sense?

## IV. AUTOMODEL FIELDS WITH DISCRETE ARGUMENT

Let $\mathbb{Z}^\nu$ be the $\nu$-dimensional integer lattice of points $t = (t^1, \ldots, t^\nu)$, $t^i = \ldots, -1, 0, 1, \ldots$ . A system of random variables $\{\xi_t, t \in \mathbb{Z}^\nu\}$ defined on a probability space (comp. sect. 2) $(W, \mathcal{B}, m)$

will be called a $\nu$-dimensional random field with discrete argument. The system of all finite dimensional probability distributions $P = \{P_{t_1,\ldots,t_r},\ t_1,\ldots,t_r \in \mathbb{Z}^\nu,\ r = 1,2,\ldots\}$ defined by the relations $P_{t_1,\ldots,t_r}(B) = m(w \in W : (\xi_{t_1}(w),\ldots,\xi_{t_r}(w)) \in B),\ B \in \mathcal{B}_{\mathbb{R}^r}$, where $P_{t_1,\ldots,t_r}$ is an element of the space $\mathcal{P}_r$ of probability measures on the $\sigma$-algebra $\mathcal{B}_{\mathbb{R}^r}$ of Borel subsets of $\mathbb{R}^r$, will be called the probability distribution of the field. The set of all probability distributions of random field with $\nu$-dimentional discrete arguments will be denoted by $\mathcal{P}(\mathbb{Z}^\nu)$. We shall introduce in $\mathcal{P}(\mathbb{Z}^\nu)$ the topology of weak convergence of finite-dimensional distributions, which is defined as the weakest of the topologies in $\mathcal{P}(\mathbb{Z}^\nu)$ for which all the transformations $\mathcal{P}(\mathbb{Z}^\nu) \to \mathcal{P}_r : P \to P_{t_1,\ldots,t_r},\ t_1,\ldots,t_r \in \mathbb{Z}^\nu,\ r = 1,2,\ldots$ are continuous if $\mathcal{P}_r$ is regarded as equipped with the usual topology of weak convergence of probability measures.

Let a parameter $\kappa \in \mathbb{R}^1$ be fixed. Using the definition introduced independently by Gallavotti and John-Lasinio [14] and Sinai [9], we shall call the set of transformations $\hat{S}_{\kappa,n} : \mathcal{P}(\mathbb{Z}^\nu) \to \mathcal{P}(\mathbb{Z}^\nu)$, $n = 1,2,\ldots$ such that $\hat{S}_{\kappa,n}$ transforms the probability distribution of the random field $\{\xi_t,\ t \in \mathbb{Z}^\nu\}$ into the probability distribution of the random field

$$\xi_t^n = n^{\kappa-\nu} \sum_{s \in V_t^n} \xi_s, \qquad t \in \mathbb{Z}^\nu \tag{4.1}$$

where $V_t^n$ is the cube with center $tn$ containing $n^\nu$ points,

$$V_t^n = \{s = (s^1,\ldots,s^\nu) \in \mathbb{Z}^\nu : t^i n - \frac{n}{2} < s^i \leq t^i n + \frac{n}{2}\},$$
$$i = 1,\ldots,\nu\},\quad t = (t^1,\ldots,t^\nu) \in \mathbb{Z}^\nu,\quad n = 1,2,\ldots \tag{4.2}$$

the discrete renorm-semigroup with parameter $\kappa$. It is evident that the probability distribution of the random field $\{\xi_t^n, t \in \mathbb{Z}^\nu\}$ is defined uniquely by the probability distribution of the random field $\{\xi_t, t \in \mathbb{Z}^\nu\}$, so that the transformations are well defined. Moreover,

$$\hat{S}_{\kappa,n_1} \hat{S}_{\kappa,n_2} = \hat{S}_{\kappa,n_1 n_2}, \quad n_1, n_2 = 1, 2, \ldots \tag{4. 3}$$

so the transformations $\hat{S}_{\kappa,n}$ form a multiplicative semigroup. A probability distribution $P \in \mathfrak{P}(\mathbb{Z}^\nu)$ is called <u>automodel with parameter</u> $\kappa \in \mathbb{R}^1$ if

$$\hat{S}_{\kappa,n} P = P, \quad n = 1, 2, \ldots \tag{4. 4}$$

We note that Sinai [9] and Gallavotti and John-Lasinio [14] use another parametrization of the renorm-semigroups and automodel distributions. Instead of the parameter $\kappa$ they consider the parameter

$$\alpha = -\frac{2\kappa}{\nu} + 2 \tag{4. 5}$$

The choice of the parametrization used in this paper is not accidental; it is founded on painful hesitation. But, of course, all reasons in such questions are very subjective. In the papers [9] and [14] the authors suppose that $1 \le \alpha < 2$, i. e., $0 \le \kappa \le \frac{\nu}{2}$. The condition $\kappa > 0$ is natural enough because there are no known examples of stationary automodel fields with discrete arguments and parameter $\kappa < 0$, and the evaluation of the mathematical expectations

$$E|\xi_t^n| \le n^{\kappa-\nu} \sum_{s \in V_t^n} E|\xi_s|$$

shows that such fields cannot exist in the class of fields with finite first moment. (Negative $\kappa$ are possible in the case of automodel fields with stationary increments; the random constants $\xi_t \equiv \xi$,

$t \in \mathbb{Z}^\nu$ are automodel fields with parameter $\kappa = 0$). However Gaussian automodel fields with discrete arguments exist for all positive $\kappa < \nu$, so the restriction $\kappa \leq \frac{\nu}{2}$ seems unnatural. The question of the existence of automodel stationary fields with discrete arguments for $\kappa > \nu$ is open.

We shall discuss now the relations between the renorm-groups and the discrete renorm-semigroups, and between automodel random fields with discrete and continuous arguments. So we shall introduce the following concept. The functional space $\mathcal{L}$ is called a discretizable space if it has, besides the properties noted in sect. 2, the following two properties:

(1) The space $\mathcal{L}$ contains the indicators $\chi_{b,c}$, $b, c \in \mathbb{R}^\nu$ of all $\nu$-dimensional rectangles

$$\Lambda_{b,c} = \{x = (x^1, \ldots, x^\nu) \in \mathbb{R}^\nu, \quad b^k < x^k \leq c^k, \quad k = 1, \ldots, \nu\},$$

$$b = (b^1, \ldots, v^\nu), \quad c = (c^1, \ldots, c^\nu) \in \mathbb{R}^\nu, \quad b^k < c^k, \quad k = 1, \ldots, \nu \tag{4.6}$$

(2) The linear space of all functions $\psi \in \mathcal{L}$ such that

$$\psi = \sum_{j=1}^{m} a_j \chi_{b_j, c_j} \tag{4.7}$$

where $b_j, c_j$, $j = 1, \ldots, m$ are vectors with rational coordinates, coefficients $a_1, \ldots, a_m \in \mathbb{R}^1$ and $m = 1, 2, \ldots,$ is dense in $\mathcal{L}$. Of course the class of discretizable spaces is very large. However it is more convenient for our aims to use a specially choosen class of function spaces $M_w(\mathbb{R}^\nu)$, where the functions $w(\lambda)$, $\lambda \in (0, \infty)$, are positive, bounded, monotonically nonincreasing, and approach to 0 as $\lambda \to \infty$. The space $M_w(\mathbb{R}^\nu)$ consists of the integrable finite range functions

$$\|\psi\| = \sup_{\lambda \in \mathbb{R}^\nu} \{ |\tilde{\psi}(\lambda)| \prod_{k=1}^{\nu} ((1 + |\lambda^k|) w(|\lambda^k|)) \} \tag{4.8}$$

where $\tilde{\psi}$ is the Fourier transform of the function $\psi$. In the spaces $M_w(S_r)$, where $M_w(S_r)$ is the set of functions $\psi \in M_w(\mathbb{R}^\nu)$ having support in the sphere $S_r \subset \mathbb{R}^\nu$ of radius r with center 0, we shall introduce the topology of the norm (4. 8). Furthermore the topology in $M_w(\mathbb{R}^\nu)$ will be introduced as the inductive limit (see [25]) of the topologies in $M_w(S_r)$, $r = 1, 2, \dots$ . (This means that the sequence $\psi_k \to \psi$ in $M_w(\mathbb{R}^\nu)$ if the supports of the functions $\psi_k$ and $\psi$ are, for large enough k, contained in a sphere $S_r$, and $\psi_k \to \psi$ in the norm (4. 8).)

<u>Proposition 4. 1.</u> <u>The space</u> $M_w(\mathbb{R}^\nu)$ <u>is discretizable.</u> <u>The set</u> $D(\mathbb{R}^\nu)$ <u>of infinitely differentiable finite range functions of</u> $x \in \mathbb{R}^\nu$ <u>is a dense subspace of the space</u> $M_w(\mathbb{R}^\nu)$. <u>The usual topology in</u> $D(\mathbb{R}^\nu)$ (see for example [18]) <u>is stronger then the topology in</u> $D(\mathbb{R}^\nu)$ <u>induced by the topology in</u> $M_w(\mathbb{R}^\nu)$.

<u>Proof.</u> The fulfillment of the continuity condition introduced in sec. 2 is evident. The fulfillment of the condition (1) introduced in this section follows from the relation

$$\tilde{\chi}_{b,c}(\lambda) = \prod_{k=1}^{\nu} (i\lambda^k)^{-1}(e^{i\lambda^k c^k} - e^{i\lambda^k b^k}),$$
$$\lambda = (\lambda^1, \dots, \lambda^\nu) \in \mathbb{R}^\nu \qquad (4. 9)$$

It is well known that the space $\mathfrak{D}(\mathbb{R}^\nu)$ is a dense subspace of the space $\mathfrak{T}(\mathbb{R}^\nu)$ of infinitely differentiable functions of rapid decrease on $\mathbb{R}^\nu$, and the topology in $\mathfrak{D}(\mathbb{R}^\nu)$ is stronger than the topology induced by the topology in $\mathfrak{T}(\mathbb{R}^\nu)$ (see [18] or [26]). Furthermore, it is known that the Fourier transform does not change the topology in $\mathfrak{T}(\mathbb{R}^\nu)$. So it is clear that $\mathfrak{D}(\mathbb{R}^\nu) \subseteq M_w(\mathbb{R}^\nu)$ and the topology in $\mathfrak{D}(\mathbb{R}^\nu)$ is stronger than the topology induced by the topology in $M_w(\mathbb{R}^\nu)$. The fact that $\mathfrak{D}(\mathbb{R}^\nu)$ is dense in $M_w(\mathbb{R}^\nu)$ can be proved as in the usual proof of the fact that $\mathfrak{D}(\mathbb{R}^\nu)$ is dense in $\mathfrak{T}(\mathbb{R}^\nu)$. The preceding remarks imply all statements of the proposition converning the space $\mathcal{D}(\mathbb{R}^\nu)$. By using the continuity of the transformations: $\mathbb{R}^\nu \to M_w(\mathbb{R}^\nu): a \to E_a \chi_{b,c}$ and $(0,\infty) \to M_w(\mathbb{R}^\nu): \lambda \to U_\lambda \chi_{b,c}$ we see that condition (2) means that $M_w(\mathbb{R}^\nu)$ coincides with the closure $M_w^1(\mathbb{R}^\nu)$ in $M_w(\mathbb{R}^\nu)$ of the subspace of all functions of the type (4. 7) where $b_j, c_j$ are arbitrary (and are not only rational). So it suffices to prove that $\mathcal{D}(\mathbb{R}^\nu) \subseteq M_w^1(\mathbb{R}^\nu)$. If $\nu = 1$ the last statement follows from the fact that if $\varphi \in \mathcal{D}(\mathbb{R}^1)$, supp $\varphi \subset [b, c)$ and

$$\varphi_m = \sum_{j=1}^{m} \varphi\left(b + \frac{c-b}{m} j\right) \chi_{b + \frac{c-b}{m}(j-1),\ b + \frac{c-b}{m} j} \tag{4. 10}$$

the sequence $\|\varphi_m - \varphi\| \to 0$ as $m \to \infty$. To establish this it is necessary to use the fact that $\varphi_m \to \varphi$ in $L_1$ and so $\tilde{\varphi}_m \to \tilde{\varphi}$ uniformly in $\lambda$, and the fact that the boundness of the derivative of the function $\varphi$ and the formula (4. 9) imply that the quantity $(1+|\lambda|)\tilde{\varphi}_m(\lambda)$ is bounded uniformly in $m$ and $\lambda$. It is easy to check that if the sequence $\psi_m \in M_w(\mathbb{R}^{\nu-1})$ goes to zero in $M_w(\mathbb{R}^{\nu-1})$, then also the sequence $\hat{\psi}_m \in M_w(\mathbb{R}^\nu)$ also goes to zero, where

$$\hat{\psi}_m(x) = \begin{cases} \psi_m(x^1, \ldots, x^{\nu-1}) & \text{if} \quad b^\nu < x^\nu \le c^\nu \\ 0 & \text{for other} \quad x = (x^1, \ldots, x^\nu) \in \mathbb{R}^\nu \end{cases} \tag{4. 11}$$

and $b, c \in \mathbb{R}^\nu$. This enables us to use induction on $\nu$. We obtain from the induction hypothesis $M_w(\mathbb{R}^{\nu-1}) = M_w^1(\mathbb{R}^{\nu-1})$ that any function

$$\psi(x) = \varphi(x^1, \ldots, x^{\nu-1}) \chi_{b^\nu, c^\nu}(x^\nu), \quad \varphi \in D(\mathbb{R}^{\nu-1}),$$
$$b^\nu, c^\nu \in \mathbb{R}^1, \quad x = (x^1, \ldots, x^\nu) \in \mathbb{R}^\nu \tag{4. 12}$$

lies in $M_w^1(\mathbb{R}^\nu)$. By changing the function $\varphi \in D(\mathbb{R}^\nu)$ to a function close to $\varphi$ which is piecewise constant in the variable, $x^\nu$ we can construct in analogy with (4. 10) a sequence of linear combination of functions (4. 12) which converges to $\varphi$. (Here it is necessary to refer, instead of to the boundedness of the derivatives, to the fact that the norms $\|\varphi_{x^\nu}\|$ are uniformly bounded on $x^\nu$, where $\varphi_{x^\nu}(x^1, \ldots, x^{\nu-1}) = \frac{\partial}{\partial x^\nu}\varphi(x^1, \ldots, x^\nu)$. So $\mathcal{D}(\mathbb{R}^\nu) \subseteq M^1_w(\mathbb{R}^\nu)$ and the proposition is proven.

Let $\mathcal{L}$ be a function space and let $\mathcal{L}^1 \supseteq \mathcal{L}$ be a discretizable space. Suppose $\mathcal{L}$ is a dense subspace of $\mathcal{L}^1$ and the topology in L is stronger than the topology induced by the topology in $\mathcal{L}^1$. (The main example of such a pair of spaces is $\mathcal{L} = \mathcal{D}(\mathbb{R}^\nu)$ and $\mathcal{L}^1 = M_w(\mathbb{R}^\nu)$). We shall say that a probability distribution $P \in \mathcal{P}^w(\mathcal{L})$ has a <u>continuation to a probability distribution</u> $P^1 \in \mathcal{P}(\mathcal{L}^1)$ if $P_\varphi = P^1_\varphi$, $\varphi \in \mathcal{L}$. It is evident that such a continuation, it it exists, is unique. The set of all probability distribution $P \in \mathcal{P}(\mathcal{L})$ having a continuation to a probability distribution $P^1 \in \mathcal{P}(\mathcal{L}^1)$ will be denoted by $\mathcal{P}^1(\mathcal{L})$. Let $P \in \mathcal{P}^1(\mathcal{L})$ be a probability distribution. The probability distribution $\hat{P} \in \mathcal{P}(\mathbb{Z}^\nu)$ of the random field

$$\xi_t = \Phi_{\chi_t}, \quad t \in \mathbb{Z}^\nu \tag{4.13}$$

where the random field $\{\Phi_\varphi, \varphi \in \mathcal{L}^1\}$ has the probability distribution $P^1$ and $\chi_t$ is the indicator of the cube

$$\Lambda_t = \{x = (x^1, \ldots, x^\nu) \in \mathbb{R}^\nu : t^i - \tfrac{1}{2} < x \leq t^i + \tfrac{1}{2},\ i = 1, \ldots, \nu\},$$
$$t = (t^1, \ldots, t^\nu) \in \mathbb{Z}^\nu \tag{4.14}$$

is called the <u>discretization</u> (with the help of the space $\mathcal{L}^1$) of the probability distribution P. Any probability distribution $\hat{P} \in \mathcal{P}(\mathbb{Z}^\nu)$ is the discretization of some probability distribution $P \in \mathcal{P}(D(\mathbb{R}^\nu))$ (with the help of the space $M_w(\mathbb{R}^\nu)$ where for example $w(\lambda) = [1 + \quad (1 + |\lambda|)]^{-1}$). In fact, if the random field $\{\xi_t, t \in \mathbb{Z}^\nu\}$ has the probability distribution $\hat{P}$ then the random field

$$\Phi_\varphi = \sum_{t \in \mathbb{Z}^\nu} \xi_t \int_{\Lambda_t} \varphi(x)dx, \qquad \varphi \in D(\mathbb{R}^\nu) \tag{4.15}$$

is well defined. Formula (4.15) also allows us to define the random field upon $M_w(\mathbb{R}^\nu)$. (The continuity of the field follows, for example, from the fact that the norm (4.8) is stronger than the norm in $L_2(\mathbb{R}^\nu)$. The probability distribution of field (4.15) is in $\mathcal{P}^1(D(\mathbb{R}^\nu))$.) The discretization of (4.15) is $\hat{P}$. We also note that if the space $D(\mathbb{R}^\nu)$ is replaced by $\mathfrak{S}(\mathbb{R}^\nu)$ the above fact becomes false.

<u>Theorem 4.2.</u> <u>Let</u> $P^1, P^2 \in \mathcal{P}^1(\mathcal{L})$ <u>be probability distributions such that for some</u> $\kappa \in \mathbb{R}^1$ <u>and</u> $n = 1, 2, \ldots$

$$P^2 = S^*_{\kappa,n} P^1 \tag{4.16}$$

<u>Then their discretizations satisfy</u>

$$\hat{P}^2 = \hat{S}_{\kappa,n} \hat{P}^1 \tag{4.17}$$

<u>The transformation</u> $\mathcal{P}^1(\mathcal{L}) \to \mathcal{P}(\mathbb{Z}^\nu) : P \to \hat{P}$ <u>transforms automodel (automodel and stationary) probability distributions with parameter</u> $\kappa$ <u>into automodel (automodel and stationary) probability distributions with the same parameter. Under this transformation different automodel probability distributions go to different ones.</u>

<u>Proof.</u> By the application of the definition (2.4) to the functions

$$\varphi = \sum_{k=1}^{r} c_k \chi_{t_k} \in \mathcal{L}^1 \tag{4.18}$$

we find that

$$S_{\kappa,n}\varphi = n^{\kappa-\nu} \sum_{k=1}^{r} c_k \sum_{t \in V^n_{t_k}} \chi_t, \qquad n = 1, 2, \ldots \tag{4.19}$$

where the cubes $V^n_t$ were introduced in (4.1). So, by comparing the definitions (4. 2), (2. 9) and (4. 13) we see that the probability distributions of the random variable $\sum_{k=1}^{r} c_k \xi_{t_k}$ defined by the joint distributions $\hat{P}^2_{t_1,\ldots,t_r}$ and $\hat{S}_{\kappa,n} \hat{P}^1_{t_1,\ldots,t_r}$ coincide. This implies that the probability distributions $\hat{P}^2_{t_1,\ldots,t_r}$ and $\hat{S}_{\kappa,n} \hat{P}^1_{t_1,\ldots,t_r}$ coincide, and establishes relation (4. 17). Relation (4. 17) shows that if the probability distribution $P \in \mathcal{P}^1(\mathcal{L})$ is automodel then $\hat{P}$ is also automodel.

In analogy with the continuous case a probability distribution of a field with discrete argument can be automodel for at most one value of $\kappa \in \mathbb{R}^1$. The only exception is the zero probability distribution (i. e. the probability distribution for which all measures $P_{t_1,\ldots,t_r}$ are concentrated at the origin), which is automodel for all $\kappa \in \mathbb{R}^1$. This implies that it suffices for the proof of the last statement of the theorem to find a contradiction in the possibility that $\hat{P}^1 = \hat{P}^2$ but $P^1$ and $P^2$ are different automodel probability distributions with the same parameter $\kappa$. Let $(P^1)^1$ and $(P^2)^1$ be the continuations of probability distributions $P^1$ and $P^2$. The equality $\hat{P}^1 = \hat{P}^2$ means that

$$(P^1)^1_\varphi = (P^2)^1_\varphi \tag{4. 20}$$

for the functions $\varphi$ of the type (4. 18). Any function $\psi$ of the type (4. 7) can be represented for some $\lambda \in (0, \infty)$ as $S_{\kappa,\lambda}\varphi$ where $\varphi$ is a function of the type (4. 18). So equation (4. 20) and the fact that the probability distributions $(P^1)^1$ and $(P^2)^1$ are automodel implies that $(P^1)^1_\psi = (P^2)^1_\psi$ for all $\psi$ of the type (4. 7). Because functions of the type (4. 7) are dense in $\mathcal{L}^1$, this implies that $(P^1)^1 = (P^2)^1$ and therefore $P^1 = P^2$.

The conjecture that any automodel probability distribution with discrete argument can be obtained by the discretization of a generalized automodel field with the appropriate choice of the spaces $\mathcal{L}$ and

$\mathcal{L}^1$ remains unproven, though confirmed in all examples. (For all known cases the choice $\mathcal{L} = \mathcal{D}(\mathbb{R}^\nu)$, $\mathcal{L}^1 = M_w(\mathbb{R}^\nu)$ for some function w is sufficient. The construction used for the proof of the last statement of theorem 4. 2 can be applied to the construction of an automodel random field defined on the set of functions of the type (4. 7) starting from the given automodel field with discrete argument, but the proof of the possibility of continuing the field to a more natural function space requires a deeper penetration into the structure of automodel fields.

Essentially the same idea can be illustrated with the help of the following simple fact*).

Proposition 4. 3. *Let $\hat{P} \in \mathcal{P}(\mathbb{Z}^\nu)$ be an automodel probability distribution and let the probability distribution P of the random field upon $M_w(\mathbb{R}^\nu)$ defined by formula (4. 15) for all $\varphi \in M_w(\mathbb{R}^\nu)$ have the large-scale limit $P^\infty$. Then for some $c > 0$ the probability distribution $\hat{P}$ is the discretization of the automodel probability distribution $T_c^* P^\infty$.*

Proof. The relation (4. 15) shows that the fact that $\hat{P}$ is automodel means that

$$P_\varphi = P_{S_{\kappa,n}\varphi} \tag{4. 21}$$

for the functions $\varphi$ of the type (4. 18). Because without loss of generality, it is possible to consider a nonzero $\hat{P}$, the arguments used in the proof of proposition 3. 1 show that the relation (4. 21) implies that the normalization $f(\lambda)$ of the limit $P^\infty$ is such that $f(\lambda) \sim c\lambda^\kappa$ as $\lambda \to \infty$, with the constant $c > 0$. So relation (4. 21) implies that

$$P_\varphi = P^\infty_{T_c\varphi} \tag{4. 22}$$

---

* It is possible to use instead of $M_w(\mathbb{R}^\nu)$ in proposition 4. 3 a wide class of other discretizable spaces. It is essential only that formula (4. 15) make sense.

for functions $\varphi$ of the type (4.18), and this implies that $\hat{P}$ is the discretization of $T_c^* P^\infty$.

Of course the fundamental difficulty mentioned above reduces now to the problem of the existence of the large-scale limit of P.

The advantages of the use of the automodel random field with continuous argument are connected with the existence in the continuous case of additional symmetries and mathematical structures. Of course, after the construction of an automodel generalized random field, the question of the existence of the continuation of the field to a discretizable function space arises. For the fields considered in sect. 7 of this paper and in the author's subsequent paper, it is easy to answer this question, though the answer is not always positive.

Let $f(n) > 0$ be a function of $n = 1, 2, \ldots$. Let $\hat{P} \in \mathcal{P}(\mathbb{Z}^\nu)$ be an automodel probability distribution of the random field $\{\xi_t, t \in \mathbb{Z}^\nu\}$ and let $\hat{P}^n$ be the probability distribution of the random field (comp. (4.1).

$$\xi_t^n = f(n) n^{-\nu} \sum_{s \in V_t^n} \xi_s, \quad t \in \mathbb{Z}^\nu \tag{4.23}$$

We shall say that a <u>probability distribution</u> $\hat{P} \in \mathcal{P}(\mathbb{Z}^\nu)$ <u>has a large-scale limit</u> $\hat{P}^\infty \in \mathcal{P}(\mathbb{Z}^\nu)$ with normalization $f(n)$ if

$$\lim_{n \to \infty} \hat{P}^n = P^\infty \tag{4.24}$$

The usual central limit theorem of probability theory (see for example [1]) for sums of random variables can be reformulated as the statement that the large-scale limit of the probability distribution of the corresponding random process coincides with the probability distribution of the automodel random process with parameter $\kappa = \frac{1}{2}$ consisting of independent Gaussian variables. More generally these variables are stable and automodel with a parameter $\kappa < \frac{1}{2}$ (see sect. 7). (More accurately, the usual formulation of the limit theorems asserts only the convergence of one-dimensional probability distributions $\hat{P}^n_t$, $t \in \mathbb{Z}^1$, but the statement about the convergence of multidimensional distributions $\hat{P}^n_{t_1,\ldots,t_r}$ follows easily from the conditions of weak dependence usually used in the limit theorems). Of course, the notion of the short-scale limit cannot be generalized to the discrete case.

The generalization to the discrete case of the general theorems of section 3 connected with the properties of the large-scale limit requires only trivial changes in their formulations and proofs.

The following fact follows immediately from the comparison of the corresponding definitions and the arguments used in the proof of the relation (4. 17).

Proposition 4. 4. *If $\mathcal{L}$ is a discretizable space and a probability distribution $P \in \mathcal{P}(\mathcal{L})$ has a large-scale limit $P^\infty$ with normalization $f(\lambda)$ then the discretization $\hat{P}$ of $P$ has a large-scale limit with the normalization $f(n)$ and this limit coincides with the discretization $\hat{P}^\infty$ of $P^\infty$.*

The conjecture, that for any $\hat{P} \in \mathcal{P}(\mathbb{Z}^\nu)$ having a large-scale limit, the probability distribution $P \in \mathcal{P}(M_w(\mathbb{R}^\nu))$ defined by the formula (4. 15) for all $\varphi \in M_w(\mathbb{R}^\nu)$ has a large-scale limit, remains unproven. The conjecture is true in all examples studied so far. This conjecture and proposition 4. 4 imply the possibility of completely reducing the study of the large-scale limits of probability distributions of fields with discrete arguments to the same problem for the fields upon $M_w(\mathbb{R}^\nu)$. Proposition 4. 3 shows that this conjecture implies that the hypothesis discussed above about the possibility of obtaining by discretization any automodel probability distribution of fields with discrete arguments. Both conjectures lead to difficulties of a similar nature. We note finally that all the previous discussion can be easily extended to the case of automodel fields and large-scale limits in the wide sense.

## V. STRUCTURE OF THE SET OF AUTOMODEL PROBABILITY DISTRIBUTIONS

The results of sect. 3 and 4 show the value of the concept of automodel probability distribution. The class $\mathcal{A}_\kappa(\mathcal{D}(\mathbb{R}^\nu))$ of all automodel probability distributions with parameter $\kappa$ is very wide (see for example [3] where it is described for the case of the usual random fields upon $\mathbb{R}^1$) and its investigation is possible only if we introduce some additional restrictions. It is easy to show that if the probability distribution $P \in \mathcal{P}(\mathcal{D}(\mathbb{R}^\nu))$ is stationary and hence by proposition 3. 2 are in the class $\mathcal{A}_\kappa^{st}(\mathcal{D}(\mathbb{R}^\nu))$. There is a mathematical and physical basis for hoping that the class $\mathcal{A}_\kappa^{st}(\mathcal{D}(\mathbb{R}^\nu))$ is not too wide. To make this statement more precise we shall first formulate some operations which enable us to construct new automodel probability distributions from the ones already found. We shall formulate all statements for the case of generalized random fields, but they are all (with the obvious exception of statement 5. 5) true also for random fields with discrete argument. All these statements follow easily from the corresponding definitions, and so we shall give them without proof.

5. 1. Multiplication by a constant. If the probability distribution $P$ is in $\mathcal{A}_\kappa^{st}(\mathcal{L})$ $(\mathcal{A}_\kappa(\mathcal{L}))$ then for any $c \in \mathbb{R}^1$ the probability distribution $T_c^* P$ is in $\mathcal{A}_\kappa^{st}(\mathcal{L})$ $(\mathcal{A}_\kappa(\mathcal{L}))$.

5. 2. Convex linear combination of random fields. If the probability distributions $P^1$, $P^2$ are in $\mathcal{A}_\kappa^{st}(\mathcal{L})$ $(\mathcal{A}_\kappa(\mathcal{L}))$ then for any $\lambda^1 \geq 0$, $\lambda^2 \geq 0$, $\lambda^1 + \lambda^2 = 1$ the probability distribution $P = \lambda^1 P^1 + \lambda^2 P^2$ defined by the formula

$$P_\varphi = \lambda^1 P^1_\varphi + \lambda^2 P^2_\varphi, \qquad \varphi \in \mathcal{L} \tag{5. 1}$$

is in $\mathcal{A}^{st}_{\kappa}(\mathcal{L})$ $(\mathcal{A}_{\kappa}(\mathcal{L}))$.

5.3. Composition of random fields. If the probability distributions $P^1$, $P^2$ are in $\mathcal{A}^{st}_{\kappa}(\mathcal{L})$ $(\mathcal{A}_{\kappa}(\mathcal{L}))$ then the probability distribution $P = P^1 * P^2$ defined by the formula

$$P_{\varphi} = P^1_{\varphi} * P^2_{\varphi}, \qquad \varphi \in \mathcal{L} \tag{5.2}$$

where $*$ in (5.2) denotes the usual convolution of measures in $\mathbb{R}^1$, is also in $\mathcal{A}^{st}_{\kappa}(\mathcal{L})$ $(\mathcal{A}_{\kappa}(\mathcal{L}))$.

5.4. Closure. The sets $\mathcal{A}^{st}_{\kappa}(\mathcal{L})$ and $\mathcal{A}_{\kappa}(\mathcal{L})$ are closed (in the topology of weak convergence of all finite-dimensional distributions).

We shall denote by $J_{\nu}$ the set of multi-indices $j = (j_1, \dots, j_{\nu})$, $j_i = 0, 1, \dots$, $i = 1, \dots, \nu$. Let $|j| = j_1 + \dots + j_{\nu}$, $j \in J_{\nu}$. The set of multi-indices $j \in J_{\nu}$ with $|j| \le q$ will be denoted by $J^q_{\nu}$, $q = 0, 1, \dots$. We shall denote by $D^j$, $j \in J_{\nu}$ the operator of partial differentiation $D^j = (D_1)^{j_1} \dots (D_{\nu})^{j_{\nu}}$, where $D_i$ is the operator of differentiation with respect to the i-th variable. The derivative of of order $j \in J_{\nu}$ of the random field $\{\Phi_{\varphi}, \varphi \in \mathcal{D}(\mathbb{R}^{\nu})\}$ is by definition the field (comp. [21], chapt. 3)

$$\Phi^j_{\varphi} = \Phi_{(-1)^j D^j \varphi}, \qquad \varphi \in \mathcal{D}(\mathbb{R}^{\nu}) \tag{5.3}$$

If the field $\{\Phi_{\varphi}, \varphi \in \mathcal{D}(\mathbb{R}^{\nu})\}$ has the probability distribution $P$, then the probability distribution of the field (5.3) will be denoted by $D^{*j}P$.

5.5. Differentiation of random fields. If the probability distribution $P \in \mathcal{A}^{st}_{\kappa}(\mathcal{D}(\mathbb{R}\ ))$ $(\mathcal{A}_{\kappa}(\mathcal{D}(\mathbb{R}^{\nu}))$ then for any $j \in J_{\nu}$ the probability distribution $D^{*j}P$ of the derivative of order $j$ of the random field having the probability distribution $P$ is in $\mathcal{A}^{st}_{\kappa-|j|}(\mathcal{D}(\mathbb{R}^{\nu}))$ $(\mathcal{A}_{\kappa-|j|}(\mathcal{D}(\mathbb{R}^{\nu}))$.

Let $\mathcal{B} \subseteq \mathcal{A}^{st}_{\kappa}(\mathcal{L})$ be a set of probability distributions. We shall say that the set $\hat{\mathcal{B}} \subseteq \mathcal{A}_{\kappa}^{st}(\mathcal{L})$ is generated (say convexly generated) by the set $\mathcal{B}$ if $\hat{\mathcal{B}}$ is the smallest closed set (closed and convex set) containing $\mathcal{B}$ such that $P^1, P^2 \in \mathcal{B}$ implies $T_c^* P^i \in \hat{\mathcal{B}}$, $c \in \mathbb{R}^1$, $i = 1, 2$ and $P^1 * P^2 \in \hat{\mathcal{B}}$. The following conjecture makes the discussion at the beginning of this section more precise. The sets $\mathcal{A}_{\kappa}^{st}(\mathcal{D}(\mathbb{R}^{\nu}))$ and $\mathcal{A}_{\kappa}^{st}(\mathcal{D}^1(\mathbb{R}^{\nu}))$, $\kappa \in \mathbb{R}^1$ are convexly generated by a finite dimensional set of probability distributions.

The problem of the complete description of the classes of random fields $\mathcal{A}_{\kappa}^{st}(\mathcal{D}(\mathbb{R}^{\nu}))$ and $\mathcal{A}_{\kappa}^{st}(\mathcal{D}^1(\mathbb{R}^{\nu}))$ seems fundamental to the theory of probability and very difficult. Some partial results in this direction are contained in another paper by this author and in sect. 6 and 7 of this paper.

## VI. MUTUAL LOCAL SINGULARITY OF AUTOMODEL PROBABILITY DISTRIBUTIONS

We can formulate the main result of this section in roughly the following way. Any two different stationary metrically transitive automodel random fields have probability distributions which are locally singular. The result illustrates the variety of automodel probability distribution, explaining the difficulty of the search.

We shall give some definitions. Let W be a set, $\mathcal{B}$, be a $\sigma$-algebra of subsets of W. Furthermore let $\{F_{\varphi}, \varphi \in \mathcal{L}\}$ be real functions of $w \in W$ measurable with respect to the $\sigma$-algebra $\mathcal{B}$ and let $m^1, m^2$ be probability measures on $\mathcal{B}$. If a set of random variables $\Phi_{\varphi}^1 \in L_0(W, \mathcal{B}, m^1)$ which coincide almost everywhere with $F_{\varphi}$ with respect to the measure $m^1$ defines a random field with probability distribution $P^2$, then we shall say that a simultaneous realization of $P^1$ and $P^2$ is defined. The Kolmogorov theorem (see for example [27], §4. 3) implies that any two probability distributions $P^1, P^2 \in \mathcal{P}(\mathcal{L})$ have a simultaneous realization, with $W = (\mathbb{R}^1)^{\mathcal{L}}$. For a wide class spaces $\mathcal{L}$ including the Schwarz spaces $\mathcal{D}(\mathbb{R}^{\nu})$, $\mathfrak{T}(\mathbb{R}^{\nu})$, a natural simultaneous realization of any two probability distributions can be constructed with the help of the states upon the space $\mathcal{L}$ (see sect. 2).

Let $V \subseteq \mathbb{R}^\nu$ be an open set. Let $\mathcal{L}_V \in \mathcal{L}$ be the set of functions $\varphi \in \mathcal{L}$ vanishing outside V. We shall assume in this section that the set of functions

$$\hat{\mathcal{L}} = \bigcup_V \mathcal{L}_V \tag{6.1}$$

where the summation is on all bounded open sets $V \subset \mathbb{R}^\nu$ is a dense subspace of $\mathcal{L}$. It is clear that this assumption is true, for example, for the spaces $\mathcal{D}(\mathbb{R}^\nu)$ and $\mathcal{S}(\mathbb{R}^\nu)$. Let a simultaneous realization of the probability distributions $P^1$, $P^2 \in \mathcal{P}(\mathcal{L})$ be given. Let $\mathcal{B}_V$ be the smallest $\sigma$-subalgebra of the $\sigma$-algebra $\mathcal{B}$ with respect to which all functions $F_\varphi$, $\varphi \in \mathcal{L}_V$ are measurable. We shall say that <u>two probability distributions $P^1$ and $P^2$ are locally mutually singular</u> (or similarly, $P^1$ <u>is locally absolutely continuous with respect to the probability distribution</u> $P^2$) if for any open set $V \subset \mathbb{R}^\nu$ and any simultaneous realization of these probability distributions, the restrictions of the measures $m^1$ and $m^2$ to the $\sigma$-algebra $\mathcal{B}_V$ are mutually singular; i. e., there exists a set $B \in \mathcal{B}_V$ such that

$$m^1(B) = 0, \quad m^2(B) = 1 \tag{6.2}$$

(or if for any open bounded set $V \subset \mathbb{R}^\nu$ the restriction of the measure $m^1$ to the $\sigma$-algebra $\mathcal{B}_V$ is absolutely continuous with respect to the restriction of the measure $m^2$ to $\mathcal{B}_V$).

Let $\{\Phi_\varphi, \varphi \in \mathcal{L}\}$ be a stationary random field defined on the probability space $(W, \mathcal{B}, m)$. Let $\mathcal{B}_{\mathbb{R}^\nu} \subset \mathcal{B}$ be the smallest $\sigma$-algebra of sets, with respect to which all the random variables $\Phi_\varphi$, $\varphi \in \mathcal{L}$ are measurable and let $L_1(m)$ be the Banach space of all real functions measurable with respect to the $\sigma$-algebra $\mathcal{B}_{\mathbb{R}^\nu}$ and integrable with respect to the measure m. It is easy to show that the subspace $\tilde{L}_1(m) \subset L_1(m)$ containing elements of the form

$$\Psi = \psi(\Phi_{\varphi_1}, \ldots, \Phi_{\varphi_k}), \quad \varphi_1, \ldots, \varphi_k \in \mathcal{L}, \quad k = 1, 2, \ldots \tag{6. 3}$$

Where $\psi: \mathbb{R}^k \to \mathbb{R}^1$ is a continuous function of compact support, is dense in $L_1(m)$. The operators

$$\hat{E}_a : \tilde{L}_1(m) \to \tilde{L}_1(m):$$

$$\hat{E}_a \Psi = \psi(\Phi_{E_a\varphi_1}, \ldots, \Phi_{E_a\varphi_k}), \quad a \in \mathbb{R}^\nu \tag{6. 4}$$

are isometries because of the stationarity of the field $\{\Phi_\varphi, \varphi \in \mathcal{L}\}$, so they can be uniquely extended to isometries from $L_1(m)$ to $L_1(m)$. These operators will again by denoted by $\hat{E}_a$ and will be called <u>shift operators.</u> The shift operators formed a $\nu$-parameter group of operators $\{\hat{E}_a, a \in \mathbb{R}^\nu\}$. We note that for any fixed $\Psi \in L_1(m)$ the transformation $\mathbb{R}^\nu \to L_1(m) : a \to \hat{E}_a \Psi$ is continuous. Indeed, for the case $\Psi \in \tilde{L}_1(m)$ the statement follows immediately from the fact (see sect. 2) that for any $\varphi \in \mathcal{L}$ the transformation: $\mathbb{R}^\nu \to \mathcal{P}_1 : a \to P_{E_a\varphi}$ is continuous (with respect to the topology of weak convergence in the space $\mathcal{P}_1$ of probability distributions on $\mathbb{R}^1$). In the general case, it follows from the fact that the transformations $\hat{E}_a$, $a \in \mathbb{R}^\nu$ are isometries.

We shall say that a stationary random field $\{\Phi_\varphi, \varphi \in \mathcal{L}\}$ is <u>metrically transitive</u> if any element of the space $L_1(m)$ which is invariant with respect to the group $\{\hat{E}_a, a \in \mathbb{R}^\nu\}$ of shift operators is a constant. We shall say that the stationary probability distribution $P \in \mathcal{P}(\mathcal{L})$ is metrically transitive if any random field having this probability distribution is metrically transitive*).

---

* It is easy to show that if a random field with a certain probability distribution is metrically transitive, then this probability distribution is metrically transitive. Similarly, if the condition (6. 2) is true for a simultaneous realization of two probability distributions then these probability distributions are mutually singular (see for example similar constructions in [28], § 1). These remarks will not be used in the following.

<u>Theorem 6.1.</u> (I) <u>Let</u> $P^1, P^2 \in \mathcal{P}(\mathcal{L})$ <u>be distinct stationary metrically transitive automodel probability distributions. Then these probability distributions are locally mutually singular.</u>

(II) <u>Let</u> $P^1, P^2 \in \mathcal{P}(\mathcal{L})$ <u>be nonzero automodel probability distributions with different automodel parameters</u> $\kappa^1$ <u>and</u> $\kappa^2$. <u>Then the probability distribution</u> $P^1$ <u>is not locally absolutely continuous with respect to the probability distribution</u> $P^2$.

<u>Proof.</u> Suppose the conditions of statement (I) of the theorem are true and let a simultaneous realization of the probability distributions $P^1$ and $P^2$ be given. We note that it will suffice to prove that for any $\epsilon > 0$ and any open $U \in \mathbb{R}^\nu$ there exists a set $A \in \mathcal{B}_U$ such that

$$m_1(A) < \epsilon, \quad m_2(A) > 1 - \epsilon \tag{6.5}$$

Because of the stationarity of the probability distributions $P^1$ and $P^2$ it is possible without loss of generality to consider only sets $U$ containing the origin $0$. We first assume that the probability distributions $P^1$ and $P^2$ have the same automodel parameter $\kappa$. The distinctness of the probability distributions $P^1$ and $P^2$ implies the existence of a function $\varphi \in L$ such that the probability distributions $P^1_\varphi$ and $P^2_\varphi$ does not coincide. So there exists a continuous function $f : \mathbb{R}^1 \to \mathbb{R}^1$ such that the mean value

$$M^1 = \int_W f(F_\varphi(w))m^1(dw) > M^2 = \int_W f(F_\varphi w))m^2(dw) \tag{6.6}$$

Since by the condition introduced earlier $\hat{\mathcal{L}}$ is dense in $\mathcal{L}$, it is possible without loss of generality to suppose that $\varphi \in \mathcal{L}_{V_s}$ for some $s \geq 0$. Here and what follows

$$V_s = \{a = (a^1, \ldots, a^\nu) \in \mathbb{R}^\nu : -s < a^i \leq s, \ i = 1, \ldots, \nu\} \tag{6.7}$$

The continuity noted earlier of the transformation $a \to \hat{E}_a \Psi$ implies that the transformation groups $\{\hat{E}_a, a \in \mathbb{R}^\nu\}$ on the spaces $L_1(m^i)$, $i = 1, 2$ are strangly measurable. (See [29], §8.7 and sect. 3.5.11). Thus the Wiener multidimensional ergodic theorem (see [29], sect. 8.7.17 and 8.5.10) can be applied here. This theorem implies that in the sense of convergence in $L_1(m^i)$, $i = 1, 2$ the limits

$$\lim_{T \to \infty} \frac{1}{\|V_T\|} \int_{V_T} f(F_{E_a \varphi}(w))da = M^i \tag{6.8}$$

exist. (Here $\|V_T\| = \int_{V_T} da$). So for any $\epsilon > 0$ there exists $T_0$ so large that if

$$A_1 = \{w \in W: \frac{1}{\|V_{T_0}\|} \int_{V_{T_0}} f(F_{E_a \varphi}(w))da < \frac{M^1 + M^2}{2}\} \tag{6.9}$$

then the probabilities

$$m^1(A_1) < \epsilon, \quad m^2(A_1) > 1 - \epsilon \tag{6.10}$$

The automodel property of $P^1$ and $P^2$ implies that if

$$A_\lambda = \{w \in W: \frac{1}{\|V_{T_0}\|} \int_{V_{T_0}} f(F_{S_{\kappa,\lambda} E_a \varphi}(w))da < \frac{M^1 + M^2}{2}\} \tag{6.11}$$

then the probabilities

$$m^1(A_\lambda) = m^1(A_1), \quad m^2(A_\lambda) = m^2(A_1), \quad \lambda \in (0, \infty) \tag{6.12}$$

The support of the function $S_{\kappa,\lambda} E_a \varphi$ lies in the cube $V_{(s+|a|)\lambda}$. So for any open $U \in \mathbb{R}^\nu$ such that $0 \in U$ the event $A_\lambda$ is in the $\sigma$-algebra $\mathcal{B}_U$ for all small enough $\lambda$. The statement of the theorem for the case under consideration is proven.

Suppose now that in the conditions of statement (I) of the theorem the automodel parameters $\kappa^1$ and $\kappa^2$ of the probability distributions $P^1$ and $P^2$ are not equal. We can suppose that $\kappa^1 > \kappa^2$. If the probability distribution $P^1$ is zero it is possible to suppose that $\kappa^1 = \kappa^2$. So we shall suppose that $P^1$ is nonzero. Then there exists a function $\varphi \in \mathcal{L}$ such that

$$M^1 = \int_W |F_\varphi(w)|_+ m^1(dw) \in (0, \infty] \tag{6.13}$$

where $|F|_+ = F$ if $F > 0$ and $|F|_+ = 0$ if $F \leq 0$. In the same way as above we can suppose without loss of generality that $\varphi \in \mathcal{L}_{V_s}$ for some $s > 0$. By using the ergodic theorem we see as before that for large enough $T_0$ and

$$A_1^1 = \{w \in W : \frac{1}{\|V_{T_0}\|} \int_{V_{T_0}} |F_{E_a\varphi}(w)|_+ da < \frac{M^1}{2}\} \tag{6.14}$$

the probability

$$m^1(A_1^1) < \epsilon \tag{6.15}$$

For large enough $C$ and

$$A_1^2 = \{w \in W : \frac{1}{\|V_{T_0}\|} \int_{V_{T_0}} |F_{E_a\varphi}(w)|_+ da > C\}$$

the probability

$$m^2(A_1^2) > 1 - \epsilon \tag{6.16}$$

We suppose now that

$$A_\lambda^1 = \{w \in W : \frac{1}{\|V_{T_0}\|} \int_{V_{T_0}} |F_{S_{\kappa^1,\lambda} E_a\varphi}(w)|_+ da < \frac{M^1}{2}\}$$

$$A_\lambda^2 = \{w \in W : \frac{1}{\|V_{T_0}\|} \int_{V_{T_0}} |F_{S_{\kappa^2,\lambda} E_a\varphi}(w)|_+ da < C\} \tag{6.17}$$

Then the automodel property of $P^1$ and $P^2$ implies that

$$m^1(A^1_\lambda) = m^1(A^1_1)$$

$$m^2(A^2_\lambda) = m^2(A^2_1), \qquad \lambda \in (0, \infty) \tag{6.18}$$

The definition of the renorm-group and linearity implies that

$$F_{S_{\kappa^2,\lambda} E_a \varphi}(w) = \lambda^{\kappa^2 - \kappa^1} F_{S_{\kappa^1,\lambda} E_a \varphi}(w) \tag{6.19}$$

for almost all $w \in W$ with respect to the measure $m^2$. Because $\kappa^1 > \kappa^2$ it follows that the event $A^2_\lambda \subset A^1_\lambda$ for small enough $\lambda$; therefore the relations (6.15), (6.16) and (6.18) imply that for such $\lambda$

$$m^2(A^1_\lambda) > 1 - \epsilon, \quad m^1(A^1_\lambda) < \epsilon \tag{6.20}$$

Again the event $A^1_\lambda \in \mathcal{B}_U$ for all small enough $\lambda$, so the condition (6.5) is checked also for $\kappa^1 \neq \kappa^2$.

The proof of statement (II) of the theorem is similar to that of statement (I) for $\kappa^1 \neq \kappa^2$. We shall give only the main differences. It suffices for the proof of statement (II) to construct for some open bounded set $U \subset \mathbb{R}^\nu$, some $\delta < 1$, and any $\epsilon > 0$, a set $A \in \mathcal{B}_U$ such that

$$m^1(A) < \delta, \quad m^2(A) > 1 - \epsilon \tag{6.21}$$

Because the probability distribution $P^1$ is nonzero, for some $T < \infty$ and $\varphi \in \mathcal{L}_{V_T}$ the measure

$$m^1(w \in W : \Phi_\varphi(w) = 0) < 1 \tag{6.22}$$

If $\kappa^1 > \kappa^2$ we set

$$A^1_\lambda = \{w \in W : |F_{S_{\kappa^1,\lambda}\varphi}(w)| < c_1\},$$

$$A^2_\lambda = \{w \in W : |F_{S_{\kappa^2,\lambda}\varphi}(w)| < c_2\}, \quad \lambda \in (0, \infty) \tag{6.23}$$

where $c_1 > 0$ is small enough and $c_2 < \infty$ is large enough. If $\kappa^1 < \kappa^2$ we set

$$A^1_\lambda = \{w \in W : |F_{S_{\kappa^1,\lambda}\varphi}(w)| < c_1 \text{ or } = 0\},$$

$$A^2_\lambda = \{w \in W : |F_{S_{\kappa^2,\lambda}\varphi}(w)| < c_2 \text{ or } = 0\}, \quad \lambda \in (0, \infty) \tag{6.24}$$

where $c_1 < \infty$ is large enough and $c_2 > 0$ is small enough. Considerations similar to the previous ones show that the conditions (6.21) are satisfied if $A = A^1_\lambda$, $0 \in U$ and $\lambda$ is small enough.

We note that the question concerning the possibility of extending statement (II) of the theorem on the case of different automodel parameters $\kappa^1$ and $\kappa^2$ is open.

## VII. AUTOMODEL FIELDS WITH INDEPENDENT VALUES

We recall (see [21], §3.4) that the random field $\{\Phi_\varphi, \varphi \in \mathcal{L}\}$ having the probability distribution P is a field with independent values if for any $\varphi_1, \varphi_2 \in \mathcal{L}$ such that the intersection of their support is empty,

$$\text{supp } \varphi_1 \cap \text{supp } \varphi_2 = \emptyset \tag{7.1}$$

the random variables $\Phi_{\varphi_1}$ and $\Phi_{\varphi_2}$ are independent or in terms of characteristic functionals

$$L^P(\varphi_1)L^P(\varphi_2) = L^P(\varphi_1 + \varphi_2) \tag{7.2}$$

We shall consider in the following random fields with independent values upon the space $\mathcal{D}(\mathbb{R}^\nu)$ (see [21], sect. 3.4), described by the characteristic functionals of the type

$$L^P(\varphi) = \exp\{\int_{\mathbb{R}^\nu} f(D^j\varphi(x),\ j \in J_\nu^q)\,dx\} \qquad (7.3)$$

We used here the denotations introduced in sect. 5 for the formulation of statement 5.5. The function $f(t_j,\ j \in J_\nu^q)$ is a complex valued function of $m_q$ variables, where $m_q$ is the number of elements in the index set $J_\nu^q$ and $q = 1, 2, \ldots$ . Furthermore, for $t = (t_j,\ j \in J_\nu^q) \in \mathbb{R}^{m_q}$

$$f(t) = it_0 a - \sum_{j,k \in J_\nu^q} b_{j,k} t_j t_k$$

$$+ \int_{\mathbb{R}^{m_q} \setminus \{0\}} (\exp\{i(t,h)\} - 1 - i(t,h) - (\alpha(h),\ t))\Lambda(dh) \qquad (7.4)$$

where the numbers $a, b_{j,k} \in \mathbb{R}^1$, $j,k \in J_\nu^q$, the matrix $(b_{j,k},\ j,k = 1,\ldots,m_q)$ is negative-definite, $\Lambda(\cdot)$ is a measure on the $\sigma$-algebra $\mathcal{B}_{\mathbb{R}^{m_q}\setminus\{0\}}$ of Borel subsets of $\mathbb{R}^{m_q} \setminus \{0\}$ such that

$$\int_{\{h \in \mathbb{R}^{m_q}: 0 < |h| \leq 1\}} |h|^2 \Lambda(dh) < \infty, \qquad \int_{\{h \in \mathbb{R}^{m_q}: |h| > 1\}} \Lambda(dh) < \infty \qquad (7.5)$$

and finally $\alpha(h)$ is a measurable function of $h \in \mathbb{R}^{m_q}$ with values in $\mathbb{R}^{m_q}$ such that the sum $ih + \alpha(h)$ is a bounded function and such that it is twice differentiable at the point $h = 0$ and vanishes at this point together with all its first partial derivatives. The formula (7.4) is of course the general formula for $m_q$ - dimensional infinitely divisible probability distributions except that in (7.4) the terms $t_j a_j$, $j \neq 0$, included in the general formula, are absent. The inclusion of these terms in (7. 1) cannot increase the set of possible characteristic functional because

$$\int_{\mathbb{R}^\nu} D^j\varphi(x)dx = 0, \quad j\in J_\nu \quad \{0\}, \quad \varphi\in\mathcal{D}(\mathbb{R}^\nu) \tag{7. 6}$$

A centrical function $\alpha(h)$ will be fixed, therefore the measure $\Lambda$ and the numbers $a$ and $b_{j,k}$, $d$, $k\in J_\nu^q$ are uniquely determined by the function $f$.

The relation (2. 10) implies that the action of the renorm-group on the functional (7. 3) is described by the relations

$$L^{S^*_{\kappa,\lambda}P}(\varphi) = \exp\Big(\int_{\mathbb{R}^\nu} f_{\kappa,\lambda}(D^j\varphi(x),\ j\in J_\nu^q)\,dx. \tag{7. 7}$$

where

$$f_{\kappa,\lambda}(t) = \lambda^\nu f(\lambda^{\kappa-\nu-|j|}t_j,\ j\in J_\nu^q),\quad t\in\mathbb{R}^{m_q} \tag{7. 8}$$

The relations (7. 5), (7. 6) imply that the representation (7. 4) holds for the function $f_{\kappa,\lambda}$ but the parameters $a, b_{j,k}$, $\Lambda$ must be changed to the parameters $a_{\kappa,\lambda}$, $b_{j,k,\kappa,\lambda}$, $\Lambda_{\kappa,\lambda}$ such that

$$a_{\kappa,\lambda} = a\lambda^\kappa + \int_{\mathbb{R}^{m_q}\ \{0\}} [\alpha_0(h\lambda^\kappa) - \lambda^\kappa\alpha_0(h)]\,\Lambda(dh),$$

$$b_{j,k,\kappa,\lambda} = b_{j,k}\lambda^{2\kappa-\nu-|j|-|k|},\quad j,k\in J_\nu^q,$$

$$\Lambda_{\kappa,\lambda}(B) = \lambda^\nu\Lambda(g_{\kappa,\lambda}(B)),\quad b\in\mathcal{B}_{\mathbb{R}^{m_q}\ \{0\}} \tag{7. 9}$$

where $g_{\kappa,\lambda}$ is the one-to-one transformation

$$g_{\kappa,\lambda} : \mathbb{R}^{m_q}\ \{0\}\to\mathbb{R}^{m_q}\ \{0\}$$

$$(h_j, j\in J_\nu^q)\to(\lambda^{-\kappa+\nu+|j|}h_j,\quad j\in J_\nu^q) \tag{7. 10}$$

and where $\alpha_0(h)$ is the component of the function $\alpha(h)$ corresponding to the multiindex $0$. We note that if the measure $\Lambda$ is given by a density $\ell$ with respect to the Lebesque measure then the measure $\Lambda_{\kappa,\lambda}$ is given by the density $\ell_{\kappa,\lambda}$ such that

$$\ell_{\kappa,\lambda}(h) = \lambda^{\nu}\Big(\prod_{j\in J_{\nu}^{q}} \lambda^{-\kappa+\nu+|j|}\Big)\ell(g_{\kappa,\lambda}(h)), \quad h\in \mathbb{R}^{m_q} \setminus \{0\} \tag{7.11}$$

The formulae (7. 8), (7. 9) imply the following proposition.

Proposition 7. 1. The probability distribution $P$ of a random field upon $\mathcal{D}(\mathbb{R}^{\nu})$, described by the characteristic functional given by (7. 3), (7. 4) will be automodel with parameter $\kappa$ iff the relations

$$a_{\kappa,\lambda} = a, \quad \lambda \in (0,\infty),$$

$$b_{j,k} = 0 \quad \text{if} \quad \kappa \neq \frac{\nu}{2} + |j| \quad \text{or} \quad \kappa \neq \frac{\nu}{2} + |k|,$$

$$\Lambda_{\kappa,\lambda} = \Lambda, \quad \lambda \in (0,\infty) \tag{7.12}$$

are satisfied.

In the case $q = 0$ the conditions (7. 12) reduce (it is easier to understand this directly using (7. 8)) to the condition that the function $f$ be the logarithm of the characteristic function of the probability distribution which is stable in the strict sense (see [23], § 6. 1) with exponent

$$\alpha = \frac{\nu}{\nu - \kappa} \tag{7.13}$$

The condition $0 < \alpha \leq 2$ corresponds to the condition

$$-\infty < \kappa \leq \frac{\nu}{2} \tag{7.14}$$

In another special case when the function $f$ depends only on the variables $t_j \in J_\nu^q$ with $|j| = q$ where $q > 0$ the condition (7. 2) reduces to the condition that f be the logarithm of the characteristic function of a probability distribution which is stable in the wide sense (see [23], § 6. 1 and 17. 11) with exponent

$$\alpha = \frac{\nu}{\nu + q - \kappa} \tag{7. 15}$$

In such a way it is easy to obtain any values of $\kappa$. The general case does not reduce to the stable probability distributions.

It is easy to check that the set of all probability distributions $P \in \mathcal{P}(\mathcal{D}(\mathbb{R}^\nu))$ of the field with independent values is closed. So the closure of the set of probability distributions described in a proposition 7. 1 consists of automodel probability distribution of random fields with independent values. This closure is infinite dimensional but it can be generated (see sect. 5) by a finite dimensional set of such probability distributions. It is not known whether this operation gives all such probability distributions. (This problem is connected with the open problem of the description of all stationary generalized fields with independent values (see [21], § 3. 4).

It is easy, by the use of the same methods, to investigate the set of automodel random fields upon the space $\mathcal{D}^1(\mathbb{R}^\nu)$ (i. e. automodel fields in wide sense, see sect. 3) with independent values. The only essential difference is that in this case we may freely set the value $a = 0$, so the first of the conditions (7. 5) is not now needed. Now all probability distributions stable in the wide sense appear even in the case $q = 0$.

Among the random fields described by proposition (7. 1), only the fields with $q = 0$, having probability distributions stable in the strict sense, have a discretization (see sect. 4). In such a situation the simplest choice of the space $\mathcal{L}^1 \supset \mathcal{D}(\mathbb{R}^\nu)$ is $L_2^1$, where $L_2^1$ is the space of square-integratable finite-range functions, regarded as the inductive limit of the Hilbert spaces $L_2(S_r)$ (see sect. 4). The explicitly known formula for stable laws shows that the characteristic functional (7. 3) can be continuously extended to the entire space $L_2^1$, and therefore the corresponding probability distribution can be continued to a probability distribution in $\mathcal{P}(\mathcal{L}^1)$. We can for example choose $w(\lambda) = [1 + \ln(1 + \lambda)]^{-1}$; then it is easy to see that $L_2^1 \supseteq M_w(\mathbb{R}^\nu)$ (see sect. 4) and that the topology in $L_2^1$ is weaker than the topology in $M_w(\mathbb{R}^\nu)$; therefore the restriction of this probability distribution to $M_w(\mathbb{R}^\nu)$ is a probability distribution in $\mathcal{P}(M_w(\mathbb{R}^\nu))$. The automodel fields with discrete argument which can be obtained by such discretization are simply composed of independent random variables stable in the strict sense. The set of all independent random variables stable in the wide sense can be obtained by the discretization of the random fields with independent values automodel in the wide sense which are continuations of the corresponding random fields upon $\mathcal{D}^1(\mathbb{R}^\nu)$.

## REFERENCES


1. I. A. Ibragimov, J. V. Linnik, Independent and Stationary Connected Variables, M. 1965, "Nauka".
2. M. Rosenblatt , Independence and Dependence, Proc. 4th. Berkeley Symp. on Math. Stat. and Prob. , Berkeley 1961, 43-47.
3. J. Lamperti, Semi-Stable Stochastic Processes, Tr. Am. Math. Soc. , 104, No. 1, 62-78, 1962.
4. M. S. Taqqu, Weak Convergence to Fractional Brownian Motion and to the Rosenblatt Process, Z. Wahrs. verf. Geb. 31, No. 4 287-302, 1975.
5. L. P. Kadanoff, Scaling Laws for Ising Models near $T_c$, Physics, 2, No. 6, 263-272, 1966.

6. K. G. Wilson, Renormalization group and critical phenomena II, Phase-Space Cell Analysis of Critical Behavior, Phys. Rev., B4, No. 9, 3184-3205, 1971.

7. S. K. Ma, Introduction to the Renormalization Group, Rev. of Mod. Phys., 45, No. 4, 589-614, 1973.

8. P. Blecher, J. Sinai, Investigation of the Critical Point in Models of the Type of Dyson's Hierarchical Models. Comm. Math. Phys. 33, No. 1, 23-42, 1973.

9. Y. G. Sinai, Automodel Probability Distributions, Theory of prob. and its appl., 21, No. 1, 63-80, 1976.

10. G. Gallavotti, H. Knops, Block-Spins Interaction in the Ising Model, Comm. Math. Phys., 36, No. 2, 171-184, 1974.

11. G. Gallavotti, A. Martin-Löf, Block-Spin Distributions for Short-Range Attractive Ising Models, Il Nuovo Cimento, 25B, No. 1, 425-441, 1975.

12. G. John-Lasinio, The renormalization Group: a probabilistic view, Il Nuovo Cimento, 26B, No. 1, 99-119, 1975.

13. M. Cassandro, G. Gallavotti, The Lavoisier Law and the critical point, Il Nuovo Cimento, 25B, No. 2, 691-705, 1975.

14. G. Gallavotti, G. John-Lasinio, Limit theorems for Multi-dimensional Markov processes, Comm. Math. Phys., 41, No. 3, 301-307, 1975.

15. B. Simon, The $P(\varphi)_2$ Euclidean (Quantum) Field Theory, Princeton Univ. Press, Princeton, 1974.

16. A. N. Kolmogorov, Wiener Spiral and Other Interesting Curves in Hilbert Space, Reports of Sovjet Acad. Sci., 26, No. 2, 115-118, 1940.

17. M. S. Pinsker, Theory of Curves in Hilbert Space with Stationary n-th Increments, Commun. (Isvestija) of Acad. Sci. USSR, ser. math., 19, No. 5, 319-345, 1955.

18. I. M. Gelfand, G. E. Shilov, Generalized Functions, Vol. I, Generalized Functions and Operations with Them. Second edit. Fismatgis, Moscow, 1968.

19. I. M. Gelfand, G. E. Shilov, Generalized Functions, Vol. II. , Spaces of Basic and Generalized Functions, Fismatgis, Moscow, 1968, Second edit.

20. J. Neveau, Bases mathématiques du calcul des probabilites, Maconet cie, Paris, 1964.

21. I. N. Gelfand, N. J. Vilenkin, Generalized Functions, Vol. IV, Some Application of Harmonic Analysis, Fismatgis, Moscow, 1961.

22. N. Bourbaki, Elements de Mathématique, Intégration, ch. IX, Intégration sur les espaces topologiques séparés, Hermann, Paris, 1969.

23. W. Feller, An Introduction to Probability Theory and Its Applications, vol. II, J. Wiley and Sons, N. Y. -London-Sydney, 1966.

24. A. M. Jaglom, Some Classes of Random Fields in n-Dimensional Space Related to Stationary Random Processes, Theory of Prob. and Its Applic. , 2, No. 3, 292-337, 1957.

25. A. P. Robertson, A. P. , W. J. Robertson, Topological Vector Spaces, Cambr. Univ. Press, Cambridge, 2nd. Ed. , 1973.

26. L. Hörmander, Linear Partial Differential Operators, Springer-Verlag, Berl. - Got. - Heid. , 1963.

27. M. Loeve, Probability Theory, D. Van Nostrand Comp. Inc. , Princeton, 1960.

28. R. L. Dobrushin, R. A. Minlos, Polynoms of Random Function, Advances (Uspechi) of Math. Sci. , 32, No. 2, 67-122, 1967.

29. N. Dunford, J. Schwartz, Linear operators, Part. I, General Theory, Intersci. Publ. , N. J.-Lnd. , 1958.

chapter 6

# NONEXISTENCE OF ONE– AND TWO–DIMENSIONAL GIBBS FIELDS WITH NONCOMPACT GROUP OF CONTINUOUS SYMMETRIES

R. L. Dobrushin and S. B. Shlosman

## I. INTRODUCTION AND STATEMENT OF RESULTS

We study two-dimensional lattice Gibbs fields with G-invariant potential, where G is a Lie group, acting on the configuration space X of our system. In [1, 2] is was shown that under certain assumptions of smoothness and decrease of the potential (in [1] a finite range potential was considered, and in [2] an exponentially decreasing potential) and compactness of the group G, it follows that any Gibbs field with that potential is G-invariant. Naturally, in the noncompact case the situation must be the same (in one and two dimensiona). But any invariant field has an invariant measure for its one-point correlation function, and in any physically interesting case of noncompact symmetry any nontrivial invariant measure is infinite. Therefore it is natural to believe that there exists no one- and two-dimensional Gibbs field at all with G-invariant potential, if G is a noncompact connected Lie group. In this note we prove this statement under general assumptions.

Let us introduce some notations. Let $(X, B, \mu)$ be a measure space with $\sigma$-algebra B and $\sigma$-finite measure $\mu$. Suppose we have a measurable action of the Lie group $G$ on $X$ (this means that $G \times X \to X$ is measurable with respect to the $\sigma$-algebras $B_G \times B$ and $B$, where $B_G$ is a Borel $\sigma$-algebra), which leaves invariant $B$ and $\mu$. Let $\mu$ satisfy the following conditions:

A. There exists a function $\varphi$ on $X$, such that:

(1) $\varphi$ is $B$-measurable,

(2) $1 > \varphi > 0$ $\mu$- almost everywhere,

(3) $\varphi$ is $\mu$-integrable,

(4) for any $x \in X$ and any $g_1, g_2, \ldots \in G$, $\rho(e, g_n) \to \infty$, we have $\varphi(g_n x) \to 0$, where $e \in G$ is the unit element and $\rho$ is any right invariant Riemann metric on $G$.

Notes. 1. Such a function always exists in each case of physical interest: namely, $\exp\{-U(x_t|\bar{y})\}$, where $t \in \mathbb{Z}^2$, $\bar{y}$ any configuration on $\mathbb{Z}^2 \setminus t$, and $U(x_t|\bar{y})$ the interaction of a particle at $t$ with the outside world (see f-la (1) below).

2. If $X = G$, $\varphi$ can be taken of the form $[\chi_G\{h \in G, \rho(e,h) \leq \rho(e,g)\}]^{-2}$, where $\chi_G$ is right-invariant Haar measure of $G$.

Let $\mathbb{Z}^2$ be the two-dimensional lattice, $A \subset \mathbb{Z}^2$ any subset, $\bar{A}$ its complement, $|A|$ the number of elements in $A$, $X^A$ the set of configurations in $A$ (= functions $x_A : A \to X$; $t \to (x_A)_t$, $t \in A$), and $B_A$ the product $\sigma$-algebra in $X^A$, $\mu_A$ the product measure on $B_A$, $|A| < \infty$.

There is an action of $G$ on $X^A$; namely, $x_A \to gx_A$, $(gx_A)_t = gx_t$.

The potential is a family of real-valued functions

$$U = \{U_A;\ U_A : X^A \to \mathbb{R}^1,\ |A| < \infty\}$$

Let the following conditions be satisfied (compare with [2]):

B. Invariance. For any $A \subset \mathbb{Z}^2$, $|A| < \infty$

$$U_A(g x_A) = U_A(x_A)$$

C. Decrease. Let $V_n \subset \mathbb{Z}^2$ be a square,

$$V_n = \{t \in \mathbb{Z}^2;\ |t_i| \leq n,\ t = (t_1, t_2)\},$$

and $x_{\mathbb{Z}^2}$ any configuration. We suppose that for $m - n > N$,

$$\sum_{\substack{A,\, A \cap V_n \neq \phi \\ A \cap V_m \neq \phi}} |U_A(x_A)| \leq \alpha n \exp\{-\beta(m - n)\}$$

uniformly in $x_{\mathbb{Z}^2}$, where $N$ is a fixed integer, $x_A = x_{\mathbb{Z}^2}\big|_A$, and $\alpha, \beta > 0$.

Note. This property is satisfied for finite range potentials and for uniformly and exponentially decreasing pair potentials. Unfortunately it is not valid in the general case of noncompact groups, when $U(x_A)$ decreases nonuniformly in $x_A$ as diam (A) goes to infinity.

For example, it is not valid in the Gaussian case. It seems, nevertheless, that our result is still valid for this particular case, but the proof must be modified.

D. Differentiability. Let $n$ be an integer and $x_{\mathbb{Z}^2}$ any configuration. Let

$$(g_n^* x_{\mathbb{Z}^2})_t = \begin{cases} g x_t, & t \in V_n \\ x_t, & t \in \bar{V}_n \end{cases}$$

Let

$$F(g\,|\,n, x_{\mathbb{Z}^2}) = \sum_{A,\, A \cap V_n \neq \phi} U_A((g_n^* x_{\mathbb{Z}^2})|_A)$$

Let $w$ be any right-invariant tangent field on $G$ of unit length. Let us suppose the function $F$ has a second derivative along $w$ and

$$|w w F(g\,|n, x_{\mathbb{Z}^2})| \leq \gamma(w)\, n$$

uniformly in g, $x_{\mathbb{Z}^2}$ with $\gamma(w)$ bounded. (If we introduce the system of local coordinates $y_1, \dots, y_m$ on G is such a way that vector fields $\partial/\partial y_k$ become right-invariant and of unit length, our condition may be rewritten in a more familiar form

$$\left|\frac{\partial^2}{\partial y_k^2} F(y_1, \dots, y_m\,|n, x_{\mathbb{Z}^2})\right| \leq \gamma(\partial/\partial_{y_k})\, n$$

Note. This condition is satisfied only for potential which increase "at infinite of G" only as a square. Our results are probably true in a more general situation, but the proof would be more complicated.

E. Convergence and measurability. Suppose the functions $U_A$ are $B_A$-measurable for any A, $|A| < \infty$. Let $V \subset \mathbb{Z}^2$, $|V| < \infty$ and let us consider the series (absolutely convergent according to condition C)

$$U(x_V|x_{\bar{V}}) = \sum_{A;\, A \cap V \neq \phi} U_A((x_{\mathbb{Z}^2}|_A) \tag{1}$$

where $x_V = x_{\mathbb{Z}^2}|_V$, $x_{\bar{V}} = x_{\mathbb{Z}^2}|\bar{V}$. Let us call the configuration $x_{\bar{V}}$ a boundary condition for volume V, and the sum $U(x_V|x_{\bar{V}})$ the energy of $x_V$ with respect to the boundary conditions $x_{\bar{V}}$. Suppose that for any V and $x_{\bar{V}}$ the integral

$$\int_{X^V} \exp\{-U(x_V|x_{\bar{V}})\}d\mu_V$$

converges.

We suppose the reader to be familiar with the definition of Gibbs state (otherwise see [1], where we define it using the same notations).

The main result of this paper is the following.

Theorem. *If the measure* $\mu$ *on* X *satisfies* A, *the potential* U satisfies B - E, and if G is noncompact and connected, then there are no Gibbs states with potential U.

Notes. 1. Without condition A our theorem is not true; moreover, it seems that the existence of a Gibbs state with potential U implies its G - invariance.

2. In case $\nu \geq 3$ the situation is just the opposite. Indeed, let $X = G = \mathbb{R}^n$, and let us consider the following translation -- invariant quadratic pair potential:

$$U_A(x_A) = \begin{cases} U(s-t)(x_s - x_t)^2 & A = \{s, t\} \\ 0 & |A| \neq 2 \quad \text{or} \quad |s-t| > R , \end{cases}$$

where $U(t) = U(-t)$, $t \in \mathbb{Z}^\nu \setminus 0$. It follows from Rosanov's results [3], that if the function

$$f(\lambda) = \left[ \sum_{t \in \mathbb{Z}^\nu \setminus 0} U(t)(e^{i\lambda t} - 1) \right]^{-1}$$

is integrable over the cube $\{\lambda = (\lambda_1, \ldots, \lambda_\nu);\ |\lambda_i| \leq \pi,\ i = 1, \ldots, \nu\}$, then the Gaussian stationary field with $f(\lambda)$ as its spectral density is Gibbs field with potential U. This condition is generally satisfied in three and more dimensions, and is never satisfied when $\nu = 1, 2$. (For a detailed description of the Gaussian case see [4]).

## II. THE PROOF OF THE THEOREM

The proof of the theorem is similar to that of [1, 2]. So we shall discuss it briefly, except for those details which are specific for the noncompact case. In particular, we will consider only the case of pair interaction of nearest neighbors. The general case of finite range potential can be treated as in [1], and the case of a potential with condition C -- as in [2].

From now on we can omit the condition C and rewrite the condition D in the following form:

D'. Let $s, t \in \mathbb{Z}^2$, $|s - t| = 1$, $x_s, x_t \in X$ and $w$ any right-invariant tangent vector field on $G$ of unit length. Suppose that

$$|w w U_{\{s,t\}}(g x_s, x_t)| \leq L(w)$$

uniformly in $g, s, t, x_s, x_t$ with $L(w)$ bounded.

Before starting the proof we state some facts about factor-measures. The factor-measure $\tilde{\mu}$ of a measure $\mu$ with respect to t the action of a group $G$ on the space $(X, B)$ is a measure on the $\sigma$-subalgebra $\tilde{B} \subset B$ of $G$-invariant elements of $B$, such that for any measurable and integrable function $f$ on $X$ we have

$$\int_X f(x)\, d\mu = \int_X \tilde{f}(x)\, d\tilde{\mu} \tag{2}$$

where

$$\tilde{f}(x) = \int_G f(gx)\, d\chi(g) \tag{3}$$

is $\tilde{B}$-measurable, and $\chi$ the right-invariant Haar measure on $G$.

In [5] there were found necessary and sufficient conditions for the existence of the factor-measure. It is easy to see that our condition A implies the existence of the factor-measure. If we replace our group $G$ by its closed noncompact subgroup $H \subset G$, then our condition A remains valid with $H$ instead of $G$. This is a consequence of the fact that any closed noncompact subgroup of $G$ "goes to infinity". Because $G$ is noncompact and connected, it is possible to find some closed subgroup which is isomorphic to $\mathbb{R}^1$. So it is enough to study the particular case $G = \mathbb{R}^1$.

Now, let $V_n \subset \mathbb{Z}^2$ be a square $(2n+1) \times (2n+1)$, centered at the origin. As in [1], let

$F_0 = V_{n-k}$, $F_i = V_{n+i-k} \setminus V_{n+i-k-1}$, $i = 1, \ldots, k$, so $V_n = \bigcup_{i=0}^{k} F_i$.

Let $x_{V_n} \in X^{V_n}$ and $x_{F_i} = x_{V_n}|F_i$. There is an action of $\mathbb{R}^{k+1}$ on $X^{V_n}$:

$$b x_{V_n} = (b_0 x_{F_0}, \ldots, b_k x_{F_k}) \tag{4}$$

where

$$b = (b_0, \ldots, b_k), \quad (b_i x_{F_i})|_t = b_i(x_{F_i}|_t)$$

Let $p(x_{V_n}|x_{\bar{V}_n})$ be the density function (according to $\mu_{V_n}$) of the conditional distribution in $V_n$ with respect to the boundary condition $x_{\bar{V}_n}$,

$$p(x_{V_n}|x_{\bar{V}_n}) = \frac{\exp\{-U(x_{V_n}|x_{\bar{V}_n})\}}{\int \exp\{-U(x_{V_n}|x_{\bar{V}_n})\} d\mu_{V_n}}$$

We are interested in the projection of this distribution into $F_0$. Its density $p(x_{F_0}|x_{V_n})$ (with respect to $\mu_{F_0}$) is given by the formula

$$p(x_{F_0}|x_{V_n}) = \int_{X^{V_n \setminus F_0}} p(x_{F_0}, x_{V_n \setminus F_0}|x_{\bar{V}_n}) \, d\mu_{V_n \setminus F_0}$$

Let us consider

$$p(b_0|x_{F_0}, x_{\bar{V}_n}) = \frac{p(b_0 x_{F_0}|x_{\bar{V}_n})}{\int_{\mathbb{R}^1} p(b_0 x_{F_0}|x_{\bar{V}_n}) \, db_0}$$

The integral in this formula converges (almost everywhere) because of the existence of the factor-measure of the measure $\mu_{F_0}$ according to the $\mathbb{R}^1$ action on $X^{F_0}$. For the same reason, for any measurable and integrable $f$ on $X^{F_0}$, we have

$$\int_{X^{F_0}} f(x_{F_0}) p(x_{F_0} | x_{V_n}) d\mu_{F_0}$$

$$= \int_{X^{F_0}} \left( \int_{-\infty}^{+\infty} f(b_0 x_{F_0}) p(b_0 | x_{F_0}, x_{V_n}) db_0 \right) \left( \int_{-\infty}^{+\infty} p(b_0 x_{F_0} | x_{\overline{V}_n}) db_0 \right) d\tilde{\mu}_{F_0} , \tag{5}$$

analogous to the full probability formula. For proving the theorem it is enough to show that for any $N_1$, $N_2$, $-\infty < N_1 \le N_2 < +\infty$

$$\int_{N_1}^{N_1+N_2} p(b_0 | x_{F_0}, x_{\overline{V}_n}) db_0 \to 0 \tag{6}$$

when $n \to \infty$, uniformly in $x_{F_0}$.

Indeed, let us take $\prod_{t \in F_0} \varphi(x_t)$, $x_t = x_{F_0}|_t$ for $f(x_{F_0})$, where the function $\varphi$ is the same as in A. According to (6) and (5)

$$\int_{X^{F_0}} \prod_{t \in F_0} \varphi(x_t) p(x_{F_0} | x_{V_n}) d\mu_{F_0} \to 0 \tag{7}$$

when $n \to \infty$. For proving (7) let us take two numbers $N$, $\epsilon > 0$ and consider three subspaces of $X^{F_0}$:

$$X_1 = \{x \in X^{F_0}, \int_{\{\prod_{t \in F_0} \varphi(b_0 x|_t) > \epsilon\}} db_0 > N\},$$

$$X_2 = \{x \in X^{F_0}; \prod_{t \in F_0} \varphi(x|_t) < \epsilon\} \cap (X^{F_0} \setminus X_1),$$

$$X_3 = X^{F_0} \setminus \{X_1 \cup X_2\} .$$

Because of the existence of the factor-measure, all $X_i$ are measurable, and $\mu_{F_0}(X_1) \to 0$ when $N \to \infty$. Thus

$$\int_{X_2} \Big( \prod_{t \in F_0} \varphi(x|_t) \, p(x|x_{\overline{V}_n}) d\mu_{F_0}(x) < \epsilon ,$$

because the measure $p(x_{F_0}|x_{V_n})$ has total mass one, also the integral over $X_3$ goes to zero when $n \to \infty$ because of (6). Taking the limit $N \to \infty$, $\epsilon \to 0$, we have (7). Now if $p$ is the Gibbs state, and $\mathcal{P}_{F_0}$ its projection on $(X^{F_0}, \mathcal{B}_{F_0})$ then

$$\int_{X^{F_0}} \Big( \prod_{t \in F_0} \varphi(x|_t) d\mathcal{P}_{F_0}(x) = 0$$

But the measure $\mathcal{P}_{F_0}$ is absolutely continuous with respect to $\mu_{F_0}$ and the function $\prod_{t \in F_0} \varphi(x_{F_0}|_t)$ is positive $\mu_{F_0}$ -- almost everywhere. This contradiction implies, that (6) implies the Theorem. For proving (6) let us consider the probability distribution

$$p(b_0, b_1, \dots, b_k | x_{F_0}, x_{F_1}, \dots, x_{F_k}, x_{\overline{V}_n}) \, db_0 \dots db_k$$

$$= p(b|x_{V_n}, x_{\overline{V}_n}) db = p(bx_{V_n}|x_{\overline{V}_n}) \Big/ \int_{\mathbb{R}^{k+1}} p(bx_{V_n}|x_{\overline{V}_n}) db),$$

where $b = (b_0, b_1, \dots, b_k)$. Let us introduce new coordinates in $\mathbb{R}^{k+1}$ by the formula

$$r_i = b_i - b_{i+1}, \qquad i = 0, 1, \dots, k-1,$$

$$r_k = b_k \tag{8}$$

Because the only interacting particles are the nearest neighbors, the $r_i$-s are (as in [1]) independent:

$$p(r_0, r_1, \dots, r_k | x_{F_0}, x_{F_1}, \dots, x_{F_k}, x_{\overline{V}_n})$$

$$= p(r_0 | x_{F_0}, x_{F_1}) p(r_1 | x_{F_1}, x_{F_2}) \dots p(r_k | x_{F_k}, x_{\overline{V}_n}) \tag{9}$$

The interaction between the two layers $F_m$ and $F_{m+1}$ is

$$H(x_{F_m}, x_{F_{m+1}}) = \sum_{\substack{t \in F_m, s \in F_{m+1}, \\ |s-t| = 1}} U_{\{s,t\}}(x_s, x_t)$$

According to A, E, the function $\exp\{-H(rx_{F_m}, x_{F_{m+1}})\}$ is integrable in $r$, $r \in \mathbb{R}^1$ and by $D^1$

$$d^2\, dr^2\, H(rx_{F_m}, x_{F_{m+1}}) \leq c_1 m,$$

where $c_1 < \infty$. Hence, the function $H(\cdot x_{F_m}, x_{F_{m+1}})$ has a global minimum. As in [1] it follows that

$$p(r_m | x_{F_m}, x_{F_{m+1}}) \leq c_2 \sqrt{m} \tag{10}$$

From (8) and (9) it follows that

$$p(\cdot | x_{F_0}, x_{\overline{V}_n}) = p(\cdot | x_{F_0}, x_{F_1}) * p(\cdot | x_{F_1}, x_{F_2}) * \ldots * p(\cdot | x_{F_k}, x_{\overline{V}_n}),$$

where $*$ means convolution. So we can apply the technique of concentration functions of independent random variables to study the distribution of $b_0$.

The concentration function of the random variable (see [6, §3, 1, 3.2], or [7]) is defined as

$$Q(\xi, \lambda) = \sup_{x \in \mathbb{R}^1} p_r\{x \leq \xi < x + \lambda\}$$

Let $Q(\xi)$ be $Q(\xi, 1)$. For the random variable $\eta$ with the distribution function $G(x)$, put

$$F(x) = \int_{-\infty}^{+\infty} G(y - x)\, dG(y),$$

(the symmetrization of $G(x)$), and

$$D^*(\eta) = \int_{|x| < 1} x^2 dF(x) + \int_{|x| \geq 1} dF(x)$$

Now if $\xi = \eta_1 + \dots + \eta_k$ ($\eta_i$ are independent), then

$$Q(\xi) \le A\left(\sum_{i=1}^{k} D^*(\eta_i)\right)^{-1/2}$$

where $A > 0$ is a constant. (Esseen [7], see also [6, §3.2]). Let us apply (11) to our case. From (10) we have

$$D^*(r_m \mid x_{F_m}, x_{F_{m+1}}) \le c_3 m^{-1}$$

where $c_3 > 0$ is a constant. Hence

$$Q(b_0 \mid x_{F_0}, x_{\overline{V}_n}) \le A\left(\sum_{m=1}^{k} c_3/m\right)^{-1/2} \le c_4(\ln n)^{-1/2} \to 0$$

when $n \to \infty$, and the proof of (6) follows.

## REFERENCES


1. R. L. Dobrushin, S. B. Shlosman, Absence of Breakdown of Continuous Symmetry in Two-Dimensional Models of Statistical Physics. Comm. Math. Phys. 42 (1975), 31-40.
2. S. B. Shlosman, Absence of Continuous Symmetry Breaking in Two-Dimensional Models of Statistical Physics. Teor. Math. Phys. 33 (1977), 86-94. (In Russian).
3. U. A. Rozanov, Gaussian Fields with Given Conditional Distributions. Teor. Prob. Appl., 12 (1967), 433-443. (In Russian).
4. R. L. Dobrushin, Gaussian Random Fields - Gibbsian Point of View (see this volume).
5. R. L. Dobrushin, R. A. Minlos, Factor-measures on measure spaces. Trudi Mosc. Math. Soc., 32 (1975), 77-92. (In Russian).

6. V. V. Petrov, Sums of Independent Random Variables, "Nauka", Moscow, 1972.

7. C. -G. Esseen, On the Concentration Function of a Sum of Independent Random Variables. Z. Wahrsheinlichkeitsteorie Verw. Geb. 9 (1968), 290-308.

chapter 7

# ANALITICITY OF CORRELATION FUNCTIONS FOR LATTICE SYSTEMS WITH NONFINITE POTENTIAL IN THE MULTIPHASED CASE

V. M. Gertzik

## I. INTRODUCTION

There are two viewpoints concerning transitions. One of them is that phase transition occurs at points where the analyticity of thermodynamic and correlation functions breaks down and the other is that it occurs at points that lie on the boundary between domains that have different structures for the set of Gibbs states.

To confirm the hypothesis about the equivalence of these points of view it is necessary to prove in particular the analiticity in the Gibbs state non-unique domain for the lattice systems with the existence of such domain proved. For the systems with a finite potential i.e., the potential with the finite radius of the interaction) the analyticity in the multiphased domain follows from the fact that the system of Minlos-Sinai correlation equations for the contour assembly has ([1, 2]) the solution for the complex values of the potential and temperature parameters.

Attempts to use correlation equation methods for models with nonfinite potentials have been unsuccessful, however, so we have to use in this case a modification of the method used by R. L. Dobrushin ([3, 4]) for one dimensional systems. This method is on estimations of mean values of exponentional functionals on random fields. In this paper we prove by this method the analiticity of the correlation functions and Gibbs' free energy in the Gibbs state non-unique domain for the classical discrete lattice systems with a potential which is the sum of a finite potential which ensures phase transition and a sufficiently small nonfinite pair potential which decreases as $r^{-(\nu + \delta + 1)}$ where r is distance, $\nu$ is the dimension of the lattice and $\delta$ is an arbitrary positive constant. The necessary condition for applicability of this method is the existence of a Peierls' transformation which removes the contour without changing the rest of the configuration.

## II. THE MAIN RESULT

Let $T^\nu = \{t : t = (t^1, \ldots, t^\nu),\ t^k = 0, \pm 1, \ldots;\ k = 1, \ldots, \nu\}$ be the $\nu$-dimensional ($\nu = 2, 3$) integer lattice with the metric $p(t_1, t_2) = \max_{1 \leq k \leq \nu} |t_1^k - t_2^k|$, and let $X^{T^\nu}$ be the set of all functions $x \equiv (x_t, t \in T^\nu)$, $x_t \in X$, where X is a finite set. Let us call a function $U \equiv U(x, t_1, t_2)$, $t_1, t_2 \in T^\nu$, $x \in \lambda T^\nu$, taking complex values and the value $+\infty + 0 \cdot i$ ("real infinity") <u>a potential in the broad sense</u> if it satisfies the following conditions:

$$U(x, t_1, t_2) = U(x, t_2, t_1) \qquad \text{for all} \quad t_1, t_2 \in T^\nu,\ x \in X^{T^\nu};$$

$$Y^u = \{x : x \in X^{T^\nu},\ |U(x, t_1, t_2)| < \infty \quad \text{for all} \quad t_1, t_2 \in T^\nu\} \neq \phi;$$

$$\sum_{t' \in T^\nu} |U(x, t, t')| < \infty \qquad \text{for all} \quad x \in Y^U,\ t \in T^\nu .$$

If a potential in the broad sense $U(x, t_1, t_2)$ depends on $t_1, t_2$ and the values $x_{t_1}, x_{t_2}$ of a function x of the points $t_1, t_2$ only, we shall call it <u>a potential in the narrow sense</u> or just <u>a potential.</u>

We put $Y_V^U(x) = \{x' : x' \in Y^U,\ x'_t = x_1 \text{ if } t \in V\}$, $x \in Y^U$, $V \subset T^\nu$, and introduce the $\sigma$-algebra $\mathcal{B}$ on $Y^u$, generated by the sets $Y_V^U(x)$, $x \in Y^U$, $V \subset T^\nu$, $|V| < \infty$ ($|A|$ is the number of elements in the set A), and the $\sigma$-algebras $\mathcal{B}_V$, $V \subset T^\nu$, $|V| < \infty$, generated by the families of sets $\{Y_{\tilde{V}}^U : x \in Y^U, \tilde{V} \subset T^\nu \setminus V, |\tilde{V}| < \infty\}$ respectively. For any $x \in Y^U$, $B \in \mathcal{B}$, $V \subset T^\nu$, $|V| < \infty$, we assume

$$U_V(x) = \sum_{t \in V} U(x, t, t) + \sum_{\substack{\{t_1, t_2\} \subset T^\nu \\ \{t_1, t_2\} \cap V \neq \phi}} U(x, t_1, t_2) \tag{2.1}$$

$$Z_V^U(x) = \sum_{\tilde{x} \in Y^U_{T^\nu \setminus V}(x)} \exp\{-\beta U_V(\tilde{x})\} \tag{2.2}$$

and if $Z_V^U(x) \neq 0$

$$q_V^U(B|x) = [Z_V^U(x)]^{-1} \sum_{\tilde{x} \in B \cap Y^U_{T^\nu \setminus V}(x)} \exp\{-\beta U_V(\tilde{x})\} \tag{2.3}$$

where $\beta$ is a parameter, $0 < \beta < \infty$ ($\beta^{-1}$ has the physical interpretation of temperature). For real potentials U (i. e. for potentials taking real values and the value $+\infty$ only) we define now Gibbs measures on $Y^U$. Let us call a probability measure $P^U$ on $\{Y^U, \mathcal{B}\}$ a Gibbs state of the lattice system with the potential U, if the conditional distributions $P_x^U\{B|\mathcal{B}_V\}$, $x \in Y^U$, $B \in \mathcal{B}$, are defined almost everywhere by the equalities

$$P_x^U\{B|\mathcal{B}_V\} = q_V^U(B|x) \tag{2.4}$$

We shall call the probability measure $P_{V,x}^U$, $V \subset T^\nu$, $v \in Y^U$, on $\{Y^U, \mathcal{B}\}$ given by the equality

$$P_{V,x}^U\{B\} = q_V^U(B|x) \tag{2.5}$$

the Gibbs distribution in the finite vessel $V$ with boundary condition $x$. The probabilities $P^U\{Y_V^U(x)\}$, $x \in Y^U$, $V \subset T^\nu$, $|V| < \infty$, are usually called correlation functions. (As a rule Gibbs measures are defined on the whole space $X^{T^\nu}$ but as shown in Lemma 1 of [5], the set $Y^U$ has measure 1 for any Gibbs state on $X^{T^\nu}$, so that our definitions differ in no essential way from the usual ones).

Introduce the group of transformations $\bar{Q}$ acting on the sets $T^\nu$ and $X^{T^\nu}$ and generated by the groups $Q$ and $M^{T^\nu}$, where $Q$ is the group of all translations $q_t$, $t \in T^\nu$ acting by the rules $q_t t' = t + t'$ and $(q_t x)_{t'} = x_{t'-t}$ for all $t, t' \in T^\nu$, $x \in X^{T^\nu}$, and the group $M^{T^\nu}$ consists of the elements $m \equiv (m_t, t \in T^\nu)$, $m_t \in M$; where $M$ is the permutation group of the set $X$, and acts according to rules $mt = t$, $(mx)_t = m_t x_t$ for all $t \in T^\nu$, $x \in X^{T^\nu}$. Let $G$ be the subgroup of the group $\bar{Q}$ containing the group $Q_n = \{q_{nt}, t \in T^\nu\} \subset Q$, where $n > 0$, $nt = (nt^1, \dots, nt^\nu)$, and the set $\{x^i, i \in I\}$, $x^i \in X^{T^\nu}$, $|I| < \infty$, is the orbit of the group $G$ (i. e. $\{gx^j, g \in G\} = \{x^i, i \in I\}$ for all $j \in I$) such that $gx^i = x^i$ for all $i \in I$, $g \in Q_n$.

Introduce the real potential $U^o$ having the following properties:

$A_1$. $U^o(x, t_1, t_2) = U^o(gx, gt_1, gt_2)$ for all $x \in X^{T^\nu}$, $t_1, t_2 \in T^\nu$, $g \in G$.

$A_2$. $U^o(x^i, t_1, t_2) = 0$ for all $i \in I$, $t_1, t_2 \in T^\nu$.

$A_3$. There exists a natural number $\bar{n} > n$ such that $U^o(x, t_1, t_2) = 0$ for all $x \in X^{T^\nu}$, $t_1, t_2 \in T^\nu$, $\rho(t_1, t_2) \geq 2\bar{n}$.

$A_4$. For all $t_1, t_2 \in T^\nu$, $1 \leq \rho(t_1, t_2) < 2\bar{n}$, $x \in X^{T^\nu}$ the function $U^o(x, t_1, t_2)$ takes values $0$ or $\infty$ and

(a) if $x, \tilde{x} \in X^{T^\nu}$ are such that $x_{t_1} = \tilde{x}_{t_1}$, $x_{t_2} = x^i_{t_2}$, $\tilde{x}_{t_2} = x^j_{t_2}$, $i, j \in I$, $i \neq j$, the equality $U^o(x, t_1, t_2) = 0$ implies $U^o(x, t_1, t_2) = \infty$;

(b) $U^o(x, t_1, t_2) = 0$ if $U^o(x, t'_j, t'_{j+1}) = 0$ for any sequence $t'_1, \dots, t'_k$, $k \leq 2\bar{n}$, $t'_1 = t_1$, $t'_k = t_2$, $\rho(t'_j, t'_{j+1}) = 1$, $j = 1, \dots, k-1$.

$A_5$. Let $X_0$ be the set of the values of the functions $x^i$, $i \in I$, and $N(x) = \{t : t \in T^\nu, x_t \in X \ X_0\}$. If $x \in Y^{U^o}$ and $|N(x)| < \infty$ then $U^o_{N(x)}(x) \geq a|N(x)|$, a is a positive constant. (This is the <u>Peierls' condition</u> with respect to "the <u>ground states</u>" $x^i$, $i \in I$.)

<u>Notation.</u> It was shown in [5] that if the finite potential of some lattice system is invariant under the action of the group G (in the sense of condition $A_1$) and the Peierls' condition analogous to $A_5$ is fulfilled with respect to the finite set of ground states, then there exists a lattice system with a $U^o$ - type potential equivalent to the original system. (The simplest examples of such systems are Ising's ferromagnet and antiferromagnet). To guild this new system we must take the lattice $S^\nu = T^\nu + \frac{e}{2}$ (e is the vector with unit co-ordinates) and assume that the generic function x on this lattice takes values $x_t$, $t \in S^\nu$ in the set of configurations of the original system in the square with center at t and with side length $2\bar{n}$, where $\bar{n}$ is the sum of the radius of the interaction and the mutual period of the potential and the ground states. Then the potential must be renormed as in [5], and the infinite values of the pair interaction are introduced for the points $t_1, t_2 \in S^\nu$ according to the intersecting squares if the values $x_{t_1}, x_{t_2}$ define the different configurations of the original system in the mutual region of these squares. The last operation gives rise to condition $A_4$. The addition of a small nonfinite pair potential which is invariant under the action of the group G and decreases as $\tau^{-(1+\nu+\delta)}$ to the original (finite) potential involves as one can easily show the addition to the potential of the new system a potnetial which is described by conditions $B_1 - B_3$ given below.

Let us call a set $V \subset T^\nu$ a connected set if for any $t$, $t' \in V$ there exists a finite set $t_j \in V$, $j = 1, \ldots, k$, such that $t_1 = t$, $t^k = t'$,

$\rho(t_j, t_{j+1}) = 1$ for all $j = 1, \dots, k-1$, and introduce for any $\epsilon > 0$, $\delta > 0$, $V \subset T^\nu$ the family $F_{\epsilon,\delta}(V)$ of complex-valued potentials in the broad sense such that any $\Phi \in F_{\epsilon,\delta}(V)$ has the following properties:

$B_1$. $\Phi(x^i, t, t) = 0$ for any $t \in T^\nu$, $i \in I$.

$B_2$. $$\sup_{x \in X^{T^\nu}} \max_{\substack{t_1, t_2 \in T^\nu \\ \rho(t_1, t_2) = \ell}} |\Phi(x, t_1, t_2)| \le \epsilon \min\{1, \ell^{-(\nu+1+\delta)}\}, \quad \ell = 0, 1, \dots .$$

$B_3$. For any connected component $\tilde{V}$ of the set $V$, any $t_1, t_2 \in \tilde{V}$, $g \in G$ such that $gt_1, gt_2 \in \tilde{V}$, and any $x \in X^{T^\nu}$

$$\Phi(x, t_1, t_2) = \Phi(gx, gt_1, gt_2)$$

We introduce also a potential $\Phi^\zeta$ depending on the complex parameter $\zeta \in W$, where $W$ is a neighborhood of the origin in n-dimensional complex space $C^n$, such that $\Phi^\zeta \in F_{\epsilon,\delta}(T^\nu)$ for any $\zeta \in W$ and for any fixed $t_1, t_2 \subset T^\nu$, $x \in X^{T^\nu}$, $\Phi^\zeta(x, t_1, t_2)$ is a holomorphic function of $\zeta \in W$, real if $\zeta$ is real and equal to zero if $\zeta = 0$ (i. e. $\zeta$ is the origin of $C^n$).

Below we shall write $U^\zeta = U^0 + \Phi^\zeta$ and for the case $U = U^\zeta$ we shall replace the upper index $U$ in the terms defined by the formulae (2. 2) - (2. 5) by the index $\zeta$, e. g. $Z_V^{U^\zeta}(x) \equiv Z_V^\zeta(x)$. Noting that $Y_V^{U^0+\Phi}(x) = Y_V^{U^0}(x)$, $Y^{U^0+\Phi} = Y^{U^0}$ for any $x \in Y^{U^0}$, $V \subset T^\nu$, $\Phi \in F_{\epsilon,\delta}(\tilde{V})$, $\tilde{V} \subset T^\nu$, we shall put $Y^{U^0} \equiv Y$, $Y_V^{U^0}(x) \equiv Y_V(x)$.

<u>Theorem</u>. <u>Let</u> $\epsilon \equiv \epsilon(a, \delta, \nu)$ <u>and</u> $\beta_0^{-1} \equiv [\beta_0(\epsilon, a, \nu)]^{-1}$ <u>be sufficiently small constants and</u> $\beta > \beta_0$. <u>Then for any</u> $i \in I$ <u>there exists a sequence of finite boxes</u> $V_k^i$, $V_k^i \subset T^\nu$. $V_k^i \subset V_{k+1}^i$, $k = 1, 2, \dots$, $\bigcup_{k=1}^\infty V_k^i = T^\nu$, <u>such that for any</u> $V \subset T^\nu$. $|\bar{V}| < \infty$, $\bar{x} \in Y$

$$q^\zeta(Y_{\bar{V}}(\bar{x})|x^i) = \lim_{k \to \infty} q^\zeta_{V_k^i}(Y_{\bar{V}}(\bar{x})|x^i) \tag{2. 6}$$

exists and is analytic on W, and the set of Gibbs states of the system with the potential $U^{\zeta}$ contains for any real $\zeta \in W$ the probability measure $P^{\zeta, i}$ such that

$$P^{\zeta, i}\{Y_{\overline{V}}(\overline{x})\} = q^{\zeta}(Y_{\overline{V}}(\overline{x})\,|x^{i}) \tag{2.7}$$

and

$$P^{\zeta, i}\{Y_{\overline{V}}(x^{i})\} \geq 1 - \alpha_{\overline{V}}(\beta) \tag{2.8}$$

where $\lim_{\beta \to \infty} \alpha_{\overline{V}}(\beta) = 0$, $\alpha_{\overline{V}}(\beta) > 0$.

In addition, for any sequence $V_k \subset T^{\nu}$, $|V_k| < \infty$, $V_k \subset V_{k+1}$, $k = 1, 2, \ldots,$ such that $\bigcup_{k=1}^{\infty} V_k = T^{\nu}$, $\lim_{k \to \infty} \frac{|\Delta_k|}{|V_k|} = 0$, $\Delta_k = \{t : t \in V_k,\ \rho(T^{\nu} \setminus V_k, t) = 1\}$ and such that the set $T^{\nu} \setminus V_k$ is connected for any $k$, the Gibbs free energy $F^{\zeta} = \lim_{k \to \infty} |V_k|^{-1} \ln Z^{\zeta}_{V_k}(x^{i}$ exists, is analytic on W and does not depend on i.

In particular for the system with the real potential $U = U^{o} + \Phi$, $\Phi \in F_{\epsilon_1, \delta}(T^{\nu})$, $\epsilon_1 < \epsilon$, the analyticity of the correlation functions and the free energy with respect to the parameter $U(\tilde{x}, t_1', t_2')$ for any fixed $\tilde{x}$, $t_1', t_2'$ is a consequence of the theorem if we put

$$\Phi^{\zeta}(x, t_1, t_2) = \begin{cases} (1+\zeta)\,\Phi(x, t_1, t_2) + \zeta U_0(x, t_1, t_2) & \\ \qquad\qquad \text{if } x = g\tilde{x},\ t_1 = gt_1',\ t_2 = gt_2',\ g \in G; & \\ \Phi(x, t_1, t_2) \qquad \text{otherwise} & \end{cases}$$

$$W = \{\zeta : \zeta \in C^{1},\ |\zeta| \leq (\epsilon - \epsilon_1)[\epsilon_1 + |U^{o}(\tilde{x}, t_1', t_2')|\,]^{-1}\}$$

The analyticity with respect to $\beta$ for this system also follows from the theorem; take

$$\Phi^{\zeta} = \beta_1^{-1}\zeta(U^{o} + \Phi) + \Phi, \quad \zeta \in C^{1}, \quad \beta_1 > \beta_0 .$$

Evidently, $\beta_1 U^{\tilde{\zeta}} = \beta U = (\beta_1 + \tilde{\zeta})U$ for $\tilde{\zeta} = \beta - \beta_1$ and analyticity takes place in the circle

$$|\zeta| \leq \min\{\beta_1 - \beta_0 ; \beta_1(\epsilon - \epsilon_1)[\epsilon_1 + \sup_{x,t} |U^0(x, t, t)|]^{-1}\}$$

With some modification of the details of our proof one can prove the analogous results for the lattice system with nonfinite potential introduced by R. L. Dobrushin in [6].

The question about analyticity (and about nonuniqueness of the Gibbs state) for lattice systems with nonfinite potentials is open for the case when the set of ground states is not an orbit of the group G and for the case when the group G includes revolutions and reflections (e. g. for the case when the potential is invariant under the action of some revolution but the ground states are not). For these cases the difficulty arises from the absence of the Peierl's transformation.

Certain essential parts of our proof are extensions of the technique used by Dobrushin for the one-dimensional case [3, 4] to the case of contour models. (In contrast to the one-dimensional case the analyticity of the correlation functions in the multiphased situation does not follow from the analyticity of the free energy since the conditions $A_1$ and $\Phi \in F_{\epsilon,\delta}(T^\nu)$ do not allow one to get the correlation functions by differentiating the free energy function as was done in [4]). The main point of the proof is Lemma 1 and its consequence, Lemma 2.

## III. THE PROOF OF THE THEOREM

The existence of Gibbs distributions $P^{\zeta i}$ satisfying conditions (2. 7) - (2. 8) can be proved by standard methods (see, e. g. [7]), so we omit that part of the proof. Most of its elements can be found in the subsequent arguments given below. By the compactness of the space of probability measures on $\{Y, \mathcal{B}\}$ we can choose, using the diagonal process for any $i \in I$, a sequence of increasing boxes $V_k^i$, $|V_k^i| < \infty$, $k = 1, 2, \ldots$, $\bigcup_{k=1}^{\infty} V_k^i = T^{\nu}$, such that the limits (2.6) exist on a countable set of values of $\zeta$ dense in the intersection of the region $W$ and the real subspace of $C^n$.

Furthermore we follow Dobrushin([3, 4]), and use the multidimensional generalization of Vitali's theorem, which follows from the compactness of the space of bounded holomorphic functions (cf. [9], Theorem 1. A. 12) and the theorem that a holomorphic function which vanishes for real values of its arguments is identically zero (cf. [10], p. 286). Vitali's theorem ensures the existence and the analiticity of the limits $q^{\zeta}(Y_{\overline{V}}(\overline{x})|x^i)$, $\zeta \in W$, $i \in I$, provided that for any $i \in I$, $\zeta \in W$, any $V \subset T^{\nu}$, $|V| < \infty$, such that $T^{\nu} \setminus V$ is a connected set and for any $\overline{x} \in Y$, $\overline{V} \subset V$, the inequalities

$$|\ln \Xi_V^{\zeta}(x^i)| < \infty, \tag{3. 1}$$

$$|q_V^{\zeta}(Y_{\overline{V}}(\overline{x})|x^i)| \leq K(\beta, |\overline{V}|) \tag{3. 2}$$

hold, where $K(\beta, |V|)$ is a real positive function of $\beta$ and $|V|$. (Without loss of generality we can assume that each $T^{\nu} \setminus V_k^i$ is a connected set). That the inequality (3. 1) implies t e existence and the analyticity of the functions $q_V^{\zeta}(Y_{\overline{V}}(\overline{x})|x^i)$ is a simple consequence of Lemma 1, which will be proved below. We shall give the proof of (3. 1) at the end of this section together with the proof of the analiticity of the free energy, but we assume for now that all of $q_V^{\zeta}(Y_{\overline{V}}(\overline{x})|x^i)$ are defined. The main part of the paper deals with the proof of (3. 2).

We define <u>the cycle of a function</u> $x \in Y$ as a connected component $\Gamma$ of the set $N(x) = \{t : t \in T^{\nu}, x_t \in X \setminus X_0\}$, $|\Gamma| < \infty$, and introduce the set $Y(\Gamma) \subset Y$ of functions having $\Gamma$ as a cycle. The set $Y(\Gamma)$ can be divided into subsets $Y_s(\Gamma)$, $s = 1, \ldots, s(\Gamma)$, each of which contains all the functions having the same values on the set $\Gamma^{(+)} = \{t : t \in T^\nu,\ \rho(t, \Gamma) \leq 1\}$. Let $v_k(\Gamma)$, $k = 0, \ldots, m(\Gamma)$, be the connected components of $T^\nu \setminus \Gamma^{(+)}$, with $|v_k(\Gamma)| < \infty$ if $k > 0$, $|v_0(\Gamma)| = \infty$. The conditions $A_4$(a), $A_2$ and the definition of a cycle imply for any $x \in Y_s(\Gamma)$, $s = 1, \ldots, s(\Gamma)$, $t \in \delta_k$, $k = 0, \ldots, m(\Gamma)$, $\delta_k = \{t : t \in T^\nu,\ \rho(t, v_k(\Gamma)) = 1\}$, the equality $x_t = x_t^{j(s,k)}$, $j(s,k) \in I$. (The indexes of the ground states $j(s,k)$ are essentially defined by the values of $x$ on the set $\Gamma$. We define the action of the group $G$ on $I$ by the rule $gi = j, g \in G$, $i, j \in I$, if $gx^i = x^j$.

Let $g_\Gamma(s,k)$ be the elements from $G$ that take the origin of $T^\nu$ into the positive orthant and such that $g_\Gamma(s,k)j(s,k) = j(s,0)$, $\rho(t, g_\Gamma(s,k)t) < n$. (The existence of such elements follows from the definitions of $\{x^i, i \in I\}$ and $G$. In particular $g_\Gamma(s,0)$ is unit of $G$.) For functions from $Y(\Gamma)$ we define the Peierls' transformation $\pi_\Gamma$ acting on $x \in Y_s(\Gamma)$, $s = 1, \ldots, s(\Gamma)$ by the rule

$$\pi_\Gamma x_t = \begin{cases} g_\Gamma(s,k)x_{g_\Gamma(s,k)^{-1}t} & \text{if } t \in g_\Gamma(s,k)v_k(\Gamma),\ k = 0, \ldots, m(\Gamma); \\ x_t^{j(s,0)} & \text{otherwise} \end{cases} \tag{3.3}$$

It follows from condition $A_4$(a) and the definition of $g_\Gamma(s,k)$ that the different sets $g_\Gamma(s,k)v_k(\Gamma)$ have empty intersections (since "the width of the walls" of $\Gamma$ dividing the sets $v_k(\Gamma)$ and $v_{k'}(\Gamma)$ with different $j(s,k)$, $j(s,k')$ is not less than $\bar{n}$); therefore the definition of $\pi_\Gamma$ is correct. The conditions $A_2, A_3, A_4$(b) imply that $\pi_\Gamma x \in Y$ for any $x \in Y(\Gamma)$. Furthermore, the mapping $Y_s(\Gamma) \to \pi_\Gamma Y_s(\Gamma)$ is one-to-one, and it follows from $A_1, A_2$ and $A_5$ that $\overset{\circ}{U}_V(x) - \overset{\circ}{U}_V(\pi_\Gamma x) \geq a|\Gamma|$ for any $V \subset T^\nu$, $\Gamma \subset V$, $x \in Y(\Gamma)$.

We put $\bar{v}_0(\Gamma) = v_0(\Gamma) \cup \delta_0(\Gamma)$ for any connected $\Gamma \subset T^\nu$, $|\Gamma| < \infty$, and introduce for any $V_1, V_2 \subset T^\nu$, $V_2 \subset V_1$, $|V_2| < \infty$, the family $A_{V_1}(V_2)$ of sets $\gamma \subset V_1$, $|\gamma| < \infty$, such that $V_2 \cap (T^\nu \setminus \bar{v}_0(\Gamma)) \neq \phi$ for any connected component $\Gamma$ of $\gamma \in A_{V_1}(V_2)$.

Let $\Gamma_1, \ldots, \Gamma_{J(\gamma)}$ be the indexed family of connected components $\gamma \in A_V(\bar{V})$. We introduce the set $Y^{\bar{V}}(\gamma)$ of all functions from $\bigcap_{n=1}^{J(\gamma)} Y(\Gamma_n)$, such that $V \subset \bar{v}_0(\Gamma)$ for any cycle $\Gamma \subset T^\nu \setminus \gamma$ of the function $x \in Y^{\bar{V}}(\gamma)$. The definition (2. 3) implies the evident inequality

$$|q_V^\zeta(Y_{\bar{V}}(\bar{x})\,|x^i)| \leq |q_V^\zeta(Y_{\bar{V}}(\bar{x}) \setminus \bigcup_{\gamma \in A_V(\bar{V})} Y^{\bar{V}}(\gamma)|x^i)| + \sum_{\gamma \in A_V(\bar{V})} |q_V^\zeta(Y^{\bar{V}}(\gamma) \cap Y_{\bar{V}}(\bar{x})\,|x^i)| \tag{3. 4}$$

(Note that some sets of functions in expressions such as (3. 4) may be empty and $q_V^U(\phi|x) = 0$ for any complex $U$ and any $x \in Y^U$.)

Let $\{s_n^j,\ n = 1, \ldots, J(\gamma)\}$, $j = 1, \ldots, s(\gamma)$ be different sequences of indexes $s_n^j = 1, \ldots, s(\Gamma_n)$. Evidently $s(\gamma) = \Pi_{n=1}^{J(\gamma)} s(\Gamma_n)$. We put

$$Y_j = Y^{\bar{V}}(\gamma) \cap Y_{\bar{V}}(\bar{x}) \cap Y_{T^\nu \setminus V}(x^i) \cap (\bigcap_{n=1}^{J(\gamma)} Y_{s_n^j}(\Gamma_n))$$

for $j = 1, \ldots, s(\gamma)$, and using the evident properties

$$q_V^\zeta(Y^{\bar{V}}(\gamma) \cap Y_{\bar{V}}(\bar{x})\,|x^i) = \sum_{j=1}^{s(\gamma)} q_V^\zeta(Y_j\,|x^i)$$

and $s(\Gamma) \leq |X \setminus X_0|^{|\Gamma|}$ for any cycle $\Gamma$ we have the estimate:

$$|q_V^\zeta(Y^{\bar{V}}(\gamma) \cap Y_{\bar{V}}(\bar{x})\,|x^i)| \leq |X \setminus X_0|^{|\gamma|} \sup_{1 \leq j \leq s(\gamma)} |q_V^\zeta(Y_j\,|x^i)| \tag{3. 5}$$

Further as in the method used by R. L. Dobrushin for the proof of analyticity in the one-dimensional case, we shall consider mean values of exponential functionals of $\exp\{-\beta\Phi_V(x)\}$ kind. We shall carry out the proof by estimating the moduli of rations of such mean values for different $\Phi$ and $V$. More precisely, for any $W \subset T^\nu$, $|W| < \infty$, $\tilde{W} \subset T^\nu$, $\Phi \in F_{\epsilon,\delta}(\tilde{W})$ and any $x \in Y$ such that $\rho(W, N(x)) > 1$ we put

$$g_W(x \mid \Phi) = \sum_{\tilde{x} \in \tilde{Y}_{T^\nu \setminus W}(x)} \exp\{-\beta\Phi_W(\tilde{x})\}\tilde{q}^0_W(Y_W(\tilde{x}) \mid x) \tag{3.6}$$

where $\tilde{Y}_{T^\nu \setminus W}(x)$ is the set of functions from $Y_{T^\nu \setminus W}(x)$ such that for any cycle $\Gamma \subset W$ of any function $\tilde{x} \in \tilde{Y}_{T^\nu \setminus W}$, the inclusion $T^\nu \setminus W \subset v_0(\Gamma)$ holds (i. e. the cycle does not "envelope the holes of the vesset" W), and where $\tilde{q}^0_W(\cdot \mid x)$ are defined by (2. 2) and (2. 3) with $U = U^0$ and $Y_{T^\nu \setminus W}(x)$ replaced by $\tilde{Y}_{T^\nu \setminus W}(x)$. (Note that the connectivity of $T^\nu \setminus V$ implies the equalities $Y_{T^\nu \setminus V}(x^i) = \tilde{Y}_{T^\nu \setminus V}(x^i)$ and $q^0_V(\cdot \mid x^i) = \tilde{q}^0_V(\cdot \mid x^i)$.) We put also $g_{\{\phi\}}(x \mid \Phi) = 1$.

Using the designations $\overline{V}_\gamma = \overline{V} \cup \{\bigcup_{n=1}^{J(\gamma)} \Gamma_n^{(+)}\}$ for any $\gamma \in A_V(\overline{V})$ and $V_\gamma = V \setminus \overline{V}_\gamma$ we write for any $j$, $1 \le j \le s(\gamma)$, such that $Y_j \ne \phi$, the following equality, which is implied by (2. 3), (3. 6) and by the definition of $U^\zeta$:

$$q^\zeta_V(Y_j \mid x^i) = q^0_V(Y_j \mid x^i)\frac{g_{V_\gamma}(x^{(j)} \mid \Phi^\zeta)}{g_V(x^i \mid \Phi^\zeta)} \times \exp\{-\beta[\Phi^\zeta_V(x^{(j)}) - \Phi^\zeta_{V_\gamma}(x^{(j)})]\} \tag{3.7}$$

where $x^{(j)} \in Y_j$ is such that $\rho(V_\gamma, N(x^{(j)})) > 1$ (the existence and the uniqueness of $x^{(j)}$ follow from the definition of $Y_j$). Analogously we have the equality

$$q_V^{\zeta}(Y_{\overline{V}}(x) \setminus \bigcup_{\gamma \in A_V(\overline{V})} Y^{\overline{V}}(\gamma) \mid x^i)$$

$$= q_V^0(\tilde{Y}_{T^{\nu} \setminus (V \overline{V})}(x^i) \mid x^i) \frac{g_{V \setminus \overline{V}}(x^i \mid \Phi^{\zeta})}{g_V(x^i \mid \Phi^{\zeta})}$$

$$\times \exp\{-\beta[\Phi_V^{\zeta}(x^i) - \Phi_{V \setminus V}^{\zeta}(x^i)]\} \tag{3. 8}$$

which is correct if $\overline{x}_t = x_t^i$ for any $t \in \overline{V}$ (in the contrary case the l left side of (3. 8) is zero). The condition $B_2$ and (2. 1) imply for any $\tilde{V} \subset T^{\nu}$, $\Phi \in F_{\epsilon, \delta}(\tilde{V})$, $V_1, V_2 \subset T^{\nu}$, $V_2 \subset V_1$, $|V_1| < \infty$, $x \in X^{T^{\nu}}$ the estimate

$$|\Phi_{V_1}(x) - \Phi_{V_2}(x)| \leq \epsilon C |V_1 \setminus V_2| \tag{3. 9}$$

where $C = 1 + 2\nu 2^{\nu - 1} \sum_{k=1}^{\infty} k^{-(2+\delta)}$. Using (3. 9) and the fact that $|\overline{V}_{\gamma}| \leq |\overline{V}| + 3^{\nu} |\gamma|$ we obtain for the moduli of the exponential terms on the right of (3. 7) and (3. 8) the upper bounds $\exp\{\beta \epsilon C(|\overline{V}| + 3^{\nu}|\gamma|\}$ and $\exp\{\beta \epsilon C |\overline{V}|\}$ respectively. Furthermore, since $q_V^0(\cdot \mid x)$ are real, it follows from Definition (2. 3) and the estimate (3. 3) that

$$q_V^0(\tilde{Y}_{T^{\nu} \setminus (V \setminus \overline{V})}(x^i) \mid x^i) \leq 1;$$

$$q_V^0(Y_j \mid x^i) \leq \frac{\sum_{x \in Y_j} \exp\{-\beta U_V^0(x)\}}{\sum_{x \in \pi_{\gamma} Y_j} \exp\{-\beta U_V^0(x)\}} \leq \exp\{-\beta a |\gamma|\} \tag{3. 10}$$

where $\pi_{\gamma} = \pi_{\Gamma_{J(\gamma)}} \circ \dots \circ \pi_{\Gamma_1}$. Combining the last four estimates we find with the help of (3. 7), (3. 8), (3. 4) and (3. 5) that

$$|q_V^\zeta(Y_{\overline{V}}(\overline{x})\,|x^i)| \le \exp\{\beta\epsilon C|\overline{V}|\}$$

$$\times\Big[\sum_{\gamma\in A_V(\overline{V})} e^{-a_1\beta|\gamma|} \sup_{j:1\le j\le s(\gamma),Y_j\ne\phi}\left|\frac{g_{V_\gamma}(x^{(j)}|\Phi^\zeta)}{g_V(x^i|\Phi^\zeta)}\right|$$

$$+\left|\frac{g_{V\setminus\overline{V}}(x^i|\Phi^\zeta)}{g_V(x^i|\Phi^\zeta)}\right|\Big] \qquad (3.11)$$

where $a_1 = a - 3^\nu\epsilon C - \beta_0^{-1}\ln|X\setminus X_0|$, $a_1 > 0$ for sufficiently small $\epsilon$ and $\beta_0^{-1}$.

Furthermore we shall need the next result, the proof of which will be given in section IV.

Lemma 1. Let $\epsilon$ and $\beta^{-1}$ be sufficiently small, $\beta > \beta_0$, $V\subset T^\nu$, $|V| < \infty$, and the function $\tilde{x}\in Y$ be such that $\rho(N(\tilde{x}), V) > 1$.

(a) If $V'\subset V$ and $\in F_{\epsilon,\delta}(V)$ then

$$\left|\ln\frac{g_{V'}(\tilde{x}|\Phi)}{g_V(\tilde{x}|\Phi)}\right| \le (\beta\epsilon C + \tfrac{1}{2})\,|V\setminus V'|$$

(b) If $V_1\subset V$, $V_2\subset T^\nu$ $V_1$, $V_1$ and $V_2\cap V$ are unions of connected components of $V$ and $\Phi^{(1)}, \Phi^{(2)}\in F_{\epsilon,\delta}(V)$ are such that inequality $\Phi^{(1)}(x, t_1, t_2) \ne \Phi^{(2)}(x, t_1, t_2)$ holds only when $t_1\in V_1$, $t_2\in V_2$, then

$$\left|\ln\frac{g_V(\tilde{x}\,|\,\Phi^{(1)})}{g_V(\tilde{x}\,|\,\Phi^{(2)})}\right| \le \beta\epsilon\,[\,3D(V_1, V_2\setminus V) + 6D(V_1, V_2\cap V)\,]$$

where for any $V', V''\subset T^\nu$

$$D(V', V'') = \sum_{t_1\in V',\, t_2\in V''} \rho(t_1, t_2)^{-(\nu+1+\delta)}$$

The statement (a) of Lemma 1 implies immediately

$$\left|\frac{g_{V\setminus\overline{V}}(x^i \mid \Phi^\zeta)}{g_V(x^i \mid \Phi^\zeta)}\right| \leq \exp\{(\beta \in C + \frac{1}{2})|\overline{V}|\} \tag{3.12}$$

Let $v_\ell$, $\ell = 0, 1, \ldots, m(\gamma)$, be numbered connected components of $V_\gamma$, $\gamma \in A_V(\overline{V})$ and $v_0 \subset \bigcap_{n=1}^{J(\gamma)} v_i(\Gamma_n)$. For any $\ell = 0, 1, \ldots, m(\gamma)$ and $n = 1, \ldots, J(\gamma)$ we introduce the numbers $k_{n,\ell} = 1, \ldots, m(\Gamma_n)$, such that $v_\ell \subset v_{k_{n,\ell}}(\Gamma_n)$. We introduce also the mappings $\pi_n = \pi_{\Gamma_0} \circ \ldots \circ \pi_{\Gamma_1}$, $n = 1, \ldots, J(\gamma)$, the identity mapping $\pi_0$, and for any $j = 1, \ldots, s(\gamma)$ such that $Y_j \neq \Phi$, the families of indexes $\{\tilde{s}^j_n$, $n = 1, \ldots, J(\gamma)\}$, $1 \leq \tilde{s}^j_n = s(\Gamma_n)$ such that $\pi_{n-1} x^{(j)} \in Y_{\tilde{s}^j_n}(\Gamma_n)$. We define the transformations $\tau(n, j)$, $n = 0, 1, \ldots, J(\gamma)$ of $V_\gamma$ by the rules $\tau(0, j)V_\gamma = V_\gamma$, $\tau(n, j)V_\gamma = \bigcup_{\ell=1}^{m(\gamma)} \tau(n, j)v_\ell$ for $n \geq 1$, where $\tau(n, j)v_\ell = g_{\Gamma_n}(\tilde{s}^j_n, k_{n,\ell}) \ldots g(\tilde{s}^j_1, k_{1,\ell})v_\ell$. Noting that $\pi_\gamma x^{(j)} = x^i$ we can write

$$\frac{g_{V_\gamma}(x^{(j)} \mid \Phi^\zeta)}{g_V(x^i \mid \Phi^\zeta)} = \frac{g_{\tau(J(\gamma), j)V_\gamma}(x^i \mid \Phi^\zeta)}{g_V(x^i \mid \Phi^\zeta)}$$

$$\times \prod_{n=0}^{J(\gamma)+1} \frac{g_{\tau(n,j)V_\gamma}(\pi_n x^{(j)} \mid \Phi^\zeta)}{g_{\tau(n+1,j)V_\gamma}(\pi_{n+1} x^{(j)} \mid \Phi^\zeta)} \tag{3.13}$$

Since it is evident that $|V \setminus \tau(J(\gamma), j)V_\gamma| = |\overline{V}_\gamma| \leq 3^\nu |\gamma| + |\overline{V}|$, the statement (a) of Lemma 1 implies the estimate

$$\left|\frac{g_{\tau(J(\gamma), j)V_\gamma}(x^i \mid \Phi^\zeta)}{g_V(x^i \mid \Phi^\zeta)}\right| \leq \exp\{(\beta \in C + \frac{1}{2})(3^\nu |\gamma| + |\overline{V}|)\} \tag{3.14}$$

To estimate the factors of the product on the right side of (3.13) we shall use Lemma 1 to prove the following result.

Lemma 2. *Let the following be given: the connected set* $\Gamma$, $|\Gamma|<\infty$, $Y(\Gamma) \neq \phi$, *the function* $x \in Y_s(\Gamma)$ *for some* $s$, $s = 1,\ldots, s(\Gamma)$, *the set* $V \subset T^\nu$, $|V| < \infty$, *such that* $\rho(N(x), V) > 1$, *and the potential in the broad sense* $\Phi \in F_{\epsilon,\delta}((T^\nu \setminus v_0(\Gamma)) \cup V)$. *We put* $\tau_\Gamma V = \bigcup_{k=0}^{m(\Gamma)} g_\Gamma(s,k)V_{(k)}$, $V_{(k)} = V \cap v_k(\Gamma)$, $k = 0,\ldots, m(\Gamma)$. *Then for sufficiently small* $\epsilon$ *and* $\beta_0^{-1}$, $\beta > \beta_0$,

$$\left|\ln\frac{g_V(\pi\,|\,\Phi)}{g_{\tau_\Gamma V}(\pi_\Gamma x\,|\,\Phi)}\right| \leq K\beta\epsilon\,|\Gamma|$$

*where* $K$ *is a positive constant.*

To prove this Lemma we use a method similar to the method applied by Dobrushin in [3], [4] and introduce the family of potentials in the broad sense $\Phi^k$, $k = 0,\ldots, m(\Gamma)+1$, where $\Phi^0 = \Phi$, and for $k \geq 1$

$$\Phi^k(x, t_1, t_2) = \begin{cases} \Phi(\pi_\Gamma x, g_\Gamma(s,k-1)t_1, g_\Gamma(s,k-1)t_2) \\ \qquad \text{for } x \in Y_s(\Gamma),\ \{t_1,t_2\} \cap v_{k-1}(\Gamma) \neq \phi, \\ \{t_1,t_2\} \subset g_\Gamma(s,k-1)^{-1}\{T^\nu \setminus \bigcup_{n<k-1} g_\Gamma(s,n)v_n(\Gamma)\}; \\ \Phi^{k-1}(x, t_1, t_2) \qquad \text{otherwise} \end{cases}$$

Evidently $\Phi_V^{m(\Gamma)+1}(x) = \Phi_{\tau_\Gamma V}(\pi_\Phi x)$ for any $x \in Y_s(\Gamma)$ and it follows from the properties of $U^0$ and $\pi_\Gamma$ that $\tilde{q}_V^0(Y_V(x)|\tilde{x}) = \tilde{q}^0_{\tau_\Gamma V}(Y_{\tau_\Gamma V}(\pi_\Gamma x)|\pi_\Gamma\tilde{x})$ for any $x \in Y_{T^\nu\setminus V}(\tilde{x})$, therefore $g_V(\tilde{x}|\ {}^{m(\Gamma)+1}) = g_{\tau_\Gamma V}(\pi_\Gamma\tilde{x}|\Phi)$ and

$$\frac{g_V(\tilde{x}\,|\,\Phi)}{g_{\tau_\Gamma V}(\pi_\Gamma \tilde{x}\,|\,\Phi)} = \prod_{k=0}^{m(\Gamma)} \frac{g_V(\tilde{x}\,|\,\Phi^k)}{g_V(\tilde{x}\,|\,\Phi^{k+1})} \tag{3.15}$$

The conditions of statement (b) of Lemma 1 are fulfilled for each factor of the right side of (3.15) if we put $\Phi^k = \Phi^{(1)}$, $\Phi^{k+1} = \Phi^{(2)}$, $V_{(k)} = V_1$, $T^\nu \setminus v_k(\Gamma) = V_2$ for $k = 0, \ldots, m(\Gamma)$. Therefore (3.15) implies

$$\left|\ln \frac{g_V(\tilde{x}\,|\,\Phi)}{g_{\tau_\Gamma V}(\pi_\Gamma \tilde{x}\,|\,\Phi)}\right| \le G\beta\epsilon \sum_{k=0}^{m(\Gamma)} D(v_k(\Gamma),\ T^\nu \setminus v_k(\Gamma)) \tag{3.16}$$

We denote $d = D(R, L)$, where $R = \{t : t \in T^\nu,\ t^1 < 0,\ t^m = 0,\ m > 1\}$ and $L = \{t \in T^\nu,\ t^1 \ge 0\}$. Since the points of $L$ which are at the distance $\ell$ from the point $t \in R$ (in the sense of the metric $\rho$) lie in the intersection of $L$ with the surface of the square (if $\nu = 2$) or the cube (if $\nu = 3$) having edge $2\ell$ and center in $t$, then

$$d = \sum_{t\in R} \sum_{\ell \ge \rho(t, L)} \{(2\ell+1)^{\nu-1} + [\ell - \rho(t, L)]\,[\,2\ell+1)^{\nu-1}$$
$$- (2\ell-1)^{\zeta-1}]\,\}\,\ell^{-(\nu+\delta+1)}$$

$$\le \sum_{m=1} \sum_{\ell=m} 2\nu(2\ell+1)^{\nu-1} \ell^{-(\nu+\delta+1)} \le 2\nu\, 3^{\nu-1} \sum_{\ell=1} \ell^{-(1+\delta)} < \infty$$

It is easy to see that for any $A \subset T^\nu$, $|A| < \infty$, the inequality $D(A,\ T^\nu \setminus A) \le 2d\,|\vartheta(A)|$ holds, where $|\vartheta(A)|$ is the area of the surface which separates $A$ from $T^\nu$ $A$ and consists of unit squares (if $\nu = 3$) or intervals (if $\nu = 2$) orthogonal to the coordinate axes and having centers at the points of the lattice $T^\nu + \frac{e}{2}$ where $e$ is a vector with unit coordinates. Since $|\vartheta(A)| \le 2\nu\,|A|$ we have from (3.16)

$$\left|\ln \frac{g_V(\tilde{x} \mid \Phi)}{g_{\tau_\Gamma V}(\pi_\Gamma \tilde{x} \mid \Phi)}\right| \leq 24\, \beta \epsilon\, d\, \nu\, |\Gamma^{(+)}| \leq 24\, \beta\, \epsilon\, d\, \nu\, 3^\nu\, |\Gamma|$$

and the statement of Lemma 2 is fulfilled for $K = 24\, d\, \nu\, 3^\nu$.

Returning to equality (3. 13) and assuming in the conditions of Lemma 2 that $\Gamma = \Gamma_{n+1}$, $V = \tau(n, j) V_\gamma$, $\tilde{x} = \pi_n x^{(j)}$, $\Phi = \Phi^\zeta$ for $n = 0, \ldots, J(\gamma) - 1$, we find with the help of (3. 14) that

$$\left|\frac{g_{V_\gamma}(x^{(j)} \mid \Phi^\zeta)}{g_V(x^i \mid \Phi^\zeta)}\right| \leq \exp\{(\beta\epsilon C + \tfrac{1}{2})\,|\bar{V}| + [\,\beta\epsilon\,(3^\nu C + K) + 3^\nu\,]\,|\gamma|\} \tag{3. 17}$$

Substituting (3. 12) and (3. 17) into (3. 11) we obtain the estimate

$$|q^\zeta(Y_{\bar{V}}(\bar{x})|x^i)| \leq \exp\{(2\beta\epsilon C + \tfrac{1}{2})|V|\} \times [1 + \sum_{\gamma \in A_V(\bar{V})} \exp\{-a_2 \beta |\gamma|\}] \tag{3. 18}$$

where $a_2 = a_1 - \epsilon\,(3^\nu (2\beta_0)^{-1}$ and $a_2 > 0$ for sufficiently small $\epsilon$ and $\beta^{-1}$.

The estimation of the sum in the square brackets of the right side of (3. 18) is carried out by standard methods, and we will only point out some of the steps. We introduce the family $A^0(t)$, $t \in T^\nu$ of the connected sets $\Gamma \subset V$, $|\Gamma| < \infty$, such that $t \in \Gamma$, and the family $A^1_V(t)$ of connected $\Gamma \subset V$, $|\Gamma| < \infty$, such that $t \quad T^\nu \setminus v_0(\Gamma)$. Let $t_1, \ldots, t_{|\bar{V}|}$ be the numbered points of $|V|$. Evidently

$$\sum_{\gamma \in A_V(\bar{V})} e^{-a_2 \beta |\gamma|} \leq \sum_{\gamma' \in A_V(\{t_1\})} e^{-a_2 \beta |\gamma'|} [1 +$$

$$+ \sum_{n=1}^{|\bar{V}|-1} \sum_{\{t_{i_1}, \ldots, t_{i_n}\} \subset \bar{V}} \prod_{n=1}^{n} (\sum_{\Gamma \in A^0(t_{i_k})} e^{-a_2 \beta |\Gamma|})]$$

Furthermore it is clear that

$$\sum_{\gamma' \in A_V(\{t_1\})} e^{-a_2\beta|\gamma'|} \leq \sum_{n=1}^{\infty} \Big( \sum_{\Gamma \in A^1_{T^\nu}(t_1)} e^{-a_2\beta|\Gamma|}\Big)^n$$

It is well known (see [6] that the quantity of connected sets including a fixed point and having volume $|\Gamma|$ does not exceed $h^{|\Gamma|}$ where $h$ is a positive constant, more then once deducing inequality (see, e. g. [5, 6])

$$\sum_{\Gamma \in A^0(t)} e^{-\beta\alpha|\Gamma|} \leq \sum_{\Gamma \in A^1_{T^\nu}(t)} e^{-\beta\alpha|\Gamma|} \leq 2\exp\{-\beta(\alpha - \frac{\ln h}{\beta_0})\} \tag{3. 19}$$

holds for any $\alpha > 0$ and sufficiently large $\beta_0 \equiv \beta_0(\alpha)$. Designating for $\alpha = a_2$, the right side of (3. 19) by $r$ we find that if $r < 1$

$$\sum_{\gamma \in A_V(\bar{V})} e^{-a_2\beta|\gamma|} \leq \frac{r}{1-r}\Big(1 + \sum_{n=1}^{|\bar{V}|-1} C^n_{|V|-1} r^n\Big) = \frac{r}{(1-r)(1+r)}(1+r)^{|\bar{V}|}$$

and thus the estimate (3. 2) is proved.

To demonstrate the analyticity of free energy we shall use again the multidimensional generalization of Vitali's theorem. The existence of the limit of the free energy for real $\zeta \in W$ follows from the well known Van-Hove's theorem, therefore we need only to prove the uniformity of the limit of $|V|^{-1}|\ln Z^{\zeta}_V(x^i)|$. Noting that

$$Z^{\zeta}_V(x^i) = Z^0_V(x^i) g_V(x^i|\Phi^{\zeta}) \tag{3. 20}$$

we find that

$$F^{\zeta} = F^0 + \lim_{k\to\infty} |V_k|^{-1} \ln g_{V_k}(x^i|\Phi^{\zeta})$$

But the statement (a) of Lemma 1 implies

$$|V|^{-1}|\ln g_V(x^i|\Phi^\zeta)| = |V|^{-1}|\ln \frac{g_V(x^i|\Phi^\zeta)}{g_{\{\phi\}}(x^i|\Phi^\zeta)}| \leq \frac{1}{2} + \beta\epsilon C$$

The existence and the analiticity of $F^\zeta$ on $W$ follow now from the existence of the limit $F^0$ and (3. 21). The inequality (3. 1) follows now from (3. 20) and (3. 21) and the proof of the theorem is complete.

## IV. THE PROOF OF LEMMA 1.

We shall prove Lemma 1 by induction. For $V = \phi$ the assertions of the Lemma are trivial. Let these be true for all $V \subset T^\nu$, $|V| < n$. We fix the vessels in the conditions of the Lemma: $V = W$, $|W| = n$, $W' = V'$, $W_1 = V_1$, $W_2 = V_2$, and begin at the proof of statement (a). According to the induction hypothesis

$$|\ln \frac{g_W(\tilde{x}|\Phi)}{g_{W'}(\tilde{x}|\Phi)}| \leq (C\beta\epsilon + \frac{1}{2})|(W\setminus t)\ W'| + |\ln \frac{g_W(\tilde{x}|\Phi)}{g_{W\setminus t}(\tilde{x}|\Phi)}| \tag{4. 1}$$

for some $t \in W\ W'$. We shall say that a cycle $\Gamma$ of a function $x \in Y$ is external in the vessel $V \subset T^\nu$, if $\Gamma \subset V$ and for any cycle $\Gamma' \neq \Gamma$, $\Gamma' \subset V$ of the function $x$ we have $\Gamma \subset v_0(\Gamma')$. We introduce the sets $Y_V^{ex}(\Gamma)$ of functions from $Y$ for which the cycle $\Gamma$ is external in the vessel $V$, and the sets $Y_V^s(\Gamma, x) = Y_s(\Gamma) \cap Y_V^{ex}(\Gamma) \cap \tilde{Y}_{T^\nu\setminus V}(x)$, $x \in Y$, $s = 1, \ldots, s(\Gamma)$. Let $S_\Gamma(V', x)$ be the set of indexes $s$, $1 \leq s \leq s(\Gamma)$, such that $Y_V^s(\Gamma, x) \neq \phi$ and let the function $x^{(s)} \in Y_V^s(\Gamma, x)$, $s \in S_\Gamma(V, x)$, be such that $\Gamma$ is the only cycle in $V$ (evidently $x^{(s)}$ exists and is unique.) For any $A$, $A' \subset T^\nu$, $A \subset A'$, $|A'| < \infty$, $x \in Y$, $\tilde{\Phi} \in F_{\epsilon,\delta}(\cdot)$ we introduce the designation

$$\varphi_{A'}(x|A, \tilde{\Phi}) = \exp\{-\beta[\tilde{\Phi}_{A'}(x) - \tilde{\Phi}_{A'\setminus A}(x)]\}$$

It follows from (3. 6) and (2. 3) that

$$g_W(\tilde{x}|\Phi) = \sum_{\Gamma \in A_W^1(t)} \sum_{s \in S_\Gamma(W, \tilde{x})} \varphi_W(x^{(s)}|\Gamma^{(+)}, \Phi)$$

$$\times g_{W \setminus \Gamma^{(+)}}(x^{(s)}|\Phi) q_W^0(Y_W^s(\Gamma, \tilde{x})|\tilde{x})$$

$$+ g_{W \setminus t}(\tilde{x}|\Phi)\varphi_W(\tilde{x}|\{t\}, \Phi)\tilde{q}_W^0(Y_W(t)|\tilde{x}) \tag{4. 2}$$

where $Y_W(t)$ is a set of function for which the external cycles in $W$ are absent. Therefore

$$\left|\frac{g_W(\tilde{x}|\Phi)}{\varphi_W(\tilde{x}|\{t\},\Phi)g_{W\setminus t}(\tilde{x}|\Phi)} - 1\right|$$

$$\leq \sum_{\Gamma \in A_W^1(t)} \sum_{s \in S_\Gamma(W, x)} \left|\frac{\varphi_W(x^{(s)}|\Gamma^{(+)}, \Phi)g_{W\setminus\Gamma^{(1)}}(x^{(s)}|\Phi)}{\varphi_W(\tilde{x}|\{t\}, \Phi)g_{\tau_\Gamma(W\setminus\Gamma^{(+)})}(\pi_\Gamma x^{(s)}|\Phi)}\right.$$

$$\left.\times \frac{g_{\tau_\Gamma(W\setminus\Gamma^{(+)})}(\tilde{x}|\Phi)}{g_{W\setminus t}(\tilde{x}|\Phi)} - 1\right| \tilde{q}_W^0(Y_W^s(\Gamma, \tilde{x})|\tilde{x}) \tag{4. 3}$$

Here we used the fact that $\pi_\Gamma x^{(s)} = \tilde{x}$ for any $\Gamma \in A_W^1(t)$, $s \in S_\Gamma(W, \tilde{x})$. It follows from (3. 9) that

$$\left|\ln \frac{\varphi_W(x^{(s)}|\Gamma^{(+)}, \Phi)}{\varphi_W(\tilde{x}|\{t\}, \Phi)}\right| < \beta\epsilon C(3^\nu|\Gamma| + 1) \tag{4. 4}$$

Using the induction hypothesis and the equality

$$\frac{g_{\tau_\Gamma(W\setminus\Gamma^{(+)})}(\tilde{x}|\Phi)}{g_{W\setminus t}(\tilde{x}|\Phi)} = \frac{g_{\tau_\Gamma(W\setminus\Gamma^{(+)})\setminus t}(\tilde{x}|\Phi)}{g_{\tau_\Gamma(W\setminus\Gamma^{(+)})\setminus t}(\tilde{x}|\Phi)} \frac{g_{\tau_\Gamma(W\setminus\Gamma^{(+)})\setminus t}(\tilde{x}|\Phi)}{g_{W\setminus t}(\tilde{x}|\Phi)}$$

we obtain from statement (a) the estimate

$$\left| \ln \frac{g_{\tau_\Gamma(W\setminus\Gamma^{(+)})}(\tilde{x}|\Phi)}{g_{W\setminus t}(\tilde{x}|\Phi)} \right| \le \left(\frac{1}{2} + C\beta\epsilon\right)(1 + 3^\nu|\Gamma|) \tag{4. 5}$$

Further we note that for the proof of Lemma 2 we needed Lemma 1 to hold for the vessel of Lemma 2 only. Therefore it follows from the induction hypothesis and Lemma 2 that

$$\left| \ln \frac{g_{W\setminus\Gamma^{(+)}}(x^{(s)}|\Phi)}{g_{\tau_\Gamma(W\setminus\Gamma^{(+)})}(\pi_\Gamma x^{(s)}|\Phi)} \right| \le K\beta\epsilon\,|\Gamma| \tag{4. 6}$$

Using the inequality $|e^z - 1| \le 2e^{|z|}$ which holds for all complex $z$, and the estimate $\tilde{q}_W^0(Y_W^s(\Gamma,\tilde{x})|\tilde{x}) \le e^{-a\beta|\Gamma|}$ analogous to (3. 10) we find from (4. 3) with the help of (4. 4) - (4. 6) that for sufficiently large $\beta_0$ and $\epsilon^{-1}$

$$\left| \frac{g_W(\tilde{x}|\Phi)}{\varphi_W(\tilde{x}|\{t\},\Phi)g_{W\setminus t}(\tilde{x}|\Phi)} - 1 \right| \le \sum_{\Gamma\in A^1_{T^\nu}(t)} 2e^{-a_3\beta|\Gamma|} \le \frac{1}{2} \tag{4. 7}$$

where $a_3 = a - (\frac{1}{2} + \ln|X\,X_0|)\beta_0^{-1} - \epsilon[K + 2C(3^\nu + 1)] > 0$.

Now the inequality $|\ln(z+1)| \le 2|z|$, which holds for $|z| < \frac{1}{2}$, and (3. 9) give

$$\left| \ln \frac{g_W(\tilde{x}|\Phi)}{g_{W\ t}(\tilde{x}|\Phi)} \right| \le \frac{1}{2} + \beta\epsilon C$$

and the statement (a) of the Lemma follows for $W$ from (4. 1).

Further we put $t_{(1)},\ldots,t_{(N)} \in W_1, \bigcup_{k=1}^N \{t_{(k)}\} = W_1$, and introduce the family $\Phi^k$, $k = 0,\ldots,N$, the complex-valued potentials in the broad sense defined by $\Phi^0 = \Phi^{(1)}$ and

$$\Phi^k(x, t_1, t_2) = \begin{cases} \Phi^2(x, t_1, t_2) \quad \text{if} \quad x \in Y^{ex}_{W_1}(\Gamma), \\ \Gamma \in A^1_{W_1}(t_{(k)}),\ t_1 \in T^\nu\ v_0(\Gamma), \\ t_2 \in W_2 \quad \text{or if} \quad x \in Y_{W_1}(t_{(k)}) \\ t_1 = t_{(k)},\ t_2 \in W_2; \\ \Phi^{k-1}(x, t_1, t_2) \quad \text{otherwise} \end{cases}$$

for $k = 1, \ldots, N$. Noting that $\Phi^N = \Phi^{(2)}$ we can write the evident equality

$$\frac{g_W(\tilde{x} \mid \Phi^{(1)})}{g_W(\tilde{x} \mid \Phi^{(2)})} = \prod_{k=1}^{N} \frac{g_W(\tilde{x} \mid \Phi^{k-1})}{g_W(\tilde{x} \mid \Phi^k)} \tag{4. 8}$$

Using the expansion (4. 2) we find that

$$J_k \equiv \left| \frac{g_W(\tilde{x} \mid \Phi^{k-1}) g_{W \setminus t_{(k)}}(\tilde{x} \mid \Phi^k) \varphi_W(\tilde{x} \mid \{t_{(k)}\}, \Phi^k)}{g_W(\tilde{x} \mid \Phi^k) g_{W \setminus t_{(k)}}(\tilde{x} \mid \Phi^{k-1}) \varphi_W(\tilde{x} \mid \{t_{(k)}\}, \Phi^{k-1})} - 1 \right|$$

$$\leq \left[ 1 - \left| \frac{g_W(\tilde{x} \mid \Phi^k)}{g_{W \setminus t_{(k)}}(\tilde{x} \mid \Phi^k) \varphi_W(\tilde{x} \mid \{t_{(k)}\}, \Phi^k)} - 1 \right| \right]^{-1}$$

$$\times \sum_{\Gamma \in A^1_{W_1}(t_{(k)})} \sum_{s \in S_\Gamma(W, \tilde{x})} \tilde{q}^0_W(Y^s_W(\Gamma, \tilde{x}) \mid \tilde{x})$$

$$\times \left| \frac{\varphi_W(x^{(s)} \mid \Gamma^{(+)}, \Phi^{k-1}) g_{W \setminus \Gamma^{(+)}}(x^{(s)} \mid \Phi^{k-1})}{\varphi_W(\tilde{x} \mid \{t_{(k)}\}, \Phi^{k-1}) g_{W \setminus t_{(k)}}(\tilde{x} \mid \Phi^{k-1})} \right|$$

$$\times \left| 1 - \frac{\varphi_W(x \mid \{t_{(k)}\}, \Phi^{k-1}) \varphi_W(x^{(s)} \mid \Gamma^{(+)}, \Phi^k)}{\varphi_W(x \mid \{t_{(k)}\}, \Phi^k) \varphi_W(x^{(s)} \mid \Gamma^{(+)}, \Phi^{k-1})} \times \frac{g_{W \setminus \Gamma^{(+)}}(x^{(s)} \mid \Phi^k) g_{W \setminus t_{(k)}}(x \mid \Phi^{k-1})}{g_{W \setminus \Gamma^{(+)}}(x^{(s)} \mid \Phi^{k-1}) g_{W \setminus t_{(k)}}(x \mid \Phi^k)} \right| \tag{4. 9}$$

It follows from (4. 7) that the first factor in the right part of (4. 9) does not exceed 2. Putting for all $k = 1, \dots, N$

$$\Phi^{k, t_{(k)}}(x, t_1, t_2) = \begin{cases} \Phi^k(x, t_1, t_2) & \text{if } x \in Y_{W_1}(t_{(k)}) \\ \Phi^{k-1}(x, t_1, t_2) & \text{if } x \in Y \setminus Y_{W_1}(t_{(k)}) \end{cases}$$

and using the evident equalities

$$g_{W \setminus t_{(k)}}(\tilde{x} \mid \Phi^k) = g_{W \setminus t_{(k)}}(\tilde{x} \mid \Phi^{k, t_{(k)}})$$

and the induction hypothesis with $V = W \setminus t_{(k)}$, $V_1 = W_2 \cap W$, $V_2 = \{t_{(k)}\}$, $\Phi^{(1)} = \Phi^{k-1}$, $\Phi^{(2)} = \Phi^{k, t(k)}$ we obtain the estimate

$$\left| \ln \frac{g_{W \setminus t_{(k)}}(x \mid \Phi^{k-1})}{g_{W \setminus t_{(k)}}(x \mid \Phi^k)} \right| \leq 3 \beta \epsilon D(\{t_{(k)}\}, W_2 \cap W) \tag{4. 10}$$

For any $\Gamma \in A^1_{W_1}(t_{(k)})$, $k = 1, \dots, N$, we introduce the following potentials in the broad sense

$$\tilde{\Phi}^{k, \Gamma}(x, t_1, t_2) = \begin{cases} \Phi^k(x, t_1, t_2) & \text{if } x \in Y^{ex}_{W_1}(\Gamma), \\ t_1 \in T^\nu v_0(\Gamma), & t_2 \in W_2 \cap W; \\ \Phi^{k-1}(x, t_1, t_2) & \text{otherwise} \end{cases}$$

$$\Phi^{k, \Gamma}(x, t_1, t_2) = \begin{cases} \Phi^k(x, t_1, t_2) & \text{if } x \in Y^{ex}_{W_1}(\Gamma), \\ t_1 \in T^\nu v_0(\Gamma), & t_2 \in W_2 \setminus W; \\ \tilde{\Phi}^{k, \Gamma}(x, t_1, t_2) & \text{otherwise} \end{cases}$$

and consider the equalities

$$\frac{g_{W\setminus\Gamma^{(+)}}(x^{(s)}|\Phi^k)}{g_{W\setminus\Gamma^{(+)}}(x^{(s)}|\Phi^{k-1})} = \left[\frac{g_{W\setminus\Gamma^{(+)}}(x^{(s)}|\Phi^k,\Gamma)}{g_{W\setminus\Gamma^{(+)}}(x^{(s)}|\tilde{\Phi}^k,\Gamma)}\right]\left[\frac{g_{W\setminus\Gamma^{(+)}}(x^{(s)}|\tilde{\Phi}^k,\Gamma)}{g_{W\setminus\Gamma^{(+)}}(x^{(s)}|\Phi^{k-1})}\right] \tag{4.11}$$

which hold for any $s \in S_\Gamma(W, \tilde{x})$. The induction hypothesis with $V = W\setminus\Gamma^{(+)}$ and, for the first and the second factors in the right side of (4.11), respectively, with $V_1 = T^\nu (v_0(\Gamma) \cup \Gamma^{(+)})$, $V_2 = W_2 \setminus W$ and $V_1 = W_2 \cap W$, $V_2 = T^\nu \setminus v_0(\Gamma)$, implies the estimate

$$\left|\ln \frac{g_{W\setminus\Gamma^{(+)}}(x^{(s)}|\Phi^k)}{g_{W\setminus\Gamma^{(+)}}(x^{(s)}|\Phi^{k-1})}\right| \leq G\beta\epsilon\, D(W_2, T^\nu\setminus v_0(\Gamma)) \tag{4.12}$$

for any $k = 1,\ldots, N$, $\Gamma \in A^1_{W_1}(t_{(k)})$, $s \in S_\Gamma(W, \tilde{x})$. It follows from condition $B_2$ and the condition

$$\Phi^k(x, t_1, t_2) = \Phi^{k-1}(x, t_1, t_2) \quad \text{if} \quad t_1 \in W_1,\ t_2 \in T^\nu\setminus W_2$$

that

$$\left|\ln \frac{\varphi_W(\tilde{x}|\{t_{(k)}\}, \Phi^{k-1})\varphi_W(x^{(s)}|\Gamma^{(+)}, \Phi^k)}{\varphi_W(\tilde{x}|\{t_{(k)}\}, \Phi^k)\varphi_W(x^{(s)}|\Gamma^{(+)}, \Phi^k)}\right|$$

$$\leq 2\beta\epsilon[D(\{t_{(k)}\}, W_2) + D(\Gamma^{(+)}, W_2)] \tag{4.13}$$

for any $k = 1,\ldots, N$, $\Gamma \in A^1_{W_1}(t_{(k)})$, $s \subseteq S\ (W, \tilde{x})$. With the help of the inequality $|e^z - 1| \leq |z|e^{|z|}$, which holds for all complex $z$, the estimates (4.10), (4.12), (4.13) and the estimate analogous to the ones used for the deduction of (4.7), we find the majorant for the left side of (4.9)

$$J_k \leq \sum_{\Gamma \in A^1_{W_1}(t_{(k)})} e^{-\beta a_3 |\Gamma|} 8\beta\epsilon [\mathcal{D}(\{t_{(k)}\}, W_2)$$

$$+ \mathcal{D}(T^\nu \setminus v_0(\Gamma), W_2)] \times \exp\{16\beta\epsilon\, \mathcal{D}(T^\nu \setminus v_0(\Gamma), W_2)\} \qquad (4.14)$$

As we have seen in the proof of Lemma 2,

$$\mathcal{D}(T^\nu \setminus v_0(\Gamma), W_2) \leq 4 d \nu 3^\nu |\Gamma| \qquad (4.15)$$

On the other hand

$$\mathcal{D}(T^\nu\ v_0(\Gamma), W_2) \leq |T^\nu \setminus v_0(\Gamma)|\, \mathcal{D}(\{t_\Gamma\}, W_2) \qquad (4.16)$$

where $t_\Gamma \in T^\nu \setminus v_0(\Gamma)$ is such that

$$\mathcal{D}(\{t_\Gamma\}, W_2) = \max_{t \in T^\nu \setminus v_0(\Gamma)} \mathcal{D}(\{t\}, W_2)$$

It follows from the triangle inequality and the inequality

$$e^{-r_1 c} \leq (1 + r_1)^{-\alpha} \leq (1 + \frac{r_1}{r_2})^{-\alpha}$$

which holds for all real $r_1 > 0$, $r_2 > 1$, $\alpha > 0$ and some $c > 0$, that

$$\mathcal{D}(\{t_\Gamma\}, W_2) = \sum_{t \in W_2} [\rho(t_\Gamma, t)]^{-(1+\delta+\nu)}$$

$$\leq e^{c|\Gamma|} \sum_{t \in W_2} [\rho(t_\Gamma, t) + |\Gamma|]^{-(1+\delta+\nu)}$$

$$\leq e^{c|\Gamma|} \sum_{t \in W_2} [\rho(t_{(k)}, t)]^{-(1+\delta+\nu)}$$

$$= e^{c|\Gamma|} \mathcal{D}(\{t_{(k)}\}, W_2)$$

for all $k = 1, \ldots, N$, $\Gamma \in A^1_{W_1}(t_{(k)})$. Using this estimate and the estimate

$$|T^\nu \backslash v_0(\Gamma)| \leq |\vartheta(T^\nu \backslash v_0(\Gamma))|^\nu \leq (3^\nu\, 2\nu)^\nu |\Gamma|^\nu$$

we find from (4. 16) that

$$\mathcal{D}(T^\nu \backslash v_0(\Gamma), W_2) \leq (2\nu\, 3^\nu)^\nu |\Gamma|^\nu e^{c|\Gamma|} \mathcal{D}(\{t_{(k)}\}, W_2) \tag{4. 17}$$

Then (4. 14), (4. 15), and (4. 17) imply

$$J_k \leq \epsilon\, \mathcal{D}(\{t_{(k)}\}, W_2) \sum_{\Gamma \in A^1_{W_1}(t_{(k)})} \beta K_1 |\Gamma|^\nu e^{-\beta \alpha_4 |\Gamma|}$$

where $K_1 = 8[1 + (2\nu\, 3^\nu)^\nu]$, $a_4 = a_3 - 64\epsilon\, d\, \nu\, 3^\nu - c_{\beta_0}$. Since $|\Gamma|^\nu \leq \exp\{\beta \frac{\nu}{\beta_0} |\Gamma|\}$ we have from (3. 19)

$$J_k \leq \epsilon\, \mathcal{D}(\{t_{(k)}\}, W_2)$$

and with the help of (4. 10), the condition $B_2$ and the estimate $|\ln(z + 1)| \leq 2|z|$ for $|z| < \frac{1}{2}$ we find for sufficiently small $\epsilon$, $\beta_0^{-1}$ that

$$\left|\ln \frac{g_W(\tilde{x} | \Phi^{k-1})}{g_W(\tilde{x} | \Phi^k)}\right| \leq 3\beta\epsilon\, \mathcal{D}(\{t_{(k)}\}, W_2 \backslash W) + G\beta\epsilon\, \mathcal{D}(\{t_{(k)}\}, W_2 \cap W)$$

The statement (b) of Lemma 1 follows now from (4. 8).

ACKNOWLEDGMENTS

I am very grateful to R. L. Dobrushin and S. A. Pirogov for their attention to this work.

## REFERENCES

1. R. A. Minlos, Ya. G. Sinai, Some New Results on First Order Phase Transitions in Lattice Gas Models. Proceedings of Moscow Mathematical Society (in Russian), 17, pp. 213-242, 1967.
2. R. A. Minlos, Ya. G. Sinai, The Phenomenon of "Phase Separation" at Low Temperatures in Some Lattice Models of a Gas. I, Mathematical Collection, (in Russian), v. 73, pp. 375-449, 1967; II, Proceedings of Moscow Mathematical Society, (in Russian), 19, pp. 113-178, 1968.
3. R. L. Dobrushin, Analyticity of Correlation Functions in One-Dimensional Classical Systems with Slowly Decreasing Potentials. Communications in Mathematical Physics, v. 32, No. 4, pp. 269-289, 1973.
4. R. L. Dobrushin, Analyticity of Correlation Functions in One-Dimensional Classical Systems with Degree Decreasing of Potential. Mathematical Collection (in Russian), v. 94, No. 1, pp. 16-48, 1974.
5. V. M. Gertzik, Conditions of the Nonuniqueness of Gibbs State for Lattice Models with Finite Potential. Izvestija Akademii Nauk SSSR Ser. Matem., (in Russian), 40, 448-462, 1976.
6. R. L. Dobrushin, Existence of Phase Transitions in Models of a Lattice Gas. Proceedings of the 5-th Berkeley Symposium on Mathematical Statistics and Probability, v. 3, pp. 73-87, 1967.
7. R. L. Dobrushin, The Problem of Uniqueness of a Gibbs Random Field and Phase Transitions. Functional Analysis and Applications (in Russian), v.1, No. 4, pp. 44-47, 1968.
8. R. G. Gunning, H. Rossi, Analytic Functions of Several Complex Variables, New York, Prentice Hall, 1965.
9. B. V. Shabat, Introcudtion to Complex Analysis, (in Russian), Sciences edit., 1969.

chapter 8

# ASYMPTOTIC PROPERTIES OF THE PRODUCT OF RANDOM MATRICES DEPENDING ON A PARAMETER

I. Ja. Goldsheid

## I. FORMULATION OF THE PROBLEM. ITS CONNECTION WITH LYAPUNOV'S INDICES THEORY. SOME PRELIMINARY RESULTS.

This work is devoted to a detailed exposition of the results announced in paper [12].

We consider the product of random matrices depending on a parameter:

$$A(n, t, \omega) = A_1(t, \omega) A_2(t, \omega) \dots A_n(t, \omega) \tag{1}$$

Here $A_i(t, \omega)$, $i = 1, 2, \dots$ is a sequence of independent random processes with identically finite-dimensional distributions taking values in the group of non-degenerate $m \times m$ complex matrices. We shall call the parameter $t$ the time of the random process, $t \in T$, $T$ is a compact metric space; $\{\Omega, \mathcal{F}, \Pr\}$ is a probability space. We suppose the matrices $A_i(t, \omega)$ to be continuous functions of $t$ and

$$E \sup_{t \in T} |\ln \|A_i(t, \omega)\| | < \infty, \quad E \sup_{t \in T} |\ln |\det A_i(t)| | < \infty$$

One of the main problems of this work is to find out the conditions under which there exists a function $\alpha(t)$, independent of $\omega$ such that with probability I for all $t \in T$

$$\lim_{n \to \infty} n^{-1} \ln \|A(n, t, \omega)\| = \alpha(t) \tag{2}$$

Convergence in (2) may be uniform in $t$ or not. Both cases are of interest to us. We shall formulate one general result from work [I] connected with the above mentioned problem. Let $B_i(\omega)$ be a stationary ergodic sequence of matrices and $E \ln \|B_i\| < \infty$. Then almost surely

$$\lim_{n \to \infty} n^{-1} \ln \|B(n, \omega)\| < \beta$$

exists and is independent of $\omega$.

For $t \in T$ the set of $\omega \in \Omega$ for which (2) holds will be denoted by $\Omega_t$. The fact that $\Pr\{\Omega_t\} = 1$ follows from the just formulated result. We shall denote by $\Omega_0$ those $\omega \in \Omega$ for which (2) holds for all $t \in T$. It is evident that $\Omega_0 = \bigcap_{t \in T} \Omega_t$. Apriori we cannot state that $\Pr\{\Omega_0\} = 1$ for the intersection is uncountable.

$$n^{-1} \sum_{i=1}^{n} \ln |\det A_i(t, \omega)| \to E \ln |\det A_1(t, \omega)| \tag{3}$$

and according to Mourier's ergodic theorem [3] (for functions taking values in a Banach space) the convergence in (3) takes place uniformly in $t$ with probability 1. Therefore it will be sufficient for this paper to consider only the case when $\det A_i(t, \omega) = 1$, and we shall always make this assumption. The motivation for the present work was the study of the asymptotic behavior of the solution of Schrödinger's equation with random potential

$$-y'' + q(x, \omega) y = \lambda y \tag{4}$$

and the analogous difference equation

$$y_{n+1} + q_n(\omega) y_n + y_{n-1} = \lambda y_n \tag{5}$$

as $x \to \pm\infty$ respectively $n \to \pm\infty$ in connection with spectral problems for such operators.

The second aim of the work is to study logarithmic asymptotics of the solutions to (4) and (5) as functions of the parameter $\lambda$ (the conditions for $q(x,\omega)$, $q_i(\omega)$ will be given below). For the solution of this problem it is necessary to have detailed knowledge about the first problem for the case of $2 \times 2$ matrices.

We concentrate on equation (5) for there is no principal difference between (4) and (5). Suppose $q_i(\omega)$, $i = 0, \pm 1, \pm 2, \ldots$ is a sequence of independent identically distributed random variables and $E \ln (|q_0| + 1) < \infty$.

If $y_0$ and $y_1$ are given the other $y_i$ $(i > 1)$ are defined from (5) recurrently. It is convenient to consider the sequence of two-dimensional vectors $(y_0, y_1)$, $(y_1, y_2), \ldots, (y_n, y_{n+1})$ where $\{y_i\}_{i=0}^{\infty}$ is the solution of (5). It is evident that

$$(y_0, y_1) A_1(\lambda, \omega) A_2(\lambda, \omega) \ldots A_n(\lambda, \omega) = (y_n, y_{n+1}) \tag{6}$$

where

$$A_i(\lambda, \omega) = \begin{pmatrix} 0 & -1 \\ 1 & -q_i(\omega) + \lambda \end{pmatrix} \tag{7}$$

From (6) we can see that the asymptotics of the solutions to (5) are closely connected with the asymptotics of (1), when $A_i(\lambda, \omega)$ is given by (7) ($\lambda$ instead of $t$).

The solutions which tend to 0 when $n \to \infty$ (or $n \to -\infty$) are most interesting from the spectral theory point of view. The existence of such solutions follows from (2) if $\alpha(t) > 0$.

At present there are some theorems containing sufficient conditions for $\alpha(t) > 0$ ($t$ is fixed). We shall suppose the conditions of Furstenberg's theorem to be fulfilled ([2], the formulation is given below). Let us consider the question of the connection of the above problem with Lyapunov's indices theory.

We will give two definitions of Lyapunov in a form convenient for us ([4]).

Definition 1. The characteristic index of a vector $Z \in C^m$ ($C^m$ is m-dimensional Euclidean space) is the value:

$$\alpha_Z = \limsup_{n \to \infty} n^{-1} \ln \|ZB_1B_2 \ldots B_n\|$$

Lyapunov proved that for any sequence of matrices $B_i$ ($\|B(n)\| < \exp(\gamma n)$) there exists a basis $Z_1, Z_2, \ldots, Z_m$ in $C^m$ such that for any other basis $\tilde{Z}_1, \tilde{Z}_2, \ldots, \tilde{Z}_m$ we have

$$\alpha_{z_i} \geq \alpha_{\tilde{z}_i} \quad \text{if} \quad \alpha_{\tilde{Z}_1} \geq \alpha_{\tilde{Z}_2} \geq \ldots \geq \alpha_{\tilde{Z}_m}$$

Definition 2. The values $\alpha_{z_1}, \alpha_{z_2}, \ldots, \alpha_{z_m}$ are called the characteristic indices of the sequence of matrices $B_i$, $i = 1, 2, \ldots$. The basis $Z_1, Z_2, \ldots, Z_m$ is called normal.

It is easy to see that with probability 1, $\lim_{n \to \infty} n^{-1} \ln \|ZB(n,\omega)\| = \beta$ if $Z$ does not belong to the random subspace generated by the vectors $Z_2, Z_3, \ldots, Z_m$.

Osseledets ([11]) and Millionschikov ([8, 9]) proved a theorem which essentially strengthend this remark. It turns out, in particular, that if $B_i(\omega)$ is a strictly stationary sequence of matrices ($\det B_i(\omega) = 1$, $E \ln \|B_i\| < \infty$) then with probability 1 for any $Z \in C^m$ $\lim_{n \to \infty} n^{-1} \ln \|ZB(n, \omega)\| = \alpha_Z(\omega)$ exists. Moreover for the

normal basis $Z_j$, $j = 1, 2, \ldots, m$, we have:
$\sum_{j=1}^{m} \alpha_{Z_j} = 0$ and the $Z_j(\omega)$ are measurable functions of $\omega (\|Z_j(\omega)\| = 1)$

From this theorem it follows that if $\beta > 0$ then with probability 1 there exists at least one vector $Z$ with $\alpha_Z < 0$. We remark that if $\beta = 0$ then the result of the theorem is a trivial one: any basis in $C^m$ is normal, the index $\alpha_Z = 0$ for any $Z$ and nothing can be told about the existence of the decreasing solution of equation (5).

Thus we see that $\alpha(t)$ is the leading Lyapunov index of the sequence of matrices $A_i(t, \omega)$. As $A_i(t, \omega)$ depends continuously on $t$ there is a relation between our problem and the problem of perturbation of $\alpha(t)$ when $A_i(s, \omega) - A_i(t, \omega) \to 0$ while $s \to t$ uniformly $i, t, \omega$ (of course not all random processes are of such kind).

Perron ([5, 6, 7]) studied the question of the stability of Lyapunov's indices. From his results it follows that there systems of linear differential equations of the form

$$\frac{dy}{dx} = \Phi(x) y \tag{8}$$

($x \geq 0$, $\Phi(x)$ is a $m \times m$ matrix, $y$ is an n-dimensional vector, $\sup_x \|\Phi(x)\| < \infty$) for which Lyapunov's indices are changed essentially for some arbitrarily small perturbations of $\Phi(x)$.

Nevertheless some classes of systems (8) are stable under perturbation of some concrete types. Millionschikov showed that the leading index of the so-called absolutely regular system is upper semi-continuous and all the characteristic indices of an absolutely regular system depend continuously on the perturbation of a white noise type (see [10], where other references can be found).

The special nature of our work consists of the following: first of all we are interested in the validity of relation (2) for all $t \in T$ almost surely; secondly the dependence on the parameter $t$ has a sufficiently concrete structure. For example in case (6)

$$A_i(\lambda + h) - A_i(\lambda) = \begin{pmatrix} 0 & 0 \\ 0 & h \end{pmatrix}$$

In spite of such a simple type of perturbation the answer in this case is not trivial.

We need the following result proved in [13].

Let $\xi_i(t, \omega)$ be a sequence of random processes, $i = 1, 2, \ldots$, $t \in T$, $T$ a compact metric space, $\omega \in \Omega$. From now on we shall assume the processes $\xi_i(t)$ to be measurable and separable and $E \sup_{t \in T} |\xi_i(t)| < \infty$. Besides that we shall suppose $\xi_i(t, \omega)$ to be strictly stationary in $i$ in the sense of the following definition.

Definition 3. The sequence of random processes $\xi_i(t, \omega)$ is called stationary in $i$, $i = 1, 2, \ldots$ if $\Pr\{\xi_{k+1} \in B_1, \xi_{k+2} \in B_2, \ldots \xi_{k+s} \in B_s\}$ is independent of $k$ for all $k \geq 0$, $s > 0$ and arbitrary Borel sets $B_1, B_2, \ldots, B_s$ belonging to the $\sigma$ algebra of the space of functions on $T$.

Theorem 1. 1. *Assume that the sequence of processes* $\{\xi_i(t, \omega)\}_{i=1}^{\infty}$ *is stationary in* $i$ *and that the shift in* $i$ *is an ergodic transformation.*

If $T = [0, 1]$ and $\xi_i(t, \omega)$ for every $t \in T$ are continuous from the right or from the left with probability 1 then almost surely for all $t \in T$

$$n^{-1} \sum_{i=1}^{n} \xi_i(t, \omega) \to E\xi_1(t) \tag{9}$$

If $T$ as any compact metric space and the $\xi_i(t, \omega)$ are continuous at every point with probability 1, then the convergence in (9) is uniform in $t$.

In some special situation we shall also use the following lemma ([13]). We set $S_n(t,\omega) = n^{-1}\sum_{i=1}^{n} \xi_i(t,\omega)$, $m(t) = E\xi(t)$.

Lemma 1.1. Let the conditions of Theorem 1.1 hold and assume that the processes $\xi_i(t,\omega)$ are continuous in the point $t_0 \in T$. Then for $\epsilon > 0$ there exists a $\delta > 0$ such that almos surely for all $t \in (t_0 - \delta, t_0 + \delta)$

$$\limsup_{n\to\infty} S_n(t,\omega) - \liminf_{n\to\infty} S_n(t,\omega) < \epsilon$$

and

$$\lim_{n\to\infty} |S_n(t,\omega) - m(t_0)| < \epsilon$$

Let $\Gamma$ be an arbitrary Borel set, $\Gamma \subset SL(m, C)$. We define the measure $\nu_t(\cdot)$ by the following formula

$$\nu_t(\Gamma) = \Pr\{A(t,\omega) \in \Gamma\} \tag{10}$$

Let $G(t)$ be the subgroup of $SL(m, C)$ obtained by taking the closure of the group generated by the support of the measure $\nu_t$: $C^m$ is m-dimensional complex space with the ordinary scalar product, the corresponding Euclidean norm will be denoted by $\|\cdot\|$. $P^{m-1}$ denotes projective space. It will be convenient for us to consider $P^{m-1}$ as the sphere $S^{m-1} \subset C^m$, modulo the identification of points x and y for which $r(x,y) = 0$ where $S^{m-1} = \{x \subset C^m : \|x\| = 1\}$ and $r(x,y) = (1 - |(x,y)|^2)^{\frac{1}{2}}$.

For any matrix $A(\det A \neq 0)$ we define the mapping $P^{m-1} \to P^{m-1}$ by the formula $x \to \|xA\|^{-1} xA (x \in P^{m-1})$.

We consider a Markov chain on $P^{m-1}$ with transition probabilities defined by

$$F_t(x, B) = \Pr\{\|xA(t,\omega)\|^{-1} xA(t) \in B\} \tag{11}$$

for $x \in P^{m-1}$ and $B$ an arbitrary Borel set of $P^{m-1}$. Let $\mu_t$ be an invariant measure of the chain:

$$\mu_t(B) = \int \mu_t(dx)\, F_t(x, B)$$

Furstenberg's theorem ([2]). Let $G(t)$ be a noncompact subgroup of $SL(m, C)$ such that each of its subgroups of finite index is irreducible.

Then for any fixed $x \in P^{m-1}$ with probability 1

$$\lim_{n\to\infty} n^{-1} \ln \|xA(n, t, \omega)\| = \alpha(t) > 0 \tag{12}$$

where

$$\alpha(t) = \iint_{P^{m-1}\times SL(m, C)} \ln \|xA\| \mu_t(dx)\nu_t(dA)$$

($\alpha(t)$ is independent of $\mu_t$).

From (12) it is easy to obtain for fixed $t \in T$ w. p. 1:

$$\lim_{n\to\infty} n^{-1} \ln \|A(n, t, \omega)\| = \alpha(t) .$$

Now we list some assumptions which are common for most of our theorems:

$A_i(t, \omega)$, $i = 1, 2, \ldots,$ is a sequence of independent random processes $(t \in T)$; (13)

$T$ is a compact metric space; (14)

$$E \sup_{t\in T} \ln \|A_i(t)\| < \infty ; \tag{15}$$

$$\det A_i(t) = 1 ; \tag{16}$$

$A_i(t, \omega)$ is continuous in $t$ w. p. 1; (17)

Conditions for Furstenberg's theorem are fulfilled for any fixed $t \in T$. (18)

## II. SUFFICIENT CONDITIONS FOR EXPONENTIAL DIVERGENCE OF THE PRODUCT OF MATRICES

Theorems 2. 2 and 2. 3 of this paragraph solve the first problem formulated in § 1 and are the main ones in this work. The main difficulty arises from the non-commutativity of the product of matrices. It is natural to reduce the question to the ergodic theorem in a function space by analogy with the case when the parameter $t$ is absent in which case the problem was solved by applying the usual ergodic theorem.

Theorem 2. 1 shows that this approach works for many cases.

The investigation of the problem mentioned in §1 shows that the answer depends on the continuity properties of trajectories of a random process $\xi(t)$ which takes values in $P^{m-1}$ and is invariant with respect to the transformation

$$\xi(t) \to \|\xi(t)A(t,\omega)\|^{-1}\xi(t)A(t,\omega)$$

The exact definition is given below.

Definition 1. A random process $\xi(t)$ is called invariant with respect to the transformation $\xi(t) \to \|\xi(t)A(t,\omega)\|^{-1}\xi(t)A(t,\omega)$, if $\xi(t)$ and $A(t,\omega)$ are independent and the finite-dimensional distributions of $\xi(t)$ coincide with the distributions of

$$\|\xi(t)A(t,\omega)\|^{-1}\xi(t)A(t,\omega)$$

From now on we shall call $\xi(t)$ an invariant process for the sake of brevity.

We shall construct an invariant process $\xi(t) = \xi(t,\omega)$ on a probability space $(\Omega^{\sim}, \mathfrak{F}^{\sim}, \Pr^{\sim})$ which is unrelated to the space $\Omega$ discussed above. This will be done by means of Kolmogorov's extension theorem [14]. For this purpose we want to find the finite dimensional distributions

$$\mu_{t_1,\dots,t_k}(B) = \Pr\{(\xi(t_1), \xi(t_2), \dots, \xi(t_k)) \in B\}$$

for B a Borel set in $\mathfrak{D}_k = P^{m-1} \times P^{m-1} \times \dots \times P^{m-1}$ (k times). The conditional distributions are defined by analogy with formula (11) in §1:

$$F_{t_1,\dots,t_k}((\xi_1,\dots,\xi_k), B) =$$
$$= \Pr\{(\|\xi_1 A(t_1)\|^{-1}\xi_1 A(t_1), \dots, \|\xi_k A(t_k)\|^{-1}\xi_k(t_k)) \in B\} \tag{1}$$

Because $\xi$ is invariant

$$\mu_{t_1,t_2,\dots,t_k}(B) = \int_{\mathfrak{D}_k} \mu_{t_1,\dots,t_k}(d(\xi_1,\xi_2,\dots,\xi_k)) F_{t_1,t_2,\dots,t_k}(\xi_1,\xi_2,\dots,\xi_k;B) \tag{2}$$

The right-hand side of (2) generates a continuous operator (in the topology of weak convergence) in the space of probability measures on the compact $\mathfrak{D}_k$. The space of measures is a compact one According to Tichonov's fixed point theorem there exists a solution $\mu_{t_1,t_2,\dots,t_k}(\cdot)$ of (2). If $B = B^{\sim} \times P^{m-1}$ where $B^{\sim} \subset \mathfrak{D}_{k-1}$ then

$$F_{t_1,t_2,\dots,t_k}(\xi_1,\xi_2,\dots,\xi_k;B) = F_{t_1,t_2,\dots,t_{k-1}}(\xi_1,\xi_2,\dots,\xi_{k-1};B^{\sim}) .$$

That is why the restriction of $\mu_{t_1,t_2,\dots,t_k}(\cdot)$ to the subspace $\mathfrak{D}_{k-1}$ satisfies (2) with the kernel

$$F_{t_1,t_2,\dots,t_{k-1}}(\xi_1,\xi_2,\dots,\xi_{k-1};B^{\sim})$$

If in addition it were known that the solution of (2) is unique Kolmogorov's consistency condition for finite dimensional distributions would be immediate (in the other case it is not clear how to choose a consistent family of distributions).

Theorem 2.1. Suppose that for any fixed $t \in T$; $\xi, \zeta \in C^m (\xi \neq 0, \zeta \neq 0)$ the angle between the vectors $\xi A(n, t, \omega)$ and $\zeta A(n, t, \omega)$ tends to zero almost surely as $n$ tends to infinity (the exceptional null set may depend on $t$, $\xi$, $\zeta$). Suppose also that assumptions (13) - (17), § 1 are fulfilled.

Then the equation (2) has a unique solution. The corresponding process exists and $E(r(\xi(t), \xi(S))) \to 0$ when $t \to S$.

Proof: Suppose (2) does not have unique solution. Let us consider a Markov chain on $\mathcal{D}_k$ with transition probabilities defined by (1); for the initial distribution we take some solution of (2). Then the chain is stationary. As the solution is not unique there exists a nontrivial decomposition of the chain into ergodic components. We consider two such ergodic components with their initial distributions $\mu_0$ and $\mu_1$ being different. Let $f(\zeta_1, \ldots, \xi_k)$ be a continuous function on $\mathcal{D}_k$ for which

$$\int_{\mathcal{D}_k} f(\xi_1, \xi_2, \ldots, \xi_k) \mu_0 (d(\xi_1, \xi_2, \ldots, \xi_k))$$

$$\neq \int_{\mathcal{D}_k} f(\xi_1, \xi_2, \ldots, \xi_k) \mu_1 (d(\xi_1, \xi_2, \ldots, \xi_k)) \tag{3}$$

We shall compare averages

$$S_n(\mu_0) = n^{-1} \sum_{i=0}^{n-1} f(\xi_1^{(i)}, \xi_2^{(i)}, \ldots, \xi_k^{(i)}) \tag{4}$$

$$S_n(\mu_1) = n^{-1} \sum_{i=0}^{n-1} f(\eta_1^{(i)}, \eta_2^{(i)}, \ldots, \eta_k^{(i)}) \tag{5}$$

where $(\xi_1^{(i)}, \xi_2^{(i)}, \ldots, \xi_k^{(i)})$ and $(\eta_1^{(i)}, \eta_2^{(i)}, \ldots, \eta_k^{(i)})$, $i = 0, 1, 2, \ldots$ are sequences of the elements of chains with initial distributions $\mu_0$ and $\mu_1$ respectively. As $\mu_0$ and $\mu_1$ generate the ergodic chains then for the initial points $(\xi_1^{(0)}, \xi_2^{(0)}, \ldots, \xi_k^{(0)})$ and $(\eta_1^{(0)}, \eta_2^{(0)}, \ldots, \eta_k^{(0)})$ typical in the sense of these measures

$$S_n(\mu_j) \int_{\mathcal{D}_k} f(\xi_1, \xi_2, \ldots, \xi_k) \mu_j (d(\xi_1, \xi_2, \ldots, \xi_k)) \tag{6}$$

with probability 1. Hence if points $(\xi_1^{(0)}, \xi^{(0)}, \ldots, \xi_k^{(0)})$, $\eta_1^{(0)}, \eta_2^{(0)}, \ldots, \eta_k^{(0)})$ are fixed, then the corresponding conditional probability of the fact, that there is a convergence of averages to the same limits, will be equal to 1.

The construction of our chain is such that $\xi_\ell^{(i)}$ can be defined by formula

$$\begin{aligned} \xi_\ell^{(i)} &= \| \xi_\ell^{(i-1)} A_i(t_\ell, \omega) \|^{-1} \xi_\ell^{(i-1)} A_{i-1}(t_\ell, \omega) \\ &= \| \xi_\ell^{(0)} A(i-1, t_\ell, \omega) \|^{-1} \xi_\ell^{(0)} A(i-1, t_\ell, \omega) \end{aligned}$$

$(\ell = 1, 2, \ldots, k)$ and similarly with $\xi$ replaced by $\eta$.

According to the conditions of the theorem the angle between the vectors $\xi_\ell^{(i)}$ and $\eta_\ell^{(i)}$ tends to zero as $i \to \infty$ almost surely for all $\ell$. This means that $r(\xi, \eta) \to 0$ as $i \to \infty$. As the function $f(\xi_1, \xi_2, \ldots, \eta_k)$ is continuous then

$$|f(\xi_1^{(i)}, \xi_2^{(i)}, \ldots, \xi_k^{(i)}) - f(\eta_1^{(i)}, \eta_2^{(i)}, \ldots, \eta_k^{(i)})| \to 0 \quad \text{as} \quad i \to \infty$$

It follows that the right parts of (4) and (5) tend to the same limit and this together with (6) contradicts (3). So the uniqueness is proved.

As was mentioned above the uniqueness allows us to apply

Kolmogorov's theorem and so the invariant process $\xi(t)$ can be constructed.

Let us show that $E(r(\xi(t), \xi(s)) \to 0$ as $t \to s$. As $r(\xi, \eta)$ is a continuous function on $\mathfrak{D}_2 = P^{m-1} \times P^{m-1}$ and $r(\xi, \xi) = 0$, it is sufficient to prove that for any continuous function $f(\xi, \eta)$ on $\mathfrak{D}_2$

$$\lim_{t \to s} E(f(\xi(t), \xi(s))) = E(f(\xi(s), \xi(s)))$$

Suppose that this fact fails. By definition of invariant measure

$$E(f(\xi(t), \xi(s))) = \int_{\mathfrak{D}_2} f(\xi, \eta)\mu_{t,s}(d(\xi, \eta)) =$$

$$= \int_{\mathfrak{D}_2} \mu_{t,s}(d(\xi, \eta)) \cdot \int_{\mathfrak{D}_2} F_{t,s}[(\xi,\eta), d(\xi_1,\eta_1)]\, f(\xi_1,\eta_1) \quad (7)$$

Assume that there exists a sequence $t_i \to s$ and function $f(\cdot\,,\cdot)$ such that

$$E(f(\xi(t_i), \xi(s))) \not\to E(f(\xi(s), \xi(s))) \quad (8)$$

As $\mathfrak{D}_2$ is a compact space we may assume that the measures $\mu_{t_i, s}(\cdot)$ converge weakly to some measure $\mu_s^{\sim}$.

If we prove that

$$\int_{\mathfrak{D}_2} F_{t,s}(\xi,\eta; d)(\xi_1,\eta_1))f(\xi_1,\eta_1) \to F_{s,s}(\xi,\eta; d(\xi_1,\eta_1))f(\xi_1,\eta_1) \quad (9)$$

uniformly in $(\xi, \eta)$ as $t \to S$, then we can take the limit in the last member of (7) inside the integral and it will follow that $\mu_s^{\sim}$ is an invariant measure for the chain with transition probabilities $F_{ss}((\xi,\eta), B)$. This contradicts (8), for the invariant measure is unique. This will complete the proof of the theorem.

The uniformity in $(\xi, \eta)$ of the convergence in (9) is a consequence of the following lemma.

<u>Lemma 2.1.</u> <u>The family of functions depending on the parameters</u> $t, s$ <u>given by the left-hand side of</u> (9) <u>is compact in the topology of uniform convergence.</u>

<u>Proof of the lemma</u>: We set $S_2 = SL(m, C) \times SL(m, C)$,

$$\nu_{t,s}(B) = \Pr\{(A(t), A(s)) \in B\} \qquad (B \text{ is a Borel set in } S_2)$$

By the definition of $F_{t,s}(\cdot\,,\cdot)$ we have

$$\int_{D_2} F_{t,s}(\xi, \eta; d(\xi_1, \eta_1)) f(\xi_1, \eta_1) =$$

$$= \int_{S_2} \nu_{t,s}(d(A_1, A_2)) \cdot f(\|\xi (A_1\|^{-1} \xi A_1, \|\eta A_2\|^{-1} \eta A_2)$$

Let us take for granted for the moment that for any $\epsilon > 0$ there exists a compact set $K_\epsilon \subset S_2$ such that

$$\sup_{t, s \in T} \nu_{t,s}(S_2 \setminus K_\epsilon) < \epsilon \tag{10}$$

The function $f(\|\xi A_1\|^{-1} \xi A_1, \|\eta A_2\|^{-1} \eta A_2)$ is uniformly continuous in $(\xi, \eta, A_1, A_2)$, if $(A_1, A_2) \in K_\epsilon$, so the function

$$\int_{K_\epsilon} \nu_{t,s}(d(A_1, A_2)) f(\|\xi A_1\|^{-1} \xi A_1, \|\eta A_2\|^{-1} \eta A_2)$$

is equicontinuous in $(\xi, \eta)$. Since

$$\int_{S_2 \setminus K_\epsilon} \nu_{t,s}(d(A_1 A_2)) f(\|\xi A_1\|^{-1} \xi A_1, \|\eta A_2\|^{-1} \eta A_2) \leq \epsilon \max_{(\xi,\eta)} f(\xi, \eta)$$

our family of functions is seen to be equicontinuous in $(\xi, \eta)$.

Let us prove that (10) is true. In the inequality

$$\Pr\{\sup_t \|A(t, \omega)\| > C\} =$$

$$= \Pr\{\sup_t \ln \|A(t, \omega)\| > \ln C\} \leq (\ln C)^{-1} E(\sup_t \ln \|A(t, \omega)\|) \tag{11}$$

We choose a constant C to make the right-hand side of (11) less than $\epsilon/2$. We set

$$K_\epsilon = \{ (a_1, a_2) : \|a_1\| < C;\ \|a_2\| < C;\ a_1, a_2 \in SL(m, C)\}$$

Then $\nu_{t,s}(S_2 \setminus K_\epsilon) = \Pr\{\|A(t)\| > C \text{ or } \|A(s)\| > C\}$

$$\leq 2\Pr\{\sup \|A(t)\| > C\} \leq \epsilon$$

The lemma is proved and so the proof of the theorem (21) is complete.

Corollary 1. An index $\alpha(t) = \iint \mu_t(d\xi)\nu_t(dA) \ln \|\xi A\|$ is a continuous function of t because the process $\xi(t)$ is stochastically continuous and $E(\sup_t \ln \|A(t, \omega)\|) < \infty$ .

Remarks. (1) The convergence in (7) while $\xi, \eta$ are fixed results from the assumption of a weak continuity of $A(t, \omega)$. That is why we could require $A(t, \omega)$ to be only stochastic continuous, while proving Theorem 2. 1. But this condition is not enough for Theorems 2. 2 and 2. 3.

(2) The conditions of Theorem 2. 1 concerning the angles between the vectors $\xi A(n, t, \omega)$ and $\eta A(n, t, \omega)$ holds if the maximal Lyapunov's index $\alpha(t)$ is strictly greater than any of the other indices (cf [11]). It always takes place in the two-dimensional case.

(3) Since the process $\xi(t)$ has been constructed by means of its finite dimensional distributions, it is natural to take its measurable and separable version. Such a version exists because the process is stochastically continuous (cf. [14]). The problem whether such a version is continuous w. p. 1 turns out to be far more difficult. As we shall see this is not always the case. If an a. s. continuous version exists Theorem 2. 2 applies.

Theorem 2. 2. Let the conditions of Theorem 2. 1 hold. If there exists a version of the process $\xi(t)$ which for every $t_0$ is almost surely continuous at $t_0$, then w. p. 1

$$n^{-1} \ln \|A(n, t, \omega)\| \to \alpha(t) \quad \text{uniformly in } t \in T \tag{12}$$

($\alpha(t)$ is defined in Furstenberg's theorem, § 1).

Proof: We define a sequence of random processes:

$$\xi_{i+1}(t, \omega) = \|\xi_i(t, \omega) A_i(t, \omega)\|^{-1} \xi_i(t, \omega) A_i(t, \omega), \qquad i = 1, 2, \ldots$$

where $_1(t) = \xi(t)$, $A_i(t, \omega)$ is our sequence of matrices. The $\xi_i(t,\omega)$ have the same continuity properties as $\xi_1(t)$, for the $A_i(t, \omega)$ are continuous. Let us consider the sequence of vector functions: $X_i(t, \omega) = \{\xi_i(t, \omega), \ A_i(t, \omega)\}$. It is easy to see that this new sequence $X_i$ is stationary in $i$ in the sense of Definition 3, § 1, for it is markovian and its initial distribution is invariant with respect to the transformation $\xi_{i-1} \to \xi_i$.

We use the following fact, analogous to the situation when there is no extra parameter

$$\ln \|\xi_1(t) A(n, t, \omega)\| = \ln \|\xi_1(t) A(n-1, t, \omega)\| + \ln \|\xi_n(t) A_n(t, \omega)\|$$

and

$$n^{-1} \ln \|\xi_1(t) A(n, t, \omega)\| = n^{-1} \sum_{i=1}^{n} \ln \|\xi_i(t) A_i(t, \omega)\|$$

We define

$$\zeta_i(t, \omega) = \ln \|\xi_i(t, \omega) A_i(t, \omega)\|$$

The following assertions are true for the sequence $\zeta_i(t, \omega)$:

(a) The sequence $\zeta_i(t, \omega)$ is stationary in $i$, for it is a function of the stationary sequence

(b) The ergodicity of the induced measure on the space of sequences $\{\zeta_i(t, \omega)\}_{i=1}^{\infty}$ results from the ergodicity of the Markov chain $X_i$ (the latter results from the uniqueness of invariant measure)

(c) The continuity of $\zeta_i(t, \omega)$ at every fixed point $t_0$ results from the conditions of Theorem 2. 2

(d) $E(\sup_t |\zeta_i(t, \omega)| \leq m \cdot E(\sup_t \ln \|A_i(t, \omega)\|) < \infty$

The properties (a) - (d) of the sequence $\zeta_i(t, \omega)$ show that the conditions of the Theorem 1. 1 are fulfilled. Hence

$$n^{-1} \sum_{i=1}^{n} \zeta_i(t, \omega) \to \alpha(t) \qquad n \to \infty$$

uniformly in t almost surely. Thus

$$n^{-1} \ln \|\xi_1(t) A(n, t, \omega)\| \to \alpha(t) \tag{13}$$

uniformly in t w. p. 1.

By consturction, the distribution of $\{X_i(t, \omega)\}_{i=1}^{\infty}$ is equivalent to the measure in the Cartesian product of the spaces $\Omega^{\sim} \times \Omega$ with the measure $\Pr^{\sim}(\cdot) \times \Pr(\cdot)$ $(\xi_0(t) = \xi(t, \omega^{\sim})$ $\omega^{\sim} \in \Omega^{\sim}$-- cf. Theorem 2. 1). Since the measure of those pairs $(\omega^{\sim}, \omega)$ for which (13) holds is equal to 1, (13) holds for almost all values of $\xi_1(t, \omega^{\sim})$.

We have to derive (12) from (13). It is evident that $\liminf_{n\to\infty} n^{-1} \ln \|A(n, t, \omega)\| \geq \alpha(t)$. The one-dimensional distributions of $\xi(t)$ coincide with the invariant measure of the chain, which has been considered in Furstenberg's theorem and the invariant measure is unique. For any open neighborhood of an arbitrary point in the support of the measure $\mu_t(\cdot)$ in $P^{m-1}$ there is a nonzero probability that the point $\xi_1(t, \omega^{\sim})$ falls in this neighborhood.

The irreducibility of the group $G(t)$ (in Furstenberg's theorem) means in particular, that the support of the measure $\mu_t(\cdot)$ in $P^{m-1}$ is m-dimensional in the following sense: there exist m points $x_1, x_2, \ldots, x_m \in P^{m-1}$ which are linearly independent (when regarded as vectors in $C^m$) and the measure $\mu_t$ of an arbitrary small neighborhood of any $x_i$ is positive.

Fix $t_0 \in T$ and choose $x_1, x_2, \ldots, x_m \in P^{m-1}$ as above and find small neighborhoods $V_i \subset P^{m-1}$ of $x_i$ such that $y_1, \ldots, y_m$ are linearly independent whenever $y_i \in V_i$, $1 \le i \le m$, and $V_i \cap V_j = \emptyset$ if $i \neq j$.

Denote by $\mathfrak{U} \subset T$ some neighborhood of $t_0$ and denote by $\Omega_i^{\sim} = \{\omega^{\sim}: \xi_0(t, \omega^{\sim}) \in V_i$ for all $t \in \mathfrak{U}\}$.

It follows from the choice of $x_i, V_i$ and the fact that $\xi_0(\cdot, \omega^{\sim})$ is continuous at $t_0$ for $Pr^{\sim}$ almost all $\omega^{\sim}$ that for some $\mathfrak{U}$

$$Pr^{\sim}\{\Omega_i^{\sim}\} > 0, \quad 1 \le i \le m$$

Fix such an $\mathfrak{U}$ and choose the trajectories $\xi_0^{(i)}(t, \omega^{\sim})$ with $\omega^{\sim} \in \Omega_i^{\sim}$. Since $\xi_0^{(i)}(t) \in V_i$ implies the linear independence of $\xi_0^{(i)}(t)$, $1 \le i \le m$, and (13) holds, we see that for $Pr$ almost all $\omega$

$$\limsup_{n \to \infty} n^{-1} \ln \|A(n, t, \omega)\|$$
$$\le \lim_{n \to \infty} n^{-1} \max_{1 \le i \le m} \ln \|\xi_0^{(i)}(t) A(n, t, \omega)\| = \alpha(t)$$

uniformly for $t \in \mathfrak{U}$. Thus (12) holds uniformly for $t \in \mathfrak{U}$ w. p. 1. Since $T$ is compact we can cover it by finitely many such neighborhoods and (12) holds uniformly in $t \in T$. The proof of Theorem 2. 2 is complete.

Corollary 2. If $T = [0, 1]$ and

$$\int_{\mathcal{D}_2} \mu_{t,s}(d(\xi, \eta))\, r^{\gamma}(\xi, \eta) < \text{const} \cdot |t - s|^{1+\epsilon} \tag{14}$$

where $\gamma > 0$, $\epsilon > 0$, then (12) holds uniformly in $t$ with probability 1.

Proof: According to Kolmogorov's criterion (14) shows that there exists a continuous version of the process $\xi(t)$ and so Theorem 2. 2 can be applied.

Remark. Theorem 2. 2 can be proved when the conditions of Theorem 2. 1 are not fulfilled but it is known that there exist an invariant process which is continuous for all $t$. In this case it is enough to consider an ergodic component of the sequence $X_i(t, \omega)$ and to note that $\alpha(t)$ does not depend on the choice of $\mu_t(\cdot)$ as shown by Furstenberg in [2]).

Theorem 2. 3. Let $T = [0,1]$ and let the conditions of Theorem 2. 1 be fulfilled and assume that the process $\xi(t)$ has a version which is continuous on the right (or left) at any fixed point $t_0 \in T$ w. p. 1. Then (12) holds almost surely simultaneously for all $t \in T$; uniform convergence in $t$ may fail.

The proof of this theorem is identical with the proof of Theorem 2. 2, it is only necessary to use the first part of Theorem 1. 1.

## III. NECESSARY AND SUFFICIENT CONDITIONS OF EXPONENTIAL DIVERGENCE OF THE PRODUCT OF TWO-DIMENSIONAL MATRICES

While proving Theorem 2. 2 we applied the ergodic theorem to the sequence of the processes $\zeta_i(t, \omega) = \ln \|\xi_i(t)A_1(t, \omega)\|$. The main condition used there was the continuity of $\xi_i(t, \omega)$. This condition being only sufficient the question is if we can omit the condition of continuity and analyze the behavior of the norm of the matrix product in some other way.

We assert that at least in the two-dimensional case the above method already gives the best possible result. From now on we consider only the two-dimensional case.

First of all we remark that the conditions of Theorem 2. 1 in this case follow from (13) - (18), § I (cf. Remark 2, § 2). Secondly, the equality $\|A\| = \|A^{-1}\|$ if $\det A = 1$ takes place in the a. s. case. Both of these facts will be used by us.

Theorem 3.1. Let conditions (13)-(18), §I hold. Then the existence of a continuous version of $\xi(t)$ is a necessary and sufficient condition for

$$n^{-1} \ln \|A(n, t, \omega)\| \to \alpha(t) \text{ uniformly in } t \in T \tag{1}$$

to be held w. p. 1. If this occurs then the sequence $A_i^{-1}(t, \omega)$ also has the corresponding property:

$$n^{-1} \ln \|A_1^{-1}(t, \omega) A_2^{-1}(t, \omega), \dots, A_n^{-1}(t, \omega)\| \to \alpha(t) \tag{2}$$

uniformly in $t \in T$ w. p. 1.

Proof: We have to prove only the necessity. We point out that the continuity of $\alpha(t)$ and its strict positivity are essential for the proof of our theorem.

First of all we will formulate three lemmas, from which the theorem will be derived.

Lemma 3.1. Let $\mathfrak{a}_i$, $i = 1, 2, \dots,$ be a sequence of unimodular matrices. Define the sequence $x_n \in P^1$ by means of the following equation:

$$\|x_n \mathfrak{a}_1 \mathfrak{a}_2 \dots \mathfrak{a}_n\| = \inf_{z : \|z\| = 1} \|z \mathfrak{a}_1 \mathfrak{a}_2 \dots \mathfrak{a}_n\| \tag{3}$$

((3) defines $x_n$ uniquely as an element of $P^1$ if $\|\mathfrak{a}_1 \mathfrak{a}_2 \dots \mathfrak{a}_n\| > 1$).

Set $\mathfrak{a}(n) = \mathfrak{a}_1 \mathfrak{a}_2 \dots \mathfrak{a}_n$. If

$$\sum_{n=1}^{\infty} \|\mathfrak{a}(n+1)\|^{-2} \|\mathfrak{a}_{n+1}\|^2 < \infty \tag{4}$$

then $x = \lim_{n \to \infty} x_n$ exists. If the matrices $\mathfrak{a}_i$ depend continuously on a parameter $t$, $\mathfrak{a}_i = \mathfrak{a}_i(t)$, and the series in (4) converges uniformly in $t$, then $x(t)$ is a continuous function of $t$.

Lemma 3.2. *Let the conditions of Lemma* 3.1 *hold and let* $\mathfrak{a}_0$ *be a matrix with* $\det a_0 = 1$. *Define* $y_n$ *by the equation*

$$\|y_n \mathfrak{a}_0 \mathfrak{a}(n)\| = \inf_{z:\,\|z\|=1} \|z\,\mathfrak{a}_0 \mathfrak{a}(n)\|$$

*and set* $y = \lim_{n\to\infty} y_n$. *Then* $y = \|x\mathfrak{a}_0^{-1}\|^{-1} x\mathfrak{a}_0^{-1}$, *where* $x$ *is defined in Lemma* 3.1.

Lemma 3.3. *Let the conditions of Theorem* 3.1 *hold, then the series*

$$\sum_{n=1}^{\infty} \|A(n+1, t, \omega)\|^{-2} \|A_{n+1}(t, \omega)\|^2 \tag{5}$$

*converges uniformly in* $t \in T$ *almost surely.*

Suppose that (1) is fulfilled. According to Lemma 3.3 the conditions of Lemma 3.1 are true for the sequence $A_n(t, \omega)$ almost surely and we obtain the continuous random process $x(t) = x(t, \omega)$; the vector $x(t, \omega)$ is defined in Lemma 3.1. Let us consider the sequence of random matrices $A_0(t, \omega)$, $A_1(t, \omega), \ldots$ . It is evident that for this sequence (where $A_0$ is identically distributed with $A_i$, $i \geq 1$, and independent of $A_1, A_2, \ldots$) the conditions of Lemma 3.3 are also fulfilled and a random process $y(t, \omega)$ can be defined by Lemma 3.1 as well. It is clear that the finite-dimensional distributions of the processes $y(t, \omega)$ and $x(t, \omega)$ coincide and by Lemma 3.2 $y(t, \omega) = \|x(t, \omega)A_0^{-1}(t, \omega)\|^{-1} x(t, \omega)A_0(t, \omega)$. Besides that $x(t, \omega)$ is independent of $A_0$. We obtain the following lemma.

Lemma 3.4. *Let* (1) *hold. Then there exists a continuous random process which is invariant with respect to the transformation*

$$x(t) \to \|x(t)A_0^{-1}(t, \omega)\|^{-1} \cdot x(t)A_0^{-1}(t, \omega) .$$

It means that Theorem 2.2 can be applied to the sequence $A_i^{-1}(t, \omega)$, i.e. the property (2) is fulfilled and the second assertion of Theorem 3. 1 is proved. From the remark that

$$E \ln \|A_1^{-1}A_2^{-1}\dots A_n^{-1}\| = E \ln \|A_n^{-1}A_{n-1}^{-1}\dots A_1^{-1}\|$$

$$= E \ln \|(A_1 A_2 \dots A_n)^{-1}\| = E \ln \|A_1 A_2 \dots A_n\|$$

(where the first equality holds because the vectors $\{A_1^{-1}, A_2^{-1}, \dots, A_n^{-1}\}$ and $\{A_n^{-1}, A_{n-1}^{-1}, \dots, A_1^{-1}\}$ are identically distributed and the other equalities are simply identities) it follows that $\alpha(t)$ is the same as in (1).

Now owing to (2) Lemma 3. 4 can be applied to the matrices $A_i^{-1}(t,\omega)$, i. e. there exists a continuous random process which is invariant with respect to the transformation $\xi(t) \to \|\xi(t)A(t,\omega)\|^{-1} \times \xi(t)A(t,\omega)$.

The theorem is proved.

The proofs of Lemmas 3. 1, 3. 2, 3. 3 are placed outside the proof of Theorem 3. 1 firstly for convenience of exposition and secondly because Lemma 3. 1 is of special interest to us.

The proof of Lemma 3. 1: Let $g$ be a matrix, $\det g = 1$, $\|g\| > 1$. The vector $x_g$ is defined by the equality

$$\|x_g g\| = \inf_{x:\ \|x\|=1} \|xg\|$$

We shall prove that

$$r(x_g, x_{gg^\sim}) < 2\|g^\sim\|^2 \|gg^\sim\|^{-2} \tag{6}$$

whenever $g^\sim$ is a matrix for which $\det g^\sim = 1$ and

$$\|g^\sim\|^2 \|g\|^{-2} \|gg^\sim\|^{-2} < 1/4 \tag{7}$$

If we take $g = a_1 a_2 \cdot \dots \cdot a_n$, $g^\sim = a_{n+1}$, then we get that (7) is fulfilled for a sufficiently big $n$ and so

$$r(x_n, x_{n+1}) \leq 2 \, \|a(n+1)\|^{-2} \, \|a_{n+1}\|^2$$

Thus by virtue of (4) $x_n$ is a cauchy sequence, so that Lemma 3.1 is a consequence of (6).

To prove (6) we shall use the method of Tutubalin [15]; we shall make more precise the principal lemma [15] for the two-dimensional case (see estimation (12) below).

Represent $g$ in the form of $g = UDV$, where U, V, are unitary matrices, D is a diagonal matrix (this is always possible), i. e.

$$g = \begin{pmatrix} u_1 & u_2 \\ -\bar{u}_2 & \bar{u}_1 \end{pmatrix} \begin{pmatrix} d & 0 \\ 0 & d^{-1} \end{pmatrix} \begin{pmatrix} v_1 & v_2 \\ -\bar{v}_2 & \bar{v}_1 \end{pmatrix}$$

$\|g\| = d$, $\bar{u}_i$ is the complex conjugate of $u_i$. Then $x_g = eU^{-1}$, where the vector $e = (0, 1)$.

Then let us present $Vg^{\sim}$ in the form of $Vg^{\sim} = KV_1$ where $K = \begin{pmatrix} k & 0 \\ h & k^{-1} \end{pmatrix}$ and $V_1$ is a unitary matrix. Then $gg^{\sim} = UDKV_1$.

Lastly $DK = U'D'V' = \begin{pmatrix} u'_1 & u'_2 \\ -\bar{u}'_2 & \bar{u}'_1 \end{pmatrix} \begin{pmatrix} d' & 0 \\ 0 & d'^{-1} \end{pmatrix} \begin{pmatrix} v'_1 & v'_2 \\ -\bar{v}'_2 & \bar{v}'_1 \end{pmatrix}$ and

$gg^{\sim} = UU'DV'V_1$. Since $x_{gg^{\sim}} = eU'^{-1}U^{-1}$, we have $r(x_g, x_{gg^{\sim}}) =$
$(I - |x_g, x_{gg^{\sim}})|^2)^{1/2} = (I - |(eU'^{-1}U^{-1}, eU^{-1})|^2)^{1/2} =$
$= (I - |(eU'^{-1}, e)|^2)^{1/2} = (I - |u'_1|^2)^{1/2} = |u'_2|$.

Thus (6) will be proved if we show that

$$|u'_2| \leq 2 \cdot \|K\|^2 \cdot d^{-2} \tag{8}$$

as $\|K\| = |g^{\sim}|$, $d' = \|gg^{\sim}\|$. From $DK(DK)^* = U'D'^2U'^{-1}$ it follows that $DKK^*DU' = U'D'^2$. From the equality of the upper left elements of the matrices in the last equation we get

$$|u_1'^{-1}u_2'| = k^{-1}|h|^{-1}|d^2k^2 - d'^2| = k|d|d'^{-2}|1-(|h|^2+k^{-2})d^{-2}d'^{-2}|^{-1}$$

(We made use of the fact that $d'^4-(d^2k^2+d^{-2}|h|^2+d^{-2}k^{-2})d'^2+1=0$, $d'^2$ can be found easily).

As $k^2+k^{-2}+|h|^2 = \|K\|^2 + \|K\|^{-2} \leq 2\|K\|^2$, we have $k|h| < \frac{1}{2}(k^2+|h|^2) \leq \|K\|^2$ and $(|h|^2+k^{-2})d^{-2}d'^{-2} \leq 2\|K\|^2 d^{-2}d'^{-2} < \frac{1}{2}$. Now we have:

$$|u_2'| \leq |u_2'u_1'^{-1}| \leq 2\|K\|^{-2}d'^{-2}$$

Lemma 3. 1 is proved.

<u>Proof of Lemma 3. 2:</u> Let $z_n = \|x_n a_0^{-1}\| x_n a_0^{-1}$. It is enough to show that $z_n \to y$. $z_n$ is convergent sequence, because $x_n \to x$. Let $\lim z_n = y^\sim \neq y$. If $n$ is big enough then $r(y_n, z_n) > \frac{1}{2}r(y, y^\sim)$ and for the same $n$, $r(\|y_n a_0\|^{-1}y_n a_0, \|z_n a_0\|^{-1}z_n a_0) = r(\|y_n a_0\|^{-1}y_n a_0, x_n) \geq \epsilon > 0$. This means that $\|y_n a_0\|^{-1}y_n a_0 = c_n x_n + c_n^\sim x_n^\sim$, where $\|x_n^\sim\| = 1$, $(x_n, x_n^\sim) = 0$, and $c_n^\sim \geq \epsilon$. (We remark that $r(\|y_n a_0\|^{-1}y_n a_0, x_n) = (1-|(c_n x_n + c_n^\sim x_n^\sim, x_n)|^2)^{1/2} = (1-|c_n|^2)^{1/2} = |c_n^\sim| \geq \epsilon$.) As $\|x_n^\sim a(n)\| = \|a(n)\|$, we have:

$$\|y_n a_0 a(n)\| = \|y_n a_0\| \|(c_n x_n + c_n^\sim x_n^\sim)a(n)\| = \|y_n a_0\| \cdot \|c_n x_n a(n) + c_n^\sim x_n^\sim a(n)\|$$

$$\geq \|a_0\|^{-1} \cdot (|c_n^\sim| \cdot \|x_n^\sim a(n)\| - \|x_n a(n)\|)$$

$$\geq \|a_0\|^{1} \cdot (\epsilon \|a(n)\| - \|a(n)\|^{-1}) \to \infty$$

as $n \to \infty$. The latter contradicts the definition of $y_n$. Lemma 3. 2 is proved.

<u>Proof of Lemma 3. 3:</u> By the uniform convergence of (1) $\alpha(t)$ is

continuous and we can choose $0 < \epsilon < \frac{1}{2} \min \alpha(t)$. Again by the uniform convergence we then have

$$\|A(n, t, \omega)\| \geq \exp(n(\alpha(t) - \epsilon)) \quad \text{for} \quad n > N(\epsilon, \omega)$$

So

$$\|A(n)\|^{-2} \|A_n\|^2 \leq \exp(-2n(\alpha(t) - \epsilon)) + 2 \ln \|A_n(t)\|)$$

$$= \exp\{-2n(\alpha(t) - \epsilon - n^{-1} \ln \|A_n(t)\|)\} \exp\{-2(\alpha(t) - \epsilon - n^{-1} \sup \ln \|A_n(t)\|)\}$$

We denote $\kappa_i(\omega) = \sup \ln \|A_i(t, \omega)\|$ and $S_n(\omega) = n^{-1} \sum_{i=1}^{n} \kappa_i(\omega)$. Since $E\kappa_i(\omega) < \infty$, $S_n(\omega) \to E\kappa_i$ as $n \to \infty$ almost surely, according to the ergodic theorem. But this means that

$$n^{-1} \kappa_n(\omega) = S_n(\omega) - n^{-1}(n-1) S_{n-1}(\omega) \to 0$$

as $n \to \infty$ almost surely.

Lemma 3. 3 is proved.

In the course of Lemma 3. 1 we proved the following estimate for $r(x_n, x)$:

$$r(x_n, x) \leq \sum_{j=n}^{\infty} r(x_j, x_{j+1}) \leq 2 \sum_{j=n}^{\infty} \|\mathfrak{a}_1 \mathfrak{a}_2 \ldots \mathfrak{a}_{j+1}\|^{-2} \|\mathfrak{a}_{j+1}\|^2$$

(here we suppose that $n$ is big enough so that (6) can be applied.

We present $x$ in the form of $x = \tilde{c}_n \tilde{x}_n + c_n x_n$, where $|\tilde{c}_n| = r(x_n, x)$, $\|\tilde{x}_n\| = \|x_n\| = 1$, $(x_n, \tilde{x}_n) = 0$, as was done in Lemma 3. 2. Then

$$\|x \mathfrak{a}(n)\| \leq |\tilde{c}_n| \cdot \|\tilde{x}_n \mathfrak{a}(n)\| + |c_n| \cdot \|x_n \mathfrak{a}(n)\|$$

$$= r(x_n, x) \|\mathfrak{a}_n\| + |c_n| \|\mathfrak{a}(n)\|^{-1}$$

$$\leq \mathfrak{a}(n)\|^{-1} + 2 \|\mathfrak{a}(n)\| \sum_{j=n}^{\infty} \|\mathfrak{a}(j+1)\|^{-2} \|\mathfrak{a}_{j+1}\|^2 \tag{9}$$

Now the following lemma can be easily proved.

Lemma 3. 5. Let $\mathfrak{a}_i$, $i = 1, 2, \ldots$ be a sequence of two-dimensional matrices, satisfying $\det \mathfrak{a}_i = 1$, $\|\mathfrak{a}_i\| < \text{const} \exp(\gamma \cdot i)$,

$$\limsup_{n\to\infty} n^{-1} \ln \|\mathfrak{a}(n)\| = \overline{\beta}, \quad \text{and} \quad \liminf_{n\to\infty} n^{-1} \ln \|\mathfrak{a}(n)\| = \beta \tag{10}$$

with

$$\overline{\beta} + 2\gamma - 2\beta < 0 \tag{11}$$

Then there exists a vector $x$ such that

$$\limsup_{n\to\infty} n^{-1} \ln \|x\,\mathfrak{a}(n)\| \leq \overline{\beta} + 2\gamma - 2\beta \tag{12}$$

and $x = \lim_{n\to\infty} x_n$, where $x_n$ is defined by the equality

$$\|x_n\,\mathfrak{a}(n)\| = \inf_{z : z \in P^1} \|z\,\mathfrak{a}(n)\|$$

Proof: It is evident that $\beta > 0$ and $\gamma < \beta$. Thus the conditions of Lemma 3. 1 are fulfilled and $x = \lim_{n\to\infty} x_n$ exists. Let $\epsilon > 0$, $\beta - \epsilon > \gamma$ and $N(\epsilon)$ be such that for $n > N(\epsilon)$

$$\exp(\beta - \epsilon)n \leq \|\mathfrak{a}(n)\| \leq \exp(\overline{\beta} + \epsilon)n$$

By using the estimate (9) we obtain

$$\|x\,\mathfrak{a}(n)\| \leq \exp\{-(\beta - \epsilon)n\}$$

$$+ 2\,\text{const} \cdot \exp(\overline{\beta} + \epsilon)n \sum_{j=n}^{\infty} \exp\{-2(j+1)(-\gamma + \beta - \epsilon)\}$$

$$\text{const} \exp\{n(\overline{\beta} + 2\gamma - 2\beta + 3\epsilon)\}$$

Hence

$$\limsup_{n\to\infty} n^{-1} \ln \|x\,\mathfrak{a}(n)\| \leq \overline{\beta} + 2\gamma - 2\beta + 3\epsilon$$

Since $\epsilon$ can be taken arbitrarily small, the lemma is proved.

Now we can see that if the conditions of the lemma hold with $\overline{\beta} = \beta$ and $\gamma$ arbitrarily small (this always takes place in the case of a stationary sequence as Lemma 3. 3 shows) then

$$\limsup_{n \to \infty} n^{-1} \ln \|x \mathfrak{a}(n)\| \leq -\overline{\beta}$$

On the other hand for any $x(\|x\| = 1)$ $\|x \mathfrak{a}(n)\| \geq \|\mathfrak{a}(n)\|^{-1}$.

So

$$\liminf_{n \to \infty} n^{-1} \ln \|\mathfrak{a}(n)\| \geq -\overline{\beta}$$

Hence

$$\lim_{n \to \infty} n^{-1} \ln \|x \mathfrak{a}(n)\| = -\overline{\beta} \tag{13}$$

Remark. In the case when $\mathfrak{a}_i(\omega)$ is a stationary sequence of matrices the almost sure existence of the vector $x(\omega)$ with property (13) follows from Osseledets' theorems (cf. [11]). Conversely for the two-dimensional case the results of [11] can be obtained from Lemma 3. 5. For further results it is important that the existence of an exponentially decreasing vector $x$ follows from property (11) only. Consequently the following corollary holds.

Corollary 1. *Let* $A_i(t, \omega)$ *be a stationary sequence such that almost surely* (1) *holds for all* $t$. *Then there exists* $x(t, \omega)$ *with* $\|x(t, \omega) = 1$ *such that for all* $t$

$$\lim_{n \to \infty} n^{-1} \ln \|x(t, \omega) A(n, t, \omega)\| = -\alpha(t) \tag{14}$$

(here we do not assume that there is uniform convergence in (1), but if the convergence in (1) is uniform, then it is uniform in (14)).

The Corollary follows from Lemmas 3. 3 and 3. 5.

Let us note that if $\overline{\beta} = 0$ then (13) holds for any $x$ and the assertion of Lemma 3. 5 is trivial.

In §II we formulated a sufficient condition of the continuity of the process $\xi(t)$ in terms of asymptotic properties of two-dimensional distributions ($T$ is a segment). Of course it is difficult to check the condition (14) in §II. In the two-dimensional case we shall therefore formulate two sufficient conditions for the existence of a continuous version of the process $\xi(t)$ (these conditions are necessary because of Theorem 3. 1).

Theorem 3. 2. *Let conditions (13) - (18), § I, hold; assume also that for some strictly increasing sequence of integers* $n_k = n_k(\omega)$ *the series*

$$\sum_{k=1}^{\infty} \|A(n_{k+1}, t)\|^{-2} \|A_{n_k+1}(t) A_{n_k+2}(t) \dots A_{n_{k+1}}(t)\|^2 < \infty \qquad (15)$$

*converges uniformly in* $t$ *almost surely. Then there exists a continuous process which is invariant with respect to the transformation* $\xi(t) \to \|\xi(t)A(t,\omega)\|^{-1} \xi(t)A(t,\omega)$, i. e. (1), (2), and (14) hold.

Proof: We set $\mathfrak{a}_k = A_{n_{k-1}} \cdot A_{n_{k-1}+1} \dots A_{n_k}$, $k = 1, 2, \dots$ ($n_0 = 1$). Then we can define a continuous function $x_\omega(t)$ by Lemma 3. 1 for all $t \in T$ (here we change the notation, for a priori it is not known that the resulting function $\omega \to x_\omega(t)$ is measurable when $t$ is fixed, as the sequence $n_k(\omega)$ was not supposed to be measurable).

Now we note, that we can define $x(t, \omega)$ almost surely for every fixed $t \in T$ using Lemma 3. 5 and Furstenberg's theorem (for this $t$). We consider a countable set $\{t_i\}_{i=1}^{\infty} \subset T$, which is everywhere dense in $T$. Due to countability we can define the random process $x(t_i, \omega)$ for all $t_i$, $i = 1, 2, \dots$ and the finite-dimensional distributions of this process, defined for $t \in \{t_i\}_{i=1}^{\infty}$ only, are invariant with respect to the transformation

$$x(t_i) \to \|xA^{-1}(t_i, \omega)\|^{-1} x(t_i)A^{-1}(t_i, \omega)$$

(This is proved in the same way as in Theorem 3. 1).

Next we remark that $x_\omega(t_i) = x(t_i, \omega)$ for all $t_i \in \{t_i\}_{i=1}^{\infty}$ (because $x_\omega(t_i) = \lim_{k\to\infty} x_{n_k}(t_i,\omega) = \lim_{n\to\infty} x_n(t_i, \omega) = x(t_i, \omega)$ by Lemma 3. 1. ) But $x(t)$ is a continuous function of $t$ and so $x(t_i,\omega)$ can be extended continuously to all $T$ with probability 1. The finite-dimensional distributions of the random process obtained by this procedure are also invariant with respect to the transformation $x(t) \to \|x(t)A^{-1}(t, \omega)\|^{-1} x(t)A^{-1}(t, \omega)$ since this process is stochastically continuous. Now we obtain the process $\xi(t)$ by applying Theorem 3. 1 to the matrices $A_i^{-1}(t, \omega)$. The theorem is proved.

Theorem 3. 3. *Suppose that the conditions (13) - (18), § I, hold. If with probability 1 there exists a continuous vector-valued function $x_\omega(t) \neq 0$ such that for all $t$*

$$\|x_\omega(t) A(n_k, t, \omega)\| \to 0 \tag{16}$$

*for some subsequence $n_k = n_k(\omega, t) \to \infty$, then there exists a continuous version of the process $\xi(t)$, and*

$$n^{-1} \ln \|x_\omega(t) A(n, t, \omega)\| \to -\alpha(t)$$

*uniformly in $t$ almost surely.*

The proof of this theorem has almost the same pattern as that of Theorem 3. 2: we construct the same process $x(t_i, \omega)$. Since

$$\|x(t_i, \omega) A(n, t, \omega)\| \to 0, \quad x(t_i, \omega) = x_\omega(t_i),$$

for there cannot exist two points in $P^1$ for which (16) is fulfilled in the case of two-dimensional and unimodular matrices. The remaining part of the proof coincides with the corresponding part for Theorem 3. 2. The theorem is proved.

Remarks. (1) The process $x(t_i, \omega)$ we constructed in Theorem

3. 2 and 3. 3 can always be formed. But as we shall see below, they do not always have a continuous extension to T.

(2) It is important to note that the uniform and exponentially fast convergence of the whole sequence to zero results from the continuity of $x_\omega(t)$ and an arbitrary rate of convergence to zero of the subsequence in (16).

In conclusion we notice that property (1) is characteristic for the Markov chain $\{\xi_i(t), A_i(t,\omega)\}$ in the following sense: if (1) is fulfilled, then there exists a process $\xi_0(t,\omega)$ such that for any bounded continuous function $f(\cdot,\cdot,\cdot)$ on $T \times P^{-1} \times SL(2, C)$ one has almost surely

$$n^{-1}\sum_{k=1}^{n} f(t, \xi_k(t), A_k(t)) \rightarrow Ef(t, \xi_0(t), A_0(t)) \tag{17}$$

uniformly in t. (17) results from the continuity of the invariant process $\xi_0(t)$ and the ergodic theorem. If (17) is fulfilled for $f = \ln \|\xi A\|$ only, then (17) is fulfilled for all continuous functions.

## IV. THE STRUCTURE OF THE SET WHERE THE INVARIANT RANDOM PROCESS IS CONTINUOUS IN THE TWO-DIMENSIONAL CASE

In this paragraph we shall prove a theorem, describing the set where the invariant random process $\xi(t)$ is continuous in the two-dimensional case. It follows from this theorem that there exists a version of $\xi(t)$ which has no random points of discontinuity.

Lemma 4. 1. Let the conditions of Theorem 2. 1 be fulfilled and let $\xi(t)$ be a separable random process which is invariant with respect to the transformation $\xi(t) \rightarrow \xi(t)A(t,\omega)/\|\xi(t)A(t,\omega)\|$. Then if the probability of $\xi(t)$ being continuous (or continuous from one side) at $t_0$ is not equal to zero, it is equal to 1.

Proof: We consider the sequence of random processes

$$\xi_{i+1}(t) = \xi_i(t)A_i(t)/\|\xi_i(t)A_i(t)\|, \quad \xi_1(t) = \xi(t)$$

If the trajectory $\xi_1(t, \omega^\sim)$ for a certain fixed $\omega^\sim$ is continuous at $t_0$, then the whole sequence $\xi_i(t)$ is continuous at $t_0$. We set $\delta_i(\omega) = \limsup_{t \to t_0} r(\xi_i(t), \xi_i(t_0))$. Since the process $\xi_1(t)$ is separable, all $\delta_i(\omega)$, $i = 1, 2, \ldots,$ are measurable random variables. By the above, if $\delta_1(\omega) = 0$, then $\delta_i(\omega) = 0$ for all $i > 1$. It means that the function $\delta_1(\omega)$ with a positive probability is invariant with respect to the shift, and hence $\delta_1(\omega) = 0$ almost surely. The proof is the same in the case of one-sided continuity.

<u>Theorem 4. 1.</u> <u>Let</u> $\xi(t)$ <u>be as in Lemma 4. 1. The set of</u> $t \in T$ <u>in which the process</u> $\xi(t)$ <u>is continuous with probability</u> 1 <u>is an open subset of the compact</u> T.

<u>Proof:</u> We consider the following stationary sequence of random processes: $\eta_i(t) = \ln \|\xi_i(t) A_i(t)\|$, $i = 1, 2, \ldots$ .

As above, if $\xi(t)$ is continuous at $t_0$ w. p. 1, when the $\eta_i(t)$ are continuous at $t_0$. Applying Lemma 1. 1 to the sequence $\eta_i(t)$, we obtain that for any $\epsilon > 0$ there exis a neighborhood of the point $t_0$ such that almost surely for $n > N(\epsilon, \omega)$ and all $t$ in this neighborhood

$$n^{-1} \sum_{i=1}^{n} \ln \|\xi_i(t) A_i(t)\| \geq \alpha(t_0) - \epsilon$$

or, equivalently

$$n^{-1} \ln \|\xi_1(t) A(n, t, \omega)\| > \alpha(t_0) - \epsilon$$

An argument similar to the one at the end of the proof of Theorem 2. 2 (i. e. the use of a big enough collection of $\xi_1(t)$ shows that

$$n^{-1} \ln \|A(n, t, \omega)\| \geq \alpha(t_0) - \epsilon$$

for $n > N(\epsilon, \omega)$ and all $t$ from our neighborhood.

If $\epsilon$ is such that $\alpha(t) - \epsilon > 0$ for the a.s. t then by Lemma 3. 3 the series

$$\sum_{n=1}^{\infty} \|A(n,t)\|^{-2} \|A_n(t)\|^2 < \infty$$

converges uniformly in t from this neighborhood.

This means that Theorem 3. 2 can be applied (if we set $n_k = k$), i. e. there exists a version of the process $\xi(t)$ which is continuous in the chosen neighborhood of the point $t_0$.

We can deal similarly with any point of continuity of the process $\xi(t)$. Since the process $\xi(t)$ may be defined constructively, i. e. as the limit of $\xi_n(t)$ (cf. Lemma 3. 1 where $x(t) = \lim_{n\to\infty} x_n(t)$ the only difference is that the process $\xi_n(t)$ is constructed from the sequence of matrices $A_1^{-1}, A_2^{-1}, \ldots, A_n^{-1}$), $\xi(t)$ is defined uniquely in the intersection of any two neighborhoods as above (if the intersection is nonempty). We get the whole set of points of continuity by taking the union of all constructed neighborhoods of the points of the continuity.

The theorem is proved.

The next theorem is proved similarly.

Theorem 4. 2. *If* $T = [0, 1]$ *and* $\xi(t)$ *is continuous from the right with probability* 1 *at the point* $t_0 \in T$, *then* $\xi(t)$ *is continuous* w. p. 1 in some right semi-neighborhood of the point $t_0$.

## V. THE DESCRIPTION OF CLASSES OF POTENTIALS. SOME PRELIMINARY INFORMATION ON DIFFERENTIAL OPERATORS

As was mentioned in §I the problems of one-dimensional spectral theory of differential operators and difference operators are the natural source of the products of random matrices depending on a parameter. We shall discuss second order operators only, for even this case is not trivial.

Let us consider the equations

$$-y'' + q(x)y = \lambda y \tag{1}$$

$$y_{n+1} + q_n y_n + y_{n-1} = \lambda y_n \tag{2}$$

where $q(x) = q(x,\omega)$ is random process ($x$ is called the "time", $-\infty < x < \infty$, $\omega \in \Omega$), $q_n = q_n(\omega)$ is sequence of random variables. We shall suppose in both cases, that the random processes are stationary or periodically stationary in the strict sense.

Definition 1. A random process is called periodically stationary in the strict sense if there exists a random partition of $(-\infty,\infty)$ into intervals $(x_i(\omega), x_{i+1}(\omega))$, $i = 0, \pm 1, \pm 2, \ldots, x_i < x_{i+1}$, $x_i(\omega)$ measurable random variables, such that the following property holds: let $\mathcal{L} = \mathcal{L}(q(\cdot,\omega), \mathfrak{a}(\omega), \mathfrak{b}(\omega))$ be an arbitrary measurable functional, taking values in a finite dimensional vector space, whose value is determined if the trajectory $q(x,\omega)$ is given for $\mathfrak{a}(\omega) \le x \le \mathfrak{b}(\omega)$ ($\mathfrak{a}(\omega)$ and $\mathfrak{b}(\omega)$ are also measurable random variables); then the sequence of the vectors $\mathcal{L}_i = \mathcal{L}(q(\cdot,\omega), x_i(\omega), x_{i+1}(\omega))$ forms a strictly stationary (in $i$) random process.

We shall always suppose that

$$E(x_{i+1}(\omega) - x_i(\omega)) < \infty \tag{3}$$

$$E \sup_{x \in [x_i, x_{i+1}]} |q(x,\omega)| < \infty \tag{4}$$

Due to Definition 1 the values in the left-hand sides of inequalities (3) and (4) do not depend on $i$.

The basic functionals for us will be the fundamental matrices of equations (1) and (2), i. e. matrices $A_i(\lambda,\omega)$ giving the transformation from the vector of initial conditions $(y(x_i), y'(x_i))$ to the vector $(y(x_{i+1}), y'(x_{i+1}))$ according to the solution of the differential equation (1) on the interval $[x_i, x_{i+1}]$. It is evident that the matrices $A_i(\lambda, \omega)$ form a stationary sequence in $i$ of random processes if $q(\cdot)$ belongs to the above described class of the potentials; here $\lambda$ is the time of the process, and an arbitrary compact to the complex plane can be taken as the compact $T$.

It is easy to construct random processes which satisfy the conditions of Definition 1.

Definition 1 and all the comments after it are applicable to equation (2) as well. The main restriction on $q_n$ is the following:

$$E \ln (|q_n(\omega)| + 1) < \infty \tag{5}$$

It is evident, that condition (5) is a weaker restriction than (4). The possibility of weakening (5) is connected with the fact that the transformation $(y_{i-1}, y_i)$ to $(y_i, y_{i+1})$ which results from equation (2) is governed by the matrix

$$A_i(\lambda, \omega) = \begin{matrix} 0 & -1 \\ 1 & -q_i(\omega) + \lambda \end{matrix}, \tag{6}$$

for which we make the following requirement

$$E(\sup_{\lambda \in T} \ln \|A(\lambda, \omega)\|) < \infty \tag{7}$$

It is enough to consider the case of a metrically transitive sequence of matrices, and in the sequel we restrict ourselves to this case.

Lastly the main assumption is that

$$n^{-1} \ln \|A_1(\lambda, \omega) \cdot A_2(\lambda, \omega) \cdot \ldots \cdot A_n(\lambda, \omega)\| \to \alpha(\lambda) > 0 \tag{8}$$

with probability 1 for every fixed $\lambda$.

As we mentioned in §II, Furstenberg's theorem contains sufficient conditions for (8) to be fulfilled. M. M. Benderskiy and L. A. Pastur (cf. [16]) verified these conditions for our case; in [17] L. A. Pastur pointed out a wide class of Markovian potentials, satisfying (8); the difference equation was also considered in [18].

We shall list some needed facts about the spectrum of random operators corresponding to equations (1) and (2):

$$H(\omega)y = -\frac{d^2y}{dx^2} + q(x,\omega)y \tag{9}$$

$$(\mathcal{H}(\omega)y)_n = y_{n+1} + q_n(\omega)y_n + y_{n-1} \tag{10}$$

By $\mathcal{L}_2(-\infty,\infty)$ we denote the Hilbert space of functions $y(x)$ $(x \in -\infty, \infty))$ satisfying $\int_{-\infty}^{\infty} |y(x)|^2 dx < \infty$; $\ell_2(-\infty,\infty)$ is the space of sequences $\{y_i\}$ satisfying $\sum_{i=-\infty}^{\infty} |y_i|^2 < \infty$. The operators (9) and (10) are usually (cf. [19]) defined on subspaces, which are everywhere dense in $\mathcal{L}_2(-\infty,\infty)$ and $\ell_2(-\infty,\infty)$.

$H_\theta(\omega)$ and $\mathcal{H}_\theta(\omega)$ denote the operators acting on the subspaces $\mathcal{L}_2(0.\infty)$ and $\ell_2(0,\infty)$ and defined by the right-hand sides of formulas (9) and (10) respectively, for functions satisfying the boundary conditions

$$y(0)\sin\theta - y'(0)\cos\theta = 0 \tag{11}$$

$$y_0 \sin\theta - y_i \cos\theta = 0 \tag{12}$$

Since condition (4) is assumed, it follows with probability 1 (for example, due to the ergodic theorem) that

$$x_i^{-1}(\omega) \sup_{x\in[x_i, x_{i+1}]} |q(x,\omega)| \to 0 \text{ as } i \to \infty \tag{13}$$

Hence the closure of the operator $H_\theta$ is selfadjoined with probability 1 (cf. for example, [19] where this fact results from the estimate $q(x) \geq -x^2$ for a sufficiently large $x$; it is clear that this estimate is implied by (13).) The fact that the closure of the operator $\mathcal{H}_\theta$ is selfadjoined may be found in [20] for example.

It is evident that the mentioned facts are also applicable to the operators $H_\theta$ and $\mathcal{H}_\theta$ on the spaces $\mathcal{L}_2(-\infty, 0)$ and $\ell_2(-\infty, 0)$ respectively. Hence the closures of the operators on the left half-axes are also selfadjoined.

It is easy to see that the closure of the operator $H(\omega)$ in the space $\mathcal{L}_2(-\infty, \infty)$ is with probability 1 a selfadjoined operator. The closure of the operator $\mathcal{H}(\omega)$ in $\ell_2(-\infty, \infty)$ is (for every $\omega$) a selfadjoined operator.

Now we list some results from [17] that we shall need later.

(a) The spectra of the operators $H(\omega)$ and $\mathcal{H}(\omega)$ are almost surely independent of $\omega$; we shall denote them by $S_H$ and $S_{\mathcal{H}}$, omitting $\omega$.

(b) The sets $S_H$ and $S_{\mathcal{H}}$ do not contain isolated points, i. e. almost surely the spectra of the operators $H$ and $\mathcal{H}$ are spectra of condensation. We shall denote the spectral function of the operator $H_\theta(\omega)$ by $\rho_\theta(\lambda, \omega)$ (the definition is given in [20]). We formulate a lemma, the proof of which can be found in [21].

<u>Lemma 5.1.</u> <u>For almost all</u> $\lambda$ <u>with respect to the spectral measure</u> $d\rho_\theta(\lambda, \omega)$ <u>the solution of equation (2) with initial conditions</u> $y_0 = \cos\theta$, $y_1 = \sin\theta$ <u>satisfies the inequality</u>

$$|y_n(\lambda)| < c(\lambda, \epsilon)\, n^{1/2+\epsilon} \tag{14}$$

<u>where</u> $c(\lambda, \epsilon)$ <u>is a constant depending on</u> $\lambda, \epsilon, \omega$.

Let us denote the spectrum of the operator $\mathcal{H}_\theta(\omega)$ by $S_\theta(\omega)$.

Further on we need

<u>Lemma 5. 2.</u> <u>With probability 1, $S_{\mathcal{H}} \subset S_{\theta}(\omega)$ and $S_{\theta}(\omega) \setminus S_{\mathcal{H}}$ consists at most of a countable set of isolated points.</u>

<u>Proof:</u> Let us add to the boundary condition (12) the following boundary condition at the point $n + 1$:

$$y_{n+1} = 0 \tag{15}$$

We denote by $\mathcal{H}_{\theta, n}(\omega)$ the operator acting by formula (10) on the space $\ell_2(0, n)$ of functions satisfying conditions (12) and (15). Let us denote by $\nu(n, \lambda, \omega)$ the number of eigenvalues of the operator $\mathcal{H}_{\theta, n}(\omega)$ which are less than $\lambda$. Now consider the function

$$N(n, \lambda, \omega) \stackrel{\text{def}}{=\!=} n^{-1} \nu(n, \lambda, \omega) \tag{16}$$

It is known that almost surely the following limit exists and is nonrandom

$$N(\lambda) = \lim_{n \to \infty} N(n, \lambda, \omega) \tag{17}$$

(Note that the relation (17) can be proved in the same way as the corresponding assertion for the differential equation was proved in [22]; the oscillation theorems used here can be found in [20]).

It is shown in [17] that $S_{\mathcal{H}}$ is the set of points of increase of the function $N(\lambda)$. The points of increase of $N(\lambda)$ are contained in $S(\omega)$, for if $\lambda_0$ is a point of increase, then $\lambda_0$ is a limit point of eigenvalues of the sequence of operators $\mathcal{H}_{\theta, n}$ (as $n \to \infty$), hence $\lambda_0 \in S_{\theta}$ (cf. [21]). So $S_{\mathcal{H}} \subseteq S_{\theta}(\omega)$ with probability 1.

On the other hand it is known ([21]), that the part of the spectrum of the operator $\mathcal{H}_{\theta}$, consisting of nonisolated points, is contained in the spectrum of the operator $\mathcal{H}$ (for all $\omega$). So $S_{\theta} \setminus S_{\mathcal{H}}$ consists of isolated points only.

The lemma is proved.

Remark. The corresponding assertion for differential equations can be proved in the same way.

## VI. THE PRODUCT OF RANDOM MATRICES WITH A PARAMETER, LYING IN THE SPECTRUM AND OUTSIDE THE SPECTRUM

We shall use the notations of § V.

Theorem 6. 1. *Let* $T$ *be an arbitrary compact of the complex plane and* $T \cap S = \emptyset$. *Let the matrices* $A_i(\lambda, \omega)$ *(corresponding to equations (1) and (2) of § V) form a sequence of independent random processes. Then almost surely*

$$n^{-1} \ln \|A(n, \lambda, \omega)\| \to \alpha(\lambda), \quad n \to \infty \tag{1}$$

*uniformly in* $\lambda \in T$.

Proof: If $\lambda \notin S$, then it is known that there exist the Weyl solutions of equations (1) and (2) of § V, i. e. solutions for which $\int_0^\infty |y(x, \lambda)|^2 dx < \infty$, respectively $\sum_{i=0}^{\infty} |y_i(\lambda)|^2 < \infty$. Since the operators $H_\theta(\omega)$ and $\mathcal{H}_\theta(\omega)$ are selfadjoined, these solutions are unique in the following sense: any other solution can be obtained by multiplying our solution with some number. So the solution of equation (1), § I possesses almost surely the property

$$n^{-1} \ln (|y(x_n, \lambda)|^2 + |y'(x_n, \lambda)|^2) \to -\alpha(\lambda) \tag{2}$$

for fixed $\lambda$. Indeed if $\lambda$ is fixed such a solution exists by virtue of property (8) of § V and Lemma 3. 5. From property (3) of § V it follows that

$$x_n^{-1}(\omega) \ln (|y(x_n, \lambda)|^2 + y'(x_n, \lambda)|^2) \to -\alpha(\lambda) \{E(x_2(\omega) - x_1(\omega))\}^{-1}$$

and hence the solution satisfying condition (2) belongs to $\mathcal{L}_2(0, \infty)$,

i. e. it coincides with the Weyl solution. The same fact is also true for the difference equation. Consider a countable everywhere dense subset $T^{\sim}$ of $T$. Since $T^{\sim}$ is countable then property (2) is fulfilled with probability 1 for all $t \in T^{\sim}$. In addition it is known that either $y'(x,\lambda)^{-1}\ y(x,\lambda)$ or $[y(x,\lambda)]^{-1} y'(x,\lambda)$ is an analytic function of the parameter because $T \cap S = \emptyset$. Hence the properties of the vector $(y(x,\lambda),\ y'(x,\lambda))$ which are used in the proof of Theorem 3. 3 are fulfilled. From Theorem 3. 1 and 3. 3 it follows that (1) is true and (2) holds almost surely uniformly in $\lambda \in T$.

The theorem is proved.

Now we shall describe the behavior of the product of random matrices for $\lambda \in S$. We impose the additional condition

$$\inf q(x,\omega) \geq q_0 > -\infty \qquad \text{(for almost all } \omega)$$

Theorem 6. 2. With probability 1, in an arbitrary neighborhood of any point $\lambda' \in S$ ($S$ is the spectrum of the operator $H(\omega)$) it is possible to find a point $\lambda_0 \in S$ such that either

$$\limsup_{n\to\infty} n^{-1} \ln \|A_1(\lambda_0,\omega)A_2(\lambda_0,\omega)\ldots A_n(\lambda_0,\omega)\|$$

$$\geq 2 \liminf_{n\to\infty} n^{-1} \ln \|A_1(\lambda_0,\omega)A_2(\lambda_0,\omega)\ldots A_n(\lambda_0,\omega)\| > 0 \tag{3}$$

or

$$\lim_{n\to\infty} n^{-1} \ln \|A_1(\lambda_0,\omega)A_2(\lambda_0,\omega)\ldots A_n(\lambda_0,\omega)\| = 0 \tag{4}$$

Remark. It is understood that $\lambda_0$ is random and varies with $\omega$. We note that there is at least a countable set of such points and that it is dense in $S$.

Proof: Let $\lambda' \in S$ and assume the assertion of the theorem is false for some neighborhood of the point $\lambda'$. Hence with probability 1

$$g(\lambda, \omega) \stackrel{\text{def}}{=} \limsup_{n\to\infty} n^{-1} \ln\|A(n, \lambda, \omega)\| - 2 \liminf_{n\to\infty} n^{-1} \ln\|A(n,\lambda,\omega)\| < 0 \tag{5}$$

in some neighborhood of the point $\lambda'$.

Equation (5) means in particular that in a $\delta$-neighborhood of the point $\lambda'$ there exists an $N$ and a subneighborhood such that

$$\|A(n, \lambda, \omega)\| > 1 \quad \text{for} \quad n \geq N$$

and for all $\lambda$ from this subneighborhood. Indeed, if this is not true, then for every subneighborhood $\mathfrak{U}$ of the $\delta$-neighborhood of $\lambda'$ and for every $n_0$ there exists an $n \geq n_0$ and $\lambda \in \mathfrak{U}$ with $\|A(n, \lambda, \omega)\| = 1$. Let us choose a point $\lambda_1$ and numbers $\delta_1 > 0, n_1$ such that

$$\|A(n, \lambda, \omega)\| \leq 2 \quad \text{for} \quad \lambda \in [\lambda_1 - \delta_1, \lambda_1 + \delta_1]$$

(this can be done because $\|A(n, \lambda, \omega)\|$ is continuous in $\lambda$). As the set of points, in which $\|A(\cdot, \cdot, \omega)\| = 1$, is everywhere dense in $[\lambda' - \delta, \lambda' + \delta]$, then there exists $n_2 > n_1$ and $\lambda_2 \in [\lambda_1 - \delta_1, \lambda_1 + \delta_1]$ such that $\|A(n_2, \lambda_2, \omega)\| = 1$, and we can again construct $\delta_2 > 0$ so that $\|A(n_2, \lambda, \omega)\| \leq 2$ when $\lambda \in [\lambda_2 - \delta_2, \lambda_2 + \delta_2] \subset [\lambda_1 - \delta_1, \lambda_1 + \delta_1]$. Repeating this procedure we obtain a sequence of nested intervals $[\lambda_i - \delta_i, \lambda_i + \delta_i]$. The intersection of these intervals is non-empty. Let us denote by $\lambda_0$ an arbitrary point of this intersection. Then it is evident that $\|A(n_i, \lambda_0, \omega)\| \leq 2$ for all $n_i$, hence

$$\lim_{i\to\infty} n_i^{-1} \ln \|A(n_i, \lambda_0, \omega)\| = 0 \tag{6}$$

(6) contradicts (5) (for $\liminf_{n\to\infty} n^{-1} \ln\|A(n, \lambda, \omega)\| > 0$).

Thus we showed that in any $\delta$-neighborhood of the point $\lambda$ in which (5) is true there exists a subneighborhood $[a, b]$ in which

$$\|A(n, \lambda, \omega)\| > 1 \tag{7}$$

for all n greater than some number $N(\delta, \lambda', \omega)$.

Let us denote by $\varphi(n, \lambda, \omega)$ an angle such that the norm of the vector

$$\| (\cos \varphi, \sin \varphi) A(n, \lambda, \omega) \| = \| A(n, \lambda, \omega) \|^{-1}$$

and

$$-\frac{\pi}{2} < \varphi < \frac{\pi}{2}$$

It follows from (7) that $\varphi(n, \lambda, \omega)$ is defined uniquely and depends continuously on $\lambda \in [a, b]$ (we define the metric in the space of angles by the formula $|\sin(\varphi_2 - \varphi_1)|$; this coincides with the above mentioned metric in $P^1$, given in angular coordinates).

From Lemma 3.5 and property (5) follows the convergence of $\varphi(n, \lambda, \omega)$ for every $\lambda$ from $[a, b]$ to some $\varphi(\lambda, \omega)$, where $\varphi(\lambda, \omega)$ satisfies

$$\limsup_{n \to \infty} n^{-1} \ln \| (\cos \varphi(\lambda, \omega), \sin \varphi(\lambda, \omega)) A(n, \lambda, \omega) \| \leq g(\lambda, \omega) \quad (8)$$

It will be important for us that according to a well known theorem of Baire the function $\varphi(\lambda, \omega)$ must have continuity points in $S \cap [a, b]$ for it is the limit of the continuous functions $\varphi(n, \lambda, \omega)$.

From (8) it follows that the solution of equation (1) of § V with the boundary conditions

$$y(0, \lambda) = \cos \varphi(\lambda, \omega), \quad y'(0, \lambda) = \sin \varphi(\lambda, \omega)$$

belongs to $\mathcal{L}_2(0, \infty)$. From this fact it also follows that the solution of equation (1) of § V with any other initial condition possesses the following property

$$|y(x_i, \lambda)|^2 + |y'(x, \lambda)|^2 > \text{const} \exp[(-g(\lambda, \omega) - \epsilon) \cdot i] \quad (9)$$

for all $\epsilon > 0$ (the const depends on the initial condition and $\epsilon$).

Up to now we did not use the fact that $\lambda' \in S$, i. e. $S \cap [a,b] \neq \emptyset$ (it is important to notice that as $S$ has no isolated points, it is rich enough).

Now we will compare inequalities (8) and (9) with the elementary corollary of Shnol's theorem (cf. for exampl [21]). Shnol showed that if $q(x) \geq q_0$ then for almost all $\lambda$ with respect to the measure $d\rho_\theta(\lambda)$ and for sufficiently big $x$

$$y(x,\lambda) < C(\lambda,\epsilon_1)x^{1/2+\epsilon_1}$$

($y(x,\lambda)$ is the solution of equation (1) of § V with the initial conditions $y(0,\lambda) = \cos\theta$, $y'(0,\lambda) = \sin\theta$).

We use the fact that with probability 1

$$|q(x,\omega)| \leq C_1(\omega)(|x|+1)$$

which is a consequence of condition (4) of § V). So

$$|y'(x,\lambda)| = |\sin\theta + \int_0^x y(\tau,\lambda)(q(\tau)-\lambda)d\tau|$$

$$\leq 1 + C(\lambda,\epsilon_1)C_1(\omega)\int_0^x \tau^{1/2+\epsilon_1}(|\lambda|+\tau)d\tau ,$$

i. e. if $x$ is large enough

$$|y'(x,\lambda)| \leq C_2(\lambda,\omega,\epsilon_1)x^{5/2+\epsilon_1}$$

It means in particular that with probability 1

$$y^2(x,\lambda) + y'^2(x,\lambda) < C_3(\lambda,\epsilon,\omega)|x|^{5+2\epsilon_1}$$

for almost all $\lambda$ with respect to measure $d\rho_\theta(\lambda,\omega)$ and $x$ large enough. Since the latter inequality contradicts inequality (9), it follows that (8) takes place for almost all $\lambda \in S \cap [a,b]$ with respect to measure $d\rho_\theta(\lambda)$. But this means that for $\lambda \in S \cap [a,b]$

all the eigenfunctions of the operator $H_\theta$ belong to $\mathcal{L}_2(0,\infty)$, and the function $\varphi(\lambda,\omega)$ intersects the level $\theta$ countably often.

Let us consider another operator $H_\psi$, $\psi \neq \theta$. By Lemma 5.2 $S \cap [a,b] \subset S_\psi$ and the analogous reasoning shows that the function $\varphi(\lambda,\omega)$ intersects the level $\psi$ countably often in $S \cap [a,b]$. Consequently, at every point $\lambda \in S \cap [a,b]$ the function $\varphi(\lambda,\omega)$ has a discontinuity of the second kind. This fact contradicts the above mentioned Baire theorem (we remark that from the beginning we consider only those $\omega$ for which $S_\theta(\omega) \supset S$ and hence $S_\psi(\omega) \supset S$).

The theorem is proved.

Corollary 1. We obtained a wide class of examples of random matrices, depending analytically on a parameter and such that the invariant process defined in §II, is in every point of $S$ discontinuous with probability 1, and this property holds whatever version of the process is chosen. This assertion results from Lemma 4.1 and Theorem 4.1, according to which from the continuity with positive probability of the invariant process in some point it follows its continuity in a neighborhood of this point with probability 1. C. Hence it follows from Theorem 2.2 that property (8) of §V holds in this neighborhood, and this contradicts Theorem 6.2.

It is clear that Theorem 6.2 is connected with Perron's theorem about the discontinuity of Lyapunov's characteristic indices, mentioned in §I: here the discontinuity holds with probability 1 and the type of perturbation is very simple.

At last we remark that for matrices corresponding to the difference equation (2) of §V the theorem analogous to Theorem 6.2 is proved by using Lemma 5.1.

ACKNOWLEDGMENTS

The author expresses his deep gratitude to the referee for his careful reading of this work; for correcting my english and misprints of mathematical character.

## REFERENCES

1. H. Furstenberg, H. Kesten, Products of random matrices, Ann. Math. Stat., No. 2, 457-469, 1960.
2. H. Furstenberg, Noncommuting random products, Trans. Amer. Math. Soc., v. 108, No. 3, 377-428, 1963.
3. E. Mourier, Elements aleatoir dans un espace de Banach, Ann. Inst. Henri Poincare, t. 13, 161, 1953.
4. A. M. Lyapunov, Collected works, v. 2, Moscow-Leningrad, Academy of Science of the USSR, 1956. (In Russian).
5. O. Perron, Die Ordnungszahlen der Differentialgleichungssysteme, Math. Z., 31, 748-766, 1929.
6. __________, Die Stabilitatsfrage bei Differentialgleichungen, Math. S., 32, 7-3-728, 1930.
7. __________, Über lineare Differentialgleichungen, bei denen die unabhängige Variable real ist, G. reine und angew. Math. 142, 254-270, 1931.
8. V. M. Millionschikov, Linear systems of differential equations, International Congress of Mathematicians in Nice. 1970. Reports of Soviet Mathematicians, Moscow, "Nauka", 1972. (In Russian).
9. __________, Statistically regular systems, Math. sbornik, 75, I, 154-165, 1968.
10. N. A. Isobov, Linear systems of ordinary differential equations, Results of science and techniques, Ser. Math., v. 12, 71-146.
11. V. I. Osseledets, The multiplicative ergodic theorem. Characteristic Lyapunov's indices of dynamical systems, Trudi Moskov. Math. Soc., v. 19, 179-210, 1968. (In Russian).
12. I. Ja. Goldsheid, Asymptotics of the product of random matrices depending on a parameter, Soviet Math. Dokl., v. 16, No. 5, 1975.

13. I. Ja. Goldsheid, The law of large numbers in some functional spaces, Uspehi Mat. Nauk, v. 31, No. 2, 1976. (In Russian).

14. I. D. Gihman, A. V. Skorohod, The theory of stochastic processes, vol. I, "Nauka", Moscow, 1971, English transl., Die Grundlehren der math. Wissenschaften, Band 210, Springer-Verlag, Berlin and New York, 1974.

15. V. N. Tutubalin, Asymptotics for the distribution of the product of complex unimodular matrices, Vestnik Moskov. Univ. Ser. Mat. Meh. 21, No. 2, 70, 1966. (In Russian).

16. M. M. Benderskiy, L. A. Pastur, About the spectrum of the one-dimensional Schrödinger's equation with a random potential, Mat. Sbornik, v. 82(124), No. 2, 273-284, 1970. (In Russian).

17. L. A. Pastur, About the spectrum of random Jacoby's matrices and of the Schrödinger's equation with a random potential on the whole axis, Physical-Engineering Institute of Low Temperatures, Academy of Sciences of the Ukrainian SSR, Kharkov, (preprint), 1974. (In Russian).

18. K. Ishii, Preprint, Reseafch Inst. Fundam. Phisics, Kioto University, Kioto, 1972.

19. M. A. Naymark, Linear differential operators, Moscow, "Nauka", 1969.

20. F. V. Atkinson, Discrete and continuous boundary problems, Academic Press, New York, London, 1964.

21. I. M. Glasman, Direct methods of the qualitative spectral analysis of sinfular differential operators, Fizmatgiz, Moscow, 1963, English transl., Israel Program for Scientific Translations, Jerusalem, 1965, Davey, New York, 1966.

22. M. M. Benderskiy, L. A. Pastur, The theory of functions, functional analysis and its applications, 20, (Kharkov), 1974. (In Russian).

chapter 9

# SYMMETRICAL RANDOM WALKS ON DISCRETE GROUPS

R. I. Grigorchuk

In this paper groups whose problem of identity is solved by Dehn's algorithm are proven to be non-amenable. The proof is by using estimates of the spectral radius of an associated random walk.

## I. INTRODUCTION

A group G is said to be amenable if in the space B(G) of all real valued bounded functions with the norm $\|f\|_\infty = \sup_{g \in G} |f(g)|$ there exists a positive linear functional $\Phi$ satisfying the condition $\Phi(1) = 1$.

A free group $F_m$ of the rank $m \geq 2$ is not amenable. The set of amenable groups is closed with respect to subgroups and factor groups (see [1]).

The conjecture that any non-amenable group contains a free subgroup with two generators is not solved up to now; in this connection an important task is to discover the criteria of amenability (or non-amenability) for groups given by a co-representation, i. e., by a collection of generators and their relations.

No difficulties arise in the case when a group G decomposes

into a free product of non-trivial groups $G_1$ and $G_2$, since the group $G \simeq G_1 \times G_2$ is non-amenable except for the case: $G_1 \simeq \mathbb{Z}_2$, $G_2 \simeq \mathbb{Z}_2$. Indeed, in that case the subgroup $[G_1, G_2]$ from $G_1 \times G_2$ generated by all commutators of the form $aba^{-1}b^{-1}$, $a \in G_1$, $b \in G_2$ is singly generated. Consequently, $[G_1, G_2$ is a free group of rank $m \geq 2$ except when $G_1 \simeq \mathbb{Z}_2$, $G_2 \simeq \mathbb{Z}_2$. For a free product of cyclic groups this result was obtained by another method in Dixmier [2].

The situation becomes much more complicated if to the set of defining words $a_1^{t_1}, \ldots, a_m^{t_m}$ of the group $\mathbb{Z}_{t_1} \times \mathbb{Z}_{t_2} \times \ldots \times \mathbb{Z}_{t_m}$ one more defining word is added: e. g., $(a_1 a_2)^{t_{m+1}}$; the purpose of this paper is to propose an approach to problems of this kind.

If $G$ is an arbitrary countable group and $p(g)$ is a distribution of probabilities on it, a bounded operator $T$ on $\ell_2(G)$ may be defined by the formula

$$(Tf)(g) = \sum_{g_1 \in G} f(g_1) p(g^{-1} g_1)$$

If the distribution $p(g)$ is symmetrical, the operator $T$ is symmetric with spectrum in the segment $[-1, 1]$, and the spectral radius $r$ coincides with that of a symmetrical random walk on the group $G$ with a matrix of transitional probabilities $\Pi = \|\pi(g, h)\|$, where $\pi(g, h) = p(g^{-1}h)$ and may be calculated by the formula

$$r = \limsup_{n \to \infty} \{P^n_{e,e}\}^{1/n}$$

where $P^n_{e,e}$ is the probability of returning into the unit element $e$ of the group $G$ at the n-th step [3].

Kesten [4], as well as Day [5], proved that the group $G$ is amenable iff $r = 1$ for an arbitrary symmetrical distribution $p(g)$

whose support generates the group G; if $r = 1$ for one such distribution of probabilities, then $r = 1$ for the rest of distributions of probabilities $p(g)$ such that $p(g) = p(g^{-1})$ and supp $p(g)$ generates the group G.

It is known [6] that any finitely generated group may be represented as a factor of a free group F with a finite number of generators by a normal subgroup. It turns out that the generating function $H(x)$ of the number of words of length n in the normal subgroup H may yield essential information on the ergodic properties of the group G, e. g , whether it is amenable or not, whether simple walk is recurrent. We are going to prove these facts in §IV. The value inverse to the radius of convergence of the series $H(x)$ is called the growth exponent of the normal subgroup H. In §III the principal properties of the growth exponent are proven; its merit consists in having an explicit formula expressing the spectral radius via the growth exponent. The rest of the paper is devoted to studying the behavior of the group G growth exponent when the length of defining relations of the group $G \simeq F/H$ increases.

## II. NOTATIONS

Let $F_m$ be a free group of rank m, $F(a_1, \ldots, a_m)$ -- a free group of rank m with m free generators $a_1, \ldots, a_m$. An element of the group $F(a_1, \ldots, a_m)$ is identified with the set of all words W such that they may be reduced to the same word by crossing out words of the type $a_\nu^\epsilon a_\nu^{-\epsilon}$, $1 \leq \nu \leq m$, $\epsilon = \pm 1$. It is known that the result of word reduction does not depend on the order of reduction [6]. As a rule, an element g of a free group will mean an irreducible word W corresponding to g. The length of the element g (of the word W), i. e., the number of symbols in W, will be denoted by $|g|$ or $|W|$. The same symbol $|A|$ denotes the set cardinality A

However, it will always be clear from the context what object we are dealing with.

Let $G \simeq F(a_1, \ldots, a_m)/H$. Then the elements $a_1, \ldots, a_m$ under the canonical homomorphism are mapped into the generating elements of the group $G$ which we also denote by the symbols $a_1, \ldots, a_m$. If $H$ is any subgroup of the group $F_m$, then $H_n$ is the set of words of the length $n$ in the group $H$, and the series $H(x) = \sum_{n=0}^{\infty} |H_n| x^n$ is called the generating function of the group $H$.

## III. PROPERTIES OF THE GROWTH EXPONENT

Let $H$ be a subgroup of the group $F(a_1, \ldots, a_m)$.

Definition. The upper limit

$$\alpha_H^{a_1, \ldots, a_m} = \limsup_{n \to \infty} |H_n|^{1/n}$$

is called the growth exponent of the group $H$ with respect to the free generators $a_1, \ldots, a_m$ of the group $F(a_1, \ldots, a_m)$. The growth exponent of the free group $F_m$ with respect to an arbitrary set of the free generators of the group equals $2m - 1$. Indeed, the number of words of the length $n$ in the group $F_m$ equals $2m(2m-1)^{n-1}$ independently of the free generators set of the group $F_m$.

The growth exponent of the group $H \subset F_m$ depends on the choice of free generators of the group $F_m$. Let, e. g., the group $H \subset F(a_1, a_2, a_3)$ be generated by the elements $a_1, a_2$. Then $\alpha^{a_1, a_2, a_3} = 3$. However, let $b_1 = a_1 a_3$, $b_2 = a_2$, $b_3 = a_3$. We wish to calculate the growth exponent of the group $H \subset F(b_1, b_2, b_3)$ generated by the elements $b_1 b_3^{-1}$, $b_2$ with respect to the free generators $b_1, b_2, b_3$ of the group $F_3$. Denote the set of words of the length $n$ of the group $H$ ending with either the word $b_1 b_3^{-1}$ or the word $b_3 b_1^{-1}$ (either with the word $b_2$ or the word $b_2^{-1}$) by $H_n^{(1)}$ ($H_n^2$) respectively. Then

$$H_n^{(1)} = H_{n-2}^{(1)} + 2H_{n-2}^{(2)}$$

$$H_n^{(2)} = 2H_{n-1}^{(1)} + H_{n-1}^{(2)}$$

Turning to the generating functions $H^{(1)}(x)$, $H^{(2)}(x)$ of the numbers $H_n^{(1)}$, $H_n^{(2)}$ we obtain the system

$$\begin{cases} H^{(1)}(x) = x^2 H^{(1)}(x) + 2x^2 H^{(2)}(x) + 2x^2 \\ H^{(2)}(x) = 2xH^{(1)}(x) + xH^{(2)}(x) + 2x \end{cases}$$

Hence

$$H(x) = H^{(1)}(x) + H^{(2)}(x) = \frac{2x(2x^2 + x + 1)}{-3x^3 - x^2 - x + 1}$$

The growth exponent is the inverse of the radius of convergence of the series $H(x)$. Since $1/3$ is not a root of the equation

$$-3x^3 - x^2 - x + 1 = 0$$

$\alpha_H^{a_1, a_2, a_3} \neq \alpha_H^{a_1 a_3, a_2, a_3}$. However, below we shall often omit the subscripts and superscripts of $\alpha$; but the context will always make it clear which group and what generators we are dealing with. From the growth exponent definition, as well as from the fact that the growth exponent of the free group $F_m$ equals $2m-1$ it follows that $\alpha_H^{a_1, \ldots, a_m} \leq 2m-1$.

Statement 3.1. The growth exponent of a non-trivial normal subgroup $H \subset F_m$ is bounded from below by the constant $\sqrt{2m-1}$, i.e.,

$$\alpha_H^{a_1, \ldots, a_m} \geq \sqrt{2m-1} \tag{3.1}$$

Proof: Let the freely reduced word W belong to H. At least $2m - 2$ words from the set of words $a_\nu^{\epsilon} W a_\nu^{-\epsilon}$, $\nu = 1, \ldots, m$, $\epsilon = \pm 1$ are freely reduced, and they all belong to the normal subgroup $H_\epsilon$. If $W_1$ is any of these words, the collection $a_\nu^{\epsilon} W_1 a_\nu^{-\epsilon}$, $\nu = 1, \ldots, m$, $\epsilon = \pm 1$ contains $2m-1$ freely reduced words, and they all also belong to H. Holding on to this strategy we obtain in H the subset $\mathfrak{E}(W)$ which contains no less than $(2m-2)(2m-1)^{n-1}$ words of the length $|W| + 2n$. Hence

$$\alpha_H^{a_1, \ldots, a_m} \geq \lim_{n \to \infty} [(2m-2)(2m-1)^{n-1}]^{1/(|W|+2n)} = \sqrt{2m-1}$$

Statement 3. 2. Let H be a non-trivial normal subgroup of the group $F_m$. Then

$$|H_n| \cdot |H_m| \leq |H_{n+m+2}| \tag{3. 2}$$

Proof: Let $U \in H_n$, $V \in H_m$. It is always possible to find such $\nu$ and $\epsilon = \pm 1$ that $a_\nu^{\epsilon} U a_\nu^{-\epsilon} V$ is a freely reduced word and $|a_\nu^{\epsilon} U a_\nu^{-\epsilon} V| = n + m + 2$. The resulting mapping $H_n \times H_m \to H_{n+m+2}$ is one-to-one, whence follows (3. 2).

Statement 3. 3. If H is a non-trivial normal subgroup of the group $F_m$ and

(a) if not all the elements in H are of an even length, there exists a limit

$$\alpha_H = \lim_{n \to \infty} |H_n|^{1/n}$$

(b) if all the elements in H are of an even length, there exists a limit

$$\alpha_H = \lim_{n \to \infty} |H_{2n}|^{1/2n}$$

Proof: (a) Let W and V be freely reduced words belonging to the group H, the length of the word W being even, and that of the word V being odd. Consider the set $\mathfrak{E}(W) \cup \mathfrak{E}(V)$ (definition of the set $\mathfrak{E}(W)$ was given while proving statement 3. 1. ). There exists such N that when $n > N$ the intersection of the set $\mathfrak{E}(W) \cup \mathfrak{E}(V)$ with the set $H_n$ is not empty, i. e. when $n > N$ the numbers $|H_n|$ do not equal zero. And at last we are to use the fact that the sequence of the numbers $|H_n|$ is semi-multiplicative, by force of statement 3. 2.

(b) is proved similarly.

Statement 3. 4. If $\alpha_H$ is the growth exponent of a normal subgroup $H \subset F_m$, then

$$|H_n| \leq \alpha_H^{n+2} \tag{3. 3}$$

Proof: Suppose the contrary: there exist $\epsilon > 0$ and N such that

$$|H_N| > (\alpha_H^2 + \epsilon)\alpha_H^N$$

Let $N_1 = 2N + 2$, $N_2 = 2N_1 + 2, \ldots, N_k = 2N_{k-1} + 2$. Then

$$|H_{N_1}| \geq |H_N|^2 > (\alpha_H^2 + \epsilon)^2 \alpha_H^{2N}$$

$$\cdots\cdots\cdots\cdots$$

$$|H_{N_k}| \geq |H_N|^{2k} > (\alpha_H^2 + \epsilon)^{2^k} \alpha_H^{2^k N}$$

Hence

$$\limsup_{k\to\infty} |H_{N_k}|^{1/N_k} \geq \limsup_{n\to\infty} [ (\alpha_H^2 + \epsilon)^{2^k} \alpha_H^{2^k N}]^{1/N_k} = (\alpha_H^2+\epsilon)^x \alpha_H^y$$

where

$$x = \lim_{k\to\infty} \frac{2^k}{N + 2^k + 2^{k-1} + \ldots + 1} = \frac{1}{N + 1 + 2^{-1} + \ldots} = \frac{1}{N+2}$$

$$y = \lim_{k\to\infty} \frac{2^k N}{2^k N + 2^k + 2^{k-1} + \ldots + 1} = \frac{N}{N+2}$$

Thus

$$\limsup_{k\to\infty} |H_{N_k}|^{1/N_k} \geq (\alpha_H^2 + \epsilon)^{1/(N+2)} \alpha_H^{N/(N+2)} > \alpha_H$$

The contradiction obtained proves (3. 3).

<u>Statement 3. 5.</u> <u>Let</u> $H^{(n)}$, $n = 1, 2, \ldots$ <u>be a sequence of normal subgroups of the group</u> $F(a_1, \ldots, a_m)$. If $H^{(1)} \subset H^{(2)} \subset \ldots \subset H^{(n)} \subset \ldots$ <u>and</u> $H = \bigcup_{n=1}^{\infty} H^{(n)}$, <u>then</u>

$$\lim_{n\to\infty} \alpha_{H^{(n)}}^{a_1, \ldots, a_m} = \alpha_H^{a_1, \ldots, a_m}$$

<u>Proof:</u> Since $\alpha_{H^{(1)}} \leq \alpha_{H^{(2)}} \leq \ldots$, the limit $\alpha_0 = \lim_{n\to\infty} \alpha_{H^{(n)}}$ exists and deos not exceed $\alpha_H$. Using (3. 3) we have

$$|H_n^k| \leq \alpha_{H^{(n)}}^{k+2} \leq \alpha_0^{k+2}$$

when all $k, n = 1, 2, \ldots$. However, beginning from a certain number $k = k(n)$ all the sets $H_n^{(k)}$ coincide with the set $H_n$. Consequently, $|H_n| \leq \alpha_0^{k+2}$, and therefore $\limsup_{n\to\infty} |H_n|^{1/n} \leq \alpha_0$, q. e. d.

<u>Remark.</u> Let $H^{(n)}$, $n = 1, 2, \ldots$ be a sequence of normal subgroups of the group $F(a_1, \ldots, a_m)$. If $H^{(1)} \supset H^{(2)} \supset \ldots \supset H^{(n)} \supset \ldots$ and $H = \bigcap_{n=1}^{\infty} H^{(n)}$, then, generally speaking, the equality

$$\lim_{n\to\infty} \alpha_{H^{(n)}}^{a_1, \ldots, a_m} = \alpha_H^{a_1, \ldots, a_m}$$

does not hold.

Example. Let $H^{(n)}$ be a normal subgroup of the group $F(a,b)$ generated by the elements $a^{2^n}, b^{2^n}$. It is obvious that $\alpha^{a,b}_{H^{(1)}} \geq \alpha^{a,b}_{H^{(2)}} \geq \dots \geq \sqrt{3}$ but $\bigcap_{n=1}^{\infty} H^{(n)} = e$.

Let $H$ be an arbitrary subgroup of the group $F(a, b)$. In what way can its growth exponent be calculated? A subgroup of a free group is known [6] to be also free, and the set $W_i(a_\nu)$ of its free generators may be chosen so that it would satisfy the two following conditions.

Let $V(W_i)$ be a freely reduced word in the symbols $W_i$.

$$V(W_i) = W_{i_1}^{\epsilon_1} W_{i_2}^{\epsilon_2} \dots W_{i_r}^{\epsilon_r}, \qquad \epsilon_j = \pm 1$$

(i) after complete reduction of the word $V(W_i(a_\nu))$ there remains at least one $a$, symbol $a_{\nu_j}^{\eta_j}$, $\eta_j = \pm 1$, from each $W_{i_j}^{\epsilon_j}$.

(ii) $a$, the length of the word $V(W_i(a_\nu))$, is no less than $a$, the length of any $a$, a symbol entering $V(W_i)$.

The set of the group $H$ generators, satisfying conditions (i) and (ii) is called a Nielsen set. Denote by $\beta(W_i, W_j)$ the number of symbols which may be omitted by reducing from the word $W_iW_j$. The beginning $S$ of the word $W$ such that $1/2\,|W| < |S| < 1/2\,|W|+1$ is called the principal beginning of the non-empty freely reduced word $W$. The principal end of the word $W$ is defined similarly. The beginning $S$ of the word $W_i$ is called isolated if it is not the beginning of any other word $W_j$. It is known [6] that if the set $\{W_i(a_\nu)\}$ of non-empty freely reduced words is a Nielsen one, the principal beginning and the principal end of any word are isolated. Therefore, the numbers $|W_k^\epsilon|$, $\beta(W_i^\mu, W_k^\epsilon)$, and $\beta(W_k^\epsilon, W_i^\mu)$ satisfy the inequalities

$$|W_k^\epsilon| \geq \beta(W_i^\mu, W_k^\epsilon), \qquad |W_k^\epsilon| \geq \beta(W_k^\epsilon, W_i^\mu)$$

The element $g \in H$ is said to end with the word $W_k^\epsilon$ if $g = W_{i_1} \dots W_{i_n} W_k^\epsilon$. Let $H_n^{k,\epsilon}$ be the set of words of length $n$ in the group $H$, ending with the word $W_k^\epsilon$. The generating function $H(x)$ of the group $H$ and generating functions $H_k^\epsilon(x)$ of the numbers $|H_n^{k,\epsilon}|$ are related by

$$H(x) = \sum_{k,\epsilon} H_k^\epsilon(x) \tag{3. 4}$$

Statement 3. 6. The functions $H_k^\epsilon(x)$ satisfy the system of equations

$$H_k^\epsilon(x) = x^{|W_k^\epsilon|} + \sum_{\substack{m=1 \\ \mu = \pm 1}} x^{|W_k^\epsilon| - \beta(W_m^\mu, W_k^\epsilon)} H_m^\mu(x) \tag{3. 5}$$

Proof: The numbers $|H_n^{k,\epsilon}|$ satisfy the relations

$$|H_n^{k,\epsilon}| = \sum_{m,\mu} \left| H^{m,\mu}_{n - |W_k^\epsilon| + \beta(W_m^\mu, W_k^\epsilon)} \right| \tag{3. 6}$$

Multiplying equalities (3. 6) by $x^n$ and summing them over $n$ yields (3. 5).

Corollary. If the group $H$ is finitely generated, $H(x)$ is a rational function.

Proof: The system (3. 5) consists of a finite number of equations. Thus, if $H$ is a finitely generated group, $\alpha_H$ is an inverse value of the minimal positive pole of, the function $H(x)$, which may be found as a root of the algebraic equation

$$k,\epsilon \;\text{- - - -}\left| \begin{matrix} & m,\mu & \\ & \vdots & \\ \delta_k^m & - & x^{|W_k^\epsilon| - \beta(W_m^\mu, W_k^\epsilon)} \end{matrix} \right| = 0 \tag{3. 7}$$

## IV. THE AMENABILITY CRITERION

Consider a random walk on the group $G \simeq F(a_1, \dots, a_m)/H$ with a matrix of transitional probabilities $P = \|p(g,h)\|$ where

$$p(g,h) = \begin{cases} 1/2m & \text{if } g^{-1}h = a_\nu^\epsilon \text{ for some } \nu : 1 \leq m \text{ and } \epsilon = \pm 1 \\ 0 & \text{otherwise} \end{cases}$$

Such a walk on the group $G$ is called simple.

Calculating transition probabilities $p(e,g,n)$ from the unit $e$ of the group $F_m$ into the element $g \in F_m$, $|g| = k$ in $n$ steps during a random walk on the group $F(a_1, \dots, a_m)$ is brought down to calculating transition probabilities $P_{0k}^n$ in $n$ steps from $0$ to $k$ during a walk on the set of integer points of the semi-axis $[0, \infty)$ with the transition matrix

$$\Pi = \begin{Vmatrix} 0 & 1 & 0 & 0 & \dots \\ q & 0 & p & 0 & \dots \\ 0 & q & 0 & p & \dots \\ \dots & \dots & \dots & \dots & \dots \\ \dots & \dots & \dots & \dots & \dots \end{Vmatrix} \tag{4.1}$$

where $p = \frac{2m-1}{2m}$, $q = \frac{1}{2m}$ by the formula

$$p(e,g,n) = \frac{P_{0k}^n}{2m(2m-1)^{k-1}}$$

Let $P_{k,m}^n$ be the transition probability from the state $k$ into the state $m$ in $n$ steps for a Markov chain on $[0, \infty)$ determined by the transition matrix (4.1), and $p \geq q$.

<u>Lemma 4.1.</u> <u>Numbers</u> $P_{k,m}^n$ <u>satisfy the relations:</u> <u>if</u> $m > 0$, <u>then</u>

$$P^n_{k,m} = (\sqrt{pq})^n \left(\sqrt{\frac{q}{p}}\right)^{k-m} \left\{ \binom{n}{\frac{n+k-m}{2}} + \frac{q}{p} \binom{n}{\frac{n+k+m}{2}} - \frac{p-q}{pq} \sum_{t=1}^{\frac{n-k-m}{2}} \left(\frac{q}{p}\right)^t \binom{n}{\frac{n+k+m+2t}{2}} \right\} \tag{4.2}$$

<u>if</u> $m = 0$, <u>then</u>

$$P^n_{k,0} = (\sqrt{pq})^n \left(\sqrt{\frac{q}{p}}\right)^k \left\{ \binom{n}{\frac{n+k}{2}} - \frac{p-q}{pq} \sum_{t=1}^{\frac{n-k}{2}} \left(\frac{q}{p}\right)^t \binom{n}{\frac{n+k+2t}{2}} \right\} \tag{4.3}$$

<u>Proof:</u> It suffices to check that

1. If $m > 1$, then $P^n_{k,m} = pP^{n-1}_{k,m-1} + qP^{n-1}_{k,m+1}$

2. If $m = 1$, then $P^n_{k,m} = P^{n-1}_{k,0} + qP^{n-1}_{k,2}$

3. If $m = 0$, then $P^n_{k,0} = qP^n_{k,1}$.

Indeed,

1. $$pP^{n-1}_{k,m-1} + qP^{n-1}_{k,m+1} = (\sqrt{pq})^n \left(\sqrt{\frac{q}{p}}\right)^{k-m} \left\{ \binom{n}{\frac{n+k-m}{2}} + \frac{q}{p} \binom{n}{\frac{n+k+m}{2}} - \frac{p-q}{pq} \sum_{t=1}^{\frac{n-k-m-2}{2}} \left(\frac{q}{p}\right)^t \binom{n}{\frac{n+k+m+2t}{2}} - \frac{p-q}{pq} \left(\frac{q}{p}\right)^{n-k-m} \right\}$$

$$= (\sqrt{pq})^n \left(\sqrt{\frac{q}{p}}\right)^{k-m} \left\{ \binom{n}{\frac{n+k-m}{2}} + \frac{q}{p} \binom{n}{\frac{n+k+m}{2}} - \frac{p-q}{pq} \sum_{t=1}^{\frac{n-k-m}{2}} \left(\frac{q}{p}\right)^t \binom{n}{\frac{n+k+m+2t}{2}} \right\} = P^n_{k,m}$$

2. $$P^{n-1}_{k,0} + qP^{n-1}_{k,2} = (\sqrt{pq})^n\left(\sqrt{\frac{q}{p}}\right)^{k-1}\left\{\frac{1}{p}\binom{n-1}{\frac{n-1+k}{2}} + \binom{n-1}{\frac{n-1+k-2}{2}} + \frac{q}{p}\binom{n-1}{\frac{n-1+k+2}{2}}\right.$$

$$\left. - \frac{p-q}{pq}\sum_{t=1}^{\frac{n-1-k}{2}}\left(\frac{q}{p}\right)^t\binom{n-1}{\frac{n-1+k+2t}{2}} - \frac{p-q}{pq}\sum_{t=1}^{\frac{n-1-k-2}{2}}\left(\frac{q}{p}\right)^t\binom{n-1}{\frac{n-1+k+2+2t}{2}}\right\}$$

$$= (\sqrt{pq})^n\left(\sqrt{\frac{q}{p}}\right)^{k-1}\left\{\binom{n}{\frac{n+k-1}{2}} + \frac{q}{p}\binom{n}{\frac{n+k+1}{2}} - \frac{p-q}{pq}\,\frac{q}{p}\left[\binom{n-1}{\frac{n-1+k+2}{2}} + \binom{n-1}{\frac{n-1+k+4}{2}}\right]\right.$$

$$\left. - \frac{p-q}{pq}\left(\frac{q}{p}\right)^2\left[\binom{n-1}{\frac{n-1+k+4}{2}} + \binom{n-1}{\frac{n-1+k+6}{2}}\right] - \dots - \frac{p-q}{pq}\left(\frac{q}{q}\right)^{(n-k-1)/2}\right\}$$

$$= P^n_{k,1}$$

3. $$qP^{n-1}_{k,1} = (\sqrt{pq})^n\left(\sqrt{\frac{q}{p}}\right)^k\left\{\binom{n-1}{\frac{n-1+k-1}{2}} + \frac{q}{p}\binom{n-1}{\frac{n-1+k+1}{2}}\right.$$

$$\left. - \frac{p-q}{pq}\sum_{t=1}^{\frac{n-1-k-1}{2}}\left(\frac{q}{p}\right)^t\binom{n-1}{\frac{n-1+k+1+2t}{2}}\right\} = (\sqrt{pq})^n\left(\sqrt{\frac{q}{p}}\right)^k\left\{\binom{n}{\frac{n+k}{2}}\right.$$

$$- \frac{p-q}{q}\,\frac{q}{p}\left[\binom{n-1}{\frac{n-1+k+1}{2}} + \binom{n-1}{\frac{n-1+k+3}{2}}\right] - \frac{p-q}{q}\left(\frac{q}{p}\right)^2\left[\binom{n-1}{\frac{n-1+k+3}{2}}\right.$$

$$\left.\left. + \binom{n-1}{\frac{n-1+k+5}{2}}\right] - \dots - \frac{p-q}{pq}\left(\frac{q}{p}\right)^{(n-k)/2}\right\} = P^n_{k,0}$$

The lemma is proven.

<u>Lemma 4. 2.</u> <u>Let</u> $r$ <u>be the spectral radius of simple random walk on</u>

the group $G \simeq F(a_1, \ldots, a_m)/H$. Then

$$r = \limsup_{n \to \infty} \{ \max_{0 \le k \le n} [P^n_{0k}(\frac{\alpha_H}{2m-1})] \}^{1/n} \tag{4.4}$$

Proof. Let $P^n_{ee}$ be the probability of returning into the unit element e of the group G at the n-th step. Then

$$P^n_{ee} = {}^{n}_{00} + \sum_{k=1}^{n} P^n_{0k} \frac{|H_k|}{2m(2m-1)^{k-1}} \tag{4.5}$$

Hence

$$\limsup_{n \to \infty} \{P^n_{ee}\}^{1/n} = \limsup_{n \to \infty} \{P^n_{00} + \sum_{k=1}^{n} P^n_{0k} \frac{|H_k|}{2m(2m-1)^{k-1}}\}^{1/n}$$

$$= \limsup_{n \to \infty} \{\mu_n\}^{1/n} \limsup_{n \to \infty} \{\frac{P^n_{00}}{\mu_n} +$$

$$+ \frac{1}{\mu_n} \sum_{k=1}^{n} P^n_{0k} \frac{|H_k|}{2m(2m-1)^{k-1}}\}^{1/n}$$

where

$$\mu_n = \max\{P^n_{00}, \max_{1 \le k \le n} [P^n_{0k} \frac{|H_k|}{2m(2m-1)^{k-1}}]\}$$

But the expression

$$\limsup_{n \to \infty} \{\frac{P^n_{00}}{\mu_n} + \frac{1}{\mu_n} \sum_{k=1}^{n} P^n_{0k} \frac{|H_k|}{2m(2m-1)^{k-1}}\}^{1/n}$$

is bounded from above by the constant $(n+1)^{1/n}$, from below by the constant 1, if $\mu_n = P^n_{00}$, and by the constant $|H_p|/\alpha_H^{p+2}$ otherwise (p is the value of k which provides the maximum).

Since

$$\limsup_{p \to \infty} [\,|H_p|/\alpha_H^{p+2}\,] = 1$$

then

$$r = \limsup_{n\to\infty} \{\mu_n\}^{1/n} = \limsup_{n\to\infty} [\max_{0\le k\le n} P^n_{0k} (\frac{\alpha_H}{2m-1})^k]^{1/n}$$

The lemma is proven.

Theorem 4. 1. If $r$ is the spectral radius of simple random walk on the group $G \simeq F(a_1, \ldots, a_m)/H$, then

$$r = \frac{\sqrt{2m-1}}{2m}\left(\frac{\alpha_H}{\sqrt{2m-1}} + \frac{\sqrt{2m-1}}{\alpha_H}\right) \tag{4. 6}$$

Proof: Let $b(n,k) = q^k \binom{n}{\frac{n+k}{2}}$ where $q > 1$ and $0 \le k \le n$. If $n-k$ is even, then

$$\frac{b(n,k+2)}{b(n,k)} = q^2 \frac{n-k}{n+k+2}$$

and therefore $\frac{b(n,k+2)}{b(n,k)} > 1$, if $k \le [\frac{q^2-1}{q^2+1}] - 1$

$$\frac{b(n,k+2)}{b(n,k)} < 1, \text{ if } k \ge [\frac{q^2-1}{q^2+1}] + 1$$

Denote $(q^2-1)/(q^2+1)$ by $\beta$ and $[(q^2-1)n/(q^2+1)]$ by $\gamma$. We have proved that when $k$ changes from $0$ to $\gamma - 1$ numbers $b(n,k)$ increase and when $k$ changes from $\gamma + 1$ to $n$ they decrease. Hence

$$\max_{0\le k\le n} b(n,k) = \max_{\gamma-1\le k\le \gamma+1} b(n,k)$$

and

$$\limsup_{n\to\infty} [\max_{0\le k\le n} b(n,k)]^{1/n} = \limsup_{n\to\infty} [\max_{\gamma-1\le k\le\gamma+1} b(n,k)]^{1/n}$$

$$= \limsup_{n\to\infty} [q^{\beta n}\binom{n}{\frac{(1+\beta)n}{2}}]^{1/n} = \limsup_{n\to\infty} [q^{\beta n}(\frac{1+\beta}{2})^{-\frac{1+\beta}{2}n}(\frac{1-\beta}{2})^{-\frac{1-\beta}{2}n}]^{1/n} = q+q^{-1}$$

Note that if $k > 0$, then

$$\frac{1}{p}(\sqrt{pq})^n\left(\sqrt{\frac{q}{p}}\right)^k\binom{n}{\frac{n+k}{2}} \geq P^n_{0k} \geq \frac{2}{p}(\sqrt{pq})^n\left(\sqrt{\frac{q}{p}}\right)^k\frac{k}{n+k+2}\binom{n}{\frac{n+k}{2}}$$

if $k = 0$, then

$$(\sqrt{pq})^n\binom{n}{\frac{n}{2}} \geq P^n_{00} \geq (\sqrt{pq})^n\frac{4}{n+2}\binom{n}{\frac{n}{2}}$$

Therefore

$$p^{-1}(\sqrt{pq})^n \max_{0 \leq k \leq n}\left[\left(\sqrt{\frac{q}{p}}\right)^k\left(\frac{\alpha_H}{2m-1}\right)^k\binom{n}{\frac{n+k}{2}}\right] \geq \max_{0 \leq k \leq n}\left[P^n_{0k}\left(\frac{\alpha_H}{2m-1}\right)^k\right]$$

$$\geq (m+3)^{-1} \cdot (\sqrt{pq})^n \max_{0 \leq k \leq n}\left[\left(\sqrt{\frac{q}{p}}\right)^k\left(\frac{\alpha_H}{2m-1}\right)^k\binom{n}{\frac{n+k}{2}}\right]$$

and therefore

$$\limsup_{n\to\infty}\left\{\max_{0 \leq k \leq n}\left[P^n_{0k}\left(\frac{\alpha_H}{2m-1}\right)^k\right]\right\}^{1/n}$$

$$= \sqrt{pq}\,\limsup_{n\to\infty}\left\{\max_{0 \leq k \leq n}\left[\left(\sqrt{\frac{p}{q}}\right)^k\left(\frac{\alpha_H}{2m-1}\right)^k\binom{n}{\frac{n+k}{2}}\right]\right\}^{1/n}$$

$$= \sqrt{pq}\left(\sqrt{\frac{p}{q}}\frac{\alpha_H}{2m-1} + \sqrt{\frac{q}{p}}\frac{2m-1}{\alpha_H}\right) \tag{4.7}$$

Substituting the values $p = (2m-1)/2m$, $q = 1/2m$ into (4.7), we obtain (4.6). The theorem is proven.

Using Kesten's criterion of group amenability and formula (4.6) we have

Corollary 1. The group $G \simeq F_m/H$ is amenable iff $\alpha_H = 2m-1$, i.e., $1/(2m-1)$ is a singular point of the function $H(x)$.

Using Kesten's result, the estimate from below for the growth exponent can be made more precise. In the paper [3] it is proven that if $G \simeq F_m/H$ and a spectral radius of a random walk on $G$ equals $\sqrt{2m-1}/m$, then $G \simeq F_m$. Using this result and formula (4.6) we have

Corollary 2. $\alpha_H > \sqrt{2m-1}$.

In the subsequent paragraphs of the paper it will be proven that there exist normal subgroups of the froup $F_m$ whose growth exponents differ as little as we wish from $\sqrt{2m-1}$.

Theorem 4.2. A simple random walk on the group $G \simeq F(a_1, \ldots, a_m)/H$ is recurrent iff

$$\sum_{n=0}^{\infty} \frac{|H_n|}{(2m-1)^n} = \infty \tag{4.8}$$

Proof: It is necessary and sufficient for a walk on the group $G$ to be recurrent that $\sum_n P_{ee}^n = \infty$, where $P_{ee}^n$ is probability of returning into the unit element $e$ of the group $G$ at the n-th step. Using formulas (4.2), (4.3), as well as (4.6) we obtain

$$\sum_{n=0}^{\infty} P_{ee}^n = \sum_{n=0}^{\infty} P_{00}^n + \sum_{n=1}^{\infty}\sum_{k=1}^{n} P_{0k}^n \frac{|H_k|}{2m(2m-1)^{k-1}} = \sum_{n=0}^{\infty} P_{00}^n + \frac{2m-1}{2m}\sum_{k=1}^{\infty} \frac{|H_k|}{(2m-1)^k}\sum_{n=1}^{\infty} P_{0k}^n$$

$$= \sum_{n=0}^{\infty} (\sqrt{pq})^n \binom{n}{\frac{n}{2}} - \frac{p-q}{p}\sum_{n=1}^{\infty} (\sqrt{pq})^n \sum_{t=1}^{n/2} \left(\frac{q}{p}\right)^t \binom{n}{\frac{n+2t}{2}} + \frac{2m-1}{2m}\sum_{k=1}^{\infty} \frac{|H_k|}{(2m-1)^k}$$

$$\times \left(\sqrt{\frac{q}{p}}\right)^{-k}\left[\sum_{n=k}^{\infty} (\sqrt{pq})^n \binom{n}{\frac{n+k}{2}} + \frac{q}{p}\sum_{n=k}^{\infty} (\sqrt{pq})^n \binom{n}{\frac{n+k}{2}} - \frac{p-q}{pq}\sum_{n=k}^{\infty} (\sqrt{pq})^n \sum_{t=1}^{\frac{n-k}{2}} \left(\frac{q}{p}\right)^t \binom{n}{\frac{n+k+2t}{2}}\right]$$

$$= \sum_{n=0}^{\infty} (\sqrt{pq})^n \binom{n}{\frac{n}{2}} - \frac{p-q}{p}\sum_{t=1}^{\infty} \left(\frac{q}{p}\right)^t \sum_{n=2t}^{\infty} (\sqrt{pq})^n \binom{n}{\frac{n+2t}{2}} + \frac{2m-1}{2m}\sum_{k=0}^{\infty} \frac{|H_k|}{(2m-1)^k}$$

$$\times \left(\sqrt{\frac{p}{q}}\right)^{k}\left[\frac{1}{p}\sum_{n=k}^{\infty} (\sqrt{pq})^n \binom{n}{\frac{n+k}{2}} - \frac{p-q}{pq}\sum_{n=2t+k}^{\infty} (\sqrt{pq})^n \binom{n}{\frac{n+k+2t}{2}}\right]$$

Note that for a Bernoulli random walk with probabilities $p, q$, $p > q$

$$G(0, m) = \sum_{n=0}^{\infty} P_n(0, m) = \left(\sqrt{\frac{p}{q}}\right)^m \sum_{n=m}^{\infty} (\sqrt{pq})^n \binom{n}{\frac{n+m}{2}} = (p-q)^{-1}$$

(see [7]). Then $\sum_{n=m}^{\infty} (\sqrt{pq})^n \binom{n}{\frac{n+m}{2}} = (p-q)^{-1}\left(\sqrt{\frac{q}{p}}\right)^m$ and

$$\sum_{n=0}^{\infty} P^n_{ee} = (p-q)^{-1} - \frac{p-q}{q(p-q)} \sum_{t=1}^{\infty} \left(\frac{q}{p}\right)^{2t}$$

$$+ \frac{2m-1}{2m(p-q)} \sum_{k=0}^{\infty} \frac{|H_k|}{(2m-1)^k} \left(\sqrt{\frac{p}{q}}\right)^k \left[ P^{-1}\left(\sqrt{\frac{q}{p}}\right)^k\right.$$

$$\left. - \frac{p-q}{pq}\left(\sqrt{\frac{q}{p}}\right)^k \sum_{t=1}^{\infty} \left(\frac{q}{p}\right)^{2t}\right] = \frac{p}{p-q} + \frac{2m-1}{2m(p-q)} \sum_{k=0}^{\infty} \frac{|H_k|}{(2m-1)^k}$$

## V. AUXILIARY RESULTS FROM COMBINATORIAL GROUP THEORY

The group $G \simeq F(a_1, \ldots, a_m)/H$ may be given by the generators $a_\nu$, $\nu = 1, \ldots, m$ and defining relations

$$R_i(a_\nu) = 1, \qquad i = 1, \ldots, I$$

where $R_i$ are words in the symbols $a_\nu^{\pm 1}$. $R_i$ are called defining words of the group $G$; denote by $M$ the set of defining words. Graphic equality, equality in a free group and equality in the group $G$ are denoted by $\cong$, $\equiv$ and $=$ respectively. In the subsequent paragraphs groups are considered whose set of defining relations $M$ satisfies the three following conditions:

(1) each word $R_i$ is reduced,

(2) $M$ is closed with respect to the operations of taking the

inverse word and taking the cyclic transposition of letters of the word $R_i$,

(3) if $R_i$ and $R_j$ are not inverse to each other, the product $R_iR_j$ being reduced, $< \lambda < 1/6$ letters of the word $R_i$ are absorbed, as well as the groups whose set of determining relations M satisfies two more conditions, except (1) and (2);

(3') if $R_i$ and $R_j$ are not inverse to each other, the product $R_iR_j$ being reduced, $< \lambda < 1/4$ letters of the word $R_i$ are absorbed,

(4) if each of the words $R_i$, $R_j$ and $R_k$ is written on each side of a triangle, reduction cannot take place on all the three vertices. Greendlinger [8, 9] has proved that the problem of words identity in such groups is solved by Dehn's algorithm. The basic results of these papers are as follows.

If a non-empty free-reduced word W equals the unit element in the group G, there exists a natural number c, words $T_1, \ldots, T_c$ and defining words $R_{i_1}, \ldots, R_{i_c}$ such that

$$W \equiv \prod_{j=1}^{c} T_j^{-1} R_{i_j} T_j \tag{5.1}$$

<u>Definition.</u> We say that S has n pieces missing if S is a subword of a freely reduced word V, V = 1, V is the reduced form of $U \cong T_1^{-1}R_1T_1 \ldots T_i^{-1}R_iT_i \ldots T_q^{-1}R_qT_q$, $R_i = XSY$ and for any mode of reduction words X and Y are absorbed from $R_i$, leaving S. n is the smallest integer such that the (n+1)-th words $SR_{\beta_1} \ldots R_{\beta_n}$, $R_{\beta_n}^{-1}R_{\alpha_n}, \ldots, R_{\beta_1}^{-1}R_{\alpha_1}$ all are defining relators, but $R_{\beta_{k+1}} \ldots R_{\beta_n} SR_{\beta_1} \ldots R_{\beta_k}$ and $R_{\beta_k}^{-1} R_{\alpha_k}$, $k = 1, \ldots, n$ are not inverse to each other.

If the group G is such that the set M of its defining relations satisfies condition (1), (2), (3), the following theorem is true:

Main Theorem. If $V = 1$ and $V$ is freely reduced and not empty, then there exists a word $U \cong T_1^{-1}R_1T_1 \ldots T_n^{-1}R_nT_n$ such that $V$ is the reduced form of $U$ and for any mode of reduction of $U$ there remains either

(a) part of one $R$ which has at most one piece missing

(b) parts of two $R$'s, each of which has at most two pieces missing

(c) parts of three $R$'s, each of which has at most two pieces missing and two of which have at most three pieces missing, or

(d) parts of four $R$'s, each of which has at most three pieces missing.

Hence, if the freely reduced and not empty word $V$ equals a unit element of the group $G$, then $V$ contains occurrence of a subword $S$ which has at most three pieces missing, therefore

$$|S| > (1 - 3\lambda)|R_i| \tag{5. 2}$$

Now let the set $M$ of the defining relations of $G$ satisfy conditions (1), (2), (3), (4). Without loss of generality the product (5. 1) may be considered to have the following properties:

1. $T_j^{-1}R_{i_j}T_j$ is freely reduced,

2. For any mode of reducing the product $\Pi T_j^{-1}R_{i_j}T_j$, the word $R_{i_p}$ reduces at most $\lambda$ letters of the word $R_{i_q}$ ($p = 1, \ldots, c$, $q = 1, \ldots, c$).

Assuming property 1 as well as words $R_{i_j}$ lacking the exponent $-1$ are admissible due to properties (1) and (2) of the defining words set. Assumption of property 2 is admissible due to property 3 of the defining words set and Theorem 1 from [8] -- the part of it which we need is given below as a statement.

Statement. Let $W = \prod_{j=1}^{c} T_j^{-1}R_{i_j}T_j$. Then there exists a product

$\prod_j (T_j^1)^{-1} R_{i_j} T_j$ equal to W in the group G and such that, independently of the mode of its reduction, if $R_i^1 \cong R_i^1 R_i^2 R_i^3$ is reduced with $R_j' \cong R_j^1 R_j^2 R_j^3$ so that $R_i^2$ absorbs $R_j^2$, then $R_i^3 R_i^1 R_i^2$ and $R_j^2 R_j^3 R_j^1$ are not inverse to each other for all i and j.

For convenience we write $\prod_{j=1}^{c} T_j^{-1} R_{i_j} T_j$ as $\prod_{i=1}^{c} R_i^{-1} R_i T_i$ supposing that all the words of such type we consider have properties 1 and 2.

Definition. If the given mode of reduction takes place (complete or partial), words $R_{i_j}$ of the word $\prod T_i^{-1} R_i T_i$ are reduced with words $R_{i_j+1}$ $(j = 1, \dots, n-1;\ 1 \leq t_i < i_2 < \dots < i_n \leq c)$, none of $R_{i_p}$ $(1 \leq p \leq n)$ is reduced with any $T_i^{-1}$ or $T_i$, and besides the above reductions with one defining word (in the case $p = 1$ or $p = n$, when $n > 1$) or two of them (in the case $1 < p < n$) $R_{i_p}$ is reduced with at most one defining word, and if after the reduction of $R_{i_p}$ there remain subwords $R_{i_p}^d$ $(1 \leq p \leq n)$), then the word $R_{i_1}^d \dots R_{i_n}^d$ is a contiguous n-tuple with respect to a mode of reduction $\prod_{i=1}^{c} T_i^{-1} R_i T_i$.

Theorem. If the freely reduced not empty word W equals a unit in the group G, there exists a product of the type $\prod_{i=1}^{c} T_i^{-1} R_i T_i$ such that for any mode of complete reduction of the product W there contains an occurrence of a contiguous n-tuple with respect to the mode of reduction.

From the definition and properties of the defining relations set M it follows that

$$|R_{i_1}^d| > |R_{i_1}| - 2\lambda |R_{i_1}|, \qquad |R_{i_n}^d| > |R_{i_n}| - 2\lambda |R_{i_n}|,$$

$$|R_{ij}^d| > |R_{i_j}| - 3\lambda |R_{ij}|$$

Therefore, if the freely reduced not empty word W equals a unit in the group G, then W contains the occurrence of such a word S that for some i and T, $ST \cong R_i$ and

$$|S| > (1 - 2\lambda)|R_i|$$

## VI. CANONICAL FORM

Let M be a set of words $R_i(a_\mu)$, $i = 1, \ldots, I$, $\mu = 1, \ldots, m$ satisfying conditions (1) and (2) of §V. Assume also that the set M consists of $n_1$ words of the length $t_1, \ldots, n_k$ words of the length $t_k$ and $2 < t_1 < \ldots < t_k$.

A block graph of order p consists of a finite set V (possibly empty) containing p vertices and a set X consisting of 9 ordered triplets (i, u, v). i is a natural number, u, v are various vertices; such a definition automatically excludes loops (edges linking the vertex with itself) and multiple (parallel) edges. The vertex $u \in V$ belongs to one of k+1 types and depending on the latter, has a form of

0. a segment dissected into $t_0 = 2$ equal parts
1. a segment dissected into $t_1$ equal parts

. . . . . . . . . . . . . . . . . . .

k. a segment dissected into $t_k$ equal parts, $\nu(u)$ denotes a type of the vertex u.

The triplet $x = (i, u, v)$, $\nu(u) = j$, $1 \leq i \leq t_j - 1$, $0 \leq j \leq k$ is called an edge of the graph; the edge x is also said to connect the i-th division point of the vertex u with the vertex v. The vertices u and v are then called contiguous, while the vertex u is said to precede the vertex v. The latter fact is denoted by the inequality $u \ll v$. The vertex u and the edge x, as well as the vertex v and the edge x, are called incident to each other. A block graph without cycles is called a block tree D. The set of vertices of the block tree may be turned into a partially ordered set with the order relation $<$ (this definition requires the partial order to be only transitive) if it is supposed that $u < v$ iff there exist a sequence $u_1, \ldots, u_p$ of vertices such that $u \ll u_1 \ll \ldots \ll u_p \ll v$. The greatest lower bounds of the maximal chains in the partially-ordered set $V(D)$ of the tree $D$ vertices are called first-level vertices of the block tree D. The least upper bounds of the maximal chains are called finite vertices of the tree D. Erase the first-level vertices and the edges incidental to them. We again get a block tree for which a set of the first-level vertices is called a set of the second-level vertices of the initial block tree. Vertices of the third level and subsequent ones are defined similarly. Let $\xi(D)$ be the maximal level of the tree D vertices. We assume that the linear order $\tau(D)$ consistent with the relation $<$ introduced earlier is given on a set of vertices V. Two block trees $D_1$ and $D_2$ are called isomorphic iff there exists a one-to-one mapping $\varphi$ of the set of vertices $V(D_1)$ onto $V(D_2)$ preserving order and contiguity. Denote the set of block trees by $\Delta$. By definition, the empty set belongs to $\Delta$.

Let the mapping $\psi$ of the set of vertices $V(D)$ into the set of words $\{a_\nu^\epsilon a_\nu^{-\epsilon}, R_i(a_\mu) | \nu, \mu = 1, \ldots, m, \epsilon = \pm 1, i = 1, \ldots, I\}$ be called the coloring of the block tree $D \in \Delta$, and if $u \in V(D)$, then

$$\psi(u) \in \{a_\nu^\epsilon a_\nu^{-\epsilon} | \nu = 1, \ldots, m, \epsilon = \quad 1\} \quad \text{if } \nu(u) = 0$$

$$\psi(u) \in \{R_i : |R_i| = t_1\} \quad \text{if } \nu(u) = 1$$

$$\cdots\cdots\cdots\cdots\cdots\cdots\cdots$$

$$\psi(u) \in \{R_i : |R_i| = |t_k\} \quad \text{if } \nu(u) = k$$

i-th $(1 \le i \le t_{\nu(u)})$ division segment of the vertex u into $t_{\nu(u)}$ equal parts is colored by the i-th letter of the word $\psi(u)$. The two colored trees $D_1, D_2$ are called isomorphic iff there exists a one-to-one mapping of the set $V(D_1)$ onto the set $V(D_2)$ preserving contiguity, order, and coloring. We call a block tree obtained after erasing the coloring, a skeleton of the colored tree. Denote the set of colored trees by $\Omega$.

Let $D \in \Delta$. Extend the order $\tau(D)$ from the set of vertices on the set of all segments of division of the tree $D$ vertices. The extension is carried out as follows. Let the vertices $u$ and $v$ be contiguous, the edge $(i, u, v)$ connects them and $\{v_i\}$ is a set of all vertices $v_i$ such that $v < v_i$, $(i = 1, \ldots, S)$ in the sense of partial order on $V(D)$. Then all the division segments of the vertex $u$ from the 1-st one to the i-th one inclusive are said to precede the division segments of the vertices $v, v_1, \ldots, v_S$, while the division segments of the vertex $u$ from the (i+1)-th one to the $t_{\nu(u)}$-th one are said to follow the division segments of the vertices $v, v_1, \ldots, v_S$. The division segments of the vertex $u$ are ordered from left to right. The new order is, as before, denoted by $\tau(D)$.

Let $D \in \Omega$. Place the division segments of the tree $D$ vertices in the order $\tau(D)$. We get a sequence $\zeta_1, \ldots, \zeta_n$ of division segments. Denote the word $\psi(\zeta_1) \ldots \psi(\zeta_n)$ by $\theta(D)$. The length of the word $\theta(D)$ is called a length of a colored tree. The canonical form $Q(W)$ of the non-empty freely reduced word $W$ equal to a unit element in the group $G$ whose set $M$ of defining words satisfies conditions (1), (2), (3) of paragraph 5 is a tree $Q(W) \in \Omega$ which is constructed on the basis of $W$ in the following manner.

As above, the word $W$ has a subword $S$ such that $R_{i_1} \cong ST$ and $|T| < 3\lambda |R_{i_1}|$. If there are several such subwords, choose the one which is to the left of the others. Substitute $R_{i_1} T^{-1}$ for $S$ in $W$. Denote the word obtained by $W'$. Introduce the vertex $u_1$ into consideration. If $|R_{i_1}| = t_{\gamma_1}$, let $\nu(u) = \gamma_1$ and color it by the word $R_{i_1}$. Reduce $W'$ without touching symbols of the marked defining word $R_i$. By crossing out $R_{i_1}$ from the obtained word $W'' \cong UR_{i_1}V$ we get the word $W''' \cong UV$. If $W'''$ is reducible, the reduction may take place only between the latter symbol of the word $U$ and the first one of the word $V$, between the last but one symbol of $U$ and the second one of $V$, etc. Suppose that $U \cong U_1 a_{i_1} \dots a_{i_p}$, $V \cong a_{i_p}^{-1} \dots a_{i_1}^{-1}$ and the word $U_1V_1$ is freely reduced. To the vertex $u_1$ join vertices $u_2, \dots, u_{p+1}$ of the type $0$, color them by the words $a_{i_p} a_{i_p}^{-1}, \dots, a_{i_1} a_{i_1}^{-1}$ and introduce also edges $(1, u_2, u_1)$, $(1, u_3, u_2), \dots, (1, u_{p+1}, u_p)$. Assume $u_{p+1} \ll u_p \ll \dots \ll u_1$. Denote the word $U_1U_2$ equal to a unit element in the group $G$ by $W_1$. Now subject $W_1$ to a procedure similar to that applied to the word $W$: select a subword $S_1$ such that $R_{i_2} \cong S_1T_1$ and $|T_1| < 3\lambda\, |R_{i_2}|$. Introduce vertices $v_{q+1} \ll v_q \ll \dots \ll v_1$, the vertex $v_1$ corresponding to the word $R_{i_2}$, $\nu(v_1) = \gamma_2$, if $|R_{i_2}| = t_{\gamma_2}$ and, at last, construct the word $W_2$. If $R_{i_1}$ was situated between symbols of the word $S_1$ in the word $W'$ and if $j_1$ symbols of $S_1$ were situated to the left of $R_{i_1}$, introduce an edge $(j_1, v_1, u_p)$. In the opposite case do not introduce new edges and assume $v_1 < u_{p+1}$ if the word $R_{i_1}$ is situated to the left of $S_1$ and $v_1 > u_{p+1}$ otherwise. Do the same with the word $W_2$: select a subword $S_2$ such that $R_{i_3} \cong S_2T_2$ and $|T_2| < 3\lambda\, |R_{i_3}|$. Introduce vertices $h_{s+1} \ll \dots \ll h_1$, the vertex $h_1$ corresponding to the word $R_{i_3}$, $\nu(h_1) = \gamma_3$ if $|R_{i_3}| = t_{\gamma_3}$ and, at last, construct the word $W_3$. If $R_{i_1}$ was situated between symbols of the word $S_2$ in $W'$, introduce an edge $(j_2, h_1, u_{p+1})$, where $j_2$ is the number of symbols in $S_2$ situated to the left of $R_{i_1}$ in $W'$. Similarly, if the word $R_{i_2}$ was situated between symbols of the word $S_2$ in $W_1'$, introduce an edge $(j_3, h_1, v_{q+1})$, where $j_3$ is the number of symbols in the word situated to the left of the word $S_2$ in $W_1'$. In the opposite case new edges are not introduced and the set of vertices obtained is ordered as in the case of the pair

$u_1, v_1$. Continue the procedure till at the step $\ell + 1$, $W_\ell$ is equal to an empty word. The above procedure completely defines the canonical form $Q(W)$ of the word $W$. By definition, an empty word is corresponded by an empty tree. $\Omega_0$ will further denote a set of canonical forms.

Remark. We have defined a canonical form in a situation when the set $M$ of the group $G$ defining words satisfies conditions (1), (2), (3) of paragraph 5. This definition is literally transferred to the case of groups whose set of defining relations satisfies conditions (1), (2), (3'), (4) of paragraph 5. The only thing to be done is to select at every step of constructing the canonical form the word $S$ such that $R_i \cong ST$ and $|T| < 2\lambda |R_i|$. Note that canonical forms have no finite vertices of zero level and $W \to Q(W)$ is a one-to-one mapping.

## VII. MAIN THEOREM

Theorem 7.1. Let $\mathcal{C}_\lambda(a_1, \ldots, a_m)$ be a set of finitely-defined groups $G$ whose set $M_0$ of defining words $R_i(a_\mu)$, $1 \le \mu \le m$ after closure with respect to the operations of taking the inverse word and cyclic transposition of the word satisfies condition (3) of paragraph 5. $r_G$ is a spectral radius of a random walk on the group $G$. Then for an arbitrary $\epsilon > 0$ and natural $N$ there exists such $T$ that if $G \in \mathcal{C}_\lambda(a_1, \ldots, a_m)$, $|M_0| < N$ and $\min |R_i(a_\mu)| > T$ then

$$\left| r_G - \frac{\sqrt{2m-1}}{m} \right| < \epsilon \tag{7.1}$$

Proof: By force of Theorem 4.1 it suffices to prove that if $H$ is a normal subgroup of the group $F(a_1, \ldots, a_m)$, generated by the set of defining words $M_0$, then for an arbitrary $\epsilon > 0$ and natural $N$ there exists such $T$ that if $G \simeq F(a_1, \ldots, a_m)/H \in \mathcal{C}_\lambda(a_1, \ldots, a_m)$

and $\min_i |R_i(a_\mu)| > T$, then

$$|\alpha_H^{a_1,\ldots,a_m} - \sqrt{2m-1}| < \epsilon$$

Let the set M be obtained from the set $M_0$ of the group G by defining words by closure with respect to the operation of taking the inverse word and cyclic transpositions. Suppose M consists of $N_1$ words of the length $t_1, \ldots, N_k$ words of the length $t_k$. Using the numbers $t_1, \ldots, t_k$ construct the set $\Delta$ of block trees, and using the set of words M construct the set $\Omega$ of colored block trees with vertices of the $k+1$ type. Let $D \in \Delta$. If $u \in V(D)$, $\nu(u) = i$, then $\eta(u) = x_i^{t_i}$ is called a weight of the vertex u. The expression $\eta(D) = \eta(u_1)\ldots\eta(u_p)$ is called a weight of the tree D; $u_1, \ldots, u_p$ are vertices of the tree D written out in an arbitrary order (such definition naturally assumes weights of separate vertices to commutate with each other). A weight of a colored tree is the weight of its skeleton. Define two formal power series

$$\Phi(x_0^{t_0}, \ldots, x_k^{t_k}) = 1 + \sum_{i_0, \ldots, i_k} C_{i_0 \ldots i_k} x_0^{t_0 i_0} \ldots x_k^{t_k i_k} \quad (7.2)$$

where $C_{i_0 \ldots i_k}$ denotes the number of block trees of the weight $x_0^{t_0 i_0} \ldots x_k^{t_k i_k}$ lacking finite zero-level vertices, and

$$\Phi_0(x_0^{t_0}, \ldots, x_k^{t_k}) = \sum_{D \in \Omega} \eta(D) = 1 + \sum_{i_0, \ldots, i_k} b_{i_0 \ldots i_k} x_0^{t_0 i_0} \ldots x_k^{t_k i_k}$$

where $b_{i_0 \ldots i_k}$ denotes the number of canonical forms of the word W equal to a unit element in the group G of the weight $x_0^{t_0 i_0} \ldots x_k^{t_k i_k}$. Our immediate task is to demonstrate that the series $\Phi(x_0^{t_0}, \ldots, x_k^{t_k})$ defines the function of $x_0, \ldots, x_k$ analytic in the neighborhood of zero.

A tree with a given order on the set of its vertices is called an ordered tree, and the order should be coordinated with the contiguity relation of the vertices as in the case of a block tree (no assumption is made concerning the tree contiguity). Similar to the way it was done for block trees, the notions of tree levels, finite vertices, etc. are introduced.

Lemma 7.1. *Let* $f(x) = \sum_{n=1}^{\infty} f_n x^n$, *where* $f_n$ *is the number of ordered trees* D *with a single first-level vertex, the trees being such that* $|V(D)| = n$. *Then*

$$f(x) = \frac{1}{2}[1 - (1 - yx)^{\frac{1}{2}}] \tag{7.3}$$

Proof: Demonstrate that the numbers $f_n$ satisfy the relation

$$f_n = \sum_{p=1}^{n-1} \sum_{\Sigma r_i = n-1} f_{r_1} \cdots f_{r_p} \tag{7.4}$$

Indeed, the first-level vertex may be contained in $p$, $(1 \leq p \leq n)$ edges. If the number $p$ is fixed, then the tree may continue independently from $p$ second-level vertices, and the sum $\sum_{\Sigma r_i = n-1} f_{r_1} \cdots f_{r_p}$ embraces all the cases of the continuation. Multiplying (7.4) by $x^n$ and summing over $n$ we have

$$f(x) = \sum_{n=1}^{\infty} f_n x^n = x + \sum_{n=2}^{\infty} \sum_{p=1}^{n-1} \sum_{\Sigma r_i = n-1} f_{r_1} \cdots f_{r_p} x^n$$

$$= x + x \sum_{p=1}^{\infty} \sum_{n=p+1}^{\infty} \sum_{\Sigma r_i = n-1} f_{r_1} \cdots f_{r_p} x^{n-1}$$

$$= x + x \sum_{p=1}^{\infty} f^p(x) = x + \frac{xf(x)}{1 - f(x)}$$

Solving the latter equation with respect to $f(x)$, we have (7. 3).

Lemma 7. 2. *Let* $g(x) = \sum_{n=0}^{\infty} g_n x^n$, *where* $g_n$ *is the number of ordered trees such that* $|V(D)| = n$. *Then*

$$g(x) = \{1 - \frac{1}{2}[1 - (1-4x)^{\frac{1}{2}}]\}^{-1} \tag{7. 5}$$

Proof: Let the number of the first-level vertices of the ordered tree $D$ equal $n$. Since the tree continues independently from all these vertices, the enumerating series of the trees having the above property equals $f^n(x)$. Therefore $g(x) = \sum_{n=0}^{\infty} f^n(x)$, q. e. d. Note that the addend 1 emerged due to including an empty tree into consideration.

Corollary 7. 1. *The numbers* $g_n$ *satisfy the inequalities*

$$g_n \leq N^*(4+\epsilon)^n, \qquad \epsilon > 0$$

($N^*$ *is a constant independent of* $n$).

Indeed, the convergence radius of the series $g(x)$ equals 1/4. Therefore $\limsup_{n\to\infty} \sqrt[n]{g_n} = 4$.

Let $D$ be an arbitrary block tree, $|V(D)| = n$, and the set $V(D)$ consist of $i_0, \ldots, i_k$ vertices of the first-level, ..., $k$-th one $(\sum_{j=0}^{k} i_j = n)$. Make the following reconstruction within the tree $D$: turn the vertices $u$ of the tree $D$ into points $\Xi(u)$, unify the beginnings of edges originating in the division points of the vertex $u$ and place them into $\Xi(u)$. Denote the tree obtained by $\Xi(D)$.

Lemma 7. 3. *There exist no more than* $[(k+1)2^{2\max_i t_i}]^n$ *block trees such that* $\eta(D) = x_0^{t_0 i_0} \ldots x_k^{t_k i_k}$ *and such that under a mapping* $\Xi$ *they are transformed into a fixed tree* $\Xi(D)$.

Proof: Thus, let $\sum_{j=0}^{x} i_j = n = |V(\Xi(D))|$. Reconstructing the tree D from the tree $\Xi(D)$ we should decide, first of all, to which type each of n vertices of $\Xi(D)$ belongs. The number of possible variants is equal to that of various permutations in n boxes of $i_0$ identical objects of zero sort, ..., $i_k$ identical objects of the k-th sort so that each box contains a ball, i. e., to the number

$$\frac{n!}{i_0!\,i_1!\ldots i_k!} \leq (k+1)^n$$

After the type of each vertex of the tree $\Xi(D)$ is indicated, the question should be solved concerning the placing of edge beginnings in the division points. Let the vertex $\Xi(u_p)$ of the tree $\Xi(D)$ be contained in $q_p$ edges and $\nu(u_p) = i(p)$. The number of placements of $q_p$ edge beginnings into the division point of the vertex $u_p$ coincides with that of integer nonnegative solutions of the equation $x_1 + \ldots + x_{t_{i(p)}-1} = q_p$, which equals the number

$$\binom{t_{i(p)} - 1 + q_p - 1}{q_p}$$

Due to $\sum_p q_p \leq n - 1$ we have that the number of various placement of edge beginnings does not exceed the number

$$\prod_p \binom{t_{i(p)} - 1 + q_p - 1}{q_p} \leq \prod_p \binom{t_{i(p)} \cdot q_p}{q_p} \leq 2^{2\sum_p t_{i(p)} \cdot q_p} \leq 2^{2(\max_i t_i)n}$$

Corollary 7.2. Coefficients $C_{i_0 \ldots i_k}$ of the enumerating series (7.2) satisfy the inequalities

$$C_{i_0 \ldots i_k} \leq [(4+\epsilon)(K+1)2^{2\max_i t_i}]^{\sum_{j=0}^{k} i_j}$$

and, therefore, the enumerating series $\Phi(x_0^{t_0}, \ldots, x_k^{t_k})$ determines the function of $x_0, \ldots, x_k$ analytic in the neighborhood of zero.

Let $D \in \Delta$. A tree consisting of the vertices from the 1-st level to the i-th one inclusive, of the tres D placed in the order $\tau(D)$ with the same relation of contiguity that in D is called an i-th level section of the tree D $(1 \leq i \leq \xi(D))$. The i-th level section forms an independent object for which the notions of weight and isomorphism are naturally defined.

Lemma 7. 4. Let $\Gamma(x_0^{t_0}, \ldots, x_k^{t_k})$ be the enumerating series for the first-level sections weights. Then

$$\Gamma(x_0^{t_0}, \ldots, x_k^{t_k}) = [1 - (x_0^{t_0} + \ldots + x_k^{t_k})]^{-1} \tag{7.6}$$

Proof: The enumerating series for the weights of the first-level sections consisting of n vertices equals $(x_0^{t_0} + \ldots + x_k^{t_k})^n$. Hence

$$\Gamma(x_0^{t_0}, \ldots, x_k^{t_k}) - \sum_{n=0}^{\infty} (x_0^{t_0} + \ldots + x_k^{t_k})^n = [1 - (x_0^{t_0} + \ldots + x_k^{t_k})]^{-1}$$

Let $\Gamma_0(x_0^{t_0}, \ldots, x_k^{t_k}) = \Gamma(x_0^{t_0}, \ldots, x_k^{t_k}) - 1$. Define the sequence of functions $\Phi_n(x_0^{t_0}, \ldots, x_k^{t_k})$, $\Psi_n(x_0^{t_0}, \ldots, x_k^{t_k})$ in the following way:

(1) $\Phi_0 = 1, \ \Psi_0 = 0$

(2) $$\Phi_{n+1} = \Gamma(x_0^{t_0} \Psi_n^{t_0 - 1}, x_1^{t_1} \Phi_n^{t_1 - 1}, \ldots, x_k^{t_k} \Phi_n^{t_k - 1})$$

$$\Psi_{n+1} = \Gamma(x_0^{t_0} \Psi_n^{t_0 - 1}, x_1^{t_1} \Phi_n^{t_1 - 1}, \ldots, x_k^{t_k} \Phi_n^{t_k - 1}) \tag{7.7}$$

It is obvious that the functions $\Phi_n, \Psi_n$ are related as $\Phi_{n-1} = \Psi_n$ and the coefficients of the unknown variables $x_0, \ldots, x_k$ in the power series $\Phi_n, \Psi_n$ are stabilized as n increases. That is to say that if $C^n_{i_0 \ldots i_k}$ is a coefficient in the power series $\Phi_n(x_0^{t_0}, \ldots, x_k^{t_k})$ under the weight $x_0^{t_0 i_0} \ldots x_k^{t_k i_k}$, when $n \geq k$. It is readily seen that k may be assumed to equal $\sum_{j=0}^{k} i_j$.

<u>Lemma 7. 5.</u> $\Phi_n \to \Phi$, $\Psi_n \to \Phi - 1$, <u>and the convergence is understood in the above meaning of stabilizing the coefficients.</u>

<u>Proof:</u> Let the tree D be such that it has no finite vertices of zero type, $u \in V(D)$ and $\nu(u) = i > 0$. The tree D may be extended from an arbitrary $(t_0 - 1)$-th point of dividing a segment u into $t_i$ equal parts, the enumerating series of the weights of all possible extensions to one level being equal to $\Gamma(x_0^{t_0}, \ldots, x_k^{t_k})^{t_0 - 1}$. The situation is different when $\nu(u) = 0$. In this case the enumerating series of the weights of all possible extensions to one level equals $\Gamma_0(x_0^{t_0}, \ldots, x_k^{t_k})$, since the vertices of zero type cannot be finite vertices of the tree D, and substituting $x_0\Gamma^{t_0 - 1}$ results in considering in the enumerating series $\Phi(x_0^{t_0}, \ldots, x_k^{t_k})$ the trees whose finite vertices may belong to zero type. Thus $\Phi_1$ is an enumerating series for non-isomorphic first-level sections; $\Phi_2$ is an enumerating series of the non-isomorphic second-level section, and $\Phi_n$ is an enumerating series of the non-isomorphic n-th level section. If $n > \xi(D)$, the n-th level section of the tree D coincides with the tree D. Therefore, $\Phi' = \lim_{n \to \infty} \Phi_n$ in recalculation takes into account weights of all block trees whose finite vertices do not contain those of zero type, and only such trees, and, therefore, $\Phi' = \Phi$. q. e. d.

<u>Lemma 7.6.</u> <u>Functions</u> $\Phi, \Psi$ <u>satisfy the system of algebraic equations</u>

$$\begin{cases} \Gamma(x_0^2\Psi,\ x_1^{t_1}\Phi^{t_1-1},\dots,x_k^{t_k}\Phi^{t_k-1}) = \Phi(x_0^2 x_1^{t_1},\dots,x_k^{t_k}) \\ \Gamma_0(x_0^2\Psi,\ x_1^{t_1}\Phi^{t_1-1},\dots,x_k^{t_k}\Phi^{t_k-1}) = \Psi(x_0^2, x_1^{t_1},\dots,x_k^{t_k}) \end{cases} \tag{7.8}$$

Proof: System (7. 8) follows from the recurrence relations (7. 7), Lemma 7. 5, and the fact that $\Phi$ and $\Psi$ are functions, and $t_0 = 2$.

Lemma 7. 7. The enumerating series $\Phi$ satisfies the equation

$$\sum_{i=1}^{k} x_i^{t_i}\Phi^{t_i} + x_0^2\Phi^2 - (x_0^2 + 1)\Phi + 1 = 0$$

Proof: The functions $\Phi$ and $\Psi$ are related as $\Phi = \Psi + 1$. Using the explicit form of the function $\Gamma$ from the first-level of system (7. 8), obtain

$$[1 - x_0^2 - x_0^2\Phi - x_1^{t_1}\Phi^{t_1-1} - \dots - x_k^{t_k}\Phi^{t_k-1}]^{-1} = \Phi$$

or

$$x_1^{t_1}\Phi^{t_1} + \dots + x_k^{t_k}\Phi^{t_k} + x_0^2\Phi^2 - (x_0^2 + 1)\Phi + 1 = 0$$

The lemma is proved.

Lemma 7. 8. Coefficients $b_{i_0\dots i_k}$ of the enumerating series $\Phi^*(x_0^{t_0},\dots,x_k^{t_k})$ of the canonical forms weights do not exceed corresponding coefficients of the series $\Phi((2m-1)x_0^2, N_1x_1^{t_1},\dots,N_kx_k^{t_k})$.

Proof: An arbitrary colored tree (including the canonical form) consists of a skeleton D and its coloring $\Psi$. Let a block tree be a skeleton of the canonical form and $\eta(D) = x_0^{2i_0}x_1^{t_1 i_1}\dots x_k^{t_k i_k}$. Estimate from above the number of colorings transforming the tree D into the canonical form. If $u \in V(D)$ and $1 \le \nu(u) \le k$, there exist $N_{\nu(u)}$ possibilities to color the vertex. Let $\nu(u) = 0$. Generally speaking, the vertex u may be colored by one of 2m symbols $a_\nu^\epsilon a_\nu^{-\epsilon}$, $\nu = 1, \ldots, m$, $\epsilon = \pm 1$. However, if the edge (i, u, v) belongs to the tree D, $\nu(u) = 0$, $\Psi(u) = x = x^{-1}$ and the vertex v is the first one in the order $\tau(D)$ among those contiguous with u and such that $u \ll v$, the symbol x cannot be inverse ot the first symbol of the word $\Psi(v)$. Since zero type vertices cannot be finite vertices of the skeleton of the canonical form, hence it follows that the number of variants of coloring such a vertex does not exceed 2m-1. Therefore, if a block tree contains exactly $p_i$ vertices of the i-th type, $0 \leq i \leq k$, this tree can be a skeleton of no more than $(2m-1)^{p_0} N_1^{p_i} \ldots N_k^{p_k}$ canonical forms. Thus, coefficients of the series $\Phi((2m-1)x_0^2, N_1 x_1^{t_1}, \ldots, N_k x_k^{t_k})$ majorize coefficients of the enumerating series $\Phi^*(x_0^{t_0}, \ldots, x_k^{t_k})$.

From now on we assume the number $\lambda$ from condition (3) of paragraph 5 to be rational: $\lambda = \frac{r}{p} < \frac{1}{6}$. (This assumption is obvious not to inflict any loss of generality on our reasoning.)

<u>Lemma 7. 9.</u> <u>Let</u> $H(x) = \sum_{n=0}^{\infty} |H_n| x^n$ <u>be a generating function of the normal subgroup</u> H <u>generated by the set</u> $M_0$ <u>of the group</u> G <u>defining words.</u> <u>Then the convergence radius of the series</u> H(x) <u>is no less than that of the series</u> $\Phi((2m-1)x^2, N_1 x^{t_1}, \ldots, N_k^{t_k})$.

<u>Proof:</u> It has been proved that coefficients of the series $\Phi((2m-1)x_0^2, N_1 x_1^{t_1}, \ldots, N_k x_k^{t_k})$ majorize corresponding coefficients of the enumerating series $\Phi^*(x_0^{t_0}, \ldots, x_k^{t_k})$ of the canonical form weights. Let W be a non-empty freely reduced word, W = 1, $|W| = n$. Recall that a length of the word $\theta(Q(W))$ is called the length of the canonical form Q(W). Let $\eta(Q) = x_0^{t_0 i_0} \ldots x_k^{t_k i_k}$. We prove that $|\theta(Q(W))| - |W| < 6\lambda \sum_s t_s i_s$. Indeed, reduce $\theta(Q(W))$ to W by

inserting and crossing out words of the type $a_\nu^\epsilon a_\nu^{-\epsilon}$ in the order inverse to the one taking place in constructing the canonical form. Since at most $3\lambda R$ symbols are crossed out from each defining word R and in every crossing out one of the symbols of the word $a_\nu^\epsilon a_\nu^{-\epsilon}$ necessarily belongs to a defining word, the number of symbols crossed out from $\theta(Q(W))$ does not exceed $6\lambda \sum_s t_s i_s$. Let

$$\Phi((2m-1)x^2, N_1 x^{(1-6\lambda)t_1}, \dots, N_k x^{(1-6\lambda)t_k}) = \sum_{\rho=0}^{\rho-1} \sum_{n=0}^{\infty} f_{n,p} x^{n+p/\rho} = \varphi(x)$$

$$\tilde{\Phi}(x) = \sum_{q=0}^{\rho-1} [\sum_{n=0}^{\infty} \sum_{j=0}^{n} f_{jq}] x^{n+q/\rho} = \frac{1}{1-x} \sum_{q=0}^{\rho-1} \sum_{n=0}^{\infty} f_{n,q} x^{n+q/\rho}$$

Since the convergence radius of the series $\varphi(x)$ does not exceed 1, it coincides with that of the series $\tilde{\Phi}(x)$. Let $W = 1$, $|W| = n$ and $\eta(Q(W)) = x_0^{t_0 i_0} \dots x_k^{t_k i_k}$. Due to $W \to Q(W)$ being a one-to-one mapping, $x^{t_0 i_0} x^{(1-6\lambda)t_1 i_1} \dots x^{(1-6\lambda)t_k i_k} = x^{i_0 t_0 + (1-6\lambda)\sum_s t_s i_s}$ and $i_0 t_0 + (1-6\lambda)\sum_s t_\sigma i_\sigma < |W| = n$, we obtain

$$|H_n| \le \sum_{j=0}^{n-1} \sum_{p=0}^{\rho-1} f_{j,q}$$

Hence, the convergence radius of the series $H(x)$ is no less than that of the series $\tilde{\Phi}(x)$, and, therefore, no less than that of the series $\varphi(x)$. The lemma is proved.

<u>Lemma 7.10.</u> <u>If there exist such constants</u> C <u>and</u> $\sigma$ <u>that</u> $N_i(t_i) < C t_i^\sigma$, <u>then the convergence radius of the series</u>

$$\varphi(x) = \Phi((2m-1)x^2, N_1(t_1)x^{(1-6\lambda)t_1}, \dots, N_k(t_k)x^{(1-6\lambda)t_k}) \qquad (7.9)$$

<u>tends to</u> $1/\sqrt{2m-1}$ <u>when</u> $t_1, \dots, t_k \to \infty$.

Proof: The series $\varphi(x)$ is defined as the root of equation

$$\sum_{s=1}^{k} N_s(t_s) x^{(1-6\lambda)t_s} \varphi^{t_s} + (2m-1)x^2\varphi^2 - [(2m-1)x^2+1]\varphi + 1 = 0$$

analytic at zero and satisfying condition $\varphi(0) = 1$. Coefficients for the powers of $x$ in the series $\varphi(x)$ are nonnegative. Therefore, in estimating the convergence radius the least positive singularity $\zeta$ of the function $\varphi(x)$ should be taken into account, and since the growth exponent of the group $F_m$ normal divisor exceeds $\sqrt{2m-1}$, it may be said that $\zeta < 1/\sqrt{2m-1}$. The set of positive singular points of the function $\varphi(x)$ is to be found among the positive solutions $x_\ell$ of the system of equations

$$\begin{cases} F_1(x,\mu) = \sum_{s=1}^{k} N_s(t_s) x^{(1-6\lambda)t_s} \mu^{t_s} + (2m-1)x^2\mu^2 - [(2m-1)x^2+1]\mu + 1 = 0 \\ F_2(x,\mu) = \sum_{s=1}^{k} t_s N_s(t_s) x^{(1-6\lambda)t_s} \mu^{t_s-1} + 2(2m-1)x^2\mu - (2m-1)x^2 - 1 = 0 \end{cases} \tag{7.10}$$

Together with system (7.10) consider the system of equations

$$\begin{cases} (2m-1)x^2\mu^2 - [(2m-1)x^2+1]\mu + 1 = 0 \\ 2(2m-1)x^2\mu - (2m-1)x^2 - 1 = 0 \end{cases} \tag{7.11}$$

The point $(1/\sqrt{2m-1}, 1)$ is its unique positive solution. We prove that all the solutions of system (7.10) belonging to a strip $Q = \{0 < x < 1/\sqrt{2m-1},\ 0 < \mu\}$ tend to the solution $(1/\sqrt{2m-1}, 1)$ when $t_1, \ldots, t_k \to \infty$. The first equation of (7.11) defines the straight line $\mu = 1$ and square hyperbola $\mu = [(2m-1)x^2]^{-1}$ on the plane of variables $x, \mu$. Denote the restriction of the curves to the region $Q$ by $\Gamma_1$ and $\Gamma_2$. The second equation of (7.11) defines on the plane $(x,\mu)$ the square hyperbola $\mu = [(2m-1)x^2+1]/2(2m-1)x^2$ whose restriction to the region $Q$ is denoted by $\Gamma_3$. The function $F_1(x,\mu)$ is convex with respect to $\mu$ when $\mu > 0$, while $F_2(x,\mu)$ increases monotonically with respect to $\mu$, $(\mu > 0)$, $x$ being fixed. Therefore, the equation $F_1(x,\mu) = 0$ defines the curves $\tilde{\Gamma}_1, \tilde{\Gamma}_2$ in the region $Q$ which are solutions of the equation, while the equation $F_2(x,\mu) = 0$ defines the curve $\tilde{\Gamma}_3$. Let $x_0$ be fixed $(0 < x_0 < 1/\sqrt{2m-1})$ and $\mu \in [1, \mu_0]$, where $\mu_0 x_0^{1-6\lambda} < 1$. Since

$$K(x_0,\mu_0) = \sum_{s=1}^{k} N_s(t_s) x_0^{(1-6\lambda)t_s} \mu^{t_s} \xrightarrow[\min_i t_i \to \infty]{} 0$$

due to $N_s(t_s) < C\, t_s^{\sigma}$ and $x_0^{1-6\lambda}\mu_0 < 1$, the least root $\mu(x_0)$ of the equation $F_1(x_0,\mu) = 0$ tends to $1$ when $\min_i t_i \to \infty$ uniformly with respect to $x_0$, $0 < x_0 < 1/\sqrt{2m-1}$, and $\mu(x_0) \to 1$ when $x_0 \to 0$. Hence $\mu(x) = \varphi(x)$. Compare solutions of the equations

$$2(2m-1)x^2\mu(x) - (2m-1)^2x^2 - 1 = 0$$

$$\Gamma_2(x,\mu) - \sum_{s=1}^{k} \iota_s N_s(\iota_s) x^{(1-6\lambda)t_s} \mu(x)^{t_s - 1} + 2(2m-1)x^2\mu(x) - (2m-1)x^2 - 1 = 0$$

Denote by $I(x)$ the value $F_2(x,\mu)$ in the point $\left(x, \dfrac{(2m-1)x^2+1}{2(2m-1)x^2}\right)$,

$$I(x) = \sum_{s=1}^{k} t_s N_s(t_s) x^{(1-6\lambda)t_s} \left[\frac{(2m-1)x^2+1}{2(2m-1)x^2}\right]^{t_s - 1}$$

and by $R(x)$ the value $\dfrac{\partial F_2(x,\mu)}{\partial \mu}$ in the same point.

$$R(x) = \sum_{s=1}^{k} t_s(t_s-1)N_s(t_s) x^{(1-6\lambda)t_s} \left[\frac{(2m-1)x^2+1}{2(2m-1)x^2}\right]^{t_s - 2}$$

Since $\sum_{s=1}^{k} t_s N_s(t_s) x^{(1-6\lambda)t_s} \mu^{t_s - 1} > 0$ when $x > 0$, $\mu > 0$, the curve $\tilde{\Gamma}_3$ is not higher than the curve $\tilde{\Gamma}_3$. Let $(x_0, \mu_0) \in \tilde{\Gamma}_3$. Compare the numbers $\mu_0$ and $[(2m-1)x_0^2 + 1]/[2(2m-1)x_0^2]$. Since the derivative $\frac{\partial F_2(x, \mu)}{\partial \mu}$ grows as $\mu$ increases, then

$$0 < \frac{(2m-1)x_0^2+1}{2(2m-1)x_0^2} - \mu_0 < \frac{I(x_0)}{R(x_0)} \leq \frac{1}{\min_i (t_i - 1)} \frac{(2m-1)x_0^2+1}{2(2m-1)x_0^2} = \delta(t_1, \ldots, t_k, x_0)$$

We have found the curve $\tilde{\Gamma}_3$ gets into the $\delta(t_1, \ldots, t_k, x)$ neighborhood of the square hyperbola $\mu = [(2m-1)x^2 + 1]/[2(2m-1)x^2]$. It has been demonstrated above that when $\min_i t_i \to \infty$ the curve $\tilde{\Gamma}_1$ tends to the curve $\{0 < x < 1/\sqrt{2m-1},\ \mu = 1\}$. Hence it follows that all the intersection points of $\tilde{\Gamma}_1$ and $\tilde{\Gamma}_3$ (which are the solutions of system (7. 10) tend to the point $(1/\sqrt{2m-1}, 1)$, q. e. d.

Now complete the proof of Theorem 7. 1. Let the set $M_0$ consist of $r_1$ words of the length $t_1, \ldots, r_k$ words of the length $t_k$, $\sum_j r_j < N$. If $N_1$ is the number of words of the length $t_1, \ldots, N_k$ is the number of words of the length $t_k$ in the set of words $M$ obtained by symmetrization of the set $M_0$, then $N_1 \leq 2r_1 t_1, \ldots, N_k \leq 2r_k t_k$. Therefore, in Lemma 7. 10, $C$, $\sigma$ may be said to equal $2 \max_j r_j$, 1. Since the radius of convergence $r$ of the series $H(x)$ is no less than the radius of convergence $r_\Phi$ of the series $\Phi((2m-1)x^2, N_1 x^{(1-6\lambda)t_1}, \ldots, N_k x^{(1-6\lambda)t_k})$, and $r_\Phi \xrightarrow[t_i \to \infty]{} 1/\sqrt{2m-1}$, then $r_\Phi^{-1} \geq r^{-1} = \alpha_H \to \sqrt{2m-1}$ since $\alpha_H > \sqrt{2m-1}$. The theorem is proved.

Corollary. Let $\{W_i(a_\mu)\}$, $i = 1, \ldots, n$, $\mu = 1, \ldots, m$ be an arbitrary collection of various freely reduced words such that no $W$-word is a power of any other W-word. Then the spectral radius of a simple walk on the group

$$G = \langle a_1, \ldots, a_m \mid W_1^{t_1} = \ldots = W_n^{t_n} = e \rangle$$

tends to $\sqrt{2m-1}/m$ when $t_1, \ldots, t_i \to \infty$.

Remark 1. Theorem 7. 1 has been proved on the assumption that the set $M_0$ of defining words of the group G after symmetrication satisfies condition (3) of paragraph 5. It is easily seen that the above reasoning is valid in a situation if the class of groups $\mathcal{C}_\lambda(a_1, \ldots, a_m)$ is replaced by the class of groups $\mathcal{G}_\lambda(a_1, \ldots, a_m)$, whose set of defining words after symmetrication satisfies conditions (3), (4) of paragraph 5.

Remark 2. Limitations imposed on the class of groups considered like conditions (3) or (3'), (4), are essential. Indeed, consider the group

$$G_n = \langle a, b \mid a^{-1}ba = b^n \rangle$$

The group $G_n$ is resolvable (and, consequently, amenable) under any natrual n, the length of the defining word $a^{-1}bab^{-n}$ grows as n increases, and the spectral radius of a simple walk on $G_n$ equals 1 (due to the group $G_n$ amenability). The group $G_n$ is obvious not to enter the class of groups considered under any n.

Theorem 7. 2. Let (1) $G \in \mathcal{C}_\lambda(a_1, \ldots, a_m)$ or (2) $G \in \mathcal{G}_\lambda(a_1, \ldots, a_m)$. If (1) takes place and the system

$$\begin{cases} \sum_{s=1}^{k} N_s(t_s) x^{(1-6\lambda)t_s} \mu^{t_s} + (2m-1)x^2\mu^2 - [\,(2m-1)x^2 + 1]\mu + 1 = 0 \\ \sum_{s=1}^{k} t_s N_s(t_s) x^{(1-6\lambda)t_s} \mu^{t_s - 1} + 2(2m-1)x^2\mu - (2m-1)x^2 - 1 = 0 \end{cases}$$

has no solutions in the same region, the group G is non-amenable.

Proof: Let, e. g., $G \in \mathcal{C}_\lambda(a_1, \ldots, a_m)$. In proving theorem 7. 1, we have stated the convergence radius r os the series H(x) to be no less than the convergence radius $r_\Phi$ of the series $\Phi((2m-1)x^2, N_1 x^{(1-6\lambda)t_1}, \ldots, N_k x^{(1-6\lambda)t_k})$. The least positive singularity of the function $\Phi$ is found among the system 7. 10 solutions $x_\ell$ such that $0 < x_\ell < 1/\sqrt{2m-1}$. Therefore, $r_\Phi > 1/(2m-1)$ and, consequently, $\alpha_H = \limsup_{n\to\infty} |H_n|^{-1/n} < 2m-1$. This completes the proof.

ACKNOWLEDGMENTS

The author wishes to express his gratitude to A. M. Styopin for his stimulating interest to the paper and valuable advice, and to A. V. Yarkho for translating the paper.

REFERENCES

1. F. P. Greenleaf, Invariant means on topological groups and their applications. New York-London-Toronto-Melbourne, 1969.
2. J. Dixmier, Les moyennes invariantes dans les semigroups et leur applications, Acta Sci. Math. (Szeged), 12 (1950), 213-227.
3. H. Kesten, Symmetric random walks on groups, Trans. Amer. Math. Soc., 92 (1959), 336-354.
4. H. Kesten, Full Banach mean values on countable groups, Math. Scand., 7 (1959), 146-156.
5. M. Day, Convolutions, means, and spectra, III. J. Math., 8, (1964), 100-111.
6. W. Magnus, A. Karrass, D. Solitar, Combinatorial group theory, Interscience Publishers, Div. of John Wiley and Sons, New York-London-Sydney, 1966.
7. F. Spitzer, Principles of random walk. Van Nostrand, Princeton, 1964.

8. M. Greendlinger, On Dehn's algorithms for the conjugacy and word problems, with applications. Comm. Pure and Appl. Math., 13, No. 4 (1960), 641-677.

9. M. Greendlinger, On the problem of words identity and conjugacy. Izx. Akad. Nauk SSSR, ser. Math., 29, No. 2, 1965, 245-268. (In Russian).

chapter 10

# STATIONARY RANDOM SEQUENCES OF MAXIMAL ENTROPY

B. M. Gurevich

## I. INTRODUCTION

In the present paper we consider discrete parameter and finite state stationary stochastic processes or in other words, shift-invariant probability measures on sequence spaces.

Let V be a finite set provided with the discrete topology, and T the shirt transformation defined on the space $X = V^{\mathbb{Z}}$ by

$$T\{x_i\} = \{x_i'\}, \quad x_i' = x_{i+1}, \qquad \{x_i\} \in X$$

X is clearly compact and T a homeomorphism.

Suppose Y is a closed subset of X invariant with respect to T, and $\mu$ a T-invariant probability Borel measure on X. If $\mu(Y) = 1$, we can assume that trajectories of the stationary process corresponding to $\mu$ belong to Y. Let $\mathcal{P}(Y)$ denote the set of all such measures. We are interested in those $\mu \in \mathcal{P}(Y)$ for which the entropy $h_\mu(T)$ of **T** with respect to $\mu$ assumes as its maximal value the topological entropy, $h(T, Y)$, of T restricted to the set Y. For short we shall refer to any measure $\mu \in \mathcal{P}(Y)$ of maximal entropy as <u>a maximal measure</u>. A compactness argument shows that a maximal measure exists for an arbitrary Y mentioned above.

Our object is to formulate some sufficient conditions for the uniqueness of such a measure. The most important of these conditions is that the set Y be sufficiently well approximated in some "entropy sense" by sets $Y_n \supset Y$ (Markov sets - see below) for each of which the maximal measure is unique (it corresponds to an n-th order Markov chain).

The uniqueness problem for measures of maximal entropy with respect to a homeomorphism has been treated by several authors. The most general uniqueness conditions were apparently formulated by R. Bowen [1]. According to his terminology, these are the "expansiveness" and "specification" conditions. The former of these is automatically fulfilled for the shift transformation of the sequence space regardless of the set Y. But the latter is not necessarily true for any set Y, considered below. The point is that the "specification" condition assures the existence of a sufficiently rich collection of periodic points, while our assumptions can be fulfilled for sets Y containing no periodic sequences (see Proposition 4. 1 below).

The uniqueness problem for maximal measures with respect to the shift transformation can also be formulated in such a way that it becomes a particular case of the phase transition problem for some lattice models of Statistical Physics. Namely, we deal with a one-dimensional lattice system of a finite number of particle types interacting by means of a singular translation-invariant potential. From this point of view the set of maximal measures is identified with that of translation-invariant equilibrium Gibbs distributions (see, for instance, [2]). No uniqueness condition for such distributions at present is applicable to our case because of the singular and multi-particle character of the interaction.

A set $Y \subset X$ is called an <u>n-th-order Markov set</u> (or, simply, n-<u>Markov set</u>) if there is a set $V^{(n)} \subset V^{n+1}$ such that a sequence

$\{x_i\} \in X$ belongs to $Y$ iff any of its sections of length $n+1$ belongs to $V^{(n)}$.

The definition immediately implies that a Markov set of any order is both T-invariant and closed, and if $Y$ is an n-Markov set, then it is an m-Markov set for any $m \geq n$.

In what follows we need some notions of graph theory*. We shall use only directed graphs referring to those simply as graphs.

<u>Definition.</u> Let $G$ be a graph with vertices $v_1, v_2, \ldots$. A sequence of vertices $v_{i_1}, v_{i_2}, \ldots, v_{i_n}$ is said to be <u>a path</u> of length $n$ in $G$ with the <u>initial vertex</u> $v_{i_1}$ and the <u>terminal vertex</u> $v_{i_n}$ (or a path leading from $v_{i_1}$ to $v_{i_n}$) if $G$ contains the edge leading from $v_{i_k}$ to $v_{i_{k+1}}$, $1 \leq k \leq n-1$. An infinite path in $G$ is defined similarly. We shall refer to a path $v_{i_1}, \ldots, v_{i_n}$ as a <u>cycle</u>, if $G$ contains the edge from $v_{i_n}$ to $v_{i_1}$.

Let $\alpha(i, j) + 1$ be the minimal length of the paths in $G$ leading from $v_i$ to $v_j$ (if no such paths exist we write $\alpha(i, j) = \infty$. The quantity

$$\alpha(G) = \max_{i,j} \alpha(i, j)$$

will be called the index of connectedness of $G$.

One can assign to $G$ a 0-1-matrix $M(G) = (m_{ij}(G))$ (the incidence matrix of $G$) by the following rule: $m_{ij}(G) = 1$ if there is an edge in $G$ from $v_i$ to $v_j$, and $m_{ij}(G) = 0$ otherwise.

Suppose that $G$ is connected or, equivalently, the matrix $M(G)$ is irreducible. Due to the Frobenius theorem (see [4]) there exists a positive number $\lambda(G)$ equal to the maximum absolute value of the eigenvalues of the matrix $M(G)$; moreover, the invariant subspace for $\lambda(G)$ is one-dimensional and contains a vector

---

* Our terminology is slightly different from the usual one (cf. [3]).

$\xi(G) = (\xi_1(G), \xi_2(G), \ldots)$ with all positive components (positive vector).

Given a 1-Markov set $Y$ one can produce a graph $G(Y)$ as follows: the vertices of $G(Y)$ are those elements of the set $V$ that occur in the sequences belonging to $Y$, and an edge from $v_i \in V$ to $v_j \in V$ is drawn whenever there exists a sequence $\{x_k\} \in Y$ containing a section of the form $v_i, v_j$. It is easy to see that we get a one-to-one correspondence between all 1-Markov subsets of $X$ and a set of directed graphs. Not it is natural to identify each 1-Markov set $Y$ with the set of all doubly infinite paths in the graph $G(Y)$.

Let $Y$ be a 1-Markov set and $G(Y)$ the corresponding graph. W. Parry [5] showed that if $G(Y)$ is connected then there is a unique maximal measure $\mu \in \mathcal{P}(Y)$ which is nothing but a Markov measure with the following transition probabilities:

$$p_{ij} = \xi_j m_{ij} / \lambda \xi_i \tag{1.1}$$

where $m_{ij} = m_{ij}(G(Y))$, $\xi_i = \xi_i(G(Y))$, $\xi_j = \xi_j(G(Y))$, $\lambda = \lambda(G(Y))$.

If $Y$ is an $n+1$-Markov set, then it can be mapped in a natural way onto a 1-Markov set $Z$. For this it is sufficient to assign to any sequence $\{y_i\} \in Y$ the sequence $\varphi^{(n)}(\{y_i\}) = \{z_i\} \in Z$, where $z_i = (y_i, y_{i+1}, \ldots, y_{i+n})$ is a section of $\{y_i\}$. Clearly, $\varphi^{(n)}$ is an invertible mapping which is continuous together with its inverse and commutes with the shift transformation. Of course, $\varphi^{(n)}$ may also be applied to a 1-Markov set, $Y$. We thereby assign to the graph $G = G(Y)$ another graph $G' = G(Z)$. It is easy to describe this assignment in terms of $G$ and $G'$ only; namely, each path $a$ in $G$ of length $n+1$ turns into a vertex $a'$ in $G'$, and for any pair of such vertices $a_1$, $a_2$, the edge from $a_1$ to $a_2$ is drawn in $G$ iff there is a path in $G$ with the head and the tail coinciding with $a_1$ and $a_2$, respectively. The mapping of the graphs just described will also be denoted by $\varphi^{(n)}$.

Since there is a natural interpretation of the equality $\varphi^{(n)} = (\varphi^{(1)})^n$, the family $\{\varphi^{(n)}, n = 1, 2, \ldots\}$ can be viewed to be a cyclic semi-group with generator $\varphi = \varphi^{(1)}$. Thus we shall write $\varphi^n$ instead of $\varphi^{(n)}$. By means of $\varphi^n$ it is not difficult to extend Parry's result to the case of n- Markov sets for each $n \geq 1$. In this case the maximal measure is n- Markov, which means that with respect to the maximal measure the coordinates $y_i$ constitute an nth-order Markov chain.

Any shift-invariant closed subset Y of X can be represented in a canonical way in the form

$$Y = \bigcap_{n=1}^{\infty} Y_n \tag{1. 2}$$

where $Y_n$ is an n-Markov set, and $Y_{n+1} \subset Y_n$, $n = 1, 2, \ldots$. The sets $Y_n$ are defined by the following requirement: given a section of length $n + 1$ in an arbitrary sequence $y \in Y_n$ there is a section in some sequence $y' \in Y$ that coincides with the previous one.

Let

$$G_n(Y) = G(\varphi^n Y_n), \quad \lambda_n(Y) = \lambda(G_n(Y)) \tag{1. 3}$$

Then

$$\ln \lambda_n(Y) = h(T, Y_n) \underset{n \to \infty}{\searrow} h(T, Y) = \ln \lambda(Y)$$

where

$$\lambda(Y) = \lim_{n \to \infty} \lambda_n(Y)$$

Further set

$$\alpha_n(Y) = \alpha(G_n(Y))$$

$$\rho_n(Y) = h(T, Y_n) - h(T, Y)$$

Clearly $\rho_n(Y) \searrow 0$ as $n \to \infty$.

The main result of the present paper can be formulated as follows.

Theorem 1.1. Suppose that (a) $\lambda(Y) > 1$ and (b) the inequality

$$\rho_n(Y) \leq [\lambda_n(Y)]^{-(16+\gamma)\alpha_n(Y)} \tag{1.4}$$

holds for some $\gamma > 0$ and infinitely many integers $n$. Then $\mathcal{P}(Y)$ contains exactly one maximal measure.

Remarks 1. Assuming that the condition (a) is fulfilled, one can replace $\lambda_n(Y)$ in (1.4) by $\lambda(Y)$. This leads to a condition which seems to be a bit weaker than (b) but actually is equivalent to it.

2. Inequality (1.4) cannot be replaced by those of the form

$$\rho_n(Y) \leq \exp[-c\alpha_n(Y)] \quad \text{or} \quad \rho_n(Y) \leq \exp(-cn)$$

where $c$ does not depend on $\lambda(Y)$. This follows from Proposition 4.4 below.

3. Let us say that $Y$ is $\kappa$-approximable if $\rho_n(Y) \leq [\lambda_n(Y)]^{-(\kappa+\gamma)\alpha_n(Y)}$ for some $\gamma > 0$ and infinitely many $n$. Let $\kappa_0$ be the greatest lower bound of the numbers $\kappa$ for which the following holds: if $Y$ is $\kappa$-approximable and $\lambda(Y) > 0$, then $\mathcal{P}(Y)$ contains exactly one maximal measure. Now Theorem 1.1 means that $\kappa_0 \leq 16$. It will be shown below (see Proposition 4.4) that $\kappa_0 \geq 2/3$. Thus we have a kind of phase transition with respect to the parameter $\kappa$. The precise value of $\kappa_0$ is unknown.

Sections II and III are devoted to the proof of Theorem 1.1. Some of the auxiliary tools used there are perhaps of independent interest. Section IV contains two examples mentioned above.

## II. LEMMAS ON EIGENVALUES AND GENERATING FUNCTIONS

Lemma 2. 1. Let $G$ be a finite connected graph with $M(G)$, $\lambda(G) = \lambda$, and $\alpha(G) = \alpha$ defined as above. Let $\xi(G) = (\xi_1, \xi_2, \ldots)$ and $\eta(G) = (\eta_1, \eta_2, \ldots)$ be positive eigenvectors of $M(G)$ and $M^x(G)$, respectively, corresponding to $\lambda(G)$. Then for any $i, j$

$$\xi_j \leq \lambda^{\alpha} \xi_i, \qquad \eta_j \leq \lambda^{\alpha} \eta_i$$

The proof of this lemma is straightforward and will be omitted.

Definition. Let $G$ be a graph and $G'$ its subgraph such that any edge of $G$ belongs to $G'$ whenever the initial point and the terminal point of this edge belong to $G'$. We shall call such $G'$ a complete subgraph of the graph $G$.

Lemma 2. 2. For any finite graph $G$ with $\lambda(G) = \lambda > 1$ and $\alpha(G) = \alpha < \infty$, and for any proper complete subgraph $G'$ of $G$, the following holds:

$$\lambda - \lambda(G') \geq (\lambda^{1/\alpha} - 1)\lambda^{-2\alpha}$$

One can consider this lemma as a modification (for the case of 0 - 1-matrices) of the well-known assertion that an increase in the elements of an irreducible nonnegative matrix gives rise to the increase of its maximal eigenvalue (see [4]).

Proof of Lemma 2. 2. Let us divide the set of vertices of the graph $G$ into subsets $V_i$, $i = 0, 1, \ldots, k$, as follows. We let $V_0$ denote the set of vertices of $G$ which do not belong to $G'$, and $V_i$, $i > 0$, the set of vertices $v$ of $G$ such that $v \notin \bigcup_{j=1}^{i-1} V_j$ and there is an edge from $v$ to a vertex $v' \in V_{i-1}$. It follows from the assumption $\alpha(G) < \infty$ that each vertex of $G$ belongs to some $V_i$. Moreover, it is easy to see that $k \leq \alpha$.

Let $\xi = (\xi_1, \ldots, \xi_s)$ be a positive eigenvector of $M(G)$. For every vertex $v_i \in V_1$, $1 \leq l \leq k$ we set

$$\theta_i = \xi_i^{-1} \lambda^{-l(1+\epsilon)+\beta} \tag{2. 1}$$

where $\epsilon > 0$, and $\beta$ is chosen so that

$$\lambda^{-\beta} \leq \xi_j \leq \lambda^{\beta}, \qquad j = 1, 2, \ldots, s \tag{2. 2}$$

Clearly, $0 < \theta_i < 1$ for any $v_i \in \bigcup_{l=i}^{k} V_i$, and hence the vector $\xi'$ with the components of the form

$$\xi_i' = \xi_i(1 - \theta_i) \tag{2. 3}$$

is positive.

The matrix $M(G')$ is obtained from $M(G)$ by deleting the rows and columns corresponding to those $v_i$ which belong to $V_0$.

Let

$$\xi_i'' = \sum_{j : v_j \notin V_0} m_{ij}(G)\, \xi_j' \tag{2. 4}$$

be the i-th component of the vector $M(G')\xi'$. By the definition of $\xi$ one has

$$\lambda\xi_i = \sum_{j : v_j \in V_0} m_{ij}(G)\xi_j + \sum_{j : v_j \in V_i} m_{ij}(G)\xi_j$$

$$+ \ldots + \sum_{j : v_j \in V_k} m_{ij}(G)\xi_j, \qquad i = 1, 2, \ldots, s \tag{2. 5}$$

If $v_i \in V_1$, then the first of the sums in the right-hand side of (2. 5) contains a non-zero summand, say, $m_{ij_0}(G)\xi_{j_0}$. Due to (2. 3), (2. 4) and the inequality $\xi_j' < \xi_j$, this implies that

$$\xi_i'' < \lambda\xi_i - \xi_{j_0}, \qquad \xi_i''/\xi_i' < (\lambda\xi_i - \xi_{j_0})/\xi_i(1-\theta_i) = \lambda - \delta_i$$

where

$$\delta_i = (\xi_{j_0} - \lambda\theta_i\xi_i)/\xi_i(1-\theta_i), \qquad v_i \in V_1 \tag{2. 6}$$

If $v_i \in V_l$ for $l > 1$, then, due to (2. 5) and the definition of $V_0$, $V_1, \ldots$, one has

$$\lambda\xi_i = \sum_{j:v_j \in V_{l-1}} m_{ij}(G)\xi_j' + \sum_{j:v_j \notin V_0 \ldots V_{l-1}} m_{ij}(G)\xi_j$$

moreover, the first sum contains a non-zero term $m_{ij_l}(G)\xi_{j_l} = \xi_{j_l}$. Hence

$$\xi_i'' = \sum_{j:v_j \in V_{l-1}} m_{ij}(G)\xi_j' + \sum_{j:v_j \notin V_0 \cup \ldots \cup V_{l-1}} m_{ij}(G)\xi_j'$$

$$\leq \lambda\xi_i - \xi_{j_l} + \xi_{j_i}' = \lambda\xi_i - \theta_{j_l}\xi_{j_l},$$

$$\xi_i''/\xi_i' \leq (\lambda\xi_i - \theta_{j_l}\xi_{j_l})/\xi_i(1-\theta_i) = \lambda - \delta_i$$

where

$$\delta_i = (\theta_{j_l}\xi_{j_l} - \lambda\theta_i\xi_i)/\xi_i(1-\theta_i), \qquad v_i \in \bigcup_{l=2}^{k} V_l \tag{2. 7}$$

Now we shall make use of the variational characterization of $\lambda(G')$ which states that

$$\lambda(G') \leq \min_{\zeta} \max_{i:v_i \notin V_0} (M(G')\zeta)_i/\zeta_i$$

where $(M(G')\zeta)_i$ is the ith component of the vector $M(G')\zeta$, and the minimum is taken over all positive vectors $\zeta$ (see [4]). Due to the above

$$\lambda - \lambda(G') \geq \min_{i: v_i \notin V_0} \delta_i$$

where $\delta_i$ is defined by (2. 6) and (2. 7). If $v_i \in V_l$, $l \geq 2$, then (2. 7), (2. 1), and (2. 2) together imply that

$$\delta_i = [\lambda^{-(l-1)(1+\epsilon)-\beta} - \lambda^{-l(1+\epsilon)-(\beta-1)}]\, \xi_i^{-1}[1-\xi_i^{-1}\lambda^{-l(1+\epsilon)-\beta}]$$

$$\geq (\lambda^{-l+1-l\epsilon+\epsilon-\beta} - \lambda^{-l-l\epsilon-\beta+1})\xi_i^{-1}$$

$$= \lambda^{-l-l\epsilon-\beta+1}(\lambda^{\epsilon}-1)\xi_i^{-1} \geq \lambda^{-l-l\epsilon-2\beta+1}(\lambda^{\epsilon}-1)$$

$$\geq \lambda^{-k-k\epsilon-2\beta+1}(\lambda^{\epsilon}-1) \geq \lambda^{-\alpha-\alpha\epsilon-2\beta+1}(\lambda^{\epsilon}-1)$$

Setting $\epsilon = \alpha^{-1}$ gives

$$\delta_i \geq \lambda^{-\alpha-2\beta}(\lambda^{1/\alpha} - 1), \qquad v_i \in \bigcup_{l=2}^{k} V_l \tag{2. 8}$$

By the same arguments (2. 8) holds for $v_i \in V_1$ as well. Finally, Lemma 2. 1 enables us to take $\beta = \alpha/2$. Substituting this equality into (2. 8) leads to the desired inequality

<u>Lemma 2. 3.</u> <u>For any finite connected graph</u> G

$$\alpha(\varphi G) = \alpha(G) + 1, \quad \lambda(\varphi G) = \lambda(G) \tag{2. 9}$$

<u>Proof.</u> The first equality follows immediately from the definition of the index of connectedness. The second one can also be proved directly. However it is easier to make use of the fact that $\log \lambda(G)$ is the topological entropy of the shift transformation on the space $Y(G)$ of infinite paths in $G$ and that the shift on $Y(G)$ is topologically conjugate to that on $Y(\varphi G)$ (see Section I). The lemma is proved.

Notations. For a finite graph G with the maximal eigenvalue $\lambda(G)$ denote

$$\tau(G) = \lambda^{-1}(G), \quad \lambda_1(G) = \max_{G'} \lambda(G'), \quad \tau_1(G) = \lambda_1^{-1}(G)$$

where the maximum is taken over all proper complete subgraphs of G.

From now on we shall denote by $s(G)$ the number of vertices of the graph G..

Lemma 2.4. *Let* G *be a finite graph with* $\alpha(G) < \infty$ *and* $\lambda(G) > 1$. *Then* $\tau_1(G) \leq 1$.

Proof. Suppose that $\tau_1(G) > 1$. Then given a vertex $v_{i_1}$ of G the complete subgraph $G_{i_1}$ obtained from G by removing $v_{i_1}$ satisfies $\lambda(G_{i_1}) < 1$. As known, $\lambda(G_{i_1})$ is no more than the minimum of the sums over the rows of the matrix $M(G_{i_1})$. The above inequality means that there is a row of $M(G_{i_1})$ consisting of nothing but zeros, or equivalently, there is a vertex $v_{i_2}$ of G which is the initial point of a unique edge, namely, of that leading to $v_{i_1}$. Since G is connected, $i_2 \neq i_1$. Applying this argument repeatedly we get a sequence of vertices $v_{i_1}, v_{i_2}, \ldots, v_{i_k}$, $k \leq s(G)$, such that all $i_l$, $1 \leq l \leq k$, are different, and for $l = 2, \ldots, k$ there is a unique edge whose initial point is $v_{i_l}$, namely the edge going from $v_{i_l}$ to $v_{i_{l-1}}$. If $k = s(G)$, then taking the next step we come to a vertex $v_{i_{k+1}}$ which must coincide with $v_{i_1}$. Hence G turns out to be a simple cycle. It is easy to see that in that case $\lambda(G) = 1$ which contradicts the assumption of the lemma. Thus the desired inequality is proved.

Notations. For a finite graph G with vertices $v_1, v_2, \ldots$ denote by $m_{ij,n}(G)$, $n \geq 1$, the $i, j$th element of the matrix $(M(G))^n$, i.e., the number of paths in G of length $n+1$ going from $v_i$ to $v_j$ (we shall write $m_{ij}(G)$ instead of $m_{ij,1}(G)$). Introduce the generating functions

$$M_{ij}(t, G) = \sum_{n=1}^{\infty} m_{ij,n}(G)t^n \tag{2.10}$$

$$F_{ij}(t, G) = \sum_{n=1}^{\infty} f_{ij,n}(G)t^n \tag{2.11}$$

$f_{ij,n}^{(n)}(G)$ being the number of paths of length $n + 1$ going from $v_i$ to $v_j$ and containing $v_j$ only as the terminal point (or also as the initial point if $i = j$).

<u>Lemma 2.5.</u> <u>$M_{ij}(t, G)$ and $F_{ij}(t, G)$ are rational functions connected by the relation</u>

$$M_{ij}(t, G) = F_{ij}(t, G)[1 + M_{jj}(t, G)] \tag{2.12}$$

<u>Proof</u> (cf. [6], Chapter 16). It is convenient to set $m_{ij,0}(G) \equiv 1$. Now

$$m_{ij,n}(G) = \sum_{k=1}^{n} f_{ij,k}(G)m_{jj,n-k}(G), \qquad n = 1, 2, \dots,$$

and

$$M_{ij}(t, G) = \sum_{n=1} \sum_{k=1}^{n} f_{ij,n}(G)m_{jj,n-k}(G)t^n$$

$$= \sum_{k=1}^{n} f_{ij,k}(G)t^k \sum_{n=k}^{\infty} m_{jj,n-k}(G)t^{n-k} = F_{ij}(t, G)[1 + M_{jj}(t, G)]$$

which gives (2.12).

To prove the first assertion of Lemma 2.5 multiply (2.10) by $tm_{ki}(G)$ and add up the result obtained over $i$ from $1$ to $s = s(G)$. One obtains

$$\sum_{i=1}^{s} tm_{ki}(G)M_{ij}(t, G) = \sum_{i=1}^{s}\sum_{n=1}^{\infty} m_{ki}(G)m_{ij,n}(G)t^{n+1}\sum_{n=1}^{\infty} t^{n+1}\sum_{i=1}^{s} m_{ki}(G)m_{ij,n}(G)$$

$$= \sum_{n=1}^{\infty} m_{ki,n+1}(G)t^{n+1} = M_{kj}(t, G) - m_{kj}(G)t, \quad k = 1, 2, \dots, s \tag{2.13}$$

For any fixed j we have a set of s linear equations (numbered by k) in the functions $M_{ij}(t, G)$, $i = 1, 2, \ldots, s$, with coefficients depending on t linearly The solution of such a system is clearly formed by rational functions of t. Hence $M_{ij}(t, G)$ are rational, and due to (2. 12) the same is true for $F_{ij}(t\ G)$. So Lemma 2. 5 is proved.

Notations. Let

$$\tilde{\Delta}(t, G) = tM(G) - I$$

(I being the identity matrix) and $\tilde{\Delta}_1(t, G)$ the matrix obtained from $\tilde{\Delta}(t, G)$ by replacing its i, 1th elements by $-tm_{i1}(G)$. Further let

$$\Delta(t, G) = \det \tilde{\Delta}(t, G), \qquad \Delta_1(t, G) = \det \tilde{\Delta}_1(t, G)$$

Finally denote by $A_{ij}(t, G)$ the cofactor of the i, jth element of $\Delta(t, G)$.

In what follows we shall often denote the derivatives with respect to t by points over letters.

Lemma 2. 6. For a finite graph G with $s(G) = s$ the following equalities hold:

$$\Delta(t, G) = -\Delta_1(t, G) - A_{11}(t, G) = -\Delta_1(t, G) - \Delta(t, G_1) \tag{2. 14}$$

where $G_1$ is the complete subgraph of G obtained from G by removing the vertex $v_1$;

$$\frac{d}{dt}\Delta(t, G) = \dot{\Delta}(t, G) = t^{-1}[s\Delta(t, G) + \sum_{k=1}^{s} A_{kk}(t, G)] \tag{2. 15}$$

$$\frac{d^2}{dt^2}\Delta(t, G) = \ddot{\Delta}(t, G) = t^{-1}[(s-1)\dot{\Delta}(t, G) + \sum_{k=1}^{s} \dot{A}_{kk}(t, G)] \tag{2. 16}$$

$$F_{11}(t, G) = 1 + A_{11}^{-1}(t, G)\Delta(t, G) \tag{2. 17}$$

$$\frac{d}{dt}F_{11}(t,G) = \dot{F}_{11}(t,G) = st^{-1}A_{11}^{-1}(t,G)\Delta(t,G)$$

$$+ \sum_{k=1}^{s} t^{-1}A_{11}^{-1}(t, G)A_{kk}(t, G) - A_{11}^{-2}(t, G)\Delta(t,G)\dot{A}_{11}(t, G) \quad (2.18)$$

$$\frac{d^2}{dt^2}F_{11}(t,G) = \ddot{F}_{11}(t, G) = A_{11}^{-1}(t, G)\Delta(t, G) - 2A_{11}^{-2}(t,G)\dot{\Delta}(t, G)\dot{A}_{11}(t, G)$$

$$- A_{11}^{-2}(t,G)\Delta(t, G)\ddot{A}_{11}(t,G) + 2A_{11}^{-3}(t,G)\Delta(t,G)[\dot{A}_{11}(t,G)]^2 \quad (2.19)$$

<u>Proof.</u> For $j = 1$ the set of equations (2.13) takes the form

$$[tm_{11}(G)-1]\,M_{11}(t,G)+tm_{12}(G)M_{21}(t,G)+\ldots+tm_{1s}(G)M_{s1}(t,G)$$

$$= -tm_{11}(G), tm_{21}(G)M_{11}(t, G)+[tm_{22}(G)-1]M_{21}(t,G)+\ldots+tm_{2s}(G),\ldots$$

$$= -tm_{21}(G), tm_{s1}(G)M_{11}(t,G)+tm_{s2}(G)M_{21}(t,G)+\ldots$$

$$+ [tm_{ss}(G)-1]\,M_{s1}(t, G) = -tm_{s1}(G)$$

(2.20)

It follows from (2.20) and the definitions of $\tilde{\Delta}(t, G)$ and $\tilde{\Delta}_1(t, G)$ that

$$M_{11}(t, G) = \Delta^{-1}(t, G)\Delta_1(t, G) \quad (2.21)$$

Expanding $\Delta(t, G)$ and $\Delta_1(t, G)$ according to the first column gives

$$\Delta(t,G) = [tm_{11}(G)-1]A_{11}(t,G)+tm_{21}(G)A_{21}(t,G)+\ldots+tm_{s1}(G)A_{s1}(t,G)$$

$$\Delta_1(t,G) = -tm_{11}(G)A_{11}(t,G)-tm_{21}(G)A_{21}(t, G)-\ldots-tm_{s1}(G)A_{s1}(t,G)$$

and this in its turn gives (2.14). Now (2.12), (2.14) and (2.20) together imply (2.17).

Further,

$$\dot{\Delta}(t, G) = \sum_{k=1}^{s} D_k(t, G) \quad (2.22)$$

where $D_k(t, G)$ is the determinant of the matrix obtained from $\tilde{\Delta}(t,G)$ by replacing all elements of the kth column by their derivatives which are equal to $m_{1k}(G), \ldots, m_{sk}(G)$, respectively.

The expansions of $\Delta(t, G)$ and $D_k(t, G)$ according to the kth column take the form

$$\Delta(t, G) = [tm_{kk}(G) - 1] A_{kk}(t, G) + \sum_{i:i \neq k} tm_{ik}(G)A_{ik}(t, G)$$

$$D_k(t, G) = m_{kk}(G)A_{kk}(t, G) + \sum_{i:i = k} m_{ik}(G)A_{ik}(t, G)$$

These equalities yield:

$$D_k(t, G) = t^{-1}[\Delta(t, G) + A_{kk}(t, G)]$$

which together with (2. 22) leads to (2. 15).

Finally, (2. 16), (2. 18) and (2. 19) follow from (2. 15) and (2. 17) by simple calculations.

Lemma 2. 7. *Let* $G$ *be a finite graph with* $s(G) = s$, $\tau(G) = \tau$, *and* $\tau_1(G) = \tau_1$. *Then*

$$\text{sign } \Delta(t, G) = \text{sign } (-1)^{s-1}, \quad \tau < t \leq \tau_1 \tag{2. 23}$$

*and for any proper complete subgraph* $G'$ *of* $G$ *with* $s'$ *vertices*

$$\text{sign det } [tM(G') - I] = \text{sign } (-1)^{s'}, \quad 0 \leq t < \tau_1 \tag{2. 24}$$

Proof: If $G'$ is a complete subgraph of $G$, then $\det[tM(G') - I]$ is a principal minor determinant of $\tilde{\Delta}(t, G)$. By the definition of $\tau_1$ it takes non-zero values, if $0 \leq t < \tau_1$. Hence its sign at every point $t$, $0 \leq t < \tau_1$, coincides with that at $t = 0$. This gives (2.24).

To prove (2. 23) one may exclude the case $\tau_1 = \tau$ (which can occur if $\alpha(G) = \infty$). Furthermore, we observe that $A_{kk}(t, G)$, $k = 1, 2, \ldots, s$, is a principal minor determinant of $\tilde{\Delta}(t,G)$ corresponding to the proper complete subgraph $G' = G$ of $G$ which is obtained by removing the kth vertex. Due to (2.15), (2.24) and the equality $\Delta(\tau, G) = 0$ one has

$$\operatorname{sign} \dot{\Delta}(\tau, G) = \operatorname{sign} (-1)^{s-1} \tag{2.25}$$

This means in particular that $\Delta(\tau, G) \neq 0$. Since $\dot{\Delta}(t, G)$ is a continuous function, it preserves its sign in a neighborhood of $\tau$, and to the right of $\tau$; in this neighborhood the sign of $\Delta(t, G)$ is the same as that of $\dot{\Delta}(t, G)$. Finally, the sign of $\Delta(t, G)$ is constant if $\tau < t < \tau_1$. Otherwise, there would be a $t$ in the above interval such that $\Delta(t, G) = 0$. Let $\tau_2$ denote the minimum of such $t$ (their number is obviously finite). Then

$$\operatorname{sign} \dot{\Delta}(t, G) = \operatorname{sign} (-1)^{s-1}, \qquad \tau \le t \le \tau_2 \tag{2.26}$$

Now $\Delta(\tau_2, G) = \Delta(\tau, G) = 0$ which implies that $\dot{\Delta}(t, G) = 0$ for some $t$ satisfying $\tau < t < \tau_2$. But this does not agree with (2.26).

<u>Lemma 2.8.</u> <u>Let</u> $G$ <u>be a finite graph with</u> $s(G) = s$, $\lambda(G) = \lambda = \tau^{-1}$, <u>and</u> $\alpha(G) = \alpha < \infty$. <u>Then</u>

$$A_{11}^{-1}(\tau, G) A_{kk}(\tau, G) \le \lambda^{2\alpha}, \qquad 1 \le k \le s$$

(See above for the definitions of $A_{kk}(t, G)$ as well as of $s(G), \lambda(G)$, and $\alpha(G)$).

<u>Proof:</u> Let $B(x)$, $-\infty < x < \infty$, denote the adjoint matrix for $M(G)$. It is known (see, for example, [4], Ch. IV, §3) that all non-zero columns of the matrix $B(\lambda)$ are eigenvectors of $M(G)$ for the eigenvalue $\lambda$. Since the space of such vectors is one-dimensional, the ith column of $B(\lambda)$ is of the form $(b_{1i}, \ldots, b_{si}) = p_{i\xi}$, $1 \le i \le s$, where $M(G)\xi = \lambda\xi$. For the same reason the jth row of $B(\lambda)$ is of

the form $(b_{j1},\ldots,b_{js}) = q_j\eta$, $1 \le j \le s$, where $\eta M(G) = \lambda\eta$. The vectors $\xi = (\xi_1,\ldots,\xi_s)$ and $\eta = (\eta_1,\ldots,\eta_s)$ can be regarded as positive. This implies that all $b_{ij}$ vanish or do not vanish simultaneously. But the former is impossible ([4], Ch. XIII, §2). Hence $p_i, q_j \neq 0$ and moreover $p_i\xi_j = q_j\eta_i$, i. e., $q_j\xi_j^{-1} = p_i\eta_i^{-1}$ for all i and j. This means that $q_j\xi_j^{-1}$ is independent of j. Denote it by c. Now $b_{ij} = c\xi_i\eta_j$ and $b_{ii}b_{jj}^{-1} = \xi_i\eta_i(\xi_j\eta_j)^{-1}$. According to Lemma 2.1 $\xi_i\xi_j^{-1} \le \lambda^\alpha$, and $\eta_i\eta_j^{-1} \le \lambda^\alpha$. Therefore $b_{ii}b_{jj}^{-1} \le \lambda^2$, and it only remains to observe that, by the definition of $B(\lambda)$, $A_{kk}(\tau, G) = (-\tau)^{s-1}b_{kk}$, $1 \le k \le s$. So the lemma is proved.

Notation. Let

$$\Phi_{11}(t, G) = [\tau(G)\dot{F}_{11}(\tau(G), G)]^{-1} - F_{11}(t, G)[t\dot{F}_{11}(t, G)]^{-1} \tag{2. 27}$$

Then

$$\dot{\Phi}_{11}(t, G) = [tF_{11}(t, G)\ddot{F}_{11}(t, G) + F_{11}(t, G)\dot{F}_{11}(t, G)\dot{F}_{11}(t, G) - t(\dot{F}_{11}(t, G))^2][t\dot{F}_{11}(t, G)]^{-2} \tag{2. 28}$$

Our aim is to estimate $\dot{\Phi}_{11}(t, G)$ near the point $\tau(G)$. It will be accomplished in several steps.

Up to the end of this section we shall deal with the graph G with the vertices $v_1,\ldots,v_s$, $s(G) = s < \infty$, $\alpha(G) = \alpha < \infty$, $\lambda(G) = \lambda = \tau^{-1}$, $\lambda_1(G) = \lambda_1 = \tau_{-1}^{-1}$ and these assumptions will not be repeated below.

Lemma 2. 9. If $\Phi_{11}(t, G)$ is not identically equal to zero then

$$t^2F_{11}(t, G)\ddot{F}_{11}(t, G) + tF_{11}(t, G)\ddot{F}_{11}(t, G) - t^2[\dot{F}_{11}(t, G)]^2 \ge t^{3\alpha}, \quad 0 \le t < \tau_1 \tag{2. 29}$$

Proof: First suppose $F_{11}(t, G)$ is not of the form $\text{const} \cdot t^{k_0}$. In that case $F_{11}(1, G) > 1$. Since $F_{11}(\tau, G) = 1$ (see (2.17)) and $F_{11}(t, G)$ increases monotonically for $t > 0$, we get that $\tau = \lambda^{-1} < 1$. Due to Lemma 2.4, $\tau_1 \leq 1$. Furthermore, let $R$ be the radius of convergence of the series (2.11) for $i = j = 1$. It is clear that $\tau_1 \leq R$ (see (2.17) again).

Henceforth we shall use the abbreviated form $f_i$ instead of $f_{11,i}(G)$. In view of what has been said above one has for $0 < t < \tau_1$ the following:

$$t^2 F_{11}(t, G)\ddot{F}_{11}(t, G) + tF_{11}(t, G)\dot{F}_{11}(t, G) - t^2[\dot{F}_{11}(t, G)]^2$$

$$= \sum_{n=1}^{\infty} f_n t^n \sum_{k=1}^{\infty} f_k k(k-1)t^k + \sum_{n=1}^{\infty} f_n t^n \sum_{k=1}^{\infty} f_k kt^k - (\sum_{n=1}^{\infty} f_n nt^n)^2$$

$$= \sum_{n,k=1}^{\infty} (f_n f_k k(k-1)t^{n+k} + f_n f_k kt^{n+k} - f_n f_k nkt^{n+k})$$

$$= \sum_{n,k=1}^{\infty} f_n f_k (k^2 - nk)t^{n+k} = \frac{1}{2} \sum_{n,k=1}^{\infty} f_n f_k (n-k)^2 t^{n+k} > \frac{1}{2} \sum_{n \neq k}^{\infty} f_n f_k t^{n+k}$$

$$= \frac{1}{2} f_1 t(f_2 t^2 + f_3 t^3 + \ldots) + \frac{1}{2} f_2 t^2 (f_1 t + f_3 t^3 + \ldots) + \ldots \tag{2.30}$$

Let $\ell'$ denote the minimum of lengths of the cycles in $G$ running through the vertices $v_1$ (obviously, $\ell' \leq \alpha(G)$). In view of the above assumption there is a cycle $a$ in $G$ of length $\ell > \ell'$ running through $v_1$ only once. Let $a = (v_{i_1}, v_{i_2}, \ldots, v_{i_{\ell'}}, \ldots, v_{i_\ell})$, where $i_1 = 1$. By the definition of $\alpha(G)$ one can find a path $b = (v_{j_1}, v_{j_2}, \ldots, v_{j_{m-1}}, v_{j_m})$ such that $j_1 = i_{\ell'}$, $j_m = 1$, $j_n \neq 1$ for $n \neq m$, and $m \leq \alpha(G) = \alpha$. It is clear that $c = (v_1, v_{i_2}, \ldots, v_{i_{\ell'}}, v_{j_2}, \ldots, v_{j_{m-1}})$ is a cycle running through $v_1$ only once. Denoting the length of $c$ by $\ell$ we see that $\ell' < \ell'' \leq \ell' + m - 1 \leq 2\alpha$, and $f_{\ell'} \geq 1$, $f_{\ell''} \geq 1$. With this in mind and due to (2.30) one can write

$$t^2F_{11}(t,G)\ddot{F}_{11}(t,G) + tF_{11}(t,G)\ddot{F}_{11}(t,G) - t^2[\dot{F}_{11}(t,G)]^2$$

$$\geq \frac{1}{2}f_{\ell'}t^{\ell'}[F_{11}(t,G) - f_{\ell'}t^{\ell'}] + \frac{1}{2}f_{\ell''}t^{\ell''}[F_{11}(t,G) - f_{\ell''}t^{\ell''}]$$

$$\geq \frac{1}{2}[f_{\ell'}t^{\ell'}f_{\ell''}t^{\ell''} + f_{\ell''}t^{\ell''}f_{\ell'}t^{\ell'}] \geq t^{\ell'+\ell''} \geq t^{3\alpha}, \quad 0 \leq t \leq \tau_1$$

It remains to consider the case $F_{11}(t,G) = \text{const}\ t^{k_0}$. Due to (2. 3), in that case $\dot{\Phi}_{11}(t,G) \equiv 0$ and hence $\Phi_{11}(t,G) \equiv \text{const}$. By definition one has $\Delta(\tau, G) = 0$. Furthermore due to (2. 17) and (2. 27), $\Phi_{11}(\tau, G) = 0$. Therefore $\Phi_{11}(t,G) \equiv 0$ which contradicts the assumptions of the lemma.

Lemma 2. 10. Denote by $G_k$ the complete subgraph of the graph $G$ obtained from $G$ by removing the vertex $v_k$, $1 \leq k \leq s$. Then for $k = 1, \ldots, s$

$$\dot{A}_{kk}(t,G) = \dot{\Delta}(t,G) = t^{-1}[(s-1)A_{kk}(t,G) + \sum_{\ell=1}^{s-1} A_{11}(t,G)] \tag{2. 31}$$

$$|A_{kk}^{-1}(t,G)\dot{A}_{kk}(t,G)| \leq (\tau_1 - t)^{-1}(s-1), \qquad 0 \leq t \leq \tau_1 \tag{2. 32}$$

Proof: The equality (2. 31) follows immediately from (2. 16). To prove (2. 32) we denote by $u_1, u_2, \ldots, u_{s-1}$ all zeros of the polynomial $A_{kk}(t,G) = \Delta(t, G_k)$ taking into account the multiplicity. It follows from the definition of $\tau_1$ that $|u_i| \geq \tau_1$, $i = 1, \ldots, s-1$. Therefore

$$|A_{kk}^{-1}(t,G)\dot{A}_{kk}(t,G)| = |\sum_{i=1}^{s-1}(t-u_1)^{-1}| \leq \sum_{i=1}^{s-1}|t-u_i|^{-1} \leq (\tau_1 - t)^{-1}(s-1),$$

$$0 \leq t \leq \tau_1$$

The lemma is proved.

Lemma 2. 11. If $\Phi_{11}(t,G)$ is not equal to zero identically, then

$$\dot{\Phi}_{11}(t,G) \geq c_1 s^{-2}\tau^{7\alpha} \tag{2. 33}$$

<u>whenever</u>

$$0 \leq t - \tau \leq s^{-1} \min(\tau_1 - \tau, \tau) \tag{2.34}$$

$c_1$ <u>being an absolute constant.</u>

<u>Proof:</u> In view of Lemma 2.9 it suffices to bound the second factor on the right-hand side of (2.28). Due to (2.18)

$$\dot{F}_{11}(t, G) = A_{11}^{-1}(t, G)\Delta(t, G)[t^{-1}s - A_{11}^{-1}(t,G)\dot{A}_{11}(t,G)] + t^{-1}\sum_{k=1}^{s} A_{11}^{-1}(t,G)A_{kk}(t,G)$$

By Lemma 2.7 the functions $\Delta(t, G)$ and $A_{kk}(t, G)$, $k = 1, \ldots, s$, have the same signs in the interval $(\tau, \tau_1)$. So in this interval

$$\dot{F}_{11}(t, G) = A_{11}^{-1}(t, G)\Delta(t, G)[t^{-1}s - |A_{11}^{-1}(t, G)\dot{A}_{11}(t, G)|]$$

$$+ t^{-1}\sum_{k=1}^{s} A_{11}^{-1}(t, G)A_{kk}(t, G) \tag{2.35}$$

To estimate $A_{11}^{-1}(t, G)A_{kk}(t, G)$ we note that $A_{kk}(t, G) = \Delta(t, G_k)$ (see the statement of Lemma 2.10 for the definition of $G_k$). Due to (2.15),

$$\dot{A}_{kk}(t, G) = \dot{\Delta}(t, G_k) = t^{-1}[(s-1)A_{kk}(t, G) + \sum_{\ell=1}^{s-1} A_{11}(t, G_k)] \tag{2.36}$$

and due to Lemma 2.7

$$\operatorname{sign} A_{ii}(t, G) = \operatorname{sign}(-1)^{s-1}, \qquad \operatorname{sign} A_{11}(t, G_k) = \operatorname{sign}(-1)^{s-2}$$

is $0 < t \leq \tau_1$, $i, \ell, k = 1, 2, \ldots, s$, $\ell \neq k$. Taking into account these relations one can write

$$\frac{d}{dt}[A_{11}^{-1}(t, G)A_{kk}(t, G)] = A_{11}^{-2}(t, G)[\dot{A}_{11}(t, G)A_{11}(t, G) - A_{kk}(t,G)\dot{A}_{11}(t, G)]$$

$$= A_{11}^{-1}(t, G)\dot{A}_{kk}(t, G) - A_{11}^{-2}(t, G)A_{kk}(t, G)\dot{A}_{11}(t, G)$$

$$= t^{-1}A_{11}^{-1}(t, G)[(s-1)A_{kk}(t, G) + \sum_{\ell=1}^{s-1} A_{11}(t, G_k)]$$

$$- A_{11}^{-2}(t, G)A_{kk}(t, G)\dot{A}_{11}(t, G) =$$

$$= A_{11}^{-1}(t, G)A_{kk}(t, G)[t^{-1}(s-1) - A_{11}^{-1}(t, G)\dot{A}_{11}(t, G)]$$

$$+ t^{-1}\sum_{\ell=1}^{s-1} A_{11}^{-1}(t, G)A_{11}(t, G_k)$$

$$\leq A_{11}^{-1}(t, G)A_{kk}(t, G)[t^{-1}(s-1) + |A_{11}^{-1}(t, G)\dot{A}_{11}(t, G)|] \tag{2. 37}$$

Now (2. 32) for $k = 1$ and (2. 37) together give

$$\frac{d}{dt}[A_{11}^{-1}(t,G)A_{kk}(t,G)] \leq A_{11}^{-1}(t,G)A_{kk}(t,G)[t^{-1}(s-1) + (\tau_1-t)^{-1}(s-1)]$$

that is,

$$[A_{11}^{-1}(t,G)A_{kk}(t,G)]^{-1}\frac{d}{dt}[A_{11}^{-1}(t,G)A_{kk}(t,G)] \leq (s-1)[t^{-1} + (\tau_1-t)^{-1}]$$

By integration of the latter inequality we get

$$\ell n[A_{11}^{-1}(t, G)A_{kk}(t,G)] - [\ell n\, A_{11}^{-1}(\tau, G)A_{kk}(\tau, G)] \leq (s-1)\ \ell n[\tau^{-1}(\tau_1-t)^{-1}t(\tau_1-\tau)],$$

$$A_{11}^{-1}(t, G)A_{kk}(t, G) \leq A_{11}^{-1}(\tau, G)A_{kk}(\tau, G)[\tau^{-1}(\tau_1-t)^{-1}t(\tau_1-\tau)]^{s-1}$$

Therefore if (2. 34) holds, then

$$A_{11}^{-1}(t, G)A_{kk}(t, G) \leq A_{11}^{-1}(\tau, G)A_{kk}(\tau, G)(1 + s^{-1})^{s-1}[1 + (s-1)^{-1}]^{s-1}$$

$$\leq cA_{11}^{-1}(\tau, G)A_{kk}(\tau, G)$$

$c$ being an absolute constant. Finally, using Lemma 2. 8 we get

$$A_{11}^{-1}(t, G)A_{kk}(t, G) \leq c\lambda^{2\alpha} \tag{2. 38}$$

Substituting (2. 32) and (2. 38) into (2. 35) (in view of (2. 34)) gives

$$\dot{F}_{11}(t, G) \leq A_{11}^{-1}(t, G)\Delta(t, G)[t^{-1}s + (\tau_1-t)^{-1}(s-1)] + ct^{-1}s\lambda^{2\alpha}$$

$$\leq A_{11}^{-1}(t, G)\Delta(t, G)s[\tau^{-1} + (\tau_1-\tau)^{-1}] + cs\lambda^{2\alpha+1} \tag{2. 39}$$

Let

$$A_{11}^{-1}(t, G)\Delta(t, G) = \gamma(t), \quad s[\tau^{-1} + (\tau_1 - \tau)^{-1}] = a, \quad cs\lambda^{2\alpha+1} = b \qquad (2.40)$$

Then (2.39) takes the form

$$\dot{\gamma}(t) \leq a\gamma(t) + b \qquad (2.41)$$

Taking into account that $\gamma(\tau) = 0$ we obtain the inequality

$$\gamma(t) \leq -ba^{-1} \exp(at)[\exp(-at) - \exp(-a\tau)]$$

which together with (2.41) shows that

$$\dot{\gamma}(t) \leq b \exp[a(t - \tau)]$$

Due to (2.40) and (2.34)

$$a(t-\tau) = s[\tau^{-1}(t-\tau) + (\tau_1 - \tau)^{-1}(t-\tau)] \leq s(s^{-1} + s^{-1}) = 2$$

and finally

$$\dot{F}_{11}(t, G) = \frac{d}{dt}[A_{11}^{-1}(t, G)\Delta(t, G)] \leq c' s\lambda^{2\alpha+1} \qquad (2.42)$$

$c'$ being an absolute constant. Substituting (2.42) and (2.29) into (2.28) in view of (2.34) leads to (2.33) and hence completes the proof.

Lemma 2.12. If $\alpha = \alpha(G) > 2$ and $\Phi_{11}(t, G)$ is not equal to zero identically, then

$$\dot{\Phi}_{11}(t, G) \geq c_2 s^{-2} \tau^{7\alpha-1} \qquad (2.43)$$

whenever

$$\tau[1 - (\alpha - 1)^{-1}] \leq t \leq \tau \qquad (2.44)$$

$c_2$ being an absolute constant.

Proof: Using the fact that $\dot{F}_{11}(t, G)$ increases for $t > 0$ we obtain from (2. 28) and (2. 29) the inequality

$$\dot{\Phi}_{11}(t, G) \geq [\dot{F}_{11}(\tau, G)]^{-2} t^{3\alpha-3}$$

the right-hand side of which (due to (2. 18)) can be rewritten as follows

$$t^{3\alpha-3}\tau^2 A_{11}^2(\tau, G)[\sum_{k=1}^{s} A_{kk}(\tau, G)]^{-2}$$

Applying to this expression Lemma 2. 8 and using (2. 44) we get

$$\Phi_{11}(t, G) \geq s^{-2} t^{3\alpha-3} \tau^{4\alpha+2}$$

$$\geq s^{-2}\tau^{7\alpha-1}[1 - (\alpha-1)^{-1}]^{3\alpha-3} \geq \text{const } s^{-2}\tau^{7\alpha-1}$$

which coincides with (2. 43).

Lemma 2. 13. If $s = s(G) > 2$ and $\lambda = \lambda(G) > 1$ then

$$\dot{\Phi}_{11}(t, G) \leq c_3(s\alpha)^2 \tau^{-4\alpha-4}(1 - \tau)^{-2} \qquad (2.\ 45)$$

whenever

$$0 \leq t - \tau \leq s^{-1}(\tau_1 - \tau) \qquad (2.\ 46)$$

$c_3$ being an absolute constant.

Proof: Since the coefficients $f_{11,n}$, $n = 1, 2, \ldots$, are nonnegative, one has $F_{11}(t, G) \leq t\dot{F}_{11}(t, G)$, $t \geq 0$, which implies (see (2. 28)) that

$$\dot{\Phi}_{11}(t, G) \leq t^{-1}[\dot{F}_{11}(t, G)]^{-2} F_{11}(t, G)\ddot{F}_{11}(t, G), \qquad t \geq 0 \qquad (2.\ 47)$$

According to (2. 17) - (2. 19),

$$[\dot{F}_{11}(t,G)]^{-2}F_{11}(t,G)\ddot{F}_{11}(t,G) = [\dot{\Delta}(t,G)A_{11}(t,G) - \Delta(t,G)\dot{A}_{11}(t,G)]^{-2}$$

$$\times \{\Delta(t,G)\ddot{\Delta}(t,G)A_{11}^{2}(t,G) - \Delta^{2}(t,G)A_{11}(t,G)\ddot{A}_{11}(t,G)$$

$$- 2\Delta(t,G)\dot{\Delta}(t,G)A_{11}(t,G)\dot{A}_{11}(t,G) + 2\Delta^{2}(t,G)[\dot{A}_{11}(t,G)]^{2}$$

$$+ \ddot{\Delta}(t,G)A_{11}^{3}(t,G) - \Delta(t,G)A_{11}^{2}(t,G)\ddot{A}_{11}(t,G)$$

$$- 2\dot{\Delta}(t,G)A_{11}^{2}(t,G)\dot{A}_{11}(t,G) + 2\Delta(t,G)A_{11}(t,G)[\dot{A}_{11}(t,G)]^{2}\} \qquad (2.48)$$

We have to estimate each summand in the right-hand side of (2. 48) obtained after removing the braces. Let us show how it is accomplished for the first and the second summands. The others can be estimated in a similar way.

According to (2. 16), (2. 15) and (2. 31),

$$[\dot{\Delta}(t,G)A_{11}(t,G) - \Delta(t,G)\dot{A}_{11}(t,G)]^{-2}\Delta(t,G)\ddot{\Delta}(t,G)A_{11}^{2}(t,G)$$

$$= t^{-1}[t^{-1}s\Delta(t,G)A_{11}(t,G) + t^{-1}\sum_{k=1}^{s} A_{11}(t,G)A_{kk}(t,G)$$

$$- t^{-1}(s-1)\Delta(t,G)A_{11}(t,G) - t^{-1}\Delta(t,G)\sum_{\ell=1}^{s-1} A_{\ell\ell}(t,G_1)]^{-2}$$

$$\times \Delta(t,G)A_{11}^{2}(t,G)[(s-1)\dot{\Delta}(t,G) + \sum_{k=1}^{s}\dot{A}_{kk}(t,G)]$$

$$= [\Delta(t,G)A_{11}(t,G) + \sum_{k=1}^{s} A_{11}(t,G)A_{kk}(t,G) + \sum_{\ell=1}^{s-1}(-\Delta(t,G))A_{\ell\ell}(t,G_1)]^{-2}$$

$$\times \{\Delta(t,G)A_{11}^{2}(t,G)(s-1)[s\Delta(t,G) + \sum_{k=1}^{s} A_{kk}(t,G)]$$

$$+ t\Delta(t,G)A_{11}^{2}(t,G)\sum_{k=1}^{s}\dot{A}_{kk}(t,G)\} \qquad (2.49)$$

Consider the first of two factors in the right-hand side of (2. 49). If $\tau \leq t < \tau_1$, then, due to Lemma 2. 7, the contents of the square brackets are given by the sum of three terms of the same sign. Hence

$$[\dot{\Delta}(t, G)A_{11}(t, G) - \Delta(t, G)\dot{A}_{11}(t, G)]^{-2}\Delta(t, G)\ddot{\Delta}(t, G)A_{11}^{2}(t, G)$$

$$\leq [\Delta(t, G)A_{11}(t, G)]^{-2}\Delta^{2}(t, G)A_{11}^{2}(t, G)s(s - 1)$$

$$+ [2\Delta(t, G)A_{11}^{2}(t, G)\sum_{k=1}^{s} A_{kk}(t,G)]^{-1}\Delta(t,G)A_{11}^{2}(t,G)(s-1)\sum_{k=1}^{s} A_{kk}(t, G)$$

$$+ [2\Delta(t, G)A_{11}^{2}(t, G)\sum_{k=1}^{s} A_{kk}(t, G)]^{-1}t\Delta(t, G)A_{11}^{2}(t, G)\sum_{k=1}^{s} A'_{kk}(t, G)$$

$$\leq s(s-1) + \frac{1}{2}(s-1) + \frac{1}{2}t \max_{1 \leq k \leq s} |A_{kk}^{-1}(t, G)\dot{A}_{kk}(t, G)|$$

$$\leq s(s-1) + \frac{1}{2}(s-1) + [2(\tau_1 - t)]^{-1}t(s - 1) \tag{2.50}$$

(we have used (2. 32) for estimating the last summand).

The second term in the right-hand side of (2. 48) can be estimated as follows:

$$|\dot{\Delta}(t, G)A_{11}(t, G) - \Delta(t,G)\dot{A}_{11}(t,G)|^{-2}|\Delta^{2}(t,G)A_{11}(t,G)\ddot{A}_{11}(t, G)|$$

$$= [\Delta(t,G)A_{11}(t,G) + A_{11}(t,G)\sum_{k=1}^{s} A_{kk}(t,G) + (-\Delta(t,G))\sum_{\ell=1}^{s=1} A_{11}(t,G_1)]^{-2}$$

$$\times t\Delta^{2}(t, G)A_{11}(t, G)[(s-2)\dot{A}_{11}(t, G) + \sum_{\ell=1}^{s-1} \dot{A}_{11}(t, G_1)]$$

$$\leq t(s-2)|A_{11}^{-1}(t,G)\dot{A}_{11}(t,G)| + t|[2\sum_{\ell=1}^{s-1} A_{11}(t,G_1)]^{-1}\sum_{\ell=1}^{s-1} \dot{A}_{11}(t,G_1)|$$

$$\leq t(s-2)(s-1)(\tau_1 - t)^{-1} + t[s(G_1)-1][2(\tau_1(G_1)-1)]^{-1}$$

$$\leq (\tau_1 - t)^{-1}(s-2)(s-1) + [2(\tau_1 - t)]^{-1}t(s-2) \tag{2.51}$$

For the other six terms in the right-hand side of (2. 48) one can obtain the following upper bounds for their absolute values, respectively:

$$(\tau_1 - t)^{-1} t(2s+1)(s-1), \quad (\tau_1 - t)^{-2} 2[t(s-1)]^2, \quad \frac{1}{2}(s-1)(s+2) + (\tau_1 - t)^{-1} t(s-1),$$

$$[2(\tau_1 - t)]^{-1} t(s-2)(2s-3), \quad (\tau_1 - t)^{-1} t(s+2)(s-1), \quad (\tau_1 - t)^{-2}[t(s-1)]^2$$

Substitution of all these bounds into (2. 47) gives

$$\begin{aligned}\dot{\Phi}_{11}(t, G) &\le t^{-1} s(s-1) + (2t)^{-1}(s-1) + (2t)^{-1}(s-1)(s+2) \\ &+ [2(\tau_1 - t)]^{-1}(s-1) + (\tau_1 - t)^{-1}(s-1)(s-2) + [2(\tau_1 - t)]^{-1}(s-2) \\ &+ (\tau_1 - t)^{-1}(2s+1)(s-1) + (\tau_1 - t)^{-1}(s-1) + (\tau_1 - t)^{-1}(s+2)(s-1) \\ &+ (\tau_1 - t)^{-2} 2t(s-1)^2 + (\tau_1 - t)^{-2} t(s-1)^2 \\ &\le (s-1)[(2t)^{-1} 3(s+1) + (\tau_1 - t)^{-1}(4s+3) + (\tau_1 - t)^{-2} 3t(s-1)] \qquad (2.\ 52)\end{aligned}$$

If (2. 46) is true, then

$$\tau_1 - t \ge s^{-1}(s-1)(\tau_1 - \tau), \qquad \tau \le t \le \tau_1$$

It follows now from (2. 52) and the assumption $s > 2$ that

$$\begin{aligned}\dot{\Phi}_{11}(t, G) &\le (2\tau)^{-1} 3(s+1)(s-1) + (\tau_1 - \tau)^{-1} s(4s+3) + (\tau_1 - \tau)^{-2} 3s^2 \tau_1 \\ &\le (2\tau)^{-1} 3s^2 + 5(\tau_1 - \tau)^{-1} s^2 + 3(\tau_1 - \tau)^{-2} \tau_1 s^2 \qquad (2.\ 53)\end{aligned}$$

By Lemma 2. 2,

$$\tau^{-1} - \tau_1^{-1} \ge \tau^{2\alpha}(\tau^{-1/\alpha} - 1)$$

Hence

$$\tau_1 - \tau \ge \tau^{2\alpha+2}(\tau^{-1/\alpha} - 1) \qquad (2.\ 54)$$

Since $\tau^{-1/\alpha} = [1 - (1-\tau)]^{-1/\alpha}$, and $0 < 1 - \tau < 1$, $\alpha \geq 1$, we have

$$\tau^{-1/\alpha} - 1 \geq [1 - \alpha^{-1}(1-\tau)]^{-1} - 1 = (\alpha - 1 + \tau)^{-1}(1-\tau) \geq \alpha^{-1}(1-\tau)$$

and due to (2. 54)

$$\tau_1 - \tau \geq \alpha^{-1}\tau^{2\alpha+2}(1-\tau) \qquad (2.\ 55)$$

Substituting this inequality into (2. 53) and taking into account that $\tau_1 \leq 1$ (see Lemma 2. 4) we finally get

$$\dot{\Phi}_{11}(t, G) \leq (2\tau)^{-1} 3s^2 + \tau^{2\alpha-2}(1-\tau)^{-1} 5\alpha s^2 + \tau^{-4\alpha-4}(1-\tau)^{-2} 3(s\alpha)^2$$

$$\leq 11(s\alpha)^2 \tau^{-4\alpha-4}(1-\tau)^{-2}.$$

Thus the lemma is proved.

<u>Lemma 2. 14.</u> <u>Under the assumptions of Lemma 2. 12 one has</u>

$$\dot{\Phi}_{11}(t, G) \leq c_4 (s\alpha)^2 \tau^{-4\alpha-4}(1-\tau)^{-2} \qquad (2.\ 56)$$

<u>whenever</u>

$$0 \leq \tau - t \leq [2(2s+1)]^{-1}\tau(\tau_1 - \tau) \qquad (2.\ 57)$$

$c_4$ <u>being an absolute constant.</u>

<u>Proof:</u> As we have seen above the proof of Lemma 2. 12 was founded upon the fact that all terms in the sum

$$\sum_1(t, G) = \Delta(t, G)A_{11}(t, G) + \sum_{k=1}^{s} A_{11}(t,G)A_{kk}(t,G) + \sum_{\ell=1}^{s-1} [-\Delta(t, G)]A_{11}(t,G_\ell)$$

had the same signs provided that $\tau \leq t < \tau_1$. If $t < \tau$, this does not remain true for $\sum_1(t, G)$ but becomes true for

$$\sum_2(t,G) = -\Delta(t,G)A_{11}(t,G) + \sum_{k=1}^{s} A_{11}(t,G)A_{kk}(t,G) + \sum_{\ell=1}^{s-1} \Delta(t,G)A_{11}(t,G_\ell)$$

(see Lemma 2. 7). With this in mind one can repeat the part of the proof of Lemma 2. 12 leading to (2. 52). Here instead of (2. 52) we obtain the following:

$$\Sigma_1^{-2}(t, G)\,\Sigma_2^{2}(t, G)\,\dot{\Phi}_{11}(t, G) \leq (s-1)[(2t)^{-1}3(s+1) + (\tau_1-t)^{-1}(4s+t) + (\tau_1-t)^{-2}3t(s-1)], \quad 0 < t \leq \tau \tag{2. 58}$$

If (2. 57) holds, then $\tau/2 \leq t \leq \tau < 1$, and the right-hand side of (2. 58) is bounded by

$$(s-1)[\tau^{-1}3(s+1) + (\tau_1-\tau)^{-1}(4s+3) + (\tau_1-\tau)^{-2}3(s-1)]$$

and hence (see (2. 55)

$$\Sigma_1^{-2}(t, G)\,\Sigma_2^{2}(t, G)\,\dot{\Phi}_{11}(t, G) \leq 10(s\alpha)^2\tau^{-4\alpha-4}(1-\tau)^{-2} \tag{2. 59}$$

Now we have to estimate $\Sigma_1^{-1}(t, G)\Sigma_2(t, G)$. For short let us set

$$\Delta(t,G) = \Delta(t),\quad A_{11}(t,G) = A(t),\quad \sum_{k=1}^{s} A_{kk}(t,G) = S(t),\quad \sum_{\ell=1}^{s-1} A_{11}(t, G) = S_1(t)$$

Then

$$\Sigma_1^{-1}(t,G)\Sigma_2(t,G) = 1 + 2\{1 - [A^{-1}(t)S^{-1}(t)S_1(t)\Delta(t) - S^{-1}(t)\Delta(t)]\}^{-1}[A^{-1}(t)S^{-1}(t)S_1(t)\Delta(t) - S^{-1}(t)\Delta(t)]$$

It suffices to show that if (2. 56) holds, then

$$|A^{-1}(t)S^{-1}(t)S_1(t)\Delta(t) - S^{-1}(t)\Delta(t)| \leq |S^{-1}(t)\Delta(t)|(|A^{-1}(t)S_1(t)| + 1) \leq 1/2 \tag{2. 60}$$

For (2. 60) implies that $\Sigma_1^{-1}(t, G)\,\Sigma_2(t, G) \geq 1/3$ which together with (2. 59) gives (2. 56).

To prove (2. 60) we shall estimate separately $S^{-1}(t)\Delta(t)$ and $A^{-1}(t)S^{-1}(t)S_1(t)\Delta(t)$. Due to (2. 15), (2. 32), (2. 56), and Lemma 2. 7,

$$\left|\frac{d}{dt}(S^{-1}(t)\Delta(t))\right| = |S^{-1}(t)\dot{\Delta}(t) - S^{-2}(t)\dot{S}(t)\Delta(t)| = |(tS(t))^{-1}s\Delta(t)$$

$$+ t^{-1} - S^{-2}(t)\dot{S}(t)\Delta(t)| \leq |S^{-1}(t)\Delta(t)|[(2\tau)^{-1}s + (\tau_1-\tau)^{-1}(s-1)]$$

$$+ (2\tau)^{-1} = S^{-1}(t)\Delta(t)[-2\tau)^{-1}s - (\tau_1-\tau)^{-1}(s-1)] + (2\tau)^{-1}$$

Thus

$$\frac{d}{dt}(S^{-1}(t)\Delta(t)) \leq aS^{-1}(t)\Delta(t) + b \tag{2. 61}$$

where

$$a = -s[(2\tau)^{-1} + (\tau_1-\tau)^{-1}], \quad b = (2\tau)^{-1} \tag{2. 62}$$

Inequality (2. 61) can be easily reduced to the form

$$\frac{d}{dt}[S^{-1}(t)\Delta(t)\exp(-at)] \leq \frac{d}{dt}[-a^{-1}b\exp(-at)]$$

By integrating from a fixed $t < \tau$ to $\tau$ and taking into account that $\Delta(\tau) = 0$ we obtain from this the following

$$0 \geq S^{-1}(t)\Delta(t) \geq a^{-1}b[\exp(-a(\tau-t)) - 1] \tag{2. 63}$$

If (2. 57) holds, then $0 \leq -a(\tau - t) < 1/2$. Hence

$$0 \leq \exp[-a(\tau - t)] - 1 \leq -2a(\tau-t)$$

and due to (2. 63)

$$0 \geq S^{-1}(t)\Delta(t) \geq -2b(\tau - t) = -\tau^{-1}(\tau-t) \tag{2. 64}$$

Now consider the function $A^{-1}(t)S_1(t)$. By definition

$$A(t) = A_{11}(t, G) = \Delta(t, G_1), \qquad S_1(t) = \sum_{j=1}^{s(G_1)} A_{jj}(t, G_1)$$

Applying (2. 15) we have

$$\dot{A}(t) = t^{-1}[(s-1)A(t) + S_1(t)]$$

and hence

$$A^{-1}(t)S_1(t) = tA^{-1}(t)\dot{A}(t) - (s-1)$$

which together with (2. 32) gives

$$|A^{-1}(t)S_1(t)| \le (s-1)[(\tau_1 - t)^{-1}t + 1] \le s[(\tau_1 - \tau)^{-1}\tau + 1] \qquad (2.\ 65)$$

Finally, due to (2. 64), (2. 65), (2. 57) and Lemma 2. 4

$$|S^{-1}(t)\Delta(t)|[1 + |A^{-1}(t)S_1(t)|] \le \tau^{-1}(\tau - t)[1 + s(1 + (\tau_1 - \tau)^{-1}\tau)]$$

$$= [2(2s+1)]^{-1}(\tau_1 - \tau + s\tau_1) < 1/2$$

Thus (2. 60) and the lemma as a whole is proved.

Summarizing the results of Lemmas 2. 11 - 2. 14 we arrive at

Lemma 2. 15. If $G$ is a finite connected graph with

$s(G) = s > 2$, $\alpha(G) = \alpha > 2$, $\lambda(G) = \lambda > 1$, $\tau(G) = \lambda^{-1} = \tau$, $\tau_1(G) = \tau_1$,

and if $\Phi_{11}(t, G)$ is not identically equal to zero then

$$c^- s^{-2} \tau^{7\alpha} \le \Phi_{11}(t, G) \le c^+ (s\alpha)^2 \tau^{-4\alpha - 4}(1 - \tau)^{-2} \qquad (2.\ 66)$$

provided that

$$|t - \tau| \le [2(2s+1)]^{-1}\tau \quad (\tau_1 - \tau) \qquad (2.\ 67)$$

where $c^+ > 1$ and $c^- < 1$ are absolute constants.

## III. A NEIGHBORHOOD OF THE MAXIMAL MEASURE. THE COMPLETION OF THE PROOF OF THEOREM 1.1

Let $X = V^{\mathbb{Z}}$ be a sequence space, and $Y \subset X$ a closed shift-invariant subset of $X$ represented according to Section I in the form $Y = \bigcap_n Y_n$, where $Y_n$ is an n-Markov set. Denote by $\mathcal{P}_{max}(Y)$ (respectively, $\mathcal{P}_{max}(Y_n)$) the family of all maximal measures belonging to $\mathcal{P}(Y)$ (respectively, $\mathcal{P}(Y_n)$) (see Section I). If the graph $G(Y_n)$ is connected then $\mathcal{P}_{max}(Y_n)$ is known to consist of exactly one measure which will be denoted by $\mu_n$. For any probability Borel measure $\nu$ on $X$ denote by $\pi^m \nu$, $m = 0, 1, \ldots,$ its restriction to the $\sigma$-algebra $\sigma^m$ generated by $x_0, x_1, \ldots, x_{m-1}$ (the m-dimensional distribution of $\nu$).

The main result of this section is the following.

<u>Lemma 3.1.</u> <u>Under the assumptions of Theorem 1.1 there exists a sequence of integers</u> $n_k$ <u>such that</u> $\lim_{k\to\infty} n_k = \infty$ <u>and for any</u> $\mu \in \mathcal{P}_{max}(Y)$

$$\lim_{k\to\infty} \mathrm{Var}\,|\pi^{n_k}\mu - \pi^{n_k}\mu_{n_k}| = 0$$

<u>where</u>

$$\mathrm{Var}\,|\pi^{n_k}\mu - \pi^{n_k}\mu_{n_k}| = \max_{C\in\sigma^{n_k}} |\pi^{n_k}\mu(C) - \pi^{n_k}\mu_{n_k}(C)|$$

Before going to the proof of Lemma 3.1 we shall deduce Theorem 1.1 from this Lemma. For this purpose it is enough to observe that if $\mathcal{P}_{max}(Y)$ contains more than one measure, then one can find two measures $\mu', \mu'' \in \mathcal{P}_{max}(Y)$, $\mu' = \mu''$, with respect to which the shift transformation is ergodic. In that case $\mu'$ and $\mu''$ are mutually singular, that is, $\mathrm{Var}\,|\mu' - \mu''| = 1$, and hence

$$\lim_{n\to\infty} \mathrm{Var}\,|\pi^n\mu' - \pi^n\mu''| = 1$$

But this contradicts Lemma 3.1 which is true for both $\mu'$ and $\mu''$.

To prove Lemma 3.1 we consider the following question: given a Markov set Z how large can the difference be between a measure $\nu \in \mathfrak{P}(Z)$ whose entropy with respect to the shift T is close to $h(T, Z)$ and the maximal measure $\nu_{max} \in \mathfrak{P}(Z)$? The results obtained can be of independent interest.

The difference between two measures belonging to $\mathfrak{P}(Z)$ will be characterized here by the variational distance between their one-dimensional distributions. Of course, if $\nu' \in \mathfrak{P}(Z)$ is an arbitrary Markov measure then there can exist a Markov measure $\nu'' \in \mathfrak{P}(Z)$, far from $\nu'$ with close to or even the same entropy. However, if $\nu' = \nu_{max} \in \mathfrak{P}(Z)$ in the case of a connected graph $G(Z)$ the situation is different. Indeed, if a sequence of measures $\nu_n \in \mathfrak{P}(Z)$ weakly converges to a measure $\nu$, then $\nu \in \mathfrak{P}(Z)$, and $h_\nu(T) \geq \limsup_{n \to \infty} h_{\nu_n}(T)$. On the other hand, is a sequence $\nu_n \in \mathfrak{P}(Z)$ satisfies $\lim_{n \to \infty} h_{\nu_n}(T) = h_{\nu_{max}}(T) = h(T, Z)$, then the sequence $\nu_n$ weakly converges to $\nu_{max}$, since due to the above each limit point must be a maximal measure which is actually unique.

The notations used below are introduced in Section I.

<u>Lemma 3.2.</u> <u>Let</u> Z <u>be a</u> 1-<u>Markov set, and</u> $G(Z) = G$ <u>the corresponding graph with</u> $\alpha(G) = \alpha < \infty$ <u>and</u> $\lambda(G) = \lambda > 1$. <u>Let</u> $\nu_{max}$ <u>be the unique maximal measure, corresponding to</u> Z <u>and</u> $\nu \in \mathfrak{P}(Z)$ <u>a measure for which</u>

$$h_\nu(T) \geq \ln \lambda - \lambda^{-2\alpha-1}(\lambda^{1/\alpha} - 1) \tag{3.1}$$

<u>Then the</u> 1-<u>dimensional distributions</u> $\pi^0\nu$ <u>and</u> $\pi^0\nu_{max}$ <u>are mutually</u> absolutely continuous.

<u>Proof:</u> Replace $\nu$ by a Markov measure (of the first order) $\nu'$ uniquely determined by $\pi^1\nu' = \pi^1\nu$, i.e., by the Markov measure

with the same 2-dimensional distributions. It is easy to see that $\nu' \in \mathcal{P}(Z)$, $h_{\nu'}(T) \geq h_{\nu}(T)$, and hence in (3.1) one can replace $\nu$ by $\nu'$. The maximal measure $\nu_{max}$ is positive on each open subset of Z. Hence $\overset{0}{\pi}\nu'$ is absolutely continuous with respect to $\overset{0}{\pi}\nu_{max}$. If these two measures are non-equivalent, then there exists a 1-Markov set $Z' \subset Z$, $Z' \neq Z$, such that $\nu'(Z') = 1$ and the corresponding graph $G' = G(Z')$ is a proper complete subgraph of G. Then $\nu' \in \mathcal{P}(Z')$ and $h_{\nu'}(T) \leq \ln (G')$. According to Lemma 2.2,

$$\ln \lambda(G') \leq \ln [\lambda - \lambda^{-2\alpha}(\lambda^{1/\alpha} - 1)] = \ln \lambda + \ln[1 - \lambda^{-2\alpha-1}(\lambda^{1/\alpha} - 1)]$$

$$< \ln \lambda - \lambda^{-2\alpha-1}(\lambda^{1/\alpha} - 1),$$

which contradicts (3.1). Thus $\overset{0}{\pi}\nu'$ and $\overset{0}{\pi}\nu_{max}$ are equivalent. Finally, it follows from the definition of $\nu'$ that $\overset{0}{\pi}\nu = \overset{0}{\pi}\nu'$. The lemma is proved.

Corollary 3.3. If (3.1) remains true after substitution of $\alpha$ for $\alpha + 1$, then $\overset{1}{\pi}\nu$ and $\overset{1}{\pi}\nu_{max}$ are mutually absolutely continuous.

To prove this corollary it suffices to replace G by $\psi G$ (see Section I).

Further arguments are essentially based on the following construction which was developed for the first time apparently in [7].

Main Construction. Let G be a finite connected graph (one can also consider a countable graph G with $\lambda(G) < \infty$) with the set of vertices V(G). Denote by Z(G) the set of all doubly infinite paths in G and by $T_G$ the shift transformation defined on Z(G). For any vertex $v \in V(G)$ consider the set $Z^v \subset Z(G)$ consisting of all paths $z = (\ldots, z_{-1}, z_0, z_1, \ldots) \in Z(G)$ such that $z_i = v$ for infinitely many positive and infinitely many negative i. The set $Z^v$ is obviously $T_G$-invariant.

Now keeping v fixed we produce a new graph $G^v$ with

possibly an infinite number of vertices. For this we number all cycles in G running through v exactly once in order of increasing length, choosing an arbitrary order for cycles of the same length (the number of cycles of any fixed length is obviously finite). Let $\ell_k$ be the length of the k-th cycle, $1 \leq k < K$, $K = K(G, v) \leq \infty$. The vertices of $G^{\cdot}$ are labelled by two indices, k and $\ell$, $1 \leq k < K$, $l = 1, 2, \ldots, \ell_k$. The edges of $G^v$ are defined as follows: the edge from $(k, \ell)$ to $(k', \ell')$ exists if and only if either $k' = k$, $\ell' = \ell + 1$, or $\ell = \ell_k$, $\ell' = 1$. It is convenient to imagine the graph G to be a family of columns (or towers) or height $\ell_1, \ell_2, \ldots$, etc. with transitions allowed from every floor (except the upper ones) to the next floor of the same column and from the upper floor of every column to the lower floor of any column.

There is a natural one-to-one mapping $\psi^v$ from $Z^v$ onto the set $Z(G^v)$ of all doubly infinite paths in the graph $G^v$. It can be informally described as follows: the ascent to the i-th column in $G^v$ corresponds to the motion along the k-th cycle in G running through the vertex v exactly once. It is easy to see that $\psi^v$ is bicontinuous and transforms $T_G$ into the shift $T_{G^v}$ defined on $Z(G^v)$.

Let $\nu \in \mathcal{P}(Z(G))$ satisfy $\nu(Z^v) = 1$. Then $\psi^v$ transforms $\nu$ into a measure $\nu^v \in \mathcal{P}(Z(G^v))$. It is easy to check that if $\nu$ is a Markov measure, then $\nu^v$ admits the following simple description: there is a probability vector $p(\nu) = \{p_k(\nu)\}$, $1 \leq k < K$, such that the 1-dimensional distribution of $\nu^v$ assigns to each vertex of $G^v$ of the form $(k, \ell)$ the probability $[\sum_i \ell_i p_i(\nu)]^{-1} p_k(\nu)$, and the transition probability from $(k, \ell)$ to $(k', \ell')$ is equal to 1, if $k' = k$, $\ell' = \ell + 1$, to $[\sum_i \ell_i p_i(\nu)]^{-1} p_k(\nu)$, if $\ell = \ell_k$, $\ell' = 1$, and to 0 in the other cases.

Notations. Denote by $\mathcal{P}^v(Z(G)) = \mathcal{P}^v$ the set of all $\nu \in \mathcal{P}(Z(G))$

such that $\nu(Z^V) = 1$ and $\nu$ can be described as above (by means of a probability vector $p(\nu)$). Denote by $\mathfrak{U}^V(Z(G)) = \mathfrak{U}^V$ the set of all Markov measures $\nu \in \mathcal{P}^V(Z(G))$. One has

$$\mathcal{P}(Z(G)) \supset \mathcal{P}^V(Z(G)) \supset \mathfrak{U}^V(Z(G)) \tag{3. 2}$$

Let

$$\mathcal{P}_\epsilon = \mathcal{P}_\epsilon(Z(G)) = \{\nu \in \mathcal{P}(Z(G)) : h_\nu(T_G) \geq \ln \lambda(G) - \epsilon\}, \quad \epsilon > 0,$$

$$\Phi^V(\nu) = |(\pi^0\nu)(\tilde{v}) - (\pi^0\nu_{max})(\tilde{v})|, \quad \nu \in \mathcal{P}_\epsilon,$$

where $\tilde{v}$ is the set of all $z = (\ldots, z_{-1}, z_0, z_1, \ldots) \in Z(G)$ such that $z_0 = v$.

<u>Lemma 3. 4.</u> <u>For any</u> $\epsilon \geq 0$ <u>satisfying</u>

$$\epsilon \leq \lambda^{-2\alpha-3}(\lambda^{1/(\alpha+1)} - 1) \tag{3. 3}$$

<u>(where</u> $\lambda = \lambda(G)$, $\alpha = \alpha(G)$) <u>there exists a measure</u> $\nu_\epsilon \in P_\epsilon \cap \mathfrak{U}^V$ <u>such that</u>

$$\Phi^V(\nu_\epsilon = \underset{\nu \in \mathcal{P}_\epsilon}{\quad} \Phi^V(\nu) = \sup_{\nu \in \mathcal{P}_\epsilon \cap \mathcal{P}^V} \Phi^V(\nu) \tag{3. 4}$$

<u>Proof:</u> First consider the functional $\Phi^V(\cdot)$ acting on the set of Markov measures $\nu \in \mathcal{P}_\epsilon$. Every Markov measure is uniquely determined by its 2-dimensional distributions, i. e., by a finite number of parameters. The condition $h_\nu(T_G) \geq \ln \lambda(G) - \epsilon$ defines a compact subset of the space of parameters, and $\Phi^V(\cdot)$ is obviously continuous on this subset. Hence the supremum of $\Phi^V(\nu)$ over the set of all Markov measures $\nu \in \mathcal{P}_\epsilon$ is attained at some Markov measure $\nu_\epsilon \in \mathcal{P}_\epsilon$. Due to (3. 3) and Corollary 3. 3, $\nu_\epsilon$ is positive on each open subset of $Z(G)$. Taking into account that $\nu_\epsilon$ is Markov

and $G$ is connected it is easy to ensure that $\nu_\epsilon(Z^V) = 1$ and hence $\nu_\epsilon \in P_\epsilon \cap \mathfrak{U}^V$. Therefore

$$V(\nu_\epsilon) = \sup_{\nu \in \mathcal{P}_\epsilon \cap \mathfrak{U}^V} \Phi^V(\nu),$$

and to prove the first of the equalities (3. 4) it suffices to note that replacing any measure $\nu \in \mathcal{P}_\epsilon$ by the Markov measure with the same 2-dimensional distribution preserves the value of $\Phi^V(\cdot)$ and does not decrease the entropy of $T_G$. In view of (3. 2) the second of equalities (3. 4) follows immediately from the first one. The lemma is proved.

Lemma 3. 5. Let $\epsilon \geq 0$ satisfy (3. 3), and $\nu_\epsilon$ denote the measure specified by Lemma 3. 4. Then the probability vector $p(\nu_\epsilon) = \{p_k(\nu_\epsilon)\}$, $k = 1, 2, \ldots,$ is of the form

$$p_k(\nu_\epsilon) = c_\epsilon t_\epsilon^{\ell_k}, \qquad k = 1, 2, \ldots, \sum_k p_k(\nu_\epsilon) = 1 \tag{3. 5}$$

where $t_\epsilon$ and $c_\epsilon$ are independent of $k$.

Proof: The construction immediately implies that if $\nu \in \mathcal{P}(Z(G))$ and $\nu(Z^V) = 1$, then the dynamical systems $(Z(G^V), \nu^V, T_{G^V})$ and $(Z(G), \nu, T_G)$ are isomorphic and consequently have the same entropy. But the entropy of the former of them can easily be determined provided that $\nu \in \mathcal{P}^V$, namely,

$$h(T_G) = h_{\nu^V}(T_{G^V}) = -[\sum_k \ell_k p_k(\nu)]^{-1} \sum_k p_k(\nu) \ln p_k(\nu) \tag{3. 6}$$

Furthermore,

$$\pi^0 \nu(\tilde{v}) = [\sum_k \ell_k p_k(\nu)]^{-1} \tag{3. 7}$$

According to Lemma 3. 4, $\nu_\epsilon$ can be obtained to be a maximum point of $\Phi^V(\cdot)$ over the set $\mathcal{P}_\epsilon \cap \mathcal{P}^V$. Instead of $\Phi^V(\cdot)$ one can maximize $[\Phi^V(\cdot)]^2$. In view of equalities (3. 6) and (3. 7) this leads to a conditional extremum problem for a function of the possibly infinite number of variables $p_k$ under constraints of the form $p_k \geq 0$, $k = 1, 2, \ldots, \sum_k p_k = \text{const}$, and $(\sum_k \ell_k p_k)^{-1} \sum_k p_k \ln p_k \geq$ const. It follows from (3. 3) and Lemma 3. 3 that $p_k(\nu_\epsilon) > 0$ for every $k \geq 1$. This ensures the smoothness of all functions involved in a neighborhood of the point $p(\nu_\epsilon) = \{p_k(\nu_\epsilon)\}$. The necessary conditions of extremum are found in the standard way and give (3. 5). The calculations are omitted.

Remark. Lemma 3. 4 shows that the supremum of $\Phi^V(\cdot)$ over $\mathcal{P}_\epsilon$ coincides with that over $\mathcal{P}_\epsilon \cap \mathcal{U}^V$. Thus one should find the maximum of a function of a finite number of variables (as variables one can take the probabilities of the 2-dimensional cylinder sets) subject to many constraints. From the general point of view the method used above consists of replacing the original finite-dimensional problem by an equivalent infinite-dimensional one which involves a much smaller number of constraints.

Lemma 3. 6. Let G be a finite connected graph with $s(G) = s > 2$, $\alpha(G) = \alpha > 2$, and $\lambda(G) = \lambda > 1$. Let $\nu_{max} \in \mathcal{P}(Z(G))$ be the maximal measure, and $\nu \in \mathcal{P}(Z(G))$ a measure such that $h_{\nu_{max}}(T_G) - h_\nu(T_G) = \epsilon \geq 0$, where $\epsilon$ satisfies

$$\epsilon < c^- \lambda^{-11\alpha - 5}(\lambda^{1/\alpha} - 1)^2[6s^2(2s + 1)]^{-1} \tag{3.8}$$

Then the following holds:

$$\mathrm{Var}\,|\pi^0\nu_{max} - \pi^0\nu| \leq (c^+)^{1/2}\epsilon^{1/2}s^2\alpha\lambda^{2\alpha+(3/2)}(1-\lambda^{-1})^{-1} \tag{3. 9}$$

where $c^-$ and $c^+$ are the constants obtained in Lemma 2. 15.

Proof: Number the vertices of G to obtain a sequence $v_1, v_2, \ldots, v_s$ and set $v_1 = v$. Then the function $F_{11}(t, G)$ (see Section II) takes the form

$$F_{11}(t, G) = \sum_k t^{\ell_k} \qquad (3.10)$$

(refer to the main construction for the definition of $\ell_k$). If $\epsilon \geq 0$ satisfies (3.8), then it satisfies (3.3) as well. Let $\nu_\epsilon$ be the measure obtained in Lemma 3.4. It follows from (3.5) - (3.7), and (3.10) that

$$(\pi^0 \nu_\epsilon)(\tilde{v}) = [t_\epsilon \dot{F}_{11}(t_\epsilon, G)]^{-1} F_{11}(t_\epsilon, G) \qquad (3.11)$$

$$h_{\nu_\epsilon}(T_G) = [t_\epsilon \dot{F}_{11}(t_\epsilon, G)]^{-1} F_{11}(t_\epsilon, G) \ln F_{11}(t_\epsilon, G) - \ln t_\epsilon \qquad (3.12)$$

It is clear that $\nu_{max} = \nu_\epsilon$ for $\epsilon = 0$. Due to (3.5), (3.10) and the equality $F_{11}(\tau(G), G) = 1$ one has

$$t_0 = \tau = \tau(G) = \lambda^{-1}, \quad c_0 = 1$$

Together with (3.11) this gives

$$\pi^0 \nu_{max}(\tilde{v}) - (\pi^0 \nu_\epsilon)(\tilde{v}) = [t_0 \dot{F}_{11}(\tau, G)]^{-1} - [t_\epsilon \dot{F}_{11}(t_\epsilon, G)]^{-1} F_{11}(t_\epsilon, G)$$

$$= \Phi_{11}(t_\epsilon, G) \qquad (3.13)$$

Let

$$\Psi_{11}(t, G) = \ln(\tau^{-1} t) - [t \dot{F}_{11}(t, G)]^{-1} F_{11}(t, G) \ln F_{11}(t, G)$$

Due to (3.12)

$$h_{\nu_{max}}(T_G) - h_{\nu_\epsilon}(T_G) = \Psi_{11}(t_\epsilon, G) \leq \epsilon \qquad (3.14)$$

Furthermore one has (see (2.27)):

$$\Psi_{11}(\tau, G) = \Phi_{11}(\tau, G) = 0 \qquad (3.15)$$

$$\dot{\Psi}_{11}(t, G) = \dot{\Phi}_{11}(t, G)(t, G) \ln F_{11}(t, G) \tag{3.16}$$

Our aim is to estimate $|\Phi_{11}(t_\epsilon, G)|$ provided that $\epsilon$ is small. Let $t$ be a point such that $F_{11}(t, G)$ is defined by (3. 10). Then

$$\begin{aligned} \ln F_{11}(t, G) &= \ln F_{11}(\tau, G) + (t-\tau) F_{11}^{-1}(\theta, G) \dot{F}_{11}(\theta, G) \\ &= (t-\tau) F_{11}^{-1}(\theta, G) \dot{F}_{11}(t, G) \end{aligned}$$

where $\theta$ is some point between $t$ and $\tau$. By the definition of $F_{11}(\cdot, \cdot)$ one has $F_{11}(\theta, G) \leq \theta \dot{F}_{11}(\theta, G)$. Hence

$$\ln F_{11}(t, G) \geq (t-\tau)\theta^{-1} \geq 1 - \tau t^{-1}, \qquad t \geq \tau \tag{3.17}$$

and

$$\ln F_{11}(t, G) \leq (t-\tau)\theta^{-1} \leq (t-\tau)\tau^{-1} < t\tau^{-1} - 1, \qquad t \leq \tau \tag{3.18}$$

Now we make use of Lemma 2. 15 which states that

$$\gamma^- \leq \dot{\Phi}_{11}(t, G) \leq \gamma^+ \tag{3.19}$$

provided that

$$\tau - \delta \leq t \leq \tau + \delta \tag{3.20}$$

where

$$\gamma^- = c^- s^{-2} \lambda^{-7\alpha}, \quad \gamma^+ = c^+ (s\alpha)^2 \lambda^{4\alpha+4} (1-\lambda^{-1})^{-2} \tag{3.21}$$

$$\delta = [2(2s+1)]^{-1} \lambda^{-1} (\lambda_1^{-1} - \lambda^{-1}), \qquad \lambda_1 = \lambda_1(G) \tag{3.22}$$

Assuming $G$ to satisfy the conditions of Lemma 2. 15 we wish to find a restriction on $\epsilon$ for $t_\epsilon$ to get into the interval (3. 20). Due to the monotonicity of $\Psi_{11}(t, G)$ separately for $t > \tau$ and $t < \tau$ it suffices for $\epsilon$ to satisfy $\Psi_{11}(\tau \pm \delta) \geq \epsilon$. Using (3. 15) - (3. 17), (3. 19) and then (3. 22) we have

$$\Psi_{11}(\tau+\delta,G)=\int_{\tau}^{\tau+\delta}\dot{\Psi}_{11}(t,G)dt\geq\gamma^{-}\int_{\tau}^{\tau+\delta}(1-t^{-1}\tau)dt$$

$$=\gamma^{-}[\delta-\tau\ln(1+\tau^{-1}\delta)]\geq\gamma^{-}[\delta-\tau(\delta\tau^{-1}-\delta^{2}(2\tau^{2})^{-1}+\delta^{3}(3\tau^{3})^{-1})]$$

$$=\gamma^{-}\tau^{-1}\delta^{2}[1/2-\delta(3\tau)^{-1}]\geq\gamma^{-}\delta^{2}(3\tau)^{-1}$$

The same arguments (only replacing (3. 17) by (3. 18)) show that

$$\Psi_{11}(\tau-\delta,G)\geq\gamma^{-}\delta^{2}(2\tau)^{-1}$$

Thus

$$\tau-\delta<t_{\epsilon}<\tau+\delta \tag{3. 23}$$

whenever

$$\epsilon<\gamma^{-}\delta^{2}(3\tau)^{-1}=c^{-}\lambda^{-7\alpha-1}(\lambda_{1}^{-1}-\lambda^{-1})^{2}[6s^{2}(s+1)]^{-1} \tag{3. 24}$$

Due to Lemma 2. 2

$$\lambda_{1}^{-1}-\lambda^{-1}\geq\lambda^{-2\alpha-2}(\lambda^{1/\alpha}-1)$$

Hence (3. 24) will be true provided that

$$\epsilon<c^{-}\lambda^{-7\alpha-1}\lambda^{-4\alpha-2}(\lambda^{1/\alpha}-1)^{-2}[6s^{2}(s+1)]^{-1}$$

$$=c^{-}\lambda^{-11\alpha-5}(\lambda^{1/\alpha}-1)^{2}[6s^{2}(2s+1)]^{-1} \tag{3. 25}$$

i. e., (3. 8) implies (3. 23).

We next estimate $\Phi_{11}(t_{\epsilon},G)$ for $\epsilon$ satisfying (3. 23). Assume for definiteness $t_{\epsilon}\geq\tau$. Due to (3. 14) - (3. 17)

$$\epsilon\geq\Psi_{11}(t_{\epsilon},G)=\int_{\tau}^{t_{\epsilon}}\dot{\Psi}_{11}(t,G)dt\geq\int_{\tau}^{t_{\epsilon}}\dot{\Phi}_{11}(t,G)(t-\tau)t^{-1}dt$$

$$\geq(2\tau)^{-1}\int_{\tau}^{t_{\epsilon}}\dot{\Phi}_{11}(t,G)(t-\tau)dt=(2\tau)^{-1}[(t_{\epsilon}-\tau)\Phi_{11}(t_{\epsilon},G)-\int_{\tau}^{t_{\epsilon}}\Phi_{11}(t,G)dt] \tag{3. 26}$$

Let $R$ denote the rectangle in the plane with base $(\tau, t_\epsilon)$ and height $\Phi_{11}(t_\epsilon, G)$. The value inside the square brackets on the right-hand side of (3. 26) is obviously the area of the part of $R$ over the graph of $\Phi_{11}(t, G)$. It follows from (3. 19) that replacing $\Phi_{11}(t, G)$ by the linear function $\gamma^+(t - \tau)$ can only decrease this area. So in view of (3. 26) we get

$$\epsilon \geq (4\tau\gamma^+)^{-1}\Phi_{11}^2(t_\epsilon, G)$$

The same inequality holds for $t_\epsilon \leq \tau$. Thus due to (3. 21)

$$|\Phi_{11}(t_\epsilon, G)| \leq 2(\epsilon\tau\gamma^+)^{1/2} = 2\epsilon^{1/2}(c^+)^{1/2} s\alpha\lambda^{2\alpha+(3/2}(1-\lambda^{-1})^{-1} \tag{3. 27}$$

Now consider the case where $G$ satisfies the assumptions of Lemma 3. 6 but not those of Lemma 2. 15. It means that $\Phi_{11}(t, G) \equiv 0$, and hence in that case (3. 27) holds as well.

Finally we note that any vertex of $G$ can be taken as $v_1$. So in view of (3. 13), (3. 27) and the definition of $\nu_\epsilon$ we have

$$\begin{aligned} \text{Var}\ |\pi^0\nu_{max} - \pi^0\nu| &< (s/2)\Phi_{11}(t_\epsilon, G) \\ &\leq 2\epsilon^{1/2}(c^+)^{1/2} s^2\alpha\lambda^{2\alpha+(3/2)}(1 - \lambda^{-1})^{-1} \end{aligned}$$

The lemma is proved.

Lemma 3. 7. *For any graph* $G$ *with* $s(G) = s < \infty$, $\alpha(G) = \alpha < \infty$, *and* $\lambda(G) = \lambda > 1$ *the following inequality holds*

$$\lambda^{\alpha+1} \geq (\lambda - 1)s \tag{3. 28}$$

Proof: Denote by $M$ the incidence matrix of $G$ and observe that in view of the definition of $\alpha$ all the elements of the matrix $M = M + M^2 + \ldots + M^\alpha$ are positive integers. Hence the greatest eigenvalue of $M$ cannot be smaller than $s$. But this eigenvalue is clearly equal to $\lambda + \lambda^2 + \ldots + \lambda^{\alpha} = (\lambda - 1)^{-1}\lambda^{\alpha+1}$ which gives (3. 29).

Proof of Lemma 3. 1. : Let $Y = \bigcap_n Y_n$ and $G_n = G_n(Y)$ be the graph associated with the 1-Markov set $\varphi^{n-1}Y_n$, $n = 1, 2, \ldots$ (see Section I). We have two measures defined on the path space $Z(G_n)$ : the former of them, say $\nu$, is the image of the given measure $\mu \in \mathcal{P}_{max}(Y)$ with respect to the mapping $\varphi^{n-1}: Y_n \to Z(G_n)$, the latter one is the maximal measure $\nu_{max} \in \mathcal{P}(Z(G_n))$. It is clear that if $\mu \in \mathcal{P}_{max}(Y)$, $\mu_n \in \mathcal{P}_{max}(Y_n)$, then

$$\operatorname{Var} |\pi^n \mu_n - \pi^n \mu| = \operatorname{Var} |\pi^0 \nu_{max} - \pi^0 \nu|, \quad n = 1, 2, \ldots \tag{3. 29}$$

Set $s_n = s(G_n)$. The assumption $\lambda(Y) > 1$ implies that $\lim_{n\to\infty} s_n = \infty$, and in view of Lemma 3. 7, $\lim_{n\to\infty} \alpha_n = \infty$ . By definition

$$\rho_n = h(T, Y_n) - h(T, Y) = h_{\nu_{max}}(T_{G_n}) - h(T_{G_n})$$

If $n$ is sufficiently large and such that (1. 4) holds, then

$$\rho_n \le \bar{c}\, \lambda_n^{-11\alpha_n - 5} (\lambda_n^{1/\alpha_n} - 1)^2\, 6 s_n^2 (2s_n + 1) \tag{3. 30}$$

Indeed, it is easy to verify that

$$\lambda_n^{1/\alpha_n} - 1 \ge \alpha_n^{-1} \ln \lambda_n \ge s_n^{-1} \ln \lambda_n$$

Hence with (1. 4) and (3. 28) in mind we have

$$\bar{c}\, \lambda_n^{-11\alpha_n - 5} (\lambda_n^{1/\alpha_n} - 1)^2\, 6 s_n^2 (2s_n + 1) \ge (\bar{c}\ 13)(\lambda_n - 1)^5 (\ln^2 \lambda_n) \lambda_n^{-16\alpha_n - 10}$$
$$\ge \lambda_n^{-(16+\gamma)\alpha_n} \ge \rho_n$$

provided that $n$ is large.

Now we shall use Lemma 3. 6. Inequality (3. 30) makes it

possible to substitute $\rho_n$ for $\epsilon$ into (3. 9), setting simultaneously $s = s_n$, $\alpha = \alpha_n$ and $\lambda = \lambda_n$. Due to (1. 4) this yields

$$\mathrm{Var}\ |\pi^0 \nu_{max} - \pi^0 \nu| \leq (c^+)^{1/2} (\lambda_n - 1)^{-4} \lambda_n^{(-7\alpha_n + 11)/2}$$

which together with (2. 29) leads to the statement of Lemma 3. 1.

## IV. TWO EXAMPLES

In this section we shall use some definitions and notations of Section I.

Proposition 4. 1. Let $Z$ be a Markov set with $h(T, Z) > 0$ and $\alpha(G(Z)) < \infty$. Then for any sequence $\{\epsilon_n\}$, $\epsilon_n > 0$, there exists a closed shift-invariant subset $Y \subset Z$ such that $h(T, Y) > 0$ and

(i) $\alpha_n(Y) \leq 2n$ for every sufficiently large $n$;

(ii) $\rho_n(Y) \leq \epsilon_n$ for infinitely many $n$;

(iii) there are no periodic sequences in $Y$.

The proof of Proposition 4. 1 consists of several steps. First of all we introduce some auxiliary concepts and notation used below.

Let $G$ be a graph and $W$ a subset of the set of its vertices. We shall denote by $G \setminus W$ the complete subgraph of $G$ containing all vertices of $G$ except those belonging to $W$. Let $a^{(1)} = (v_1^{(1)}, \ldots, v_n^{(1)})$ and $a^{(2)} = (v_1^{(2)}, \ldots, v_m^{(2)})$ be two paths in $G$ such that $(v_n^{(1)}, v_1^{(2)})$ is also a path in $G$. Then $(v_1^{(1)}, \ldots, v_n^{(1)}, v_1^{(2)}, \ldots, v_m^{(2)})$ is a path in $G$ which will be denoted by $a^{(1)}a^{(2)}$. The notation $a^{(1)}a^{(2)} \ldots a^{(k)}$ will have the analogous interpretation. If $a^{(i)} = a$, for $i = 1, 2, \ldots, k$ we shall write $a^k$ instead of $a^{(1)} \ldots a^{(k)}$ and call $a^k$ the k-power of the path $a$. Let $a^{(1)} = (v_1^{(1)}, \ldots, v_n^{(1)})$ and $a^{(2)} = (v_1^{(2)}, \ldots, v_m^{(2)})$ be two paths in $G$. We shall say that $a^{(2)}$ is contained in $a^{(1)}$ (or $a^{(2)}$ is a segment of $a^{(1)}$) is one can find $i$ and $j$, $1 \leq i < j = i + m \leq n$, such that $v_1^{(2)} = v_{i+1}^{(1)}$, $v_2^{(2)} = v_{i+2}^{(1)}, \ldots,$ $v_m^{(2)} = v_j^{(1)}$. If $i = 1$ (respectively $j = n$) we refer to $a^{(2)}$ as an initial (respectively, terminal) segment of $a^{(1)}$. We let $\ell(a)$ denote the length of a path a, i. e., the number of vertices in a.

Lemma 4. 2. Let G be a finite connected graph with $\lambda(G) = \lambda > 1$. Let us fix for $k = 1, 2, \ldots$ an arbitrary path $a^{(k)}$ in G of length $\ell_k$ so that $\ell_{k+1} > \ell_k$. Then

$$\lim_{k \to \infty} \lambda(\varphi^{\ell_k} G \setminus a^{(k)}) = \lambda(G) \tag{4. 1}$$

Proof: We make use of a result of V. A. Malyshev ([8], Theorem 4). The corollary to this result which we need here can be formulated as follows. Consider a stationary finite Markov chain of positive entropy and with only one class of states. Let $a^{(k)}$ be a sequence of states of length $\ell_k$, $k = 1, 2, \ldots$, and suppose that $\ell_{k+1} > \ell_k$ and the probability of $a^{(k)}$ is positive. Denote by $p(a^{(k)}, n)$ the probability that $a^{(k)}$ does not appear within the time interval $[1, n]$. Then

$$\lim_{k \to \infty} \lim_{n \to \infty} n^{-1} \ln p(a^{(k)}, n) = 0 \tag{4. 2}$$

To prove the lemma consider the maximal measure $\nu_{max} \in P(Y)$ corresponding to the Markov set Y(G). According to Section I, $\nu_{max}$ generates the Markov chain with the transition probabilities given by (1. 1). Clearly, one can apply (4. 2) to this Markov chain. It follows from (1. 1) that for any path c in G of length n

$$c'\lambda^{-n} \leq \nu_{max}(\tilde{c}) \leq c''\lambda^{-n} \tag{4. 3}$$

where $\tilde{c}$ is an n-dimensional cylinder set corresponding to c and $c', c''$ are positive constants (independent of c and n).

Let $B(n, a^{(k)})$, $n \geq \ell_k$, be the set of paths in G of length n none of which contains $a^{(k)}$, and $N(n, a^{(k)})$ the number of elements in $B(n, a^{(k)})$. It is not hard to show that

$$\lambda(\varphi^{\ell k} G \setminus a^{(k)}) = \lim_{n \to \infty} n^{-1} \ln N(n, a^{(k)}), \quad k = 1, 2, \ldots \tag{4.4}$$

Due to (4. 3)

$$c' \lambda^{-n} N(n, a^{(k)}) \leq \sum_{c \in B(n, a^{(k)})} \nu_{max}(\tilde{c}) \leq c'' \lambda^{-n} N(n, a^{(k)}) \tag{4.5}$$

By (4. 2) it follows that

$$\lim_{k \to \infty} \lim_{n \to \infty} n^{-1} \sum_{c \in B(n, a^{(k)})} \nu_{max}(\tilde{c}) = 0 \tag{4.6}$$

Finally, (4. 4) - (4. 6) imply (4. 1), so Lemma 4. 2 is proved.

Definition. Let $a$ be an initial segment of a path $b$ in $G$. We shall say that $a$ is a determining initial segment of $b$ provided that any path in $G$ of the form $aa'$ with $\ell(aa') = \ell(b)$ coincides with $b$. Similarly, a terminal segment $\bar{a}$ of $b$ will be called a determining terminal segment of $b$, if any path in $G$ of the form $\bar{a}'\bar{a}$ with $''(\bar{a}'\bar{a}) = \ell(b)$ coincides with $b$.

Lemma 4. 3. Let $G$ be a finite graph with $\alpha(G) < \infty$, $\lambda(G) > 1$. Let $a$ be a cycle in $G$ of length $\ell(a) = \ell$ which is a power of no other cycle. Then for

$$k \geq 6 + \ell^{-1} \alpha(G) \tag{4.7}$$

there exists a subgraph $G'$ of the graph $\varphi^{k\ell} G$ such that $a^k$ is not a vertex of $G'$ and

$$\lambda(G') = \lambda(\varphi^{k\ell} G \setminus a^k) \tag{4.8}$$

$$\alpha(G') \leq (k+1)\ell + \alpha(G) \tag{4.9}$$

<u>Proof:</u> $1^{\circ}$. Let $k$ satisfy (4. 7), and $A$ be the family of paths in $G$ consisting of $a^{k}$ and all paths $b$ of the form $b = a'a^{k-1}a''$, $\ell(b) = k\ell$, where at least one of the paths $a'$, $a''$ is a determining (initial or terminal, respectively) segment of $a$. We define $G'$ to be the complete subgraph of $\varphi^{k\ell}G$ containing all vertices of $\varphi^{k\ell}G$ except those belonging to $A$ (one should recall that by definition any path in $G$ of length $k\ell$ is a vertex of $\varphi^{k\ell}G$ and vice versa). Evidently any doubly infinite path in $\varphi^{k\ell}G$ containing a vertex from $A$ also contains $a^{k}$. It follows from this by entropy arguments that

$$\lambda(G') = \lambda(\varphi^{k\ell}G\setminus A) = \lambda(\varphi^{k\ell}G\ a^{k})$$

which gives (4. 8). Now proceed to (4. 9).

$2^{\circ}$. We shall say that a path $c$ in $G$ satisfies condition $(A_{in})$ (respectively, $(A_{term})$) if no initial (respectively, terminal) segments of $c$ has the form $a'a^{2}$ (respectively, $a^{2}a''$), where $a'$ (respectively, $a''$) is a terminal (respectively, initial) segment of $a$.

We shall say that a path $c$ in $G$ satisfies condition (B) is no path belonging to $A$ is contained in $c$.

Let $c^{(o)}, c^{(1)}$ be two paths in $G$ such that $c^{(o)}$ satisfies $(A_{term})$, $c^{(1)}$ satisfies $(A_{in})$ and both satisfy (B). Due to the condition $\alpha(G) < \infty$ there is a path $\overline{c}$ in $G$ such that $\ell(\overline{c}) \leq \alpha(G)$ and $c^{(o)}\overline{c}c^{(1)}$ is a path in G. This path satisfies (B). To prove this, suppose that there exists a path $b \in A$ contained in $c^{(o)}\overline{c}c^{(1)}$. Taking into account the conditions on $c^{(o)}, c^{(1)}$ and the definition of $A$ it is easy to verify that $b$ settles down into $c^{(o)}\overline{c}c^{(1)}$ in such a way that the length of the intersection of $b$ with each of $c^{(o)}, c^{(1)}$ is less than $3l$. Then $k\ell = \ell(b) < 6l + \ell(\overline{c}) \leq 6l + \alpha(G)$ which contradicts (4. 7).

$3^{\circ}$. Henceforth we shall use the following fact. If $\lambda(G) > 1$,

$\alpha(G) < \infty$, and if a is a cycle in G then a has a non-determining initial segment and a non-determining terminal segment. To show this one can first consider the case where there exists a vertex v of G which a doesn't pass through. In this case it suffices to find a path leading from v to some vertex v which a passes through and also a path from v to v. Such paths exist because of the connectivity of G. In the opposite case our assertion follows from the condition $\lambda(G) > 1$.

Now suppose a path c in G satisfies (B) but does not satisfy $(A_{term})$. Then $c = d'a^m d$, where d is an initial segment of a, and $2 \leq m \leq k - 1$. (Suppose m to be chosen as large as possible). If d is a non-determining initial segment of a, then there is a path e in G such that de is a path in G, de $\neq$ a, and $\ell(de) = \ell(a) = \ell$. Consider the path $ce = d'a^m de$ whose length is clearly no more than $\ell(c) + \ell - 1$. This path satisfies $(A_{term})$. Indeed, otherwise ce would contain a segment f of the form $f = a^2$ which would in its turn contain a segment coinciding with a and being non-initial and non-terminal with respect to f. But this is impossible unless a is a power of another cycle. For the same reason ce satisfies (B). If d is a determining initial segment of a, then necessarily $m \leq k - 2$ (otherwise $c \in A$). As has been proved above, the cycle a has an initial segment, say d", which is non-determining (clearly, d" is an initial segment of d). Hence one can continue c to get the path $c' = d'a^{m+1}d''$ which clearly satisfies (B) and whose length is no more than $\ell(c) + \ell - 1$. Now applying to c' the above arguments give a path c'e', $\ell(c'e') \leq \ell(c) + 2l - 2$, satisfying both (B) and $(A_{term})$.

Thus it is proved that for any path c in G satisfying condition (B) there is a path $c^{(o)}$ in G satisfying both $(A_{term})$ and (B) such that $\ell(c^{(o)}) \leq \ell(c) + 2l - 2$, c is an initial segment of $c^{(o)}$. The same arguments show that given a path c' in G satisfying (B) there is a path $c^{(1)}$ in $G$ satisfying both $(A_{in})$ and (B) and such that $\ell(c^{(1)}) \leq \ell(c') + 2l - 2$, and $c'$ is a terminal segment of $c^{(1)}$. Using the result of $2^o$ one can find a path $\bar{c}$ in $G$ such that $c^{(2)} = c^{(o)}\bar{c}\,c^{(1)}$ is a path in $G$ satisfying (B) and

$$\ell(c^{(2)}) \leq \ell(c) + \ell(c') + \alpha(G) + 4l - 4$$

In particular, if $c$ and $c'$ are any two paths in $G$ of length $k\ell$ (i. e., any two vertices of $\varphi^{k\ell}G$) satisfying (B) (that is, $c, c' \in A$) then there exists a path $c^{(2)}$ in $G$ satisfying (B) and such that $c$ is an initial segment of $c^{(2)}$, $c'$ is a terminal segment of $c^{(2)}$, and

$$\ell(c^{(2)}) \leq (2l + 1)\ell + \alpha(G) - 4$$

Taking the successive segments of $c^{(2)}$ gives a path in $\varphi^{k\ell}G\setminus A$ leading from $c$ to $c'$ whose length is no more than $(2k+1)\ell + \alpha(G) - 4 - k\ell + 1 < (k + 1)\ell + \alpha(G)$. Hence (4. 9) and so the proof of the lemma is complete.

<u>Proof of Proposition 4. 1.</u>: We shall construct the set $Y \subset Z$ to be the intersection of Markov sets $Y_n$ of increasing order, $Y \subseteq Y_n \subseteq Y_{n-1} \subseteq Z$ [see (1. 2)], where the sequence $Y_n$ is uniquely determined by $Y$ and vice versa. Let $G_n(Y)$ be defined as in (1. 3). It is easy to see that $G_{n+1}(Y)$ is a subgraph of $G_n(Y)$. For our purpose it is enough to produce for any sequence $\{\delta_n\}$ of positive numbers a sequence of graphs $\{G_n\}$, $n = 0, 1, \ldots$, such that: (0) $G_o = G(Z)$; (1) $G_{n+1}$ is a complete (not necessarily proper) subgraph of $G_n$, $n = 0, 1, \ldots$; (2) $\alpha(G_n) \leq 2n$ for sufficiently large $n$; (3) $\lambda(G_n) - \lim_{i\to\infty} \lambda(G_i) \leq \delta_n$ for infinitely many $n$; (4) given any cycle $c$ in $G$ there is a number $n$ such that $c^n$ is not a vertex of $G_n$;

(5) $\lim_{i\to\infty} \lambda(G_i) > 1$.

For any cycle $c$ in $G_o$ denote by $C(c)$ the collection of all cycles in $G_o$ which can be obtained from c by permutations. If c is a power of no other cycle, then so does every $c' \in C(c)$. Hence we have a property of $C = C(c)$ which we shall refer to as simplicity. Number all simple C in such a way that along the sequence $\{C_n\}$ obtained the length $\ell_n$ of the cycles $c \in C_n$ does not decrease.

Now we shall construct a subsequence $\{G_{n_k}\}$, $k = 0, 1, \ldots$, $n_o = 0$, of the sequence of graphs we wish to obtain, and set

$$G_n = \varphi^{n-n_{k-1}} G_{n_{k-1}}, \quad n_{k-1} \leq n < n_k, \quad k \geq 1.$$

Our aim will be to satisfy the following conditions for $k = 1, 2, \ldots$.

$(1_k)$ $G_{n_k}$ is a complete subgraph of $\varphi^{n_k - n_{k-1}} G_{n_{k-1}}$;

$(2_k)$ $\alpha(G_{n_k}) \leq 2n_k$;

$(3_k)$ $\lambda(G_{n_{k-1}}) - \lambda(G_{n_k}) \leq \min(\delta_1/2^{k-1}, \ldots, \delta_k/2)$;

$(4_k)$ given any $q \leq k$ there exist $r > 0$ and $c \in C_q$ such that $r\ell_q \leq n_k$, and $c^r$ is not a vertex of $G_{r\ell_q}$.

It is easy to check that $(1_k) - (4_k)$, $k = 1, 2, \ldots$, imply (0) - (4) above.

The sequence $G_{n_k}$ will be constructed by induction. Clearly, we may assume $\delta_n$ to satisfy $\sum_{n=1}^{\infty} \delta_n < \lambda(Z) - 1$ which implies (5). Suppose for some $k > 0$, $G_{n_k}$ is defined satisfying $(1_k) - (4_k)$. Let q' denote the least of the integers subject to the restriction that there is an integer p such that $p\ell_q \geq n_k$ and $c^p$ is a vertex of the graph $\varphi^{p\ell_q - n_k} G_{n_k}$ for every $c \in C_q$. Let p' denote the least p for which the above is true with $q = q'$. It follows from $(4_k)$ that $q' \geq k + 1$. Fix an arbitrary $c' \in C_{q'}$. Due to $(1_k)$ and the definition of q' the consecutive segments of length $(\ell_{q'})^p$ in the path $(c')^{p'+1}$ constitute a cycle $\bar{c}'$ of length $\ell_{q'}$ in the graph $G'_{n_k} = \varphi^{p'\ell_q - n_k} G_{n_k}$ which is a power of no other cycle.

Now apply Lemma 4.2 to the graph $G'_{n_k}$ and the sequence $\bar{c}'$, $(\bar{c}')^2, (\bar{c}')^3, \ldots$ of its paths. According to this lemma, if m is large enough, then

$$\lambda(\varphi^{m\ell_{q'}} G'_{n_k}) - \lambda(\varphi^{m\ell_{q'}} G'_{n_k} \setminus (\bar{c}')^m) \leq$$

$$\leq \min(\delta_1/2^{k+1}, \delta_2/2^k, \ldots, \delta_k/2^2, \delta_{k+1}/2) \qquad (4.10)$$

Using Lemma 4.3 one can find for sufficiently large m a complete subgraph G' of $\varphi^{m\ell_{q'}} G'_{n_k}$ such that

$$\lambda(G') = \lambda(\varphi^{m\ell_{q'}} G'_{n_k} \setminus (\bar{c}')^m \qquad (4.11)$$

$$\alpha(G') \leq (m+1)\ell_{q'} + \alpha(G'_{n_k}) \qquad (4.12)$$

and $(c')^m$ is not a vertex of G'. Let $n_{k+1} = (m+p')\ell_{q'}$, $G_{n_{k+1}} = G'$. We state that $G_{n_{k+1}}$ satisfies $(1_{k+1}) - (4_{k+1})$. The properties indicated in $(1_{k+1})$ and $(4_{k+1})$ follow immediately from the construction. To check $(2_{k+1})$ and $(3_{k+1})$ we shall use the equalities

$$\lambda(\varphi^{m\ell_{q'}} G'_{n_k}) = \lambda(G'_{n_k}) = \lambda(G_{n_k}) \qquad (4.13)$$

$$\alpha(G'_{n_k}) = \alpha(G_{n_k}) + p'\ell_{q'} - n_k \qquad (4.14)$$

Noe $(3_{k+1})$ follows from (4.10), (4.11), (4.13), and $(2_{k+1})$ follows from (4.12) and (4.14). The proof of Proposition 4.1 is complete.

<u>Proposition 4.4.</u> <u>Given a constant</u> $a > 0$ <u>there is a closed shift-invariant subspace</u> Y <u>of a sequence space</u> X <u>such that</u> $\lambda(Y) \geq a$, $\alpha_n(Y) \leq 3n$, $\rho_n(Y) \leq \text{const}\,(a^{-2n})$, <u>and</u> P(Y) <u>contains exactly two ergodic maximal measures.</u>

<u>Proof:</u> One may obviously assume that a is an integer. Set $s = 2a$,

and $X = V^{\mathbb{Z}}$, where $V$ consists of $s$ elements labelled as follows: $v_1^{(1)}, v_2^{(1)}, \ldots, v_a^{(1)}, v_1^{(2)}, v_2^{(2)}, \ldots, v_a^{(2)}$. Denote by $Y^{(i)}$, $i = 1, 2$, the set of all sequences in $X$ containing $v_1^{(i)}, \ldots, v_a^{(i)}$ only, and by $Y^{(1,2)}$ the set of all sequences $y = \{y_i\}_{-\infty}^{\infty}$ of the form $y_i = v_1^{(1)}$ for $i \leq i_o$, $y_i = v_1^{(2)}$ for $i > i_o$ or $y_i = v_1^{(2)}$ for $i \leq i_o$, $y_i = v_1^{(1)}$ for $i > i_o$, $-\infty < i_o < \infty$. Set $Y = Y^{(1)} \cup Y^{(2)} \cup Y^{(1,2)}$.

It is easy to see that $Y$ is closed and shift-invariant. Moreover $h(T, Y) = \log a$ and there are exactly two ergodic measures $\mu^{(1)}, \mu^{(2)} \in P(Y)$ such that $h_{\mu^{(i)}}(T, Y) = h(T, Y)$, $i = 1, 2$, namely, $\mu^{(i)}$, $i = 1, 2$, is the Bernoulli measure on $Y^{(i)}$ with the uniform 1-dimensional distribution. Further $\alpha_n(Y)$ is equal to $3n-1$ and is attained, for instance, on the following pair of vertices of $G_n(Y)$: $(v_2^{(1)})^n$ and $(v_2^{(2)})^n$.

We now turn to estimation of $\lambda_n(Y)$. Let $G^{(i)}$, $i = 1, 2$, be the complete graph with the vertices $v_1^{(i)}, v_2^{(i)}, \ldots, v_a^{(i)}$. Denote by $v$ the vertex of $\varphi^{n-1}G^{(1)}$ (and simultaneously of $G_n(Y)$) of the form $(v_1^{(1)})^n$. Let $f_k^{(1)}$ be the number of the cycles of length $k$ in $\varphi^{n-1}G^{(1)}$ starting with $v$ and running through $v$ exactly once, $m_k^{(1)}$ the number of all cycles of length $k$ in $\varphi^{n-1}G^{(1)}$ starting with $v$, and $f_k$ the number of the cycles of length $k$ in $G_n(Y)$ starting with $v$ and running through $v$ exactly once. Let

$$F_n^{(1)}(t) = \sum_{k=1}^{\infty} f_k^{(1)} t^k, \quad M_n^{(1)}(t) = \sum_{k=1}^{\infty} m_k^{(1)} t^k, \quad F_n(t) = \sum_{k=1}^{\infty} f_k t^k$$

One has

$$f_k = \begin{cases} f_k^{(1)} & \text{for } 1 \leq k \leq 2n-2, \\ f_k^{(1)} + 1 & \text{for } k = 2n-1 \text{ or } 2n, \\ f_k^{(1)} + m_{k-2n}^{(1)} & \text{for } k > 2n \end{cases} \tag{4.15}$$

To prove this it suffices to observe that $\varphi^{n-1}G^{(1)}$ and $\varphi^{n-1}G^{(2)}$ can be viewed as isomorphic (in a natural sense) subgraphs of $G_n(Y)$ and that for $k = 1, 2, \ldots, 2n-1$ every cycle of length $k$ in $G_n(Y)$ running through $v$ is a cycle in $\varphi^{n-1}G^{(1)}$ as well, while for $k = 2n-1$ and $k > 2n$ there are also cycles of the above type, containing the vertices $(v_1^{(2)})^{n-1}v_1^{(1)}$ and $(v_1^{(2)})^n$, respectively. Using (4. 15) and Lemma 2. 5 we get:

$$F_n(t) = F_n^{(1)}(t) + t^{2n-1}(1 + t + tM_n^{(1)}(t))$$

$$= F_n(t) = F_n^{(1)}(t) + t^{2n-1}[1+t+tM_n^{(1)}(t)]$$

$$= F_n^{(1)}(t) + t^{2n-1}[1+t+\frac{tF_n^{(1)}(t)}{1-F_n^{(1)}(t)}] \qquad (4.\ 16)$$

As can be checked immediately $m_k^{(1)} = 1$ for $1 \leq k \leq n$, and $m_k^{(1)} = a^{k-n}$ for $k \geq n + 1$. Hence

$$M_n^{(1)}(t) = t[1 - at + (a-1)t^n]/(1 - t)(1 - at)$$

and

$$F_n^{(1)}(t) = M_n^{(1)}(t)[1 + M_n^{(1)}(t)]^{-1}$$

$$= t[1-at+(a-1)t^n]/[1-at + (a-1)t^{n+1}] \qquad (4.\ 17)$$

It is known ([7]) that $\lambda_n^{-1}(Y)$ coincides with the unique positive root of the equation $F_n(t) = 1$. We denote this root by $\tau(n)$. Due to (4. 16) and (4. 17), the above equation takes the form

$$[1 - F_n^{(1)}(t)]^2 - t^{2n-1}[1 - F_n^{(1)}(t)] - t^{2n} = 0.$$

Evidently, $0 < F_n^{(1)}(\tau(n)) < 1$. Hence

$$1 - F_n^{(1)}(\tau(n)) = \frac{1}{2}(\tau(n))^{2n-1} + [\frac{1}{4}(\tau(n))^{4n-2} + (\tau(n))^{2n}]^{1/2}$$

$$< (\tau(n))^n + (\tau(n))^{2n-1} <> (\tau(n))^n \qquad (4.18)$$

It follows from entropy arguments that

$$\tau(n) < a^{-1}, \qquad \lim_{n \to \infty} \tau(n) = a^{-1} \qquad (4.19)$$

With this in mind one obtains from (4. 18) and (4. 17) by simple transformations that

$$(1 - \tau(n))(1 - a\tau(n)) < 2(\tau(n))^n (1 - a\tau(n)) + 2(a-1)(\tau(n))^{2n+1}$$

This inequality and (4. 19) together imply that

$$a^{-1} - \tau(n) < c(\tau(n))^{2n} < ca^{-2n}$$

where $c$ is an absolute constant. So the proof is complete.

## REFERENCES

1. R. Bowen, Some systems with unique equilibrium states, Math. System Theory, 8, No. 3, 193-202, 1975.
2. D. Ruelle, Statistical Mechanics, New York-Amsterdam, Benjamin, 1969.
3. O. Ore, Theory of Graphs, American Mathematical Society Colloquium Publications, Vol. XXXVIII, Amer. Math. Soc., Providence, R. I., 1962.
4. F. R. Gantmacher, The Theory of Matrices, Chelsea, New York, 1959.
5. W. Parry, Intrinsic Markov chains, Trans. Amer. Math. Soc., 112, No. 1, 55-65, 1964.

6. W. Feller, An Introduction to Probability Theory and its Applications, Volume 1, New York, Wiley, 1964.

7. B. M. Gurevic, Shirt entropy and Markov measures in the path space of a denumerable graph, Soviet Math. Dokl. 11, No. 3, 744-747, 1970.

8. V. A Malysev, Litovsk. Mat. Sbornik. 5, No. 4, 585-591 (1965). (In Russian).

chapter 11

# MARKOV PARTITIONS FOR RATIONAL ENDOMORPHISMS OF THE RIEMANN SPHERE

M. V. Jakobson

## I. DEFINITIONS, STATEMENT OF RESULTS

1. The method of Markov partitions proposed by Ya. G. Sinai [1,2], and developed by R. Bowen [3,4], reduces the study of hyperbolic diffeomorphisms of smooth manifolds (in particular Anosov diffeomorphisms [5]) to the study of quotient systems of finite Markov chains.

A modification of this method adapted to the case of noninvertible maps was used to prove the structural stability of a certain class of polynomial mappings of the Riemann sphere [6, 7].

We shall study, by a similar method, the topological structure of rational mappings (endomorphisms) of the Riemann sphere.

The results of this paper were announced in [8].

2. Dynamical systems generated by rational endomorphisms (in other words the iteration of rational functions) were studied in detail by G. Julia [9] and P. Fatou [10]. The reader is referred to the book of P. Montel [11] for a survey of many of their results.

We state some definitions and results of [9, 10].

Let $z \to R(z)$ be a rational endomorphism, $x_1, x_2, \ldots, x_n$ - a periodic orbit. This orbit is a sink if $|DR^n(x_k)| = \Pi_{i=1}^n |DR(x_i)| < 1$, and a source if $|DR^n(x_k)| > 1$.

A domain of attraction of a periodic sink $\alpha^* = (\alpha_1, \ldots, \alpha_n)$ is an open set $\Delta_{\alpha^*} = \{x : \omega(x) = \alpha^*\}$ ($\omega(x) = \omega$ - limit set of trajectory of x).

A local domain of attraction of $\alpha^* = (\alpha_1, \ldots, \alpha_n)$ is a set $\mathfrak{D}_{\alpha^*}$ which is the union of n components of the points $\alpha_i$ in $\Delta_{\alpha^*}$. One has $\Delta_{\alpha^*} = \bigcup_{n=0}^{\infty} R^{-n}(\mathfrak{D}_{\alpha^*})$.

A critical point of an endomorphism $R(z)$ is a point y such that $DR(y) = 0$.

Fatou proved that any local domain of attraction contains a critical point of an endomorphism $R(z)$ (cf. [11]). This implies that the total number of periodic sinks of all periods for an endomorphism of degree d is no more than $2(d - 1)$.

We denote by $\mathfrak{F} = \mathfrak{F}(R)$ the closure of periodic sources of $R(z)$. $F(R)$ is the perfect set invariant under R and $R^{-1}$.

Using the theory of normal families of analytic functions Fatou proved the equivalence of the following two assumptions about $R(z)$:

$1^a$. <u>There exists</u> k, <u>such that</u> $|DR^k(z)| |\mathfrak{F} > c_0 > 1$;

$1^b$. <u>Each critical point of</u> $R(z)$ <u>belongs to the domain of attraction of some periodic sink</u>.

The hypotheses $1^{a,b}$ imply that the union of the domains of attraction of all sinks of $R(z)$--the set $\bigcup_{\alpha^*} \Delta_{\alpha^*}$-- is an open dense set of the Riemann sphere $S^2$, and $\mathfrak{F} = S^2 \bigcup_{\alpha^*} \Delta_{\alpha^*}$.

The nonwandering set of an endomorphism $R(z)$ satisfying the hypotheses $1^{a,b}$ consists of $\mathfrak{F}$ and a finite number of periodic sinks. We call such endomorphisms hyperbolic.

We denote by $H_d$ the space of all rational endomorphisms of degree d. We obtain an isomorphism of $H_d$ with an open dense subset of $P^{2d+1}(C)$ by identifying the point of projective space with homogeneous coordinates $(a_0, \ldots, a_d, b_0, \ldots, b_d)$ with

$$R(z) = \frac{a_0 + a_1 z + \ldots + a_d z^d}{b_0 + b_1 z + \ldots + b_d z^d}$$

We consider $H_d$ as a metric space with the metric induced by this isomorphism.

We shall prove that hyperbolic endomorphisms are $\Omega$-stable in the space $H_d$ (Theorem 1), and with some additional assumptions on the location of critical points, that they are structurally stable (Theorem 2).

3. Let $s : \Omega_d \to \Omega_d$ be the left semishift of the space of right d-symbol sequences (Bernoulli semishift).

The proof of Theorems 1 and 2 is based on the topological conjugacy of the dynamical system $R(z)|\mathcal{F}$ to some quotient system $(\Omega'_d, s)$ of $(\Omega_d, s)$. Let $\pi : \Omega_d \to \Omega'_d$ be the quotient map. As it was proved in [7], the quotient map $\pi$ can be choosen for a polynomial $P(z)$, so that $\pi$ identifies a finite number of periodic sequences and the inverse images of these sequences. Such identifications we shall call simple identifications. In [12] it was stated that for every rational endomorphism $R(z)$ satisfying the assumptions $1^{a,b}$, there exists an isomorphism $R(z)|\mathcal{F} \approx (\Omega'_d, s)$ with simple identifications.

The examples in Section 3 prove this statement to be wrong. These examples show that the identifications $\pi : \Omega_d \to \Omega'_d$ for rational endomorphisms are more complicated than for polynomials. However for rational mappings as well as for polynomials we have $\underline{\text{card}}\ \pi^{-1}(\omega') < c$ for any $\omega' \in \Omega'_d$ (Theorem 3, analogous to one of the results of [4]).

We note that the hypothesis of Fatou about the density of the set of $1^{a,b}$ -- endomorphisms in $H_d$ is still unproved even for polynomials (cf. [7]).

## II. SYMBOLIC DYNAMICS FOR HYPERBOLIC ENDOMORPHISMS Ω- STABILITY AND STRUCTURAL STABILITY THEOREMS

1. The following properties of any hyperbolic endomorphism follow from the hypotheses $1^{a,b}$ and the representation $S^2 = \bigcup_{\alpha^*} \Delta_{\alpha^*} \cup F$.

(a) There exist a neighborhood $\mathfrak{U} = \mathfrak{U}(R)$ of $\mathfrak{F} = \mathfrak{F}(R)$ and a $k = k(R)$ such that:

($a_1$) $R^{-k}(\mathfrak{U}) \subset \mathfrak{U}$;

($a_2$) $|DR^k(Z)| > c > 1$ for each $z \in U$;

($a_3$) Let $A = \bigcup_i \alpha_i$ be the union of all periodic sinks. Then for each compact set $M$ which verifies the condition $M \cap A = \phi$, there exists $n_0 = n_0(M)$ such that $R^{-n}(M) \subset U$ for $n > n_0$ (this implies $R^{-n}(M) \Rightarrow \mathfrak{F}$ for $n \to \infty$).

(b) Each point $x \in \mathfrak{F}$ is the limit point of the inverse images $R^{-n}(z)$ of any point $z \in S^2$ (with the exception of not more than two points, for instance $0, \infty$ for $R(z) = z^n$ or $\infty$ for polynomials).

(c) There exists $\gamma > 0$ such that from $x, y \in \mathfrak{F}$, $x \neq y$ follows $\rho(R^n(x), R^n(y)) > \gamma$ for some $n$ ($\gamma$ is an expansive constant).

2. We consider an endomorphism $R(z)$ of degree $d$. Let $x \in S^2$, and let $x_1, \ldots, x_d \in R^{-1}(x)$ be inverse images of $x$, and $\ell_1, \ldots, \ell_d$ be smooth curves joining $x$ to $x_i$. By the hypotheses $1^{a,b}$ an arbitrarily small neighborhood of the set $A$ of periodic sinks contains, for sufficiently large $n$, all of the iterates $R^n(y)$ of any critical point $y$. This implies that one can choose the curves $\ell_1, \ldots, \ell_d$ so that they do not contain either critical points and their iterations or sinks.

Let $R_i^{-1}$ be the branches of $R^{-1}$ defined by $R_i^{-1}(x) = x_i$.

We define inductively the curves $\ell_{i_1 i_2 \dots i_n}$. Let $\ell_{i_1 i_2 \dots i_{n-1}}$ be defined for each sequence of length $n-1$. By definition $\ell_{i_1 i_2 \dots i_n} = \ell_{i_1} \cup (\ell_{i_2 \dots i_n})_{i_1}^{-1}$, $(\ell_{i_2 \dots i_n})_{i_1}^{-1} = R_{i_1}^{-1}(\ell_{i_2 \dots i_n})$, where the branch $R_{i_1}^{-1}$ is well defined on $\ell_{i_2 \dots i_n}$. Thus $\ell_{i_1 i_2 \dots i_n} = \ell_{i_1} \cup \ell_{i_2}^{-1} \cup \ell_{i_n}^{-(n-1)}$ where each $\ell_{i_k}^{-(k-1)}$ is the uniquely determined inverse image of $\ell_{i_k}$.

We denote by $z_{i_1 \dots i_n}$ the end point of $\ell_{i_1 \dots i_n}$. It follows from (a) that the lengths of $\ell_{i_k}^{-(k-1)}$ decrease exponentially. Therefore the curve $L_{i_1 \dots i_n \dots} = \ell_{i_1} \cup \ell_{i_1}^{-1} \cup \dots \cup \ell_{i_n}^{-(n-1)} \cup \dots$ has finite length and unique end point, $-z_{i_1 i_2 \dots i_n \dots} = \lim_{n \to \infty} z_{i_1 i_2 \dots i_n}$, which belongs to F because of $(a_3)$.

To any sequence $I = (i_1 i_2 \dots i_n \dots) \in \Omega_d$ corresponds uniquely a point $z_{i_1 i_2 \dots i_n \dots} \in \mathcal{F}$. Thus we have defined a map $\pi : \Omega_d \to \mathcal{F}$, and by construction, $\pi \circ s(i_1 i_2 \dots i_n \dots) = \pi(i_2 \quad i_n \dots) - R(\pi(i_1 i_2 \dots i_n \dots))$. It follows from (a) that $\pi$ is continuous. Using the diagonal process we deduce from (b) that $\pi$ is surjective. Identifying for each $x \in F$ the sequences $I \in \pi^{-1}(x)$ we obtain the quotient space $\Omega_d' \approx \mathcal{F}$ and the isomorphism: $R(z)|\mathcal{F} \approx (\Omega_d', s)$.

3. _Theorem 1. Hyperbolic endomorphisms are_ $\Omega$_-stable in_ $H_d$.

_Proof._ Let $R = R(z)$ be a hyperbolic endomorphism. We denote by $R_\delta$ an endomorphism lying in the ball of radius $\delta$ in $H_d$ centered at R.

The hypothesis $1^b$ is true for sufficiently small perturbation of the coefficients of $R(z)$. Therefore $R_\delta$ remains hyperbolic for small $\delta$.

From $1^{a,b}$ follows

Proposition 1. For any $\epsilon > 0$ there exists $\delta_0 > 0$ such that for $\delta < \delta_0$ we can take in the formulation of properties (a)--(c) for $R_\delta$:

$$\mathcal{U}_\delta = \mathcal{U}, \quad k_\delta = k, \quad |c_\delta - c| < \epsilon, \quad |\gamma_\delta - \gamma| < \epsilon .$$

Having decreased $c$ and $\gamma$ we can set $c_\delta = c$, $\gamma_\delta = \gamma$.

Moreover, for small $\delta_0$ we can choose the curves $\ell_{\delta i} = (x, x_{\delta i}^{-1})$, which we use in the construction of the isomorphism $R_\delta | \mathcal{F}_\delta \approx (\Omega'_d, s)$, close to $\ell_i$.

Such a choice of $\delta_0$ implies the existence of $N$ such that for any $\delta < \delta_0$, $U$ contains the curves $\ell^{-N}_{\delta i_{N+1}} \cup \ell^{-(N+1)}_{\delta i_{N+2}} \cup \ldots$ and the length of $\ell^{-N}_{\delta i_{N+1}} \cup \ell^{-(N+1)}_{\delta i_{N+2}} \cup \ldots$ is less than $\gamma/6$.

Let $z_{\delta i_1 \ldots i_N}$ be the end point of the curve $\ell_{\delta i_1} \cup \ell^{-1}_{\delta i_2} \cup \ldots \cup \ell^{-(N-1)}_{\delta i_N}$

Having decreased $\delta_0$ is necessary, we obtain $\rho(z_{\delta i_1 \ldots i_N}, z_{i_1 \ldots i_N}) < \gamma/6$ for each sequence $I_N = (i_1 \ldots i_N)$.

<u>Proposition 2.</u> If $\delta < \delta_0$ then the quotient space $\Omega'_{\delta d}$ is isomorphic to $\Omega'_d$.

Let $I = (i_1 \ldots i_n \ldots)$, $J = (j_1 \ldots j_n \ldots)$, $\pi(I) = \pi(J) = x \in \mathcal{F}$. We prove that $\pi_\delta(I) = \pi_\delta(J)$, where $\pi_\delta : \Omega_d \to F_\delta$ is the quotient map for $R_\delta$.

We have $\rho(z_{i_1 \ldots i_N}, z_{j_1 \ldots j_N}) < \gamma/3$ because of the choice of $N$. This implies $\rho(z_{\delta i_1 \ldots i_N}, z_{\delta j_1 \ldots j_N}) < 2\gamma/3$ and $\rho(z_{\delta I}, z_{\delta J}) < \gamma$, where $z_{\delta I} = \pi_\delta(I)$, $z_{\delta J} = \pi_\delta(J)$. The same arguments for $\pi \circ s^k(I) = \pi \circ s^k(J) = R^k(x)$ imply $\rho(z_{\delta s^k(I)}, z_{\delta s^k(J)}) < \gamma$ for all $k$. As $\gamma$ is the expansive constant we obtain $\pi_\delta(I) = \pi_\delta(J)$. Because of the choice of $\delta_0$ the proof of the inverse implication: $\pi_\delta(I) = \pi_\delta(J) \Rightarrow \pi(I) = \pi(J)$ is similar.

It follows from Proposition 2, that the homeomorphism $\varphi_\delta = \pi_\delta \circ \pi^{-1} : F(R) \to F(R_\delta)$ is well defined, and $R_\delta \circ \varphi_\delta = \varphi_\delta \circ R$. We finish the proof of Theorem 1 by noticing that $\varphi_\delta \to \mathrm{Id}$ for $\delta \to 0$.

4. <u>Definition</u> (cf. [13]). The point $z_1$ is an associate of the point $z_2$ if $R^p(z_1) = R^k(z_2)$ for some $p, k \in Z$.

Let $R(z)$ be an endomorphism satisfying the following assumption:

5. <u>The critical points</u> $y_1, y_2, \ldots, y_{2d-1}$ <u>of the endomorphism</u> $R(z)$ <u>are not associates of each other.</u>

We note that the set of hyperbolic endomorphisms satisfying 2, is open in $H_d$. If the assumption 2 holds for $R(z)$ and $\delta_0$ is small, then for every $R_\delta$ with $\delta < \delta_0$ there exists a homeomorphism $\psi_\delta : \bigcup_{\alpha^*} \Delta_{\alpha^*} \to \bigcup_{\alpha^*_\delta} \Delta_{\alpha^*_\delta}$ close to Id such that $R_\delta \circ \psi_\delta = \psi_\delta \circ R$ on the union of all domains of attraction (cf. [7]).

For $\delta$ sufficiently small $\varphi_\delta : \mathcal{F} \to \mathcal{F}_\delta$ and $\psi_\delta : \bigcup_{\alpha^*} \Delta_{\alpha^*} \to \bigcup_{\alpha^*_\delta} \Delta_{\alpha^*_\delta}$ agree at periodic sources $t_{\alpha^*}$ lying on the boundaries of $\mathcal{D}_{\alpha^*}$ and at their inverse images $t^{-n}_{\alpha^*}$. It follows from $\bigcup_{\alpha^*} \bigcup_{n=0}^{\infty} t^{-n}_{\alpha^*} = \mathcal{F}$ that $\varphi_\delta$ and $\psi_\delta$ agree on $\mathcal{F}$. Thus

$$x_\delta(z) = \begin{cases} \varphi_\delta(z), & z \in \mathcal{F} \\ \psi_\delta(z), & z \in \bigcup_{\alpha^*} \Delta_{\alpha^*} \end{cases}$$

is a well-defined homeomorphism close to $I_d$ on $S^2$ satisfying the property $x_\delta \circ R = R_\delta \circ x_\delta$.

Thus we have proved the following

<u>Theorem 2.</u> <u>The hyperbolic endomorphisms, satisfying the assumption</u> 2 <u>are structurally stable in</u> $H_d$.

6. The map $\pi : \Omega_d \to \mathcal{F}$ which we used in the proof of theorems concerning $\Omega$-stability and structural stability verifies the following property (see [4] for the case of Markov partitions for diffeomorphisms):

Theorem 3. card $\pi^{-1}(x) < c$.

Proof: There exists $h > 0$ such that for any $x \in U$, $\rho(x_i^{-1}, x_j^{-1}) > h$, where $x_i^{-1}$, $x_j^{-1}$ are the preimages of $x$.

Let $N$ be such that $\mathcal{U}$ contains any curve of the form $\ell_{i_{N+1}}^{-N} \cup \ell_{i_{N+2}}^{-(N+1)} \cup \ldots$ and the length of such a curve is less than $h/2$. We will show that one can set $c = d^N$.

If card $\pi^{-1}(x) > d^N$ then there exists $d^N + 1$ different sequences $I^1, I^2, \ldots, I^{d^N+1} \in \pi^{-1}(x)$. Let $M$ be such that the strings $I_M^k = [i_{k1}, i_{k2}, \ldots, i_{kM}]$, $1 \leq k \leq d^N + 1$ in the sequences $I^k$ are pairwise distinct (obviously $N < M$). Since $d^N <$ card $\{I_M^k\}_{k_1} = d_{k_2}^N + 1$ we can choose $k_1, k_2$ so that the last $N$ coordinates in $I_M, I_M$ agree. Let $k_1 = 1$, $k_2 = 2$.

Then $i_{1M-N+1} = i_{2M-N+1}, i_{1M-N+2} = i_{2M-N+2}, \ldots, i_{1M} = i_{2M}$. Suppose that $i_{1M-N} \neq i_{2M-N}$. Let $z$ be the end point of the curve $\ell = \ell_{1M-N+1} \cup \ell_{1M-N+2}^{-1} \cup \ldots \cup \ell_{1M}^{-(N-1)}$. The end points of the curves $\ell_{1M-N} \cup \ell^{-1}$ and $\ell_{2M-N} \cup \ell^{-1}$ are respectively $z_1 \in R^{-1}(z)$ and $z_2 \in R^{-1}(z)$. As $z_1 \neq z_2$ we have $\rho(z_1, z_2) > h$. On the other hand, by the choice of $N$, $\rho(z_1, z_{I_1'}) < h/2$, $\rho(z_2, z_{I_2'}) < h/2$ where $I_1' = (i_{1M-N}, i_{1M-N+1}, \ldots)$, $I_2' = (i_{2M-N}, i_{2M-N+1}, \ldots)$. From $z_{I_1'} = R^{M-N-1}(x) = z_{I_2'}$ we get $\rho(z_1, z_2) < h$, which is a contradiction.

Thus $i_{1M-N} = i_{2M-N}$. Similarly $i_{1M-N-1} = i_{2M-N-1}, i_{1M-N-2} = i_{2M-N-2}, \ldots, i_{11} = i_{21}$.

Thus $i_{1s} = i_{2s}$ for all $s \in [1, M]$ which contradicts the inequality $I_M^1 \neq I_M^2$, and we are done.

Theorem 3 implies that the preimages $\pi^{-1}(z)$ of periodic points $z = R^k(z) \in \mathcal{F}$ are periodic sequences.

## III. EXAMPLES OF HYPERBOLIC ENDOMORPHISMS WITH A COMPLICATED STRUCTURE OF THE SET $\mathfrak{F}$

1. Example 1. $R(z) = 1 + (z - 1)^2 \cdot \frac{5z - 32}{27z}$

The critical points are: 1, 0, $w = 32/5$, $\infty$ (all simple). The fixed sinks are: $1 = R(1)$, $\infty = R(\infty)$. Since $R(0) = \infty$, $R(w) = 1$ (see the graph of $R(x)$ in Figure 1), the domains of attraction $\Delta_1, \Delta_\infty$ contain all of the critical points of $R(z)$, and by the theorem of Fatou $R(z)$ does not have any other periodic sinks.

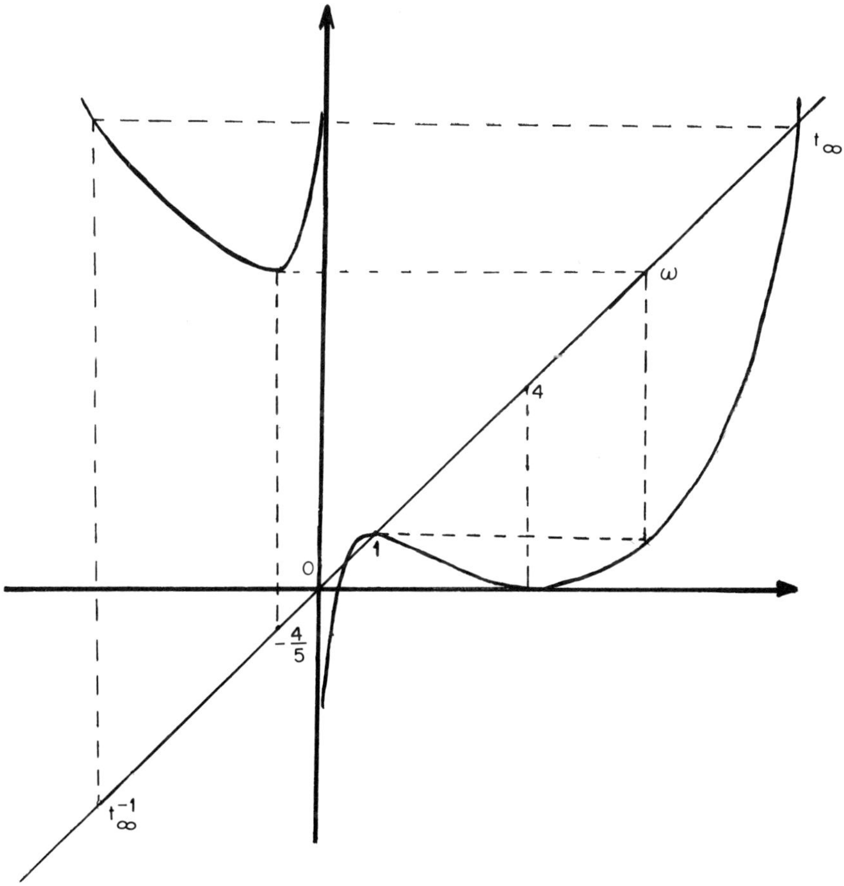


Figure 1

It can be easily shown that the local domain of attraction $\mathfrak{D}_\infty$ does not contain the point $0 \in R^{-1}(\infty)$ and similarly $w \notin \mathfrak{D}_1$. This implies that the domains $\mathfrak{D}_\infty, \mathfrak{D}_1$ and all their inverse images are simply connected.

Since the coefficients of $R(z)$ are real, all domains of attraction are symmetric with respect to the real axis. Hence $\mathfrak{D}_\infty \cap (t_\infty^{-1}, t_\infty) = \phi$ (see Fig. 1).

Two inverse images of the interval $[1, w]$, stick together at the points 1 and $y = -4/5$ and form a closed curve $\Gamma$ which is symmetric with respect to the x-axis and intersects it only at the points 1, y. From $[1, w] \subset (t_\infty^{-1}, t_\infty)$ we conclude that $\mathfrak{D}_\infty \cap \Gamma = \phi$.

Using the isomorphisms $R(z) | \mathfrak{D}_1 \approx z \to z^2 \, | \, |z| = 1 \approx R(z) | \mathfrak{D}_\infty$, and taking into account the location of the preimages of the half-plane $\operatorname{Im} z > 0$, one can prove that the boundaries of the domains $\mathfrak{D}_1$ and $\mathfrak{D}_\infty$ intersect at the points $p, q = \bar{p}$, forming a source of period two.

The location of the domains $\mathfrak{D}_\infty$, $\mathfrak{D}_1$ and some of their inverse images is shown in Figure 2.

We notice that the points of intersection of the boundaries of $\mathfrak{D}_1^{-k}$, $\mathfrak{D}_\infty^{-n}$ $(k, n = 1, 2, \ldots)$ are everywhere dense on each of these boundaries.

Since the boundaries of $\mathfrak{D}_1$ and $\mathfrak{D}_\infty$ intersect at the source of period two (and not at the fixed source), it is easily shown that $R | \mathfrak{F}$ is not isomorphic to any quotient system $(\Omega_3', s)$ with simple identifications.

There exists however a system of the cuts joining the critical points of $R(z)$ (see Figure 2) which allows us to construct an isomorphism of $R | \mathfrak{F}$ with a quotient system of the topological Markov

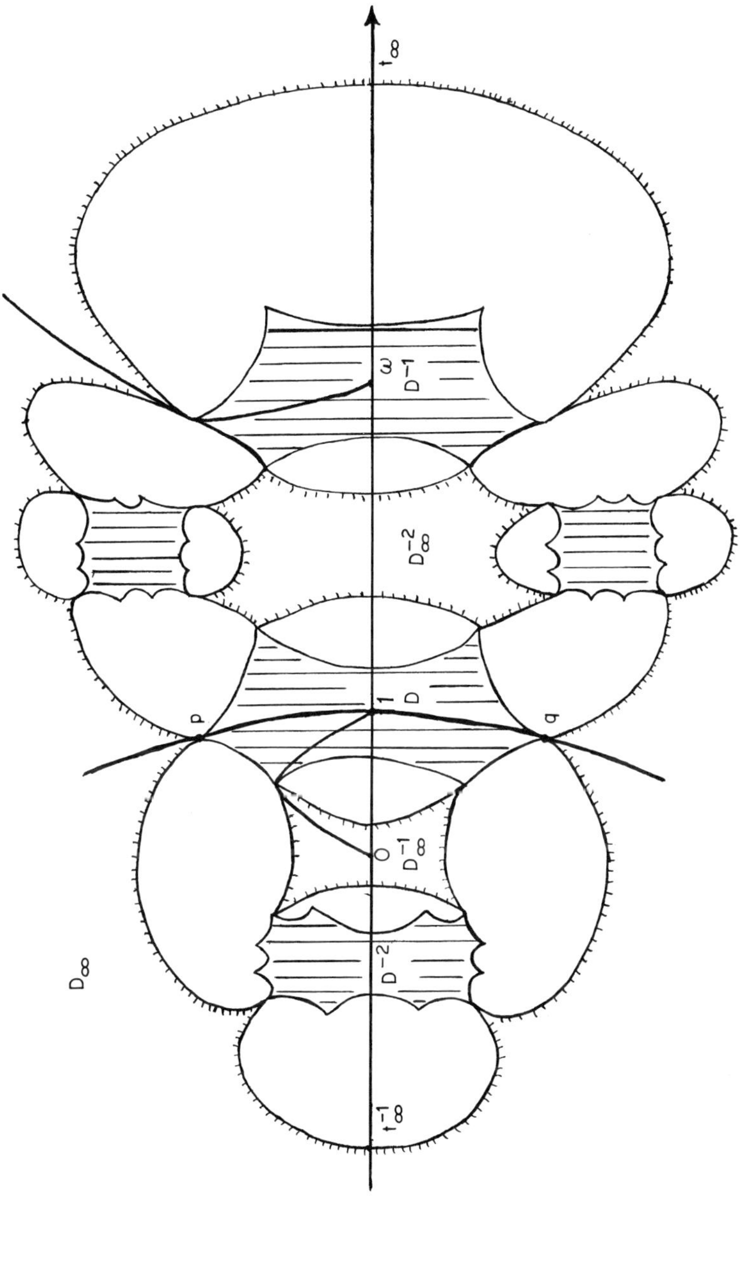


Figure 2

chain with simple identifications. The matrix of this Markov chain is

$$A = \begin{pmatrix} & & & 1 & 1 & 1 \\ & & & 1 & 1 & 1 \\ 1 & 1 & 1 & & & \\ & & & 1 & 1 & 1 \\ 1 & 1 & 1 & & & \\ 1 & 1 & 1 & & & \end{pmatrix}$$

Here the inverse images of the cuts divide $\mathcal{F}$ into the sets $\mathcal{F}_i$, $i = 1, \ldots, 6$, and $a_{ij} = 1$ corresponds to $\mathcal{F}_i \cap \mathcal{F}_j^{-1} \neq \phi$. The quotient map $\pi : \Omega_6 \to \Omega_6'$ identifies the pairs of sequences $(ijij \ldots) \sim (k\ell k\ell \ldots)$ and $(jiji \ldots) \sim (\ell k\ell k \ldots)$ and their preimages.

2. <u>Example 2.</u> $R(z) = \dfrac{z^2(z - w)(z - h)}{(z - z_1)(z - z_2)}$

One can prove that $R(z)$ with the graph $R(x)$ represented in Figure 3 exists, although $w$, $h$, $z_1$, $z_2$ are not computed explicitly.

The critical points are: $0, \infty$ simple, $z, w$ of multiplicity two. We have $DR(y_1) = DR(y_2) = 0$, $z_1 = R(y_1)$, $w = R(y_2)$. It follows from $R(0) = 0$, $R(\infty) = \infty$, $R(w) = 0$, $R(z_1) = \infty$ that $0$ and $\infty$ are the only periodic sinks. One can show as in example 1 that the domains $\mathcal{D}_0$ and $\mathcal{D}_\infty$ are simply connected.

However the closure of $\mathcal{D}_0$ is not simply connected (see Figure 4). The point $t_0$ on the boundary of $\mathcal{D}_0$ plays for $R|\mathcal{D}_0$ the role of the source of period two (the points $a, b$ in Figure 4 are the inverse images of $t_0$. The domains $\mathcal{D}_0^{-k}$ accumulate at the boundary of $\mathcal{D}_\infty$ and the domains $\mathcal{D}_\infty^{-n}$ fill up the space between $\mathcal{D}_0^{-k}$.

As in example 1, the map $R|\mathcal{F}$ is not isomorphic to the quotient system $(\Omega_4', s)$ with simple identification. We notice that any cut, joining $0$ to $\infty$ crosses $\mathcal{F}$ at an infinite number of points (there exists a cut, which crosses $\mathcal{F}$ only at the points of the form $t_0^{-k}$.)

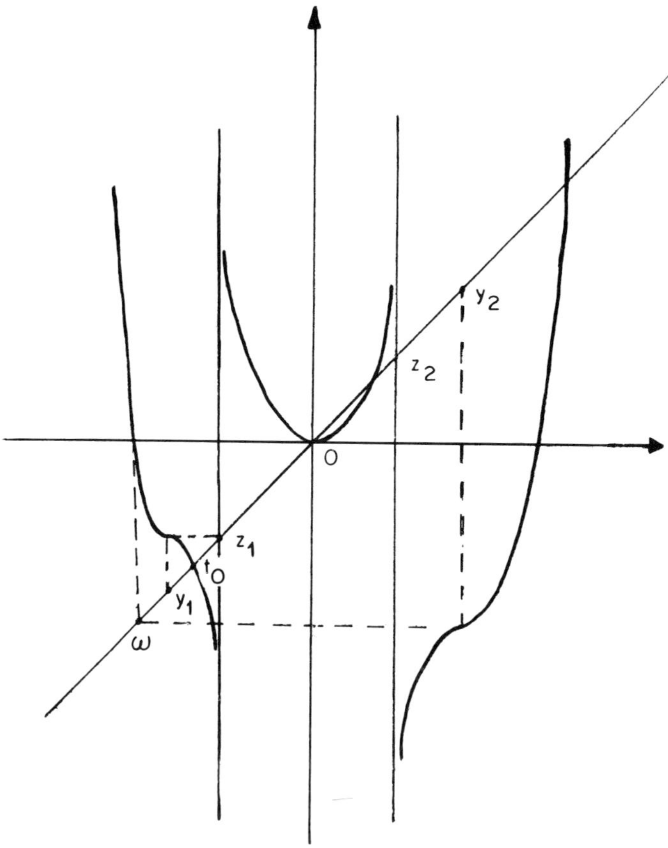


Figure 3

An example of an endomorphism with the property that F and any cut joining the critical points intersect along a nondenumerable set of points is $R(z) = \dfrac{(z-3)^3}{27(z-1)}$.

3. If we study a dynamical system generated by some rational endomorphism R(z) we may suppose that R(z) is hyperbolic (although the hypothesis about the density of the set of hyperbolic endomorphisms in $H_d$ is not proved) and proceed in the following way.

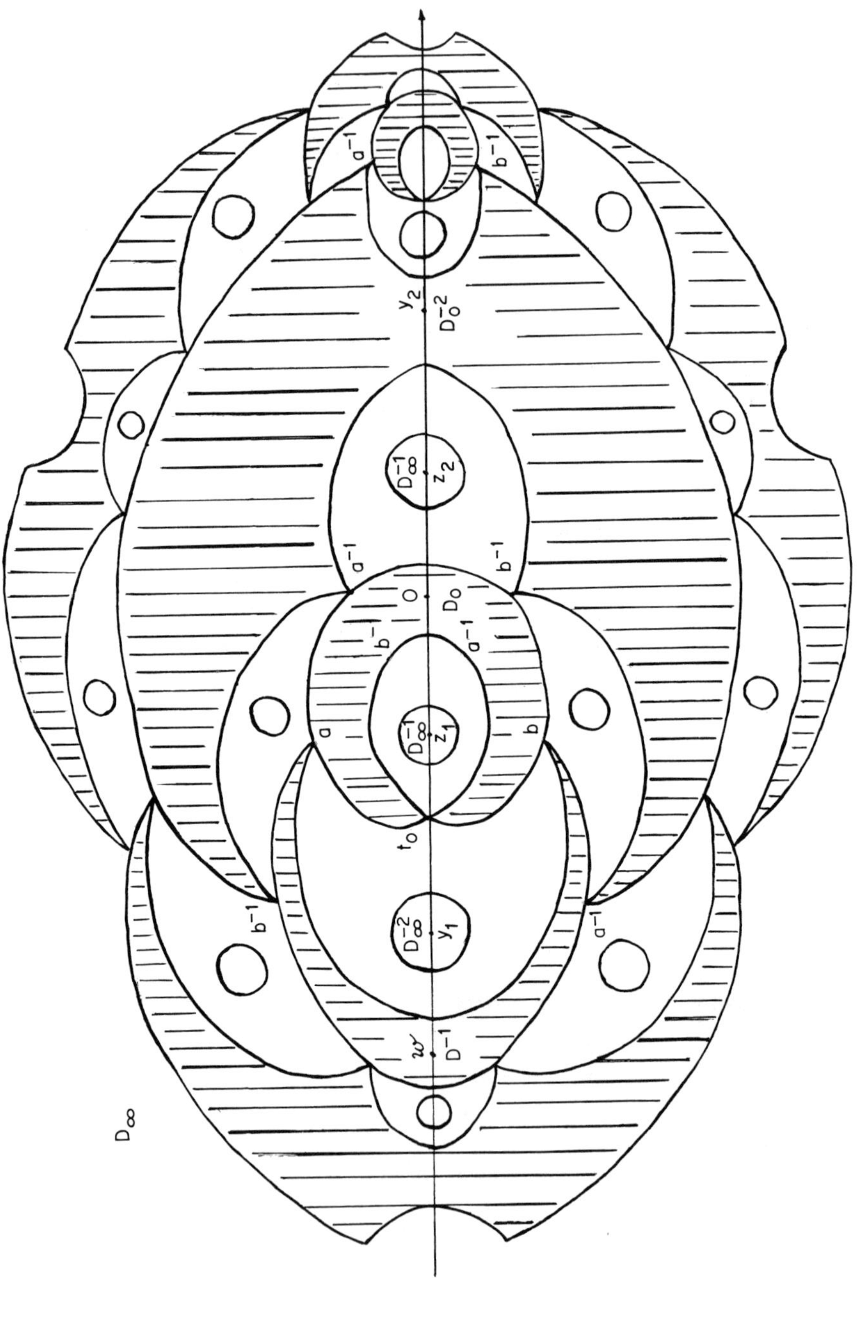

$a^{-1}$
$b^{-1}$
$y_2$
$D_0^{-2}$
$D_\infty^{-1}$
$z_2$
$a^{-1}$
$b^{-1}$
O
$D_0$
$b^{-1}$
$a^{-1}$
a
$D_\infty^{-1}$
$z_1$
b
$t_0$
$b^{-1}$
$D_\infty^{-2}$
$y_1$
$a^{-1}$
$w$
$D^{-1}$
$D_\infty$


Figure 4

(a) We find the critical points of $R(z)$, follow their iterates, and find the periodic sinks they are attracted to.

(b) For any periodic sink $\alpha^*$ we choose a little neighborhood $Q(\alpha^*)$ such that $R^{-1}(Q(\alpha^*)) \supset Q(\alpha^*)$ and we consider the corresponding increasing sequence of inverse images $R^{-k}(Q(\alpha^*))$ which converges to $\mathfrak{D}_{\alpha^*}$. This gives, at least in principle, the possibility to determine step by step, with any degree of precision, the local domains of attraction of all periodic sinks and a finite number of their inverse images.

(c) It is more complicated to determine the points of intersection of the boundaries of $\mathfrak{D}_{\alpha^*}(p, q)$ in example 1, and $t_0$ in example 2). It is helpful to use here the curves $\ell_{i_1} \cup \ell_{i_2}^{-1} \cup \ldots \cup \ell_{i_n}^{-(n-1)}$, as in the proof of the isomorphism $R|\mathfrak{F} \approx (\Omega_d', s)$.

## REFERENCES


1. Ya. Sinai, Markov partitions and C-diffeomorphisms, Func. Anal. and its Appl. 2 (1968), No. 1, 64-89.
2. Ya. Sinai, Construction of Markov partitions. Func. Anal. and its Appl. 2 (1968), No. 3, 70-80.
3. R. Bowen, Markov partitions for axiom A diffeomorphisms, Amer. J. Math. 92 (1970), No. 3, 725-747.
4. R. Bowen, Markov partitions and minimal sets for axiom A diffeomorphism, Trans. AMS 164 (1972). No. 2, 323-331.
5. D. Anosov, Geodesic flows on closed Riemann manifolds with negative curvature, Proc. Steklov Inst. Math. No. 90 (1967).

6. M. Jakobson, Structure of polynomial mappings on a singular set, Math. Sbornik, 77 (1968), No. 1, 105-124.

7. M. Jakobson, On the classification of polynomial endomorphisms of the plane, Math. Sbornik 80 (1969), No. 3, 365-387.

8. M. Jakobson, On the topological classification of rational mappings of the Riemann sphere, Uspehi Math. Nauk 28 (1973), No. 2, 247-248.

9. G. Julia, Mémoire sur l'iteration des fonctions rationnelles, J. Math Pures et Appl. 8 (1918), No. 1, 47-245.

10. P. Fatou, Sur les équations fonctionnelles, Bull. Soc. Math. Fr. 47 (1919), 161-271; 48 (1920), 33-94; 208-314.

11. P. Montel, Lecons sur les familles normales de fonctions analytiques et leurs applications, Paris, Gauthier-Villars (1927).

12. J. Guckenheimer, Endomorphisms of the Riemann sphere, Proc. of Symposia in Pure Math. 14 (1970), 95-123.

13. J. Ritt, On the iteration of rational function, Trans. AMS 21 (1920), No. 3, 348-356.

chapter 12

# THE EXISTENCE OF THE LIMITING DISTRIBUTIONS IN SOME NETWORKS OF COMMUNICATION OF MESSAGES

M. J. Kel'bert

## I. INTRODUCTION

The investigation of the probabilistic character of processes involving the communication of messages in networks is an important problem of science and technology. The first attempts to investigate these characteristics consisted in sequentially applying (from node to node) formulae of queuing theory under the simplest assumptions concerning the processes of message formation, address distributions and characteristics of transmitting equipment. Generally, these assumptions are not fulfilled in practice and the formulae obtained become very cumbersome as the number of nodes increases. Besides, note the following point: every message appearing in the network has a certain length which defines duration of its transmission from an arbitrary node of the network to the next one. Numerical experiments show that application of the formulae of queue theory obtained, under the assumption that the durations of transmission (time of service) of a given message in different nodes are independent, may produce a considerable error. Therefore, the Monte-Carlo method remains the most popular approach to the determination of probability characteristics of processes of communication of messages in networks. Another approach to approximate determination of these characteristics is the asymptotic approach. This approach was suggested in paper [1] (in connection with networks of linear structure); in this paper the limiting characteristics of processes of entry of messages in the nodes of network is studied.

In the paper mentioned above a discrete model of the linear network of communication of messages with a large number of nodes is considered, and conditions for existence of the limiting process of entry of messages when $N \to \infty$, $t \to \infty$ ($N$ is the number of nodes in the network, $t$ is the discrete time) are formulated. In this model $N$ sequentially connected nodes are enumerated as $1, 2, \ldots, N$, and the information is transmitted at times $t = 1, 2, 3, \ldots$, only in the direction of increasing node number. At the initial moment $t = 0$ there are no messages in the network. At times $t = 1, 2, 3, \ldots$, and in the n-th node ($n = 1, 2, \ldots, N$) suppose messages appear, belonging to $K$ different types (no more than $L$ messages of each type) and join the queue, waiting for their turn to be transmitted into the $(n+1)$-th node. Suppose that the messages waiting in the queue more than $T$ moments are lost. If a message has been transmitted into the $(n+1)$-th node, it is to be transmitted into the $(n+2)$-nd node with a certain probability $q$ which is fixed for a given model. This assumption means, in fact, that every message formed in the network is characterized, by its type $k$ and by its address $d$, the difference between the ordinal number of addressee and source of information. The address distribution is the geometric distribution with the parameter $q$, independently of the type of message and the number of the information source. Also, it is assumed that the duration of transmission from any node into the next one is no more than some constant $H$.

In the present paper the existence of the limiting process of reception of messages is proved for a more general model. In particular, the assumption that the messages waiting for service more than T moments are lost can be dropped. Instead of the assumption that the decision about the further transmission of any message is made independently in every node, we suppose that the probabilities of formation of the messages with a large addressee decrease rapidly enough. The assumption, that in each node there can appear simultaneously no more than KL messages, is replaced by restrictions on the average intensity of their arising. Also, the existence of the limiting process of entry of messages in a cyclic network with the fixed number of nodes when $t \to \infty$ is proved, and the relationship between these processes is discussed.

## II. FORMAL DESCRIPTION OF THE MODEL AND FORMULATION OF THE MAIN RESULTS

Let the following systems of random variables:

$$\{\zeta_{n,t}(\omega)\}, \quad \{\tau^{(k)}_{n,t,j}(\omega)\}, \quad \{d^{(k)}_{n,t,j}(\omega)\}, \quad \{\psi^{(k)}_{n,t}(\omega)\},$$

$$\{\theta_{n,t}(\omega)\}, \quad \{\chi_{n,t}(\omega)\}, \qquad n, t, j, k = 1, 2, 3, \ldots$$

be defined on a certain probability space. All random variables under study are independent; $\zeta_{n,t}(\omega)$ have the same distribution function; the distribution functions of $\tau^{(k)}_{n,t,j}(\omega)$, of $d^{(k)}_{n,t,j}(\omega)$ and of $\psi^{(k)}_{n,t}(\omega)$ depend on k only; the random variables $\theta_{n,t}(\omega)$ all have the same distribution function; and the random variables $\chi_{n,t}(\omega)$ all have the same distribution function.

The value of random variables $\zeta_{n,t}(\omega)$ is a $2\nu$-dimensional vector $(\nu = 1, 2, 3\ldots)$, with positive integer coordinates $(k_1, \ldots, k_\nu, z_1, \ldots, z_\nu)$ satisfying the condition $k_1 < k_2 < \ldots < k_\nu$, or else the vector (0, 0). Let $P\{\zeta_{n,t}(\omega) = (0,0)\} = 1 - p$, $0 < p < 1$. We suppose that $\zeta_{n,t}(\omega) \equiv (0,0)$ if $t \leq 0$, $n = 1, 2, 3, \ldots$.

Intuitively, $\zeta_{n,t}(\omega)$ describes the set of messages formed in the n-th node at the moment t. We suppose that in any node there can arise simultaneously only a finite number of messages. Denote by $\nu = \nu(\omega)$ the number of types represented in the set, and by $z_i = z_i(\omega)$ the number of messages of the $k_i$-th type.

The random variables $d^{(k)}_{n,t,j}(\omega)$ have integer positive values. Suppose that $P\{d^{(k)}_{n,t,j}(\omega) \geq m + 1\} \leq qP\{d^{(k)}_{n,t,j}(\omega) \geq m\}$, $m = 1,2,3,\ldots$ where q does not depend on k, $0 < q < 1$.

Intuitively, if $z_i$ messages $(1 \leq i \leq \nu(\omega))$ of the $k_i$-th type were formed at the moment t in the n-th node then the j-th one of them has the duration $\tau^{(k_i)}_{n,t,j}(\omega)$, and leaves the system of communication after being received in the $(n + d^{(k_i)}_{n,t,j}(\omega))$-th node, independently of the time of waiting for its transmission in the intermediate nodes. Therefore, each message is characterized by its type, duration and address. It is natural to assume that the duration of transmission of a message from an arbitrary node into the next one is a random variable depending on its type and duration; this assumption is formalized by using random variables $\psi^{(k)}_{n,t}(\omega)$.

Denote by $\mathbb{Z}^1_+$ the set of nonnegative integers. Assume that there are given non-empty sets A, $B \subset \mathbb{Z}^1_+$, and function $\varphi(a, b)$ : $A \times B \to \hat{H}$, where $\hat{H} = \{1, 2, \ldots, H\}$. Suppose that the values of random variables $\tau^{(k)}_{n,t,j}(\omega)$ belong to the set A and the values of random variables $\psi^{(k)}_{n,t}(\omega)$ to the set B.

An intuitive interpretation is, that if the j-th one of the messages of type k appearing at time $t_0$ in the $n_0$-th node of communication begins its transmission from the n-th node of communication into the (n+1)-th one, $(n \leq n_0 + d^{(k)}_{n_0,t_0,j}(\omega) - 1)$, at the moment t, then the duration of the transmission is equal to $\varphi(\tau^{(k)}_{n_0,t_0,j}(\omega), \psi^{(k)}_{n,t}(\omega))$.

Remark. The model of the linear network of communication of messages, considered in [1], imbeds in our scheme if function $\varphi(a, b)$ depends on the second argument only. Note also that all considerations could easily be generalized for the case when the type of message $k$ takes the values from an arbitrary measurable set.

The random variables $\theta_{n,t}(\omega)$ take the values 0 with probability $1 - q$, and 1 with probability $q$, while the random variables $\chi_{n,t}(\omega)$ take the values 0 with probability $r$ and 1 with probability $1 - r$, $r = p + 2q - 2pq$. They are used to prove auxiliary lemmas.

Construct a system of random variables $\{\xi_{n,t}(\omega)\}$, $n, t = 1, 2, 3, \ldots$ . The random variables $\xi_{n,t}(\omega)$ will take the values $0 = (0, 0, 0, 0, 0)$ and $(k, d, \tau, h, s)$, where $k, d, \tau, h, s$ are integers, $k > 0$, $d > 0$, $\tau \in A$, $0 < h \leq H$, $s \geq 0$.

Intuitively, the random variables $\xi_{n,t}(\omega)$, $t = 1, 2, 3, \ldots$ describe the process of message entry into the $n$-th node of communication. The equality $\xi_{n,t}(\omega) = (k, d, \tau, h, s)$ means that at the moment $t$, the $n$-th node of communication received a message, of the type $k$ and length $\tau$, which waited for transmission at the $(n-1)$th node to the $n$-th one for $h$ moments, and will leave the system after it is received in the $(n+d-1)$-th node. The equality $\xi_{n,t}(\omega) = 0$ means that at the moment $t$ no message entered the $n$-th node from the $(n-1)$-th one.

Let the function $f = f(\hat{x}, \overline{u}^{(0)}, \overline{u}^{(1)}, \ldots, \overline{u}^{(s)}, \ldots)$ be given with values 0 and $(k, d, \tau, h, s)$, where $k, d, \tau, h, s$ satisfy the restrictions $k > 0$, $d > 0$, $\tau \in A$, $0 \leq h \leq H$, $s \geq 0$, which will not be specified later. Here $\hat{x} = (\hat{k}, \hat{d}, \hat{\tau}, \hat{h}, \hat{s})$, $\overline{u}^{(s)} = \{u^{(s)}_{k,d,\tau,h}, k > 0, d > 0, \tau \in A, 0 \leq h \leq H\}$ is a collection of nonnegative integers, and only a finite number of them are different from zero. If $u^{(s)}_{k,d,\tau,h} = 0$ for all $k > 0$, $d > 0$, $\tau \in A$, $0 \leq h \leq H$ then we shall write $\overline{u}^{(s)} = \overline{0}$. Suppose that for only a finite number of $s$ values $\overline{u}^{(s)} \neq \overline{0}$.

Let the function f satisfy the following restrictions:

1. $f(\hat{x}, \bar{u}^{(0)}, \bar{u}^{(1)}, \ldots) = 0$ is equivalent to $\bar{u}^{(0)} = \bar{u}^{(1)} = \ldots = \bar{0}$.
2. $f(\hat{x}, \bar{u}^{(0)}, \bar{u}^{(1)}, \ldots) = (k, d, \tau, h, s)$ implies that $u^{(s)}_{k,d,\tau,h} \neq 0$.

Intuitively, the function f determines the discipline of choosing from the queue a message for the next transmission, $\hat{x}$ is the vector of characteristics of the message just transmitted, and $h (1 \leq h \leq H)$ is duration of transmitting a message from the previous node into the given one. For the messages formed in the given node define $h = 0$. $u^{(s)}_{k,d,\tau,h}$ is the number of messages in the queue waiting to be transmitted for s moments and having the characteristics $k, d, \tau, h$.

We thus consider all possible disciplines of uninterrupted service which satisfy restrictions 1 and 2 and effectuate the choice of a message from the queue for the next transmission proceeding from the types, lengths, the numbers of nodes to be passed before reaching the addressee, durations of transmissions from the previous node into the given one and the waiting times of the messages in the queue. It is also possible to consider disciplines which take into account the number of the nodes already passed or some other additional characteristics of messages, as well as the disciplines of interrupted service. All the reasoning may be extrapolated to them without any essential change.

Suppose that in a certain way there is given the condition at the entrance, i. e. a system of random variables $\{\xi_{1,t}(\omega)\}$, $t = 1, 2, 3, \ldots$ and the initial condition, i. e. a system of random variables $\{\xi_{n,1}(\omega)\}$, $n = 1, 2, 3, \ldots$ . Suppose the system of random variables be already constructed for $n = 1, 2, \ldots, n_0 + 1$, $t = 1, 2, \ldots, t_0$. We define the random variable $\xi_{n_0+1, t_0+1}(\omega)$.

Introduce the random variable $\gamma = \gamma_{n_0+1, t_0+1}(\omega) = \max\{t \leq t_0 : \xi_{n_0+1, t}(\omega) \neq 0\}$. Denote the components $\xi_{n_0, \gamma-\mu}(\omega)$, $\mu = 0, 1, 2, \ldots$

by $k_\mu(\omega)$, $d_\mu(\omega)$, $\tau_\mu(\omega)$, $h_\mu(\omega)$, $s_\mu(\omega)$. For each nonnegative $\mu$ consider the following system of random variables:

$$\bar{\delta}^{(\mu)}_{n_o, t_o}(\omega) = \{\delta(k-k_\mu(\omega))\delta(d-d_\mu(\omega)+1)\delta(\tau-\tau_\mu(\omega))\delta(h-h_\mu(\omega)),$$

$$k > 0,\ d > 0,\ \tau \in A,\ 0 \leq h \leq H\},$$

if $d_\mu(\omega) > 1$, and $\bar{\delta}^{(\mu)}_{n_o, t_o}(\omega) = \bar{0}$ in the opposite case. Here $\delta$ is defined by $\delta(x) = 1$ when $x = 0$, and $\delta(x) = 0$ when $x \neq 0$.

Intuitively, $\gamma_{n_o+1, t_o}(\omega)$ is the latest, not surpassing $t_o$, moment of time when the node $n_o+1$ received a message, i. e. when the transmitting device became unengaged in the node $n_o$. If at the moment $\gamma_{n_o+1, t_o}(\omega) - \mu$ no transmitted message entered the node $n_o$, then $\bar{\delta}^{(\mu)}_{n_o, t_o}(\omega) = \bar{0}$; in the opposite case $\bar{\delta}^{(\mu)}_{n_o, t_o}(\omega)$ contains a unique non-zero element.

Consider also the system of random variables

$$\bar{w}^{(\mu)}_{n_o, t_o}(\omega) = \{w^{(\mu)}_{k, d, \tau, h}(\omega),\quad k > 0,\ d > 0,\ \tau \in A,\ 0 \leq h \leq H\},$$

where $\mu = 0, 1, 2, \ldots$ . Let $2\nu = 2\nu(\omega)$ be the dimension of $\zeta_{n_o, \gamma-\mu}(\omega)$; $k_i(\omega)$, $z_i(\omega)$, $i = 1, \ldots, \nu$ are the components of $\zeta_{n_o, \gamma-\mu}(\omega)$. Let

$$w^{(\mu)}_{k, d, \tau, h}(\omega) = \sum_{i=1}^{\nu(\omega)} \delta(k_i(\omega)-k) \sum_{j=1}^{z_i(\omega)} \delta(\tau^{(k_i(\omega))}_{n_o, \gamma-\mu, j}(\omega)\delta(d^{(k_i(\omega))}_{n_o, \gamma-\mu, j}(\omega)-d)$$

if $h = 0$, $\zeta_{n_o, \gamma-\mu}(\omega) \neq (0, 0)$; $w^{(\mu)}_{k, d, \tau, h}(\omega) = 0$ in the opposite case.

Intuitively, $w^{(\mu)}_{k, d, \tau, 0}(\omega)$ is equal to the number of messages with characteristics $k, d, \tau$, in the collection formed at the moment $\gamma_{n_o+1, t_o}(\omega)$ in the node $n_o$.

At last, introduce the system of random variables

$\bar{v}^{(\mu)}_{n_o, t_o}(\omega) = \{v^{(\mu)}_{k, d, \tau, h}(\omega),\ k > 0,\ d > 0,\ \tau \in A,\ 0 \leq h \leq H\}$ where $\mu = 1, 2, \mu, \ldots$ .

Let $k_i(\omega)$, $d_i(\omega)$, $\tau_i(\omega)$, $h_i(\omega)$, $s_i(\omega)$ be the components of $\xi_{n_o+1,\gamma-i}(\omega)$, $i = 0,1,2,\ldots,\mu-1$. Let

$$v^{(\mu)}_{k,d,\tau,h}(\omega) = \sum_{i=0}^{\mu-1} \delta(k_i(\omega)-k)\delta(d_i(\omega)-d)\delta(\tau_i(\omega)-\tau)\delta(s_i(\omega)+h_i(\omega)+i-\mu)$$

Also let $\bar{v}^{(\mu)}_{n_o,t_o}(\omega) = \bar{0}$.

Intutitively, $\bar{v}^{(\mu)}_{n_o,t_o}(\omega)$ describes the messages appearing in the node $n_o$ (or those entering it) at the moment $\gamma_{n_o+1,t_o}(\omega)-\mu$, and transmitted into the $(n_o+1)$-th node up to the moment $\gamma_{n_o+1,t_o}(\omega)$ inclusive.

The next message is now chosen from the queue in the node $n_o$ at the moment $\gamma = \gamma_{n_o+1,t_o}(\omega)$ in accordance with the value taken by the function f. Namely, let

$$f(\xi_{n_o+1,\gamma}(\omega);\ \bar{w}^{(\mu)}_{n_o,t_o}(\omega) + \bar{\delta}^{(\mu)}_{n_o,t_o}(\omega) - \bar{v}^{(\mu)}_{n_o,t_o}(\omega),\ \mu = 0,1,2,\ldots)$$

$$= (k^o(\omega),\ d^o(\omega),\ \tau^o(\omega),\ h^o(\omega),\ s^o(\omega)).$$

When $k^o(\omega) \neq 0$ let

$$\xi_{n_o+1,t_o+1}(\omega) = (k^o(\omega), d^o(\omega), \tau^o(\omega), t_o+1-\gamma_{n_o+1,t_o}(\omega), s^o(\omega))$$

if $\varphi(\tau^o(\omega), \psi^{(k^o(\omega))}_{n_o,\gamma}(\omega)) = t_o+1-\gamma_{n_o+1,t_o}(\omega)$, and $\xi_{n_o+1,t_o+1}(\omega) = 0$ in the opposite case.

When $k^o(\omega) = 0$ or the random variable $\gamma = \gamma_{n_o+1,t_o}(\omega)$ is not defined, let $\gamma_{n_o+1,t_o}(\omega) = 0$ where it has not been defined earlier, and introduce the random variable. $\hat{\gamma} = \hat{\gamma}_{n_o,t_o}(\omega) =$ $\min\{t : \gamma_{n_o+1,t_o}(\omega) < t \le t_o,\ (\zeta_{n_o,t}(\omega) \neq (0,0))$ or $(d_{n_o,t}(\omega) > 1)\}$, where $d_{n_o,t}(\omega)$ is a component of $\xi_{n_o,t}(\omega)$.

Intuitively, $\hat{\gamma}_{n_o,t_o}(\omega)$ is the minimum of the moments of time which are greater than $\gamma_{n_o+1, t_o}(\omega)$ and do not surpass $t_o$, at which messages are formed in the node $n_o$ or a transmitted message enters it.

We conclude the construction. If the moment $\hat{\gamma}$ is not defined, let $\xi_{n_o+1, t_o+1}(\omega) = 0$. In the opposite case define $\hat{\delta}^{(\mu)}_{n_o, t_o}(\omega)$, $\hat{w}^{(\mu)}_{n_o, t_o}(\omega)$, $\hat{v}^{(\mu)}_{n_o, t_o}(\omega)$ as $\bar{\delta}^{(\mu)}_{n_o, t_o}(\omega)$, $\bar{w}^{(\mu)}_{n_o, t_o}(\omega)$, $\bar{v}^{(\mu)}_{n_o, t_o}(\omega)$ replacing $\gamma$ by the random variable $\hat{\gamma}$. In the case when

$$f(0, \hat{w}^{(\mu)}_{n_o, t_o}(\omega) + \hat{\delta}^{(\mu)}_{n_o, t_o}(\omega) - \hat{v}^{(\mu)}_{n_o, t_o}(\omega),\ \mu = 1,2,3,\dots)$$

$$= (k'(\omega),\ d'(\omega),\ \tau'(\omega),\ h'(\omega),\ s'(\omega)),$$

we define $\xi_{n_o+1, t_o+1}(\omega) = (k'(\omega), d'(\omega), \tau'(\omega), t_o+1-\hat{\gamma}_{n_o, t_o}(\omega), s'(\omega))$ if $\varphi(\tau'(\omega), \psi^{(k'(\omega))}_{n_o, \gamma}(\omega)) = t_o + 1 - \hat{\gamma}_{n_o, t_o}(\omega)$, and $\xi_{n_o+1, t_o+1}(\omega) = 0$ in the opposite case. Thus, the system of random variables $\{\xi_{n,t}(\omega)\}$, $n, t = 1, 2, 3, \dots$ is constructed. Note that random variables $\{\xi_{n,t}(\omega)\}$ are the functions of entrance condition, initial condition and given random variables $\{\zeta_{n,t}(\omega)\}$, $\{\tau^{(k)}_{n,t,j}(\omega)\}$, $\{d^{(k)}_{n,t,j}(\omega)\}$, $\{\psi^{(k)}_{n,t}(\omega)\}$, $n, t, j, k - 1, 2, 3, \dots$ .

Construct also an auxiliary system of random variables $\{\ell_{n,t}(\omega)\}$, $n, t = 1, 2, 3, \dots$ . Let $\zeta_{n,t}(\omega) = (k_1(\omega), \dots, k_\nu(\omega), z_1(\omega), \dots, z_\nu(\omega)) \neq (0, 0)$ and $k_t(\omega)$, $d_t(\omega)$, $\tau_t(\omega)$ be the components of $\xi_{n,t}(\omega)$. Define the total length of the queue as

$$m_{n,t}(\omega) = \sum_{i=1}^{\nu(\omega)} \sum_{j=1}^{z_i(\omega)} \varphi(\tau^{(k_i(\omega))}(\omega),\ \psi^{(k_i(\omega))}_{n,t}(\omega)),$$

and $m_{n,t}(\omega) = 0$ if $\zeta_{n,t}(\omega) = (0, 0)$. Denote $P\{m_{n,t}(\omega) = m \mid m_{n,t}(\omega) \neq 0\}$ by $f_m$.

The random variable $m_{n,t}(\omega)$ imitates the full duration of transmitting all the messages from the group described by the random variables $\zeta_{n,t}(\omega)$. Note that if the $j$-th message of the $k_i$-th type formed in the n-th node at the moment t starts to be transmitted into the (n+1)-th node at the moment $g_j(\omega)$, then the total duration of transmitting the messages equals

$$\tilde{m}_{n,t}(\omega) = \sum_{i=1}^{\nu(\omega)} \sum_{j=1}^{z_i(\omega)} \varphi(\tau_{n,t,j}^{(k_i(\omega))}(\omega),\ \psi_{n,g_j(\omega)}^{(k_i(\omega))}(\omega))$$

The distributions of the random variables $\tilde{m}_{n,t}(\omega)$ and $m_{n,t}(\omega)$ coincide, but $m_{n,t}(\omega)$ is measurable with respect to the natural $\sigma$-algebra generated by the given random variables $\{\zeta_{n,s}(\omega)\}$, $\{\tau_{n,s,j}^{(k)}(\omega)\}$, $\{\psi_{n,s}^{(k)}(\omega)\}$ for $s \leq t$.

Let $\ell_{n,o}(\omega) \equiv 0$, $\ell_{n,t}(\omega) = (\ell_{n,t-1}(\omega) + m_{n,t}(\omega) - 1$

$$+ (1 - \delta(k_t(\omega))(1 - \delta(d_t(\omega) - 1))\varphi(\tau_t(\omega), \psi_{n,t}^{k_t(\omega)}(\omega)))^+ ,$$

$n, t = 1, 2, 3, \ldots$ . Here $x^+ = \max(x, 0)$.

The system of random variables $\{\ell_{n,t}(\omega)\}$, $n, t = 1, 2, 3, \ldots$ may be interpreted as that of virtual times of waiting. If, beginning from the moment t, no new messages arise in the (n-1)-th node, and no messages come there from the n-th node, the distribution of time necessary to make the n-th node free from messages coincides with the distribution of $\ell_{n,t}(\omega)$.

We can now formulate the basic Theorem for a linear net of message communication.

Theorem 1. Let $\sum_{m=1}^{\infty} mf_m < (1 - Hq)p^{-1}$, $\sum_{m=1}^{\infty} m^2 f_m < \infty$, and the system of random variables $\{\xi_{n,t}(\omega)\}$, $n, t = 1, 2, 3, \ldots$ be built under the entrance condition $\xi_{1,t}(\omega) \equiv 0$, $t = 1, 2, 3, \ldots$ and initial condition $\xi_{n,1}(\omega) \equiv 0$, $n = 1, 2, 3, \ldots$ . Then, if $r = p + 2q - 2pq$ is sufficiently small, there exists a system of random variables $\{\bar{\xi}_n(\omega)\}$, $n = 1, 2, 3, \ldots$ such that

$$\lim_{\substack{N \to \infty \\ t \to \infty}} P\{\xi_{n_1+N,t}(\omega) = y_1, \ldots, \xi_{n_\ell+N,t}(\omega) = y_\ell\} = P\{\bar{\xi}_{n_1}(\omega) = y_1, \ldots, \bar{\xi}_{n_\ell}(\omega) = y_\ell\}.$$

$\ell = 1, 2, 3, \ldots, n_1, \ldots, n_\ell$ <u>are integers.</u> <u>The system of random variables</u> $\{\overline{\xi}_n(\omega)\}$ <u>is stationary in the sense that for any integer</u> $\overline{n}$

$$P\{\overline{\xi}_{n_1+\overline{n}}(\omega) = y_1, \ldots, \overline{\xi}_{n_\ell+\overline{n}}(\omega) = y_\ell\} = P\{\overline{\xi}_{n_1}(\omega) = y_1, \ldots, \overline{\xi}_{n_\ell}(\omega) = y_\ell\}.$$

This statement is an immediate consequence of Lemma 3 for the random variables $\{\xi_{n,t}(\omega)\}$, $n, t = 1, 2, 3, \ldots$ built under an arbitrary admissible entrance condition (see sec. 3).

<u>Remark</u>. Together with describing the processes of messages entering the communication nets we could also construct inductively a complete description of the queues at the moments $t = 1, 2, 3, \ldots$ and prove a theorem on the existence of corresponding limiting distributions. In order to make the paper not too cumbersome we confine ourselves to describing the random variables $\{\xi_{n,t}(\omega)\}$. When the communication net is cyclic, a complete description of the queues is necessary to formulate Lemma 4. For the sake of brevity we formulate all the theorems only for "one of the components" of the complete description, namely, for the processes of the messages reception $\{\xi_{n,t}(\omega)\}$. Theorems 1, 2, 3 in the formulations related to the complete description of the queues and processes of message entrance are proved in a similar way.

Consider the model of a cyclic communication net containing N nodes of communication. Intuitively this model differs from that of a linear net section since messages are transmitted from the N-th node to the first one. Besides, in the former case it is natural to assume the existence of a constant $\overline{m} < N$ such that the address of any message does not surpass $\overline{m}$. The condition $\overline{m} < N$ means that message cannot be transmitted twice via the same node, though this is unessential for the theorems proved below. If the initial condition, i. e. the random variable $\xi_1(\omega) = \{\xi_{1,1}(\omega), \ldots, \xi_{N,1}(\omega)\}$ is given, it is not difficult to construct inductively the system of random variables $\{\xi_{n,t}(\omega)\}$, describing the process of messages entering communication nodes.

Theorem 2. Let $\sum_{m=1}^{\infty} mf_m < (1 - Hq)p^{-1}$, $\sum_{m=1}^{\infty} m^2 f_m < \infty$ and the system of random variables $\{\xi_{n,t}(\omega)\}$, $n = 1, 2, \ldots, N$ be built under the initial condition $\xi_1(\omega) = (0, 0, \ldots, 0)$. Let there exist $\overline{m}$ such that $P\{d^{(k)}_{n,t,j}(\omega) = m\} = 0$ when $m > \overline{m}$ and for all $k = 1, 2, 3, \ldots$ . Then there exists a system of random variables $\{\overline{\xi}_n(\omega)\}$, $n = 1, 2, \ldots, N$ such that

$$\lim_{t\to\infty} P\{\xi_{1,t}(\omega) = y_1, \ldots, \xi_{N,t}(\omega) = y_N\} = P\{\overline{\xi}_1(\omega) = y_1, \ldots, \overline{\xi}_N(\omega) = y_N\}.$$

The distribution $P\{\overline{\xi}_1(\omega) = y_1, \ldots, \overline{\xi}_N(\omega) = y_N\}$ is invariant under rotation.

This statement, as well as the existence of the limiting distributions for the complete description all the queues of the net, and the distributions being invariant under the rotation, follows from Lemma 4 under arbitrary admissible initial condition (see sec. 3).

Assume that for any $N = 2, 3, 4, \ldots$ the initial condition $\xi_1(\omega) = (0, 0, \ldots, 0)$, and that the system of initial random variables $\{\zeta^{(N)}_{n,t}(\omega)\}$, $\{\tau^{(k,N)}_{n,t,j}(\omega)\}$, $\{d^{(k,N)}_{n,t,j}(\omega)\}$, $\{\psi^{(k,N)}_{n,t}(\omega)\}$, $n = 1, 2, \ldots, N$; $t, j, k = 1, 2, 3, \ldots$ are given on a cyclic net. Since all our reasoning concerns only finite-dimensional distributions, without loss of generality we may assume that, for any $N$ the collections of the given random variables are defined on the same probability space and, moreover, that there exists a system of random variables $\{\zeta_{n,t}(\omega)\}$, $\{\omega^{(k)}_{n,t,j}(\omega)\}$, $\{d^{(k)}_{n,t,j}(\omega)\}$, $\{\psi^{(k)}_{n,t}(\omega)\}$, $n,t,j,k = 1, 2, 3, \ldots$, such that for any $N$, the system of the initial random variables for a cyclic net of $N$ nodes is its restriction. Suppose there exists a constant $\overline{M}$ such that $P\{d^{(k)}_{n,t,j}(\omega) = m\} = 0$ when $m > \overline{M}$, $k = 1, 2, 3, \ldots$ . Let the system of given random variables provide us with coupled systems of random variables $\{\xi_{n,t}(\omega)\}$, $n = 1, 2, 3, \ldots$ and $\{\xi_{n,t}^{(N)}(\omega)\}$, $n = 1, \ldots, N$; $t = 1, 2, 3, \ldots$ describing the processes of messages entrance in the linear nets of message communication and in the cyclic net of $N$ nodes. According to Theorem 1, when $r$ is sufficiently small, there exists a limit process of message entrance into the linear net $\{\bar{\xi}_n(\omega)\}$, $n = 1, 2, 3, \ldots$. Since we have assumed that there exists the maximal possible value of the address of messages $\bar{M}$ thus, according to Theorem 2, for any $N$ there exist the limit process of message entrance into the cyclic net $\{\bar{\xi}_n^{(N)}(\omega)\}$, $n = 1, 2, \ldots, N$. The following result is a consequence of Lemmas 3 and 4.

Theorem 3. Let $\sum_{m=1}^{\infty} m f_m < (1 - Hq)p^{-1}$, $\sum_{m=1}^{\infty} m^2 f_m < \infty$ and $r = p + 2q - 2pq$ be sufficiently small. Assume that the system of random variables $\{\xi_{n,t}(\omega)\}$, $n, t = 1, 2, 3, \ldots$ is constructed under zero entrance condition and zero initial condition from the system of given random variables $\{\zeta_{n,t}(\omega)\}$, $\{\tau_{n,t,j}^{(k)}(\omega)\}$, $\{d_{n,t,j}^{(k)}(\omega)\}$, $\{\psi_{n,t}^{(k)}(\omega)\}$, $n, t, j, k = 1, 2, 3, \ldots$, and $P\{d_{n,t,j}^{(k)}(\omega) = m\} = 0$ when $m > \bar{M}$ for all $k = 1, 2, 3, \ldots$. Suppose that for any $N$ the systems of random variables $\{\zeta_{n,t}^{(N)}(\omega)\}$, $\{\tau_{n,t,j}^{(k,N)}(\omega)\}$, $\{d_{n,t,j}^{(k,N)}(\omega)\}$, $\{\psi_{n,t}^{(k,N)}(\omega)\}$, $n = 1, 2, \ldots, N$; $t = 1, 2, 3, \ldots$ are given coinciding in their distribution with the random variables given for a linear net, and the system of random variables $\{\xi_{n,t}^{(N)}(\omega)\}$, $n = 1, 2, \ldots, N$; $t = 1, 2, 3, \ldots$ is built under zero initial condition.

Let $\{\bar{\xi}_n(\omega)\}$, $n = 1, 2, 3, \ldots$ be the limit process of message entrance for a linear net, and $\{\bar{\xi}_n^{(N)}(\omega)\}$, $n = 1, 2, \ldots, N$ be the limit process for a cyclic net. Then

$$\lim_{N \to \infty} P\{\bar{\xi}_{n_1}^{(N)}(\omega) = y_1, \ldots, \bar{\xi}_{n_\ell}^{(N)}(\omega) = y_\ell\} = P\{\bar{\xi}_{n_1}(\omega) = y_1, \ldots, \bar{\xi}_{n_\ell}(\omega) = y_\ell\}$$

where $\ell = 1, 2, 3, \ldots$, $n_1, \ldots, n_\ell$ are integers.

## III. AUXILIARY LEMMAS

The sequence of random variables $\{\ell_{n,t}(\omega)\}$, $t = 0, 1, 2, \ldots$ where $\ell_{n,o} \equiv 0$ for any fixed $n$ defines a random walk on $\mathbb{Z}^1_+$. Let $\sigma_i = \sigma_{n,i}(\omega)$ be the sequence of moments when the walk returns to the point 0:

$$\sigma_{n,o}(\omega) \equiv 0, \quad \sigma_{n,i}(\omega) = \inf\{t > \sigma_{n,i-1}(\omega) : \ell_{n,t}(\omega) = 0\},$$

$\sigma_{n,o}(\omega) = \infty$ if such $t$ do not exist, $i = 1, 2, 3, \ldots$ .

__Lemma 1.__ __If__ $\sum_{m=1}^{\infty} mf_m < (1 - Hq)p^{-1}$, $\sum_{m=1}^{\infty} m^2 f_m < \infty$ __then__

$$\sup_{n,i} P\{\sigma_{n,i}(\omega) < \infty, \sup_{\sigma_{n,i}(\omega) \le t < \sigma_{n,i+1}(\omega)} \ell_{n,t}(\omega) > L\} < \epsilon(L)$$

__where__ $\epsilon(L) \to 0$ __when__ $L \to \infty$.

__Proof:__ Consider the sequence of the random variables

$$y^t = y_{n,t}(\omega) = m_{n,t}(\omega) - 1 + H\theta_{n,t}(\omega), \quad y^o_t = y^o_{n,t}(\omega) = m_{n,t}(\omega)$$
$$-1 + (1 - \delta(k_{n,t}(\omega)))(1 - \delta(d_{n,t}(\omega) - 1))\varphi(\tau_{n,t}(\omega), \psi_{n,t}^{(k_{n,t}(\omega))}(\omega))$$

(the random variables $\theta_{n,t}(\omega)$ are introduced at the beginning of sec. 2). For any $\omega$,

$$y^o_{n,t}(\omega) \le m_{n,t}(\omega) - 1 + H(1 - \delta(d_{n,t}(\omega))(1 - \delta(d_{n,t}(\omega) - 1))$$

since the conditions $k_{n,t}(\omega) \ne 0$ and $d_{n,t}(\omega) \ne 0$ are equivalent. On the set $\{\sigma_{n,i}(\omega) < \infty\}$ define the following random variables:

$$S_{i,t} = S_{n,i,t} = \sum_{j=1}^{t} y_{\sigma_{n,i}(\omega)+j}(\omega), \quad S^o_{i,t} = S^o_{n,i,t}(\omega)$$

$$= \sum_{j=1}^{t} y^o_{\sigma_{n,i}(\omega)+j}(\omega), \quad Y_{n,i}(\omega) = \sup_{t>0}(S_{i,t})^+, \quad Y^o_{n,i}(\omega) = \sup_{t>0}(S^o_{i,t})^+$$

Obviously, $P\{\sigma_i < \infty,\ y^o_{\sigma_i+j} > x\} \leq P\{\sigma_i < \infty,\ y_{\sigma_i+j} > x\}$. Hence, as in the proof of Lemma 4, §23, [2], we have that

$$EF(y^o_{\sigma_{n,i}(\omega)+j}(\omega))\chi_{\{\sigma_{n,i}(\omega)<\infty\}} \leq EF(y_{\sigma_{n,i}(\omega)+j}(\omega))\chi_{\{\sigma_{n,i}(\omega)<\infty\}}$$

for an arbitrary convex nondecreasing function $F$ such that there exists $EF(y_{\sigma_i+j})\chi_{\{\sigma_i<\infty\}}$, i. e. $y^o_{\sigma_{n,i}(\omega)+j}(\omega)$ is a convex decrease $y_{\sigma_{n,i}(\omega)+j}(\omega)\ (y^o_{\sigma_i+j} \prec y_{\sigma_i+j})$ on the set $\{\sigma_{n,i}(\omega) < \infty\}$. Copying the proof of the comparison theorems (§23, [2]), it can be shown that

$$S^o_{i,t} \prec S_{i,t},\ \max(0, S^o_{i,1}, S^o_{i,2}, \ldots, S^o_{i,t}) \prec \max(0, S_{i,1}, S_{i,2}, \ldots, S_{i,t}),$$

$$Y^o_{n,i}(\omega) \prec Y_{n,i}(\omega)$$

on the set $\{\sigma_{n,i}(\omega) < \infty\}$. Hence it follows that

$$\sup_{n,i} P\{\sigma_{n,i}(\omega) < \infty,\ \sup_{\sigma_{n,i}(\omega) \leq t < \sigma_{n,i+1}(\ )} \ell_{n,t}(\omega) > L\}$$

$$\leq \sup_{n,i} P\{\sigma_{n,i}(\omega) < \infty,\ Y^o_{n,i}(\omega) > L\} \leq \sup_{n,i} E(Y^o_{n,i}(\omega) - L + 1)^+ \chi_{\{\sigma_{n,i}(\omega)<\infty\}}$$

$$\leq \sup_{n,i} E(Y_{n,i}(\omega) - L + 1)^+ \chi_{\{\sigma_{n,i}(\omega)<\infty\}}$$

It remains to remark that $E(Y_{n,i}(\omega) - L + 1)^+ \chi_{\{\sigma_{n,i}(\omega)<\infty\}} < E(Y_{n,o}(\omega) - L + 1)^+$, $i = 1, 2, 3, \ldots$ and $E(Y_{n,o}(\omega) - L + 1)^+ < \epsilon(L)$ where $\epsilon(L) \to 0$ when $L \to \infty$, since $EY_{n,o}(\omega) < \infty$ (see, e. g., Theorem 4.13, [3]).

<u>Lemma 2.</u> <u>If</u> $\sum_{m=1}^{\infty} mf_m < (1 - Hq)p^{-1}$ <u>then for any</u> $L > 0$ <u>there exist</u> $M = M(L)$, $\beta = \beta(L)$, $T = T(L)$ <u>such that for any</u> $t > T$

$$\inf_{n, t_o} P\{\ell_{n, t_o+t}(\omega) < M | \ell_{n, t_o}(\omega) < L\} \geq \beta > 0$$

Proof: Consider an auxiliary system of the random variables $\{\tilde{\ell}_{n,t}(\omega)\}$, $n, t = 1, 2, 3, \ldots$ given by the recursive formula

$$\tilde{\ell}_{n,o}(\omega) \equiv 0, \; \tilde{\ell}_{n,t}(\omega) = (\tilde{\ell}_{n,t-1}(\omega) + m_{n,t}(\omega) - 1 + H\theta_{n,t}(\omega))^+, \quad n, t = 1, 2, 3, \ldots$$

For any fixed $n$ the system of the random variables $\{\tilde{\ell}_{n,t}(\omega)\}$, $t = 0, 1, 2, \ldots$ forms a Markov chain, and from the results of [4], it follows that the chain is ergodic. Consequently, there exist $M, \beta$, $T$ such that when $t > T$,

$$\inf_{n, t_o} P\{\tilde{\ell}_{n,t_o+t}(\omega) < M | \tilde{\ell}_{n, t_o}(\omega) < L\}$$
$$\geq \inf_{n, t_o} P\{\tilde{\ell}_{n, t_o+t}(\omega) < M | \tilde{\ell}_{n, t_o}(\omega) = L\} \geq \beta > 0$$

It remains to show that for arbitrary $n$, $t_o$,

$$P\{\ell_{n, t_o+t}(\omega) \geq M | \ell_{n,t_o}(\omega) = L\} \leq P\{\tilde{\ell}_{n,t_o+t}(\omega) \geq M | \tilde{\ell}_{n,t_o}(\omega) = L\}$$

When $t = 1$, this inequality is obvious, and the proof is completed by the induction with respect to $t$.

Now we pass to the proof of the basic Lemma 3. Let the coupled processes $\{\xi_{n,t}(\omega)\}$ and $\{\tilde{\xi}_{n,t}(\omega)\}$, $n = 1, 2, 3, \ldots$ be built (by way of the construction described above) from the same collections $\{\zeta_{n,t}(\omega)\}$, $\{\tau^{(k)}_{n,t,j}(\omega)\}$, $\{d^{(k)}_{n,t,j}(\omega)\}$, $\{\psi^{(k)}_{n,t}(\omega)\}$, $n, t, j, k = 1, 2, 3, \ldots$ under different entrance conditions $\{\xi_{1,t}(\omega)\}$ and $\{\tilde{\xi}_{1,t}(\omega)\}$, $t = 1, 2, 3, \ldots$ and zero initial conditions. Then, according to Lemma 3, for sufficiently small $r = p + 2q - 2pq$,

$$P\{\xi_{n,t}(\omega) \neq \tilde{\xi}_{n,t}(\omega)\} \to 0$$

when $n, t \to \infty$. Now we pass to an exact formulation.

The entrance condition $\{\xi_{1,t}(\omega)\}$, $t = 1, 2, 3, \ldots$ defined on the original probability space and independent of the systems of random variables $\{\zeta_{n,t}(\omega)\}$, $\{\tau^{(k)}_{n,t,j}(\omega)\}$, $\{d^{(k)}_{n,t,j}(\omega)\}$, $\{\psi^{(k)}_{n,t}(\omega)\}$, $n, t, j, k = 1, 2, 3, \ldots$ is said to be admissible if the random variables $\xi_{1,t}(\omega) = (k_t(\omega), d_t(\omega), \tau_t(\omega), h_t(\omega), s_t(\omega))$ taking the values $0$ and $(k, d, \tau, h, s)$ where $k > 0$, $d > 0$, $\tau \in A$, $0 < h \leq H$, $s \geq 0$ are such that $P\{d_t(\omega) \geq m + 1\} \leq qP\{d_t(\omega) \geq m\}$, $m, t = 1, 2, 3, \ldots$ .

__Lemma 3.__ __Let the systems of random variables__ $\{\xi_{n,t}(\omega)\}$ __and__ $\{\tilde{\xi}_{n,t}(\omega)\}$ __be built under zero initial conditions from the same collections__ $\{\zeta_{n,t}(\omega)\}$, $\{d^{(k)}_{n,t,j}(\omega)\}$, $\{\tau^{(k)}_{n,t,j}(\omega)\}$, $\{\psi^{(k)}_{n,t}(\omega)\}$, $n, t, j, k = 1, 2, 3, \ldots$ but under different admissible entrance conditions $\{\xi_{1,t}(\omega)\}$ and $\{\tilde{\xi}_{1,t}(\omega)\}$, $t = 1, 2, 3, \ldots$ . Let

$$\sum_{m=1}^{\infty} mf_m < (1 - Hq)p^{-1}, \quad \sum_{m=1}^{\infty} m^2 f_m < \infty$$

__Suppose that for some__ $n \geq 2$ __and__ $T_o \geq 0$, $P\{\xi_{n,t}(\omega) \neq \tilde{\xi}_{n,t}(\omega)\} \leq \alpha$ __for all__ $t > T_o$. __Then for arbitrary__ $\epsilon > 0$ __there exists__ $\overline{T} = \overline{T}(\epsilon)$ __such that for all__ $t > \hat{T} = T_o + T$, $P\{\xi_{n+1,t}(\omega) \neq \tilde{\xi}_{n+1,t}(\omega)\} \leq \rho\alpha + \epsilon$ __where__ $\rho = \rho(r) < 1$ __for sufficiently small__ $r = p + 2q - 2pq$.

__Proof:__ Let $d_{n,t}(\omega)$ be the component of $\xi_{n,t}(\omega)$ and $\tilde{d}_{n,t}(\omega)$ that of $\tilde{\xi}_{n,t}(\omega)$. Let $\overline{d}_u(\omega) = \max\{d_{n,u}(\omega), \tilde{d}_{n,u}(\omega), d_{n,u-1}(\omega), \tilde{d}_{n,u-1}(\omega), \ldots, d_{n,u-M}(\omega), \tilde{d}_{n,u-M}(\omega)\}$ and consider the following collection of random variables and events:

$$t_1(\omega) = \max\{u < t'_o(\omega) : (\zeta_{n,u}(\omega) = \ldots = \zeta_{n,u-M}(\omega) = (0,0))$$

$$\text{and } (\overline{d}_u(\omega) \leq 1)\},$$

$$t'_1(\omega) = t_1(\omega) - M, \quad A^1_t = \{\omega : t_1(\omega) = t\};$$

$$\overline{t}_1(\omega) = \max\{u < \overline{t}'_o(\omega) : \chi_{n,u}(\omega) = \ldots = \chi_{n,u-M}(\omega) = 1\},$$

$$\bar{t}_1'(\omega) = \bar{t}_1(\omega) - M,\ \bar{A}_t^1 = \{\omega : \bar{t}_1(\omega) = t\};$$

. . . . . . . . . . . . . . . . . . . . . .

. . . . . . . . . . . . . . . . . . . . . .

$$t_\ell(\omega) = \max\{u < t'_{\ell-1}(\omega) : (\zeta_{n,u}(\omega) = \dots = \zeta_{n,u-M}(\omega) = (0,0)$$
$$\text{and } (\bar{d}_u(\omega) \le 1)\},$$

$$t'_\ell(\omega) = t_\ell(\omega) - M,\ A_t^\ell = \{\omega : t_\ell(\omega) = t\};$$

$$\bar{t}_\ell(\omega) = \max\{u < \bar{t}'_{\ell-1}(\omega) : \chi_{n,u}(\omega) = \dots = \chi_{n,u-M}(\omega) = 1\},$$

$$\bar{t}'_\ell(\omega) = \bar{t}_\ell(\omega) - M,\ \bar{A}_t^\ell = \{\omega : \bar{t}_\ell(\omega) = t\},$$

where $t'_o(\omega) = \bar{t}'_o(\omega) = t - H$, $t > \hat{T} = T_o + \bar{T}$, while the positive integers $\bar{T}$, M and $\ell$ will be chosen later.

Intuitively, $[t'_i(\omega), t_i(\omega)]$ is the i-th period of time of length M starting from the moment t - H "backwards" from which no new message formed in the n-th node appeared and no transmitted message entered in it from the (n-1)-th node.

The sequence $(t_1, \dots, t_\ell)$ of length $\ell$ will be called admissible if $t - H > t_1 > t_2 > \dots > t_\ell > T_o + M$, $t_i - t_{i+1} \ge M$, $i = 1,2,\dots, \ell-1$. Denote the set of all admissible collections $(t_1, \dots, t_\ell)$ by $D(\ell)$. Consider the following events:

$$E = E_{\bar{T},\ell,M} = \{\omega : t'_\ell(\omega) > T_o\},\ \bar{E} = \bar{E}_{\bar{T},\ell,M} = \{\omega : \bar{t}'_\ell(\omega) > T_o\}$$

Then $(t_1(\omega), \dots, t_\ell(\omega)) \in D(\ell)$ when $\omega \in E$, $(\bar{t}_1(\omega), \dots, \bar{t}_\ell(\omega)) \in D(\ell)$ when $\omega \in \bar{E}$. It may easily be shown for arbitrary $\ell$, M, $\bar{T}$ may be chosen so that $P(E) \ge P(\bar{E}) \ge 1 - \epsilon/4$.

Let $\bar{B}_j^1 = \{\omega : t'_{j-1}(\omega) = t_j(\omega) + 1\}$. On the complement of the set $\bar{B}_j^1$ consider the random variables

$$\xi^j(\omega) = (\xi_{t_j(\omega)+1}(\omega), \dots, \xi_{t'_{j-1}(\omega)-1}(\omega)),\ \tilde{\xi}^j(\omega) = (\xi_{t_j(\omega)+1}(\omega), \dots, \tilde{\xi}_{t'_{j-1}(\omega)}(\omega))$$

and let $B_i^2 = \{\omega : (t'_{j-1}(\omega) \neq t_j(\omega)+1)$ and $(\xi^j(\omega) = \tilde{\xi}^j(\omega) = \tilde{\xi}^j(\omega))\}$. Let $\bar{B}_j = \bar{B}_j^1 \cup B_j^2$, $j = 1, 2, \ldots, \ell$, $B_i = \bigcap_{j=1}^{i} \bar{B}_j$, $i = 1, 2, \ldots, \ell$. Let $B_i^C$ be the complement of the set $B_i$, $B_\ell^C = B^C$.

Intuitively the event $\bar{B}_j$ means that between the j-th period and the (j-1)-th one with length M, when no new message arose in the n-th node and no transmitted message entered it from the (j-1)-th node, the coupled processes of the reception of messages $\xi_{n,t}(\omega)$ and $\tilde{\xi}_{n,t}(\omega)$ coincided identically.

Consider also the events

$$\bar{K}_j^1, \bar{K}_j^2, j = 1, 2, \ldots, \ell, \bar{K}_j = \bar{K}_j^1 \cup \bar{K}_j^2, K_j = \bar{K}_j \,|\, \bigcup_{i=1}^{j-1} \bar{K}_i, \quad K = \bigcup_{j=1}^{\ell} K_j \,;$$

$K^C$ is a complement of $K$ where $\bar{K}_j^1 = \bigcup_{t>0} \{(t'_j(\omega) \leq \gamma_{n+1,t}(\omega) < t'_{j-1}(\omega))$ and (at the moment $\gamma_{n+1,t}(\omega)$ there is no more than one message in the queue at the node n, i. e. $\bar{u}_{n,t}^{(\mu)}(\omega) + \bar{\delta}_{n,t}^{(\mu)}(\omega) - \bar{v}_{n,t}^{(\mu)}(\omega)$ for all $\mu = 0, 1, 2, \ldots$ except possibly one value $\mu_o$, and $\bar{u}_{n,t}^{(\mu_o)}(\omega) + \bar{\delta}_{n,t}^{(\mu_o)}(\omega) - \bar{v}_{n,t}^{(\mu_o)}(\omega)$ has no more than one nonzero component)$\}$, $\bar{K}_j^2 = \{$there exists no $t > 0$ for which the moment $\gamma_{n+1,t}(\omega)$ is defined and $t'_j(\omega) \leq \gamma_{n+1,t}(\omega) < t'_{j-1}(\omega)\}$.

Now we can estimate $P\{\xi_{n+1,t}(\omega) \neq \tilde{\xi}_{n+1,t}(\omega)\}$ when $t > \hat{T}$.

$$P\{\xi_{n+1,t}(\omega) \neq \tilde{\xi}_{n+1,t}(\omega)\} \leq \frac{\epsilon}{4} + \sum_{D(\ell)} P\{\xi_{n+1,t}(\omega) \neq \tilde{\xi}_{n+1,t}(\omega), A_{t_1}^1, \ldots, A_{t_\ell}^\ell\}$$

as $P(E) \geq 1 - \frac{\epsilon}{4}$.

$$\sum_{D(\ell)} P\{\xi_{n+1,t}(\omega) \neq \tilde{\xi}_{n+1,t}(\omega), A_{t_1}^1, \ldots, A_{t_\ell}^\ell\} \leq P(E \cap K^C)$$

$$+ \sum_{j=1}^{\ell} \sum_{D(\ell)} P\{\xi_{n+1,t}(\omega) \neq \tilde{\xi}_{n+1,t}(\omega), A_{t_1}^1, \ldots, A_{t_\ell}^\ell, K_j\} \leq 2qH\alpha$$

$$+ P(E \cap K^C) + \sum_{j=1}^{\ell} \sum_{D(\ell)} P\{\xi_{n+1,t}(\omega) \neq \tilde{\xi}_{n+1,t}(\omega), A_{t_1}^1, \ldots, A_{t_\ell}^\ell, K_j, B_j^C\}$$

as

$$\sum_{j=1}^{\ell} \sum_{D(\ell)} P\{\xi_{n+1,t}(\omega) \neq \tilde{\xi}_{n+1,t}(\omega),\ A^1_{t_1}, \ldots, A^{\ell}_{t_{\ell}},\ K_j,\ B_j\} < 2qH\alpha$$

since the random variables $\xi_{n+1,t}(\omega)$ and $\tilde{\xi}_{n+1,t}(\omega)$ coincide on the set $K_j \cap B_j$ because of their construction, if there exists no $t_o \in [t - H, t]$ such that $\max(d_{n,t_o}(\omega), \tilde{d}_{n,t_o}(\omega)) > 1$, where $d_{n,t_o}(\omega)$ is a component of $\xi_{n,t_o}(\omega)$, and $\tilde{d}_{n,t_o}(\omega)$ is that of $\tilde{\xi}_{n,t_o}(\omega)$. However,

$$\sum_{j=1}^{\ell} \sum_{D(\ell)} P\{\xi_{n+1,t}(\omega) \neq \tilde{\xi}_{n+1,t}(\omega),\ A^1_{t_1}, \ldots, A^{\ell}_{t_{\ell}},\ K_j, B^c_j\}$$

$$\leq \sum_{j=1}^{\ell} \sum_{D(\ell)} P\{A^1_{t_1}, \ldots, A^{\ell}_{t_{\ell}},\ K_j,\ B^c\}$$

$$\leq \sum_{D(\ell)} P\{A^1_{t_1}, \ldots, A^{\ell}_{t_{\ell}}, B^c\} \leq \sum_{D(\ell)} P\{A^1_{t_1}, \ldots, A^{\ell}_{t_{\ell}} \mid B^c\} P(B^c)$$

$$\leq \alpha \sum_{D(\ell)} (t - H - t'_{\ell} - M\ell) P\{A^1_{t_1}, \ldots, A^{\ell}_{t_{\ell}} \mid B^c\}$$

since $P(B^c) \leq \alpha(t - H - t'_{\ell} - M\ell)$ according to the Lemma condition. Consider the random variables $\Gamma(\omega) = t - H - t'_{\ell}(\omega)$, $\overline{\Gamma}(\omega) = t - H - \overline{t}'_{\ell}(\omega)$. Then

$$\sum_{D(\ell)} (t - H - t'_{\ell} - M\ell) P\{A^1_{t_1}, \ldots, A^{\ell}_{t_{\ell}} \mid B^c\} = E(\Gamma(\omega) - M\ell \mid B^c)$$

$$\leq E(\overline{\Gamma}(\omega) - M\ell) = \sum_{k=M\ell}^{\infty} (k - M\ell) g_k ,$$

where $g_k = P\{\overline{t}'_{\ell}(\omega) = t - H - k\}$. It remains to remark that

$$\sum_{k=M\ell}^{\infty} (k - M\ell) g_k = \rho(r) < 1 \text{ for sufficiently small } r = p + 2q - 2pq.$$

To conclude the proof of the Lemma it remains to estimate $P(E \cap K^c)$. We shall explain informally why the probability $P(E \cap K^c)$ is small. Since the event $E \cap K^c$ had been realized, it means that in the interval of time $[t'_{\ell}(\omega), \hat{T} - H]$ the "virtual waiting time" never turned into 0, and at the beginning of every time interval of length M, when no new message appeared in the n-th node and no messages came from the (n-1)-th node, that the "virtual waiting time" surpassed M. Select a sequence of "control" time intervals of length M $i_1, \ldots, i_j$ so that sufficient time elapsed between the moment $t'_{i_1}(\omega), \ldots, t'_{i_j}(\omega)$. According to Lemma 1, we may assume that the "virtual waiting time" in the interval of time $[t'_\ell(\omega), \hat{T} - H]$ does not surpass a certain constant $\bar{L}$, and according to Lemma 2, the conditional probability of the "virtual waiting time" surpassing M at the beginning of any "control" period does not surpass $1 - \beta$.

Formally the proof can be finished in the following way: choose $\bar{L}$ so that $\epsilon(\bar{L}) < \epsilon/4$, choose $M = M(\bar{L}) > H$, $T = T(\bar{L})$, $\beta = \beta(\bar{L})$ according to the formulations of Lemmas 1 and 2. Let $\ell_1 = [T/M] + 1$, $\ell_2 = [\ln \epsilon/4/\ln(1-\beta)] + 2$, $\ell = \ell_1 \cdot \ell_2$. Consider the events

$$D = \bigcup_{i=1}^{\infty} \{\omega : \sigma_{n,i-1}(\omega) < t'_\ell(\omega),\ \sigma_{n,i}(\omega) > t'_o(\omega)\}$$

and

$$K = K_{\ell_1} = \bigcap_{j=1}^{\ell_2} \{\omega : \ell_{n, L'_{j,\ell_1}(\omega)}(\omega) > M\}$$

Then

$$P(E \cap K^C) \leq P(D \cap K) \leq \epsilon/4 + (1-\beta)^{\ell_2 - 1} \leq \epsilon/2$$

From all the obtained estimates we find that when $t > T$

$$P\{\xi_{n+1,t}(\omega) \neq \tilde{\xi}_{n+1,t}(\omega)\} \leq \rho\alpha + \epsilon$$

where $\rho = \rho(r) < 1$ for small $r$. The Lemma is proved.

Consider a cyclic model of message communication consisting of N nodes of communication. The proof of existence of a limit process of message reception in a cyclic net with a fixed number of nodes is similar to that for a linear net but is much more cumbersome. For this reason we shall make it informal, because no new ideas are necessary to reformulate it in strict mathematically terms.

In the case of a cyclic net, in order to formulate the basic Lemma, one has to change the definition of the initial condition by including in it, together with the description of the process of message entrance at the zero moment of time, the complete description of the queues of messages to be transmitted from N nodes with $t = 0$. Formally this is realized by giving for every $n = 1, 2, \ldots, N$ an array of random variables with random dimensions, which describes the state of the queue when $t = 0$. We shall say that zero initial conditions are given if, at $t = 0$ there are no messages in the net.

The initial condition will be called admissible if there exists $\bar{m}$ such that $P\{d(\omega) = m\} = 0$, when $m > \bar{m}$ and $P\{d(\omega) \geq m+1\} \leq qP\{d(\omega) \geq m\}$, $m = 1, 2, \ldots, \bar{m} - 1$, where $d(\omega)$ is an address of any message contained in any queue of the given initial condition. Obviously zero initial condition is admissible .

We may define the total length of each of N queues given at time zero in a manner similar to the definition of the random variable $m_{n,t}(\omega)$ as the total length of the messages from the group described by the random variable $\zeta_{n,t}(\omega)$. We say that the initial condition is limited by the constant $\bar{L}$ if the total length of each of N queues does not surpass L when $t = 0$.

Let the systems of random variables $\{\zeta_{n,t}(\omega)\}$, $\{\tau^{(k)}_{n,t,j}(\omega)\}$, $\{d^{(k)}_{n,t,j}(\omega)\}$, $\{\psi^{(k)}_{n,t}(\omega)\}$, $n = 1, 2, \ldots, N$; $t, j, k = 1, 2, 3, \ldots,$ independent of the initial condition, be given on the given probability space. In the case of a cyclic net their description differs from that of a linear net only by the assumption of existence of a constant $\bar{m}$ such that

$$P\{d^{(k)}_{n,t,j}(\omega) = m\} = 0 \quad \text{when} \quad m > \bar{m} \qquad (*)$$

and

$$P\{d^{(k)}_{n,t,j}(\omega) \geq m+1\} \leq qP\{d^{(k)}_{n,t,j}(\omega) \geq m\}, \quad m = 1, 2, \ldots, \bar{m} - 1$$

for all $k = 1, 2, 3, \ldots$ . After the initial condition and the system of given random variables, and the function f determining the choice of a message for the next transmission are given, we can inductively build a complete description of the queues and processes of message entrance in all the nodes of the net, i. e. the system of random variables $\{\Xi_{n,t}(\omega)\}$, $n = 1, 2, \ldots, N$; $t = 1, 2, 3, \ldots$ . Denote the collection of random variables $(\Xi_{1,t}(\omega), \ldots, \Xi_{N,t}(\omega))$ by $\Xi_t(\omega)$.

Lemma 4. Let the collections of random variables $\{\Xi_{n,t}(\omega)\}$ and $\{\tilde{\Xi}_{n,t}(\omega)\}$ be constructed by means of the same collections $\{\zeta_{n,t}(\tau)\}$, $\{\tau^{(k)}_{n,t,j}(\omega)\}$, $\{d^{(k)}_{n,t,j}(\omega)\}$, $\{\psi^{(k)}_{n,t}(\omega)\}$, $n = 1, 2, \ldots, N$; $t, j, k = 1, 2, 3, \ldots$ (the collection of random variables $\{d^{(k)}_{n,t,j}(\omega)\}$ satisfies the conditions (*)) but under different admissible initial conditions limited by the constant $\bar{L}$ . Let

$$\sum_{m=1}^{\infty} mf_m < p^{-1}(1 - Hq), \quad \sum_{m=1}^{\infty} m^2 f_m < \infty$$

Suppose that for all $t > T_o$

$$P\{\Xi_t(\omega) \neq \tilde{\Xi}_t(\omega)\} \leq \rho\alpha + \epsilon$$

Then for the arbitrary $\epsilon > 0$ there exists such $\bar{T} = \bar{T}(\epsilon, \bar{L})$ that for all $t > \hat{T} = T_o + \bar{T}$

$$P\{\Xi_t(\omega) \neq \tilde{\Xi}_t(\omega)\} \leq \rho\alpha + \epsilon$$

where $\rho < 1$.

It is of importance that, in contrast to Lemma 3, we do not assume that $r = p + 2q - 2pq$ is small.

Below we illustrate in an informal way how the Lemma is proved. For an arbitrary $\epsilon > 0$ it is possible to choose such constant L that the probability of the "queue length" in each node at an arbitrary moment not surpassing L is estimated from below as $1 - \epsilon$. Let $M = \overline{m}LH$. Select $\overline{T}$ such that with a positive probability $1 - \rho$ the event $B = \{\omega$ : in the interval of time $[T_0, \hat{T}]$ there exist M sequential moments of time when the net "stands still", i. e. none of the nodes has new messages and all the messages transmitted from a certain node to the next one, leave the system of communication$\}$ be fulfilled. Let $B^C$ be the complement to the set B. If by the beginning of the period when "the net stands still" the lengths of all the queues do not surpass L, every message within the net will apply for transmission no less than $\overline{m}$ times and, consequently, by the end of the period it will leave the system of communication, and the net will be completely free from the messages.

Using the intuitive considerations discussed above the following estimate may be given accurate meaning:

$$P\{\Xi_{t+\overline{T}}(\omega) \neq \tilde{\Xi}_{t+\overline{T}}(\omega)\} \leq \epsilon$$

$$+ P\{\Xi_{t+\overline{T}}(\omega) \neq \tilde{\Xi}_{t+\overline{T}}(\omega) \mid B^C, \Xi_t(\omega) \neq \tilde{\Xi}_t(\omega)\} P\{B^C, \Xi_t(\omega) \neq \tilde{\Xi}_t(\omega)\}$$

$$\leq \epsilon + P\{B^C \mid \Xi_t(\omega) \neq \tilde{\Xi}_t(\omega)\} P\{\Xi_t(\omega) \neq \tilde{\Xi}_t(\omega)\} \leq \rho\alpha + \epsilon$$

where $t \quad T_0$, $\rho < 1$.

The author is grateful to R. L. Dobrushin under whose guidance the paper was written, and A. V. Jarkho for translating the paper.

## REFERENCES

1. R. L. Dobrushin and V. V. Prelov, Asymptotic approach to the research of linear nets of message commutation with a large number of nodes. Problemy peredachi informatsii (in press).

2. A. A. Borovkov, Probabilistic processes in the queueing theory, M. "Nauka", 1972. (In Russian).

3. Y. Chow, H. Robbins, and D. Siegmund, Great Expectations: The Theory of Optimal Stopping. Houghton Mifflin Comp. , Boston, 1971.

4. V. A. Malyshev, Classification of two-dimensional positive random walks and almost linear semi-martingales. Dokl. AN SSSR, 202, 3, 1972, pp. 526-528.

chapter 13

# THE SPECTRUM OF SMALL RANDOM PERTURBATIONS OF DYNAMICAL SYSTEMS

Yu. I. Kifer

Abstract. This paper studies the asymptotic behavior of the spectra of generators and transition operators of some Markov processes with a small parameter tending to zero. In the case of Markov processes with continuous time this problem is connected with the study of spectra of differential operators with a small parameter in the highest derivatives.

## INTRODUCTION

Consider an n-dimensional closed Riemannian manifold $M$ on which there is given a nondegenerate elliptic differential operator $L$ of second order and a vector field $B$ which generates a dynamical system $S^t$ satisfying the relation:

$$\left.\frac{d(S^t x)}{dt}\right|_{t=0} = B(x), \qquad x \in M.$$

A small random perturbation of the flow $S^t$ is a diffusion Markov process $x_t^\epsilon$ with generator (see [1]) having the form: $L^\epsilon = \epsilon^2 L + B$.

* This paper is an extended and more detailed version of the author's communication at the Fourth International Symposium on Information theory (see [2]).

In a similar way if $V$ is a homeomorphism of the manifold $M$ its small random perturbation is a Markov chain $y_n^\epsilon$ with transition density of the form: $p^\epsilon(x, y) = q^\epsilon(Vx, y)$, where $q^\epsilon(x, y)$ tends weakly to $\delta_x(y)$ as $\epsilon \to 0$ and $\delta_x(y)$ is the $\delta$-function concentrated at the point $x$. A transition operator of the Markov chain $y_n^\epsilon$ is denoted by $P^\epsilon$.

The present paper is devoted to the study of eigenvalues of the operators $L^\epsilon$ and $P^\epsilon$ as $\epsilon \to 0$. This question is interesting within the framework of the following general problem: what can one know about a dynamical system, when its parameters are transmitted with a small random noise. Due to the essential nonself-adjointedness of the operators $L^\epsilon$ and $P^\epsilon$ the study of the asymptotics of their spectra involves some difficulties. In light of the general problem formulated above, one can consider also results of the paper [3], where we found, for hyperbolic dynamical systems, the invariant measures stable under random perturbations.

This paper consists of three sections. In § I we consider perturbations of dynamical systems with a discrete spectrum and prove the convergence of the spectrum of the operator $L^\epsilon$ to the spectrum of the operator $B$. In doing this, we use a method which can be applied also for solving more general problems in perturbation theory (see §§I.IV, I. V). In § II we write the perturbation series for the eigenfunction and eigenvalues of $L^\epsilon$ and construct an example, which shows that these series, in general, diverge. §III is devoted to the study of the spectrum of the operator $P^\epsilon$, for the case of a dynamical system $\{V^m, m = \ldots, -1, 0, 1, \ldots\}$ having a continuous spectrum. We construct a model of perturbations, in which the spectrum of the operator $P^\epsilon$ consists just of one point, 0, on a subspace orthogonal to the constants. We wish to note that an example of this kind was first constructed in reference [4].

## I. PERTURBATIONS OF THE DISCRETE SPECTRUM

I. I. A dynamical system $S^t$ has pure discrete spectrum, provided there exists a set of real numbers $\{\gamma_k\}$ and an orthonormal basis $\{f_k\}$ in $L^2$-space with respect to $S^t$-invariant measure such that

$$f_k(S^t x) = e^{i\gamma_k t} f_k(x), \qquad i = \sqrt{-1}$$

If the $f_k$ are smooth functions, then (see [5, 6]) there exists a diffeomorphism $V$ which maps the manifold $M$ on the n-dimensional torus $T^n$ and

$$VS^t = R^t V ,$$

where $R^t$ is the quasiperiodic motion on the torus. In view of this fact it is reasonable to limit oneself to the case, when $M = T^n$, $n \geq 1$

$$B = (\omega_1 \frac{\partial}{\partial\varphi_1}, \ldots, \omega_n \frac{\partial}{\partial\varphi_n})$$

is a vector field with constant coefficients on $T^n$,

$$L = \sum_{k,\ell \leq n} d^{k\ell}(\varphi) \frac{\partial^2}{\partial\varphi_k \partial\varphi_\ell} + \sum_{k \leq n} b^k(\varphi) \frac{\partial}{\partial\varphi_k}, \quad \varphi = (\varphi_1, \ldots, \varphi_n)$$

is a nondegenerate elliptic differential operator on $T^n$ with $2\pi$-periodic coefficients.

Let a function $v(\varphi)$ on $T^n$ have the Fourier representation

$$v(\varphi) = \sum_q v_q e^{i(q,\varphi)}$$

where $q = (q_1, \ldots, q_n)$ is a set of integers and $(q, \varphi) = q_1\varphi_1 + \ldots + q_n\varphi_n$. The family of norms $\|\cdot\|_\rho$ is defined in the following way:

$$\|v\|_\rho^2 = \frac{1}{2}\sum_q |v_q|^2(1 + |q|^{2\rho})$$

where $|q|^{2\rho} = (q_1^2 + \ldots + q_n^2)^\rho$, and $0 = 1$. The closure of the set of all trigonometrical polynomials with respect to the norm $\|\cdot\|_\rho$ is called the Sobolev space $H^\rho$. The inner product in this space is given by the formula:

$$(v, w)_\rho = \frac{1}{2}\sum_q v_q \overline{w}_q(|q|^{2\rho} + 1)$$

where $\overline{w}$ is the complex conjugate of $w$: sometimes $(v, w)_0$ is denoted simply by $(v, w)$.

The operator $B$ has eigenvalues of the form $\lambda_m = i(m, \omega)$, where $m = (m_1, \ldots, m_n)$ and $\omega = (\omega_1, \ldots, \omega_n)$, and corresponding eigenfunctions $r_m(\varphi) = \exp\{i(m, \varphi)\}$, $\varphi = (\varphi_1, \ldots, \varphi_n)$. The frequencies $\omega = (\omega_1, \ldots, \varphi_n)$ are called rationally independent, if the equality $(m, \omega) = k$ for $k$ an integer and a set of integers $m = (m_1, \ldots, m_n)$ can be fulfilled only if $k = m = 0$.

Let the functions $a^{k\ell}(\varphi)$ and $b^k(\varphi)$, $k, \ell = 1, \ldots, n$ belong to the $H^\alpha$ space, $\alpha > 0$, and the frequencies $(\omega_1, \ldots, \omega_n)$ be rationally independent, then the following theorem is true:

Theorem 1. (a) For any set of integers $m = (m_1, \ldots, m_n)$ and $\epsilon$ small enough there exist an eigenfunction $r_m^\epsilon \in H^{2+\alpha}$ and an eigenvalue $\lambda_m^\epsilon$ of the operator $L^\epsilon$ such that

$$(r_m^\epsilon, r_m) = 1, \quad \|r_m^\epsilon\|_{2+\alpha} \le C_m, \quad \|r_m^\epsilon - r_m\|_{2+\alpha-\beta} \le C_{m,\beta}(\epsilon),$$

$$0 < \beta < \alpha, \quad |\lambda_m^\epsilon - \lambda_m| \le C_m \epsilon^2,$$

where $C_m$ depends on $|m|$ and the constant of ellipticity of the operator $L$ and $C_{m,\beta}(\epsilon) \to 0$ as $\epsilon \to 0$; the order of smallness of

$C_{m,\beta}(\epsilon)$ depends on approximative properties of the frequencies $(\omega_1, \ldots, \omega_n)$. Provided $\epsilon$ is small enough, the function $r^\epsilon_m$ and the number $\lambda^\epsilon_m$ are defined uniquely by these conditions;

(b) There exist constants $C^{(1)}$ and $C^{(2)} > 0$ and a natural number $\nu$, such that for an arbitrary eigenvalue $\lambda^\epsilon$ of $L^\epsilon$ satisfying the condition $\mathrm{Re}\,\lambda^\epsilon \geq - C^{(1)}\epsilon^2$, there exists $q$, such that

$$|\mathrm{Im}\,\lambda^\epsilon - (q, \omega)| \leq C^{(2)} \cdot \epsilon^2 ,$$

with $|q| \leq \nu$. Among such eigenvalues there exist $n$ numbers $\lambda^\epsilon_{m_1}, \ldots, \lambda_{m_n}$ satisfying the inequalities:

$$|\lambda^\epsilon_{m_k} - \omega_k| \leq C^{(2)} \cdot \epsilon^2, \quad k = 1, \ldots, n$$

Statement (b) of Theorem 1 gives an approximation method for the determination of generators of the spectrum of $B$ by the spectrum of $L^\epsilon$.

I. II. Proof. Let there exist an eigenfunction $r^\epsilon_m$ and an eigonvaluc $\lambda^\epsilon_m$ of $L^\epsilon$ satisfying the conditions

$$L^\epsilon r^\epsilon_m = \lambda^\epsilon_m r^\epsilon_m \quad \text{and} \quad (r^\epsilon_m, r_m) = 1$$

then

$$\lambda^\epsilon_m = i(m, \omega) + \epsilon^2 (L r^\epsilon_m, r_m) \tag{1}$$

and

$$\epsilon^2 L r^\epsilon_m + B r^\epsilon_m - i(m, \omega) r^\epsilon_m = \epsilon^2 (L r^\epsilon_m, r_m) r^\epsilon_m \tag{2}$$

When finding a solution of equation (2) the principal role belongs to an operator

$$F^\epsilon_m : H^{2+\alpha} \to H^{2+\alpha}$$

acting by the formula $F_m^\epsilon f = g$, where

$$\epsilon^2 D\Delta g + Bg - i(m,\omega)g = \epsilon^2(D\Delta - L)f + \epsilon^2(Lf, r_m)f \tag{3}$$

$\Delta = \sum_{k \le n} \frac{\partial^2}{\partial\varphi_k^2}$ is the Laplacian and $D > 0$ is a constant, which will be chosen later.

It is clear that $r_m^\epsilon$ is the fixed point of the operator $F_m^\epsilon$ and if $|m| \neq 0$, then the following implication is true:

$$\text{if} \quad (f, r_m) = 1, \quad \text{then} \quad (F_m^\epsilon f, r_m) = 1 \tag{4}$$

In what follows we will denote by $A(\varphi)$ the matrix $(d^{k\ell}(\varphi))$ and by $b(\varphi)$ the vector $(b^k(\varphi))$. Since the operator $L$ is a non-degenerate, there exist a constant $d > 0$, such that for the arbitrary complex vector $x = (x_1, \dots, x_n)$ the following inequality is fulfilled:

$$d^{-1}(x, x) \ge \operatorname{Re}(A(\varphi)x, x) \ge d(x, x) \tag{5}$$

where we set $(x, y) = \sum_{k \le n} x_k \bar{y}_k$, if $x = (x_1, \dots, x_n)$ and $y = (y_1, \dots, y_n)$. Let $A(\varphi) = \sum_q A_q e^{i(q,\varphi)}$ and $b(\varphi) = \sum_q b_q e^{i(q,\varphi)}$, where the $A_q$ are matrices and the $b_q$ are vectors. The coefficients of $L$ belong to the space $H^\alpha$ and so one can choose $N > 2$, such that

$$\| Lf - \hat{L}f \|_\alpha \le \frac{d}{8} \| f \|_{2+\alpha} \tag{6}$$

where the operator $\hat{L}$ has the coefficients $\hat{A}(\varphi) = (\hat{a}^{k\ell}(\varphi))$ and $\hat{b}(\varphi) = (\hat{b}^k(\varphi))$ having the form:

$$\hat{A}(\varphi) = \sum_{q:\,|q| \le N} A_q e^{i(q,\varphi)} \quad \text{and} \quad \hat{b}(\varphi) = \sum_{q:\,|q| \le N} b_q e^{i(q,\varphi)}$$

If $N$ is large enough, then, along with (6), in view of (5), the following inequality holds:

$$2d^{-1}(x,x) \geq \mathrm{Re}(\hat{A}(\varphi)x,x) \geq \frac{d}{2}(x,x) \tag{7}$$

for any complex vector $x = (x_1, \ldots, x_n)$.

Let

$$Z = n\left(\max_{k,\ell \leq n} \|a^{k\ell}(\varphi)\|_\alpha + \max_{k \leq n} \|b^k(\varphi)\|_\alpha\right)$$

It is clear that

$$\|Lf\|_\alpha \leq Z\|f\|_{2+\alpha},\ \|\hat{L}f\|_\alpha \leq Z\|f\|_{2+\alpha} \quad \text{and} \quad |(Lf, r_m)| \leq Z\|f\|_2 \tag{8}$$

Let $G_m^\epsilon : H^\alpha \to H^{2+\alpha}$ be the linear operator equal to $(\epsilon^2 D\Delta + B - i(m,\omega))^{-1}$ on a subspace orthogonal to $r_m$ and such that $G_m^\epsilon r_m = 0$. Provided $(f, r_m) = 1$ then in view of (3) and (4),

$$F_m^\epsilon f = r_m + \epsilon^2 G_m^\epsilon (D\Delta - \hat{L})f + \epsilon^2 G_m^\epsilon (\hat{L} - L)f + \epsilon^2 (Lf, r_m) G_m^\epsilon f \tag{9}$$

It is easy to see from (6) that

$$\epsilon^2 \|G_m^\epsilon (L - \hat{L})f\|_{2+\alpha} \leq \frac{d}{8D}\|f\|_{2+\alpha} \tag{10}$$

The estimations of other terms in (9) are contained in the following lemma.

<u>Lemma.</u> <u>One can choose</u> $D > 0$ <u>and</u> $\epsilon_o > 0$ <u>such that for any</u> $f \in H^{2+\alpha}$ <u>and</u> $\epsilon \leq \epsilon_o$ <u>the following inequalities are true:</u>

$$\epsilon^2 \|G_m^\epsilon (D\Delta - \hat{L})f\|_{2+\alpha} \leq \left(1 - \frac{d}{4D}\right)\|f\|_{2+\alpha} \tag{11}$$

<u>and</u>

$$\epsilon^2 \|G_m^\epsilon f\|_{2+\alpha} \leq \frac{\|f\|_{2+\alpha} \cdot d^2}{256 D^2 \cdot Z \cdot |m|^{2+\alpha}} \tag{12}$$

Proof: Let the operator $\Pi_M$ map an arbitrary function $f \in H^\alpha$, which has the Fourier representation

$$f(\varphi) = \sum_q f_q e^{i(q,\varphi)}$$

to the function $\Pi_M f$ with the Fourier representation $\Pi_M f(\varphi) = \sum_{q:\,|q| \geq M} f_q e^{i(q,\varphi)}$. It is clear that

$$G_m^\epsilon(D\Delta - \hat{L})f = G_m^\epsilon \hat{\Pi}_M(D\Delta - \hat{L})f + G_m^\epsilon \Pi_M \hat{\Gamma}\Pi_M f$$
$$+ G_m^\epsilon \Pi_M(D\Delta - \hat{L})\hat{\Pi}_M f + G_m^\epsilon \Pi_M(D\Delta - \hat{L} - \hat{\Gamma})\Pi_M f \tag{13}$$

where $\hat{\Gamma} = \sum_{k \leq n} \hat{b}^k(\varphi)\frac{\partial}{\partial\varphi_k}$, $\hat{\Pi}_M f = f - \Pi_M f$. Let

$$\delta(R) = \min_{q:\,0 < |q| \leq R} |(q,\omega)|$$

Provided the frequencies are rationally independent, $\delta(R)$ is a positive function, which is monotone nonincreasing in $R$.

It is easy to see that the following inequalities are true

$$\epsilon^2 \|G_m^\epsilon \hat{\Pi}_M(D\Delta - \hat{L})f\|_{2+\alpha} \leq \frac{\epsilon^2(D+Z)(M+N)^2}{\delta(M+|m|)}\|f\|_{2+\alpha} \tag{14}$$

$$\epsilon^2 \|G_m^\epsilon \Pi_M(D\Delta - \hat{L})\hat{\Pi}_M f\|_{2+\alpha} \leq \frac{\epsilon^2(D+Z)M^2}{\delta(M+N+|m|)}\|f\|_{2+\alpha} \tag{15}$$

$$\epsilon^2 \|G_m^\epsilon \Pi_M \hat{\Gamma}\Pi_M f\|_{2+\alpha} \leq \frac{Z}{M}\|f\|_{2+\alpha} \tag{16}$$

If

$$h = \Pi_M(D\Delta - \hat{L} - \hat{\Gamma})\Pi_M f \tag{17}$$

and the Fourier representation of a function $h$ has the form

$$h(\varphi) = \sum_q h_q e^{i(q,\varphi)}$$

then

$$h_q = - D|q|^2 f_q + \sum_{p:\, |q-p| \le N,\ |p| \ge M} f_p (A_{q-p} p, p)$$

if $|q| \ge M$ and $h_q = 0$, if $|q| < M$.

Let us put

$$V_{q,N} = \sum_{p:\, |q-p| \le N,\ |p| \ge M} \bar{f}_p (\bar{A}_{q-p} p, p)$$

where $\bar{A}_q$ is a matrix with elements which are complex conjugates to the corresponding elements of $A_q$, then one can write

$$\|h\|_\alpha^2 = D^2 \|\Pi_M f\|_{2+\alpha}^2 + \sum_{q:\, |q| \ge M} |V_{q,N}|^2 \cdot |q|^{2\alpha}$$

$$- D \sum_{q:\, |q| \ge M} |q|^{2+2\alpha} (f_q V_{q,N} + \bar{f}_q \bar{V}_{q,N}) \tag{18}$$

It is clear that the inequality below follows from the definition of the number Z

$$\sum_{q:\, |q| \ge M} |V_{q,N}|^2 \cdot |q|^{2\alpha} \le Z^2 \|\Pi_M f\|_{2+\alpha}^2 \tag{19}$$

Further

$$W_{M,N} = \sum_{q:\, |q| \ge M} |q|^{2+2\alpha} f_q \cdot V_{q,N}$$

$$= \sum_{q:\, |q| \ge M} |q|^{2+2\alpha} f_q \cdot \sum_{p:\, |q-p| \le N,\ |p| \ge M} \bar{f}_p (\bar{A}_{q-p} p, p)$$

$$= W^{(1)} + W^{(2)} + W^{(3)}$$

where

$$W^{(1)} = \sum_{p,q:\, |q-p| \le N,\ |p| \ge M,\ |q| \ge M} f_q \cdot \bar{f}_p \cdot |q|^{1+\alpha} |p|^{1+\alpha} (\bar{A}_{q-p} q, p)$$

$$= \int_{T^n} (\hat{\bar{A}}(\varphi) v(\varphi), v(\varphi)) d\varphi$$

where $v(\varphi) = (v_1(\varphi), \ldots, v_n(\varphi))$ and

$$v_k(\varphi) = \sum_{q:\, |q| \geq M} f_q \cdot |q|^{1+\alpha} q_k e^{i(q,\varphi)}, \quad q = (q_1, \ldots, q_n)$$

In view of (7),

$$\operatorname{Re} W^{(1)} \geq \frac{d}{2} \|\Pi_M f\|^2_{2+\alpha} \tag{20}$$

By $W^{(2)}$ and $W^{(3)}$ we have denoted the expressions

$$W^{(2)} = \sum_{p,q:\, |q-p| \leq N,\, |p| \geq M,\, |q| \geq M} f_q \bar{f}_p |q|^{1+\alpha} (|q|^{1+\alpha} |p|^{1+\alpha})(\bar{A}_{q-p} p, p)$$

and

$$W^{(3)} = \sum_{p,q:\, |q-p| \leq N,\, |p| \geq M,\, |q| \geq M} f_q \cdot \bar{f}_p \cdot |q|^{1+\alpha} |p|^{1+\alpha} (\bar{A}_{q-p}(p-q), p)$$

which are evaluated easily:

$$|W^{(2)}| \leq 4N^{n+3} Z M^{-1} \|\Pi_M f\|^2_{2+\alpha} \tag{21}$$

$$|W^{(3)}| \leq 4N^{n+3} Z M^{-1} \|\Pi_M f\|^2_{2+\alpha} \tag{22}$$

Let

$$D = 4Z^2 d^{-1} \tag{23}$$

and

$$M = [\, Z(8N^{n+3} + 1)^{-1}] + 1, \quad \epsilon_1 = \frac{\delta(M+N+|m|)D\lambda)^{\frac{1}{2}}}{8(D+Z)(M+N)^2} \tag{24}$$

where the square brackets mean the whole part of a number. Then, by virtue of the inequality

$$\epsilon^4 \|G^\epsilon_m \Pi_M (D\Delta - \hat{L} - \hat{I})\Pi_M f\|^2_{2+\alpha} \leq \frac{\|h\|^2_\alpha}{D^2}$$

where $h$ is defined by the formula (17), and from the relations (13)--(24), the inequality (11) takes place, provided $\epsilon \leq \epsilon_1$. Since

$$\epsilon^2 \|G_m^\epsilon f\|_{2+\alpha} \leq \epsilon^2 \|G_m^\epsilon \hat{\Pi}_{M_2} f\|_{2+\alpha} + \epsilon^2 \|G_m^\epsilon \Pi_{M_2} f\|_{2+\alpha}$$

$$\leq \frac{\epsilon^2 \|f\|_{2+\alpha}}{\delta(M_2+|m|} + \frac{\|f\|_{2+\alpha}}{DM_2^4}$$

the inequality (12) holds, provided

$$M_2 = [4\sqrt{2}\, D^{\frac{1}{4}} Z^{\frac{1}{4}} |m|^{\frac{2+\alpha}{4}} d^{-\frac{1}{2}}] + 1, \quad \epsilon_2 (\frac{\delta(M_2+|m|)}{2^9 Z |m|^{2\ \alpha}})^{\frac{1}{2}} \cdot \frac{d}{D}$$

and $\epsilon \leq \epsilon_2$. Thus the lemma is proved with $D$ being given by the formula (23) and $\epsilon_o = \min(\epsilon_1, \epsilon_2)$.

From the relations (8) - (12) it follows, provided $\epsilon \leq \epsilon_o$, that

$$\|F_m^\epsilon f\|_{2+\alpha} \leq R_m, \text{ if } \|f\|_{2+\alpha} \leq R_m \tag{25}$$

where $R_m = 16|m|^{2+\alpha} D \cdot d^{-1}$, i. e. the operator $F_m^\epsilon$ maps a ball $S_m$ of the radius $R_m$ in the space $H^{2+\alpha}$ into itself.

Let now $f$ and $g \in S_m$ and $(f, r_m) = (g, r_m) = 1$, then by (8)--(12)

$$\|F_m^\epsilon f - F_m^\epsilon g\|_{2+\alpha} \leq \epsilon^2 \|G_m^\epsilon (D\Delta - \hat{L})(f-g)\|_{2+\alpha}$$

$$+ \epsilon^2 \|G_m^\epsilon (L - \hat{L})(f-g)\|_{2+\alpha} + \epsilon^2 |(L(f-g),\ r_m)| \cdot \|G_m^\epsilon f\|_{2+\alpha}$$

$$+ \epsilon^2 |(Lg,\ r_m)| \cdot \|G_m^\epsilon (f-g)\|_{2+\alpha} \leq (1 - \frac{d}{8D}) \|f - g\|_{2+\alpha}$$

i. e. $F_m^\epsilon$ is the contraction operator in the ball $S_m$. Thus $F_m^\epsilon$ has in $S_m$ the unique fixed point $r_m^\epsilon$ and

$$\|r_m^\epsilon\|_{2+\alpha} \leq R_m \tag{26}$$

and, consequently, by virtue of (1), (8) and (26),

$$|\lambda^{\epsilon} - i(m, \omega)| \le \epsilon^2 Z R_m \tag{27}$$

It is not difficult to see that, in fact, the operator $F_m^{\epsilon}$ has a single fixed point in a ball of a radius $R_m(\epsilon)$, where $R_m(\epsilon) \to \infty$ as $\epsilon \to 0$.

By the analogy with the proof of the lemma it is easy to show, using (9), that

$$\|r_m^{\epsilon} - r_m\|_{2+\alpha-\beta} \le \frac{\epsilon^2 (M_3^{2-\beta}(\epsilon)(D + Z)R_m + ZR_m^2)}{\delta(M_3(\epsilon) + |m|)} + 2R_m (M_3(\epsilon))^{-2\beta}$$

One can choose $M_3(\epsilon)$ such that

$$\delta(M_3(\epsilon) + |m|) \ge \epsilon^{\frac{1}{2}}, \quad M_3^{2-\beta}(\epsilon)(D + Z)R_m + ZR_m^2 \le \epsilon^{\frac{1}{2}}$$

and $M_3(\epsilon) \to \infty$ as $\epsilon \to 0$. The speed of increase of $M_3(\epsilon)$ in $\epsilon$ as $\epsilon \to 0$ depends on the speed of decrease of $\delta(M)$ in $M$ as $M \to \infty$, i.e., on approximate properties of frequencies $(\omega_1, \ldots, \omega_n)$. Statement (a) of Theorem 1 is proved.

I. III Let now the eigenvalue $\lambda^{\epsilon}$ of the operator $L^{\epsilon}$ have the form $\lambda^{\epsilon} = \alpha^{\epsilon} + i\beta^{\epsilon}$, where $\alpha^{\epsilon}$ and $\beta^{\epsilon}$ are real numbers and $0 \ge \alpha^{\epsilon} \ge - C^{(1)} \cdot \epsilon^2$, $C^{(1)} \ge 0$. We introduce the operator $U^{\epsilon} : H^{2+\alpha} \to H^{2+\alpha}$ acting by the formula $U^{\epsilon} f = g$, where

$$\epsilon^2 D\Delta g + Bg - i\beta^{\epsilon} g = \epsilon^2 (D\Delta - L)f + \alpha^{\epsilon} f$$

Choosing D, N and M, as well as in the proof of (a) and putting $M_2 = [(32C^{(1)} d^{-1})^{1/4}] + 1$ and $\nu = \max(M_2, M)$ it is not difficult to prove that

$$\|U^{\epsilon} f\|_{2+\alpha} \le \left(1 - \frac{d}{16D}\right) \|f\|_{2+\alpha} \tag{28}$$

provided

$$\min_{q:|q| \le \nu} |\beta^{\epsilon} - (q, \omega)| \ge \epsilon^2 \left(\frac{8(D+Z)(M+N)^2}{Dd} + \frac{32DC^{(1)}}{d}\right) = C^{(2)} \epsilon^2 \tag{29}$$

It follows from (28) that there is no solution of the equation $U^\epsilon f^\epsilon = f^\epsilon$ except zero, provided (29) is valid. Thus for some $q$, such that $|q| \leq \nu$, the inequality $|\beta^\epsilon - (q, \omega)| \leq C^{(2)}\epsilon^2$ is true and this fact together with (27) proves statement (b) of Theorem 1.

Remark 1. Theorem 1 remains true, if the operator $L$ has complex coefficients but is a strong elliptic one, i. e. the inequality (5) holds.

Remark 2. If, by chance, the frequencies $(\omega_1, \ldots, \omega_n)$ are in resonance the situation differs from that being considered in Theorem 1. First, the infinitely many eigenvalues of the operator $L^\epsilon$ may converge to one eigenvalue of the operator $B = (\omega_1 \frac{\partial}{\partial\varphi_1}, \ldots, \omega_n \frac{\partial}{\partial\varphi_n})$, for example, in the case of the operator $L$ having constant coefficients. Second, as one can see in the case of the operator

$$L^\epsilon = \epsilon^2(2 + e^{-i\varphi_2})\frac{\partial^2}{\partial\varphi_1^2} + \epsilon^2(2 + e^{-i\varphi_1})\frac{\partial^2}{\partial\varphi_2^2} + \frac{\partial}{\partial\varphi_1}$$

all its eigenfunctions $f^\epsilon$ satisfying the condition $(f^\epsilon, e^{i(m\varphi_1 + n\varphi_2)}) = 1$ tend to infinity in $H^2$ space with the speed equal to $\text{const}\cdot\epsilon^{-1}$. Really, if

$$f^\epsilon(\varphi, \psi) = \sum_{k,\ell=-\infty}^{\infty} f_{k,\ell}\, e^{i(k\varphi + \ell\psi)}, \quad f_{m,n} = 1 \text{ and } L^\epsilon f^\epsilon = \lambda^\epsilon f^\epsilon, \text{ then}$$

$$f_{k,\ell} = \frac{-\epsilon^2(k^2 f_{k,\ell+1} + \ell^2 f_{k+1,\ell})}{2\epsilon^2(k^2+\ell^2) - i(k-m) - 2\epsilon^2(m^2+n^2) - \epsilon^2 m^2 f_{m,n+1} - \epsilon^2 n^2 f_{m+1,n}}$$

If $\|f\|_2 \leq C_1\epsilon^{-1}$, then $|f_{m+1,\ell}| \leq C_2\epsilon$ for any $\ell$. Therefore, in view of the equality $f_{m,n} = 1$ and the recurrence formula

$$f_{m,n+\nu+1} = f_{m,n+\nu}\left[f_{m,n+1} - \frac{2\nu(2n+\nu)}{m^2} + \frac{n^2}{m^2} f_{m+1,n}\right] - \frac{(n+\nu)^2}{m^2} f_{m+1,n+\nu}$$

one can find $\nu \geq 0$ such that $f_{m,n+\nu+1}$ has the order $\epsilon^{-1}$.

I. IV. The method of proof of Theorem 1 can be used in more general problems. On the Cartesian product of the n-dimensional torus $T^n$ and $\nu$-dimensional Euclidian space $R^\nu$ let there be given a nondegenerate elliptic differential operator:

$$L^\epsilon = \epsilon^2(L_\varphi + L_{\varphi,x} + L_x) + B_\varphi + B_x, \quad \varphi \in T^n, \; x \in R^n$$

where

$$L_\varphi = \sum_{k,\ell \leq n} a^{k\ell}(\varphi, x)\frac{\partial^2}{\partial\varphi_k \partial\varphi_\ell} + \sum_{k \leq n} b^k(\varphi, x)\frac{\partial}{\partial\varphi_k}, \quad \varphi = (\varphi_1, \ldots, \varphi_n)$$

$$L_{\varphi,x} = \sum_{k\leq n, \ell \leq \nu} b^{k\ell}(\varphi, x)\frac{\partial^2}{\partial\varphi_k \partial x_\ell}, \quad x = (x_1, \ldots, x_\nu), \; B_\varphi = \sum_{k\leq n} \omega^k(\varphi, x)\frac{\partial}{\partial\varphi_k}$$

$$L_x = \sum_{k,\ell\leq\nu} c^{k\ell}(\varphi, x)\frac{\partial^2}{\partial x_k \partial x_\ell} + \sum_{\ell\leq\nu} d^\ell(\varphi, x)\frac{\partial}{\partial x_\ell}, \; B_x = \sum_{k\leq\nu} \zeta^k(\varphi, x)\frac{\partial}{\partial x_k}$$

and each coefficient of $L^\epsilon$ is a smooth function; and for some point $O \in R^n$, which is not supposed to be unique, the following holds:

$$\sum_{k\leq n, \ell\leq\nu} |b^{k\ell}(\varphi, x)| + \sum_{\ell\leq\nu} |d^\ell(\varphi, x)| \leq C|x - O|, \quad \sum_{k,\ell\leq\nu} |c^{k\ell}(\varphi, x) - c^{k\ell}$$

$$+ \sum_{k\leq n} |\omega^k(\varphi, x) - \omega^k| \leq C|x - O|^{1+\gamma}, \quad \sum_{k\leq\nu} |\zeta^k(\varphi, x) - \sum_{\ell\leq\nu} \zeta^{k\ell}(x_\ell - O_\ell)|$$

$$\leq C|x - O|^{2+i}$$

where $C > 0$; $(c^{k\ell})$ is a matrix whose elements are constants and without loss of generality we suppose it to be the unit one: $\omega = (\omega^1, \ldots, \omega^n)$ is a constant vector with rationally independent components, $\zeta = (\zeta^{k\ell})$ is a matrix with eigenvalues, which have negative real parts and $\gamma = 0$, if n equals to 0 or 1, and $\gamma = 1$

if $n > 1$. In order to use the same words in the statements (a) and (b) of Theorem 2, formulated below, we suppose that $\zeta = (\zeta^{k\ell})$ is a normal matrix although this condition is not necessary for (a) of Theorem 2 to be true.

Let us denote by $\Delta_\varphi$ and $\Delta_x$ the Laplacians on $T^n$ and $R^\nu$, respectively. Let $H_\epsilon$ be the Hilbert space with the inner product:

$$(f, g) = \int_{T^n \times R^\nu} f(\varphi, x) \cdot \overline{g(\varphi, x)} p^\epsilon (x) d\varphi dx, \qquad \|f\| = (f, f)^{-\frac{1}{2}}$$

where

$$p^\epsilon (x) = \text{const. } \exp(- \frac{\zeta(x - O), (x - O))}{2\epsilon^2})$$

where $\text{const} > 0$ is a normalization factor. It is easy to show that an operator

$$D^\epsilon = \epsilon^2 Q \Delta_\varphi + \sum_{k \leq n} \omega^k \frac{\partial}{\partial \varphi_k} + \epsilon^2 \Delta_x + \sum_{k, \ell \leq \nu} \zeta^{k\ell} (x_\ell - O_\ell) \frac{\partial}{\partial x_k}$$

has in $H_\epsilon$ the complete set of orthogonal eigenfunctions of the form:

$$r^\epsilon_{m, \ell} (\varphi, x) = e^{i(m, \varphi)} \cdot f^\epsilon_\ell (x), \quad m = (m_1, \ldots, m_n), \quad \ell = 0, 1, \ldots,$$

where $f^\epsilon_\ell (x)$ is an eigenfunction of the operator

$$D^\epsilon_x = \epsilon^2 \Delta_x + \sum_{k, \ell \leq \nu} \zeta^{k\ell} (x_\ell - O_\ell) \frac{\partial}{\partial x_k}$$

with the eigenvalue $\lambda_\ell$, not depending on $\epsilon$. The functions $f^\epsilon_\ell (x)$ can evidently be written in terms of the polynomials of Hermite and Laguere. Let us introduce a Hilbert space $H^2_\epsilon$ with the norm

$$\|f\|_2 = \frac{1}{2} [ (f, f) + ((\Delta_\varphi + D^\epsilon_x) f, (\Delta_\varphi + D^\epsilon_x) f)]$$

We consider an operator $F^\epsilon_{m, \ell} : H^2_\epsilon \to H^2_\epsilon$ acting by the formula:

$$F^{\epsilon}_{m,\ell}f = r^{\epsilon}_{m,\ell} + \epsilon^2 G^{\epsilon}_{m,\ell}(Q\Delta_{\varphi} - L_{\varphi})f + \epsilon^2 G^{\epsilon}_{m,\ell}(\Delta_x - L_x)f$$

$$+ \epsilon^2 G^{\epsilon}_{m,\ell} L_{\varphi,x} f + \sum_{k \le n} (\omega^k - \omega^k(\varphi, x)) \frac{\partial f}{\partial \varphi_k}$$

$$+ \sum_{k \le \nu} (\zeta^{k\ell} \cdot (x_\ell - O_\ell) - \zeta^k(\varphi, x)) \frac{\partial f}{\partial x_k}$$

$$+ ((L^{\epsilon} - D^{\epsilon} + \epsilon^2 Q\Delta_{\varphi}) f, r^{\epsilon}_{m,\ell}) G^{\epsilon}_{m,\ell} f ,$$

where $G^{\epsilon}_{m,\ell}$ is the linear operator, which equals to $(D^{\epsilon} - i(m, \omega) - \lambda_\ell)^{-1}$ on a subspace, orthogonal to $r^{\epsilon}_{m,\ell}$ and such that $G^{\epsilon}_{m,\ell} r^{\epsilon}_{m,\ell} = 0$ and $Q$ is large enough. In the case of $n = 0$ the operator $L^{\epsilon}$ acts in a space of functions on $R^{\nu}$ and a corresponding operator $F^{\epsilon}_{m,\ell}$ is denoted by $F^{\epsilon}_{\ell}$.

Using the method of proof of theorem 1 and the following inequalities:

$$(|x - O|^{k_2}, |x - O|^{k_2}) \le \text{const} \cdot \epsilon^k; \quad \| \frac{\partial h}{\partial x_\ell} \| \le \text{const} \cdot \epsilon^{-1} \cdot \|h\|_2;$$

$$\left\| \frac{\partial^2 h}{\partial x_k \partial x_\ell} \right\| \le \text{const} \cdot \epsilon^{-2} \cdot \|h\|^2$$

it is easy to show, provided $\epsilon$ is small enough, that for any natural number $\ell$ and for any set of integers $m = (m_1, \ldots, m_n)$ the operators $F^{\epsilon}_{\ell}$ and $F^{\epsilon}_{m,0}$ are contracting in some ball of the space $H^2_{\epsilon}$ and their fixed points are eigenfunctions of $L^{\epsilon}$ satisfying the conditions of the following theorem.

<u>Theorem 2.</u> (a) <u>If $n \ge 1$ then for any set of integers</u> $m = (m_1, \ldots, m_n)$, <u>provided $\epsilon$ is small enough, there exist an eigenfunction</u> $e^{\epsilon}_m(\varphi, x)$ <u>and the corresponding eigenvalue</u> $\lambda^{\epsilon}_m$ <u>of the operator</u> $L^{\epsilon}$ <u>such that</u>

$$(e^{\epsilon}_m, r^{\epsilon}_{m,0}) = 1, \quad \|e^{\epsilon}_m - r^{\epsilon}_{m,0}\| \to 0$$

<u>and</u>

$$|\lambda_m^\epsilon - i(m, \omega)| \to 0 \text{ as } \epsilon \to 0$$

(b) If $n = 0$ then for any natural number $\ell$, provided $\epsilon$ is small enough, there exists an eigenfunction $e_\ell^\epsilon(x)$ and the corresponding eigenvalue $\lambda_\ell^\epsilon$ of the operator $L^\epsilon$ such that

$$(e_\ell^\epsilon, f_\ell^\epsilon) = 1, \quad \|e_\ell^\epsilon - f_\ell^\epsilon\| \to 0$$

and

$$|\lambda_\ell^\epsilon - \lambda_\ell| \to 0 \quad \text{as} \quad \epsilon \to 0$$

Statement (a) of theorem 2 can be interpreted as the spectral stability of quasiperiodic motion on the topologically stable torus under the perturbation by small diffusion given in some neighborhood of this torus. In consequence of this proposition it is interesting to understand the physical and probabilistic sense of eigenvalues and eigenfunctions of the nonselfadjoint operators under consideration.

I V. The method used in a proof of theorem 1 can be applied in the investigation of the asymptotic behavior of the spectrum of differential operators having a small parameter on the highest derivatives.

On the n-dimensional Riemannian manifold M of class $C^{k+1}$ let there be given a differential operator $L^\epsilon = \epsilon^2 L_1 + L_2$ with complex coefficients of class $C^{2\nu}$, where $2\nu \leq k$. We assume that $L_1$ is a strong elliptic operator (see [7]) of order $2\nu$ and $L_2$ is a differential operator of order $\eta < 2\nu$.

Let $\mu$ be a normalized measure on M with a density relative to the Riemannian volume on M satisfying a Hölder condition with the index $\alpha > 0$. We consider two cases:

(a) M is a closed manifold and we study the spectrum of the operator $L^\epsilon$ acting in a Hilbert space H which is the closure of the set of all functions of class $C^{k+\alpha}$ on $M$ with respect to the norm $\|f\| = (f, f)^{1/2}$, where

$$(f, g) = \int_M f(x)\overline{g(x)}\mu(dx)$$

(b) $M$ is the manifold with the border $\partial M$ and we study the spectrum of the operator $L^\epsilon$, acting in a Hilbert space $H_o$, which is the closure of the set of all finite functions of class $C^{k+\alpha}$ on $M$ with respect to the same norm as in (a).

Theorem 3. Let the operator $L_2$ have in $H$ (respectively in $H_o$) a complete set of orthonormal eigenfunctions $r_o = 1, r_1, r_2, \ldots$ coinciding with the set of eigenfunctions of some nondegenerate self-adjoint in $H$ (respectively in $H_o$) strong elliptic differential operator of order $2\nu$ with coefficients of class $C^{2\nu}$ and: (1) each eigenvalue of the operator $L_2$ has a finite multiplicity and lies on the imaginary axis; (2) eigenvalues $\{\rho_k, k = 0, 1, \ldots\}$ of the operator $L_o$ are nonpositive and $0 = |\rho_o| \le |\rho_1| \le \ldots \le |\rho_\ell| < \ldots$ and $|\rho_\ell| \to \infty$ as $\ell \to \infty$. Then for any simple eigenvalue $\lambda_k$ and the corresponding eigenfunction $r_k$ of the operator $L_2$ such that $L_o r_k = \rho_k r_k$ and $\rho_k \ne 0$ there exist the eigenvalue $\lambda_k^\epsilon$ and the eigenfunction $r_k^\epsilon$ of the operator $L^\epsilon$ such that

$$|\lambda_k^\epsilon - \lambda_k| \le C_k \epsilon^2, \quad \|r_k^\epsilon - r_k\| \to 0$$

as $\epsilon \to 0$ and $c_k > 0$ does not depend on $\epsilon$.

Proof: Let $r_\ell$ be an eigenfunction of the operator $L_o$ with an eigenvalue $\rho_\ell$, $\ell = 0, 1, \ldots$ each eigenvalue counting as many times as its multiplicity. We introduce a norm

$$|||f||| = \left[\frac{1}{2} \sum_{\ell=0}^{\infty} |f_\ell|^2 (|\rho_\ell|^2 + 1)\right]^{1/2}$$

where $f(x) = \sum_{\ell=0}^{\infty} f_\ell \cdot r_\ell(x)$ is the expansion of the function f with respect to the basis $\{r_k\}$. By virtue of Barding's inequality (see [7]) the norm $|||\cdot|||$ is equivalent to the usual Sobolev's norm [7].

As in the proof of Theorem 1 we are looking for the function $r_k^\epsilon$ as a fixed point of the operator $F_k^\epsilon$ acting by the formula $F_k^\epsilon f = g$, where

$$\epsilon^2 DL_o g + L_2 g - \lambda_k g = \epsilon^2 (DL_o - L_1)f + \epsilon^2 (L_1 f,\ r_k)f$$

for some $D > 0$ and

$$\lambda_k^\epsilon = \lambda_k + \epsilon^2 (L_1 r_k^\epsilon,\ r_k) \tag{30}$$

As in §I. II, we introduce the linear operator $G_k^\epsilon$ equal to $(\epsilon^2 DL_o + L_2 - \lambda_k)^{-1}$ on the subspace orthogonal to $r_k$ and such that $G_k^\epsilon r_k = 0$. If

$$\Pi_M f(x) = \Pi_M(\sum_{\ell=0}^{\infty} f_\ell \cdot r_\ell(x)) = \sum_{\ell=0}^{M} f_\ell r_\ell(x) \quad \text{and} \quad \Pi^M f = f - \Pi_M f\ ,$$

then, provided $(f,\ r_k) = 1$, the equality below takes place:

$$F_k^\epsilon f = r_k + \epsilon^2 G_k^\epsilon (DL_o - L_1)f + \epsilon^2 (L_1 f,\ r_k) G_k^\epsilon f \tag{31}$$

In view of Garding's inequality [7]

$$(L_1 f,\ L_o f) \geq c_1 |||f|||^2 - c_2 \|f\|^2 ; \qquad c_1,\ c_2 > 0 \tag{32}$$

provided $|||f||| < \infty$. Therefore, if constants D and M are chosen large enough, then by analogy with the proof of theorem 1 one can find $d > 0$ such that

$$\epsilon^2 |||G_k^\epsilon (DL_o - L_1)\Pi^M f||| \leq (1 - 0,5dD^{-1})\, |||f||| \tag{33}$$

for any $\epsilon > 0$. Further, provided M is fixed, one can choose N large enough such that for any $\epsilon > 0$

$$\epsilon^2 |||G_k^\epsilon \Pi^N (DL_o - L_1)\Pi_M f||| \le 0, \;125dD^{-1} |||f||| \tag{34}$$

Now, for M and N being fixed, one can find $\epsilon_o$ small enough such that for $\epsilon \le \epsilon_o$ the inequalities below are true:

$$\epsilon^2 |||G_k^\epsilon \Pi_N (DL_o - L_1)\Pi_M f||| \le 0, \quad 125dD^{-1} |||f||| \tag{35}$$

and

$$\epsilon^2 |(L_1 f, r_k)| \cdot |||G_k^\epsilon f||| \le 0, \; 015625 |||f|||^2 d^2 D^{-2} |\rho_k|^{-1} \tag{36}$$

From the relations (33)--(36) and (31) it follows that the operator $F_k^\epsilon$ maps a set of functions

$$W = \{f : |||f||| \le 8D|\rho_k|d^{-1}, \quad (f, r_k) = 1\}$$

into itself and for arbitrary $f, g \in W$

$$|||F_k^\epsilon f - F_k^\epsilon g||| \le (1 - 0, \; 25dD^{-1}) |||f - g|||$$

i. e., $F_k^\epsilon$ is the contraction operator in $W$. Thus $F_k^\epsilon$ has in $W$ the single fixed point $r_k^\epsilon$ and $\|r_k^\epsilon\| \le 8Dd^{-1} \cdot |\rho_k|$. Using this and the relations (30) and (31), the statement of Theorem follows.

## II. THE PERTURBATION SERIES

II. I. Let the operator L have infinitely differentiable coefficients. One can look for the eigenvalue $\lambda_m^\epsilon$ and the eigenfunction $r_m^\epsilon$ of the operator $L^\epsilon$ in the form:

$$\lambda_m^\epsilon = i(m, \omega) + \epsilon^2 \lambda^{(1)} + \epsilon^4 \lambda^{(2)} + \dots$$

and

$$r_m^\epsilon = e^{i(m, \varphi)} (1 + \epsilon^2 r^{(1)} + \epsilon^4 r^{(2)} + \dots)$$

The formal substitution of these expressions into the equation

$L^\epsilon r^\epsilon_m = \lambda^\epsilon_m r^\epsilon_m$ and the equating of terms with the same power of $\epsilon$ leads to equations of the form

$$Br^{(k)} = \lambda^{(k)} + \lambda^{(k-1)} r^{(1)} + \ldots + \lambda^{(1)} r^{(k-1)} - e^{-i(m,\varphi)} L(e^{i(m,\varphi)} \cdot r^{(k-1)}) \tag{37}$$

After integration of the left and right sides of (37) we obtain

$$\lambda^{(k)} = \int_{T^n} e^{-i(m,\varphi)} \cdot L(e^{i(m,\varphi)} r^{(k-1)}) d\varphi$$

since, as it is easy to see, each $r^{(\ell)}$ is orthogonal to constants.

Equations (37) do not always have solutions in the space $H^0$ even in the case of rationally independent frequencies $(\omega_1, \ldots, \omega_n)$. However, if there exists $\mu > 0$ such that for any $\ell$

$$\max_{q: |q| \le \ell} |(q, \omega)|^{-1} < \text{const} \cdot \ell^\mu$$

then, in view of [8], each equation in (37) has a solution. One can see from (37) that the definition of $r^{(k)}$ contains derivatives of order $2 \cdot (k - 1)$ of coefficients of the operator $L$. Since the norm of these derivatives has, in general, the order $(2 \cdot (k-1))!$ there is no reason to expect the series for $\lambda^\epsilon_m$ and $r^\epsilon_m$ to converge. Below we construct an example, which seems to illustrate the general case, where $r^\epsilon_m$ is not an analytic function of $\epsilon^2$, although, in view of remark 1 of §I, this example satisfies the conditions of Theorem 1.

II. II. Let $L^\epsilon$ be a strong elliptic operator on the circle of the form:

$$L^\epsilon = \epsilon^2 (2 + \epsilon^{-in\varphi}) \frac{d^2}{d\varphi^2} + \frac{d}{d\varphi}$$

and $r^\epsilon_m$ be an eigenfunction satisfying the condition:

$$\int_0^{2\pi} r_m^\epsilon(\varphi) e^{-im\varphi}\, d\varphi = 1$$

where m and n are integers. If $m \neq \ell n$ with $\ell$ being an integer, then a direct substitution of the function

$$r_m^\epsilon(\varphi) = \alpha + \sum_{k=0} \beta_k e^{i(m-k_n)\varphi}, \quad \beta_o = 1$$

into the equation $L^\epsilon r_m^\epsilon = \lambda^\epsilon r_m^\epsilon$ leads to the formulas:

$$\lambda^\epsilon = im - 2\epsilon^2 m^2, \quad \alpha = 0, \quad \beta_o = 1,$$

$$\beta_k = \frac{\epsilon^{2k} \prod_{\ell=1}^{k} (m - (\ell - 1)n)^2}{\prod_{\ell=1}^{k} (-i\ell n + 2m\ell n\epsilon^2 - 2\ell^2 n^2 \epsilon^2)}, \quad k = 1, 2, \ldots \tag{38}$$

<u>Proposition.</u> <u>There exists no</u> $R > 0$ <u>such that the function</u> $r_m^\epsilon$ <u>has a power series in</u> $\epsilon^2$ <u>which converges in</u> $H^o$ <u>for</u> $\epsilon^2 \leq R$.

<u>Proof:</u> By virtue of (38)

$$|\beta_k|^2 = \frac{\epsilon^{4k} \prod_{\ell=1}^{k} (m - (\ell - 1)n)^4}{n^{2k}(k!)^2 \prod_{\ell=1}^{k} (4\epsilon^2(\ell n - m)^2 + 1)}$$

Let

$$f_k(\epsilon^2) = [\prod_{\ell=1}^{k} (4\epsilon^2(\ell n - m)^2 + 1)]^{-1}$$

It is easy to verify, differentiating $f(\epsilon^2)$ $\nu$ times in $\epsilon^2$, that

$$|f_k^{(\nu)}(0)| > (\nu - 1)!\, (kn - m)^{2\nu} \tag{39}$$

If $g_k(\epsilon^2) = |\beta_k|^2$, then, in view of (39),

$$|g_k^{(\nu)}(0)| > (\nu - k - 1)!\,(kn-m)^{2(\nu-k)} n^{-2k} (k!)^{-2} \prod_{\ell=1}^{k} (m-(\ell-1)n)^4$$

where $g_k^{(\nu)}$ is the $\nu$'th derivative in $\epsilon^2$ of the function $g_k$. From here it follows by the Cauchy-Hadamard formula that

$$R_k = \lim_{\nu \to \infty} \left(\frac{g_k^{(\nu)}(0)}{\nu!}\right)^{-1/\nu-} < \frac{1}{(kn-m)^2} \tag{40}$$

where $R_k$ is the radius of convergence of a Taylor series in $\epsilon^2$ for the function $g_k(\epsilon^2) = |\beta_k|$. If the function $r_m^\epsilon$ could be expanded into a power series in $\epsilon^2$ which converges in the space $H^O$ with $\epsilon^2 \leq R$ and $R > 0$, then $\beta_k = (r_m^\epsilon, e^{i(k,\varphi)})$ and therefore $|\beta_k|^2$ would also expand into a power series in $\epsilon^2$ with $\epsilon^2 \leq R$ which contradicts (40).

It is easy to see that expansions in fractional powers of $\epsilon^2$ also diverge.

## §III. ON PERTURBATION OF THE CONTINUOUS SPECTRUM

III. I. Let $V$ be a hoemomorphism of the manifold $M$ and let $\mu$ be a measure, which is invariant with respect to $V$. The operator $U$, acting by the formula

$$Uf(x) = f(V^{-1/x}) \tag{41}$$

is unitary in the Hilbert space $H$, which is generated by the inner product

$$(f,g) = \int_M f(x)\overline{g(x)}\,\mu(dx) \tag{42}$$

The spectrum of the dynamical system $\{V^n, n = \ldots, -1, 0, 1, \ldots\}$ is defined as the spectrum of the operator $U$. Let now $Q^\epsilon$ be a completely continuous selfadjoint operator in $H$ with a positive continuous kernel $q^\epsilon(x,y)$, which is a member of continuous one-parameter semigroup $Q_t$, $t = \epsilon^2$, of hermitian operators $(Q_t Q_s = Q_{t+s},\ Q_o = 1)$ preserving function 1 invariant, and has, in accordance with the Hilbert-Schmidt theorem, the complete orthonormal system of real eigenfunctions $f_o = 1, f_1, f_2, \ldots$ and corresponding eigenvalues $e^{\epsilon^2 \lambda_o}, e^{\epsilon^2 \lambda_1}, \ldots$, with $\lambda_o = 0$, $\lambda_k \; 0$ for $k \neq 0$ and $\lambda_k \to -\infty$ as $k \to \infty$. It is obvious that the kernel $q^\epsilon(x,y)$ can be represented in the form:

$$q^\epsilon(x, y) = \sum_{k=0}^{\infty} e^{\epsilon^2 \lambda_k} f_k(x) f_k(y) \tag{43}$$

As the random perturbation of the homeomorphism $V_\epsilon$ we consider a Markov chain $y_n^\epsilon$ with the transition operator $P^\epsilon$ having the kernel $p^\epsilon(x, y) = q^\epsilon(Vx, y)$. The spectrum of $P^\epsilon$ can be studied as follows. Let $g(x) \in H$ and $g(x) = \sum_{k=0}^{\infty} g_k f_k(x)$, then

$$P^\epsilon g(x) = \int_M p^\epsilon(x, y) g(y) \sum_{k=0}^{\infty} e^{\epsilon^2 \lambda_k} g_k f_k(Vx) \tag{44}$$

Let

$$\gamma_{mn}^\epsilon = e^{\epsilon^2 \lambda_n} \int_M f_m(x) f_n(Vx) \mu(dx$$

and $\Gamma^\epsilon$ be the infinite matrix with elements $\gamma_{mn}^\epsilon$. If

$$P^\epsilon g(x) = \lambda^\epsilon g(x) \tag{45}$$

then, by virtue of (44), the equality (45) is equivalent to the relation

$$\Gamma^\epsilon g = \lambda^\epsilon \bar{g} \tag{46}$$

where $\bar{g} = (g_o, g_1, \ldots)$ is the vector, the elements of which are the coefficients of the expansion of the function $g$ relative to the basis $\{f_o, f_1, \ldots\}$.

III. II. The problem of the determination of eigenvalues of the matrix $\Gamma^{\epsilon}$ does not seem to be easy, in general, but under some conditions it can be solved.

Theorem 4. Let the following conditions hold: (a) the operator $U$, defined by the formula (41), has continuous spectrum on the subspace orthogonal to constants, i. e., the family of the projectors $F(\theta)$ in the spectral representation of the operator $U$ (see [7])

$$U = \int_0^{2\pi} e^{i\theta} \, dF(\theta)$$

depend on the parameter $\theta$ continuously;

(b) there exists a one to one correspondence $t$ of integers such that

$$f_k(Vx) = \gamma_k f_{t(k)}(x), \quad t(0) = 0, \quad |\gamma_k| = 1$$

Then the spectrum of the operator $P^{\epsilon}$ on the subspace orthogonal to constants consists of one point $0$ for any $\epsilon > 0$.

Proof: The operator $P^{\epsilon}$ is completely continuous, and so (see [7]) it has pure point spectrum apart perhaps from $0$. Let $\lambda$ be an eigenvalue of $P^{\epsilon}$ and $P^{\epsilon} g = \lambda g$, $(g, g) = 1$, with $g(x) = \sum_{k=0}^{\infty} g_k f_k(x)$ and $g_0 = 0$. Then, in view of the condition (b), for any integers $k$ and $m$ the following holds:

$$\lambda^m \cdot g_k = g_{t^{-m}(k)} \exp\{\epsilon^2 \sum_{s=1}^{m} \lambda_{t^{-s}(k)}\} \prod_{s=1}^{m} \gamma_{t^{-s}(k)} \tag{47}$$

As the operator $U$ and all its powers have, by virtue of the condition (a), no eigenfunctions, then for any $k \neq 0$

$$t^{-s}(k) \to \infty \quad \text{as} \quad s \to \infty \tag{48}$$

Hence $\lambda$ can not differ from zero since, by the assumption, $\lambda_\ell \to -\infty$ as $\ell \to \infty$. From the equality (47) one can see that the point 0 also is not an eigenvalue of $P^\epsilon$. In order to answer the question of whether 0 belongs to the spectrum, it is necessary to consider an equation $P^\epsilon g = h$, which is equivalent to the system

$$h_k = g_{t^{-1}(k)} \exp\{\epsilon^2 \lambda_{t^{-1}(k)}\} \gamma_{t^{-1}(k)}, \quad k = 0, 1, \ldots, \tag{49}$$

where

$$h(x) = \sum_{k=0}^{\infty} h_k f_k(x) \quad \text{and} \quad g(x) = \sum_{k=0}^{\infty} g_k f_k(x)$$

It follows from (49) that

$$|g_k| = h_{t(k)} | \exp\{\epsilon^2 \lambda_k\}$$

and, since $\lambda_k \to -\infty$ as $k \to \infty$, there exist $h \in H$ for which the function $g$ does not belong to $H$. Therefore the point 0 belongs to the limit spectrum of $P^\epsilon$.

The spectrum of $P^\epsilon$ has the form indicated in Theorem 4 only in consequence of the coordination condition (b) and, in general, the behavior of the spectrum as $\epsilon \to 0$ seems to be different. The instability of the spectral behavior depicted in Theorem 4 is corroborated by the following circumstance: for any $\lambda$ such that $0 \leq |\lambda| < 1$:

$$\|(P^\epsilon - \lambda)^{-1}\| \to \infty \quad \text{as} \quad \epsilon \to 0$$

The latter formula follows from the evident representation of a solution $r(x)$ of the equation $P^\epsilon r(x) - \lambda r(x) = g(x)$, which has the form $r(x) = \sum_{k=0}^{\infty} r_k f_k(x)$, where

$$r_k = -\frac{1}{\lambda}\left[g_k + \sum_{\ell=1}^{\infty} \left(\prod_{m=1}^{\ell} \gamma_{t^{-m}(k)}\right) \frac{g_{t^{-\ell}(k)}}{\lambda^\ell} \exp\left(\epsilon^2 \sum_{m=1}^{\ell} \lambda_{t^{-m}(k)}\right)\right]$$

and

$$g(x) = \sum_{k=0}^{\infty} g_k f_k(x)$$

The natural field of applications of Theorem 4 is the case of affine transformations of compact Abelian groups with the function $q^\epsilon(x, y)$ which is invariant under shifts. Let, for instance, M be the n-dimensional torus $T^n$, V is an automorphism, which is given by a matrix with integer elements and determinant equal to one and $q^\epsilon(x, y) = q(x - y)$. In the case, when the eigenvalues of V differ from 1 in module (hyperbolic case), and also in the case, when there are eigenvalues of each type, i. e. , which have module greater than 1, less than 1 and equal to 1, but not roots of unity, (partially hyperbolic case) the corresponding unitary operator has a continuous spectrum (see [9]) and, consequently, all conditions of Theorem 4 hold. In the case of an automorphism of the two-dimensional torus the result of Theorem 4 was obtained in [4] by another method using Fourier transformation.

Theorem 4 can be applied in the case of transformations with quasi-discrete spectrum, provided we consider them on a subspace which is orthogonal to eigenfunctions.

## REFERENCES


1. E. B. Dynkin, Markov processes. Moscow: FM 1963 (English translation: Springer 1965).
2. Yu. I. Kifer, On spectrum of small random perturbations of the dynamical systems. In: Proceedings of the Fourth International symposium on Information theory, pp. 78-80. Moscow-Leningrad. 1976 (In Russian).
3. ____________, On small random perturbations of some smooth dynamical systems, Izv. Akad. Nauk SSSR Ser. Math. 38, 1091-1115 (1974) = Math. USSR Izvestija 8, 1083-1107 (1974).

4. M. B. Dunskaja, On spectrum of the perturbed automorphism of the torus, Uspehi Mat. Nauk 26, 215(1971). (In Russian).

5. D. A. Lind, Locally compact measure preserving flows, Advances in Math. 15, 175-193 (1975).

6. ________, Spectral invariants in smooth ergodic theory, Lecture Notes in Phis., 38, 296-308 (1975).

7. K. Yosida, Functional analysis, Math. Wissenschaften b. 123. Berlin-Göttingen - Heidelberg : Springer 1965.

8. I. Brezin, R. Ellis, L. Shapiro, Recognizing G-induced flows, Israel J. of Math. 17, 56-65 (1974).

9. M. I. Brin, Ja. B. Pesin, Partially hyperbolic dynamical system, Izv. Akad. Nauk SSSR Ser. Mat. 38, 170-212 (1974)= Math. USSR Izvestija 8 (1974).

chapter 14

# REVERSIBLE MARKOV CHAINS WITH LOCAL INTERACTION

O. Kozlov and N. Vasilyev

## I. INTRODUCTION

We consider Markov processes with discrete time on the space of configurations of states on a locally finite graph; each site of the graph has a finite set of states and chooses its new state with given probability depending only on the states of its neighbors at the previous moment of time. Such countable systems of stochastic elements with synchronous local interaction have been considered by many authors (see [1-- 7]). Here we describe all of the transition functions for which there exist time reversible stationary chain and study their limiting behavior and, in particular, their ergodicity. It proves to be convenient to consider the probability distribution of the chain for two successive time moments as a random field on a "doubled" graph. We show in Section II that the random field must be Gibbsian with a pair potential on the doubled graph and that the transition function also must have a Gibbsian form.

In Section III some examples of translation invariant transition functions of reversible processes are shown, including a simple example of a nonergodic strictly positive transition function on the two-dimensional lattice. The main theorem about the convergence of a Gibbsian state of translation invariant processes with a transition function is formulated. It implies that a process which started in the infinite past must either be stationary and reversible, or periodic with period 2.

Section IV contains the proof of the convergence theorem based on the specific free energy decrease principle, known to us, thanks to the work of R. Holley [7]. The other condition of ergodicity which applies in the case of an ordered space of states and a monotone transition function is treated in Section V. In particular, for the one-dimensional lattice, the transition functions are ergodic in the class of all (not necessarily translation invariant) probability measures. A similar result is proved in another of Holley's works [8], dealing, as does [7], with continuous time processes.

## II. REVERSIBILITY CONDITIONS

Let $V$ denote a countable or finite set of elements, $Y$ a finite set of states, $X = Y^V$ the set of all $x = (x_i)$, $i \in V$ where $x_i \in Y$, $\mathfrak{K}$ the class of all finite $K \subset V$, $x_K$ the restriction of $x \in X$ on $K \subset V$ and $\mathfrak{U} = \mathfrak{U}(X)$ the space of probability measures on the natural $\sigma$-algebra $\mathfrak{B} = \mathfrak{B}(X)$ of subsets of $X$ generated by cylinder sets $\{x \in X: x_K = y_K\}$, $K \in \mathfrak{K}$, $y_K \in Y^K$ We denote by $\mu(y_K)$ the value of $\mu$ on the cylinder set, and by $\mu(y_K | y_J)$ the ratio $\mu(y_{K \cup J})/\mu(y_J)$ (whenever $K \cup J = \emptyset$, $\mu(y_J) \neq 0$). The measure in $\mathfrak{U}$ corresponding to a random field $\eta$ on $V$ with values in $Y$ will often be denoted by the same letter $\eta$, i. e. $\eta(y_K)$ means $\Pr(\eta_k = y_k, k \in K)$ where $\eta$ is an X-valued random variable. We consider Markov stochastic processes $\xi = (\eta^t)$ on $(X, \mathfrak{B})$ with discrete time $t$ of the following special form. Let $\Gamma = (V, B)$ be a graph with the site set $V$ and bond set $B$, where each set $N(k) = N_\Gamma(k) = \{j \in V: (k, j) \in B\}$

("neighborhood of $k$") is finite. Let $N(K) = \bigcup_{k \in K} N(k)$ for $K \subset V$.

Definition 1. A processes $\xi^t$ is called a synchronous process on $(X, \mathcal{B})$ with graph $\Gamma = (V, B)$ or, simply a process on $(\Gamma, Y)$ if for any $t$, $K \in \mathcal{K}$, $x \in X$, $y_K \in Y^K$

$$\Pr(\xi_K^{t+1} = y_K \mid \xi^t = x) = \prod_{k \in K} \Pr(\xi_k^{t+1} = y_k \mid \xi_{N(k)}^t = x_{N(k)}) \tag{1}$$

For example if $V = Z$ and $\Gamma$ is the graph pictured in Fig. 1, then the state $\xi_i^{t+1}$ of $i \in Z$ at the time moment $t + 1$ depends only on the state $\xi^t_{\{i-1, i, i+1\}}$ of its three neighbors (including $i$ itself) at time $t$.

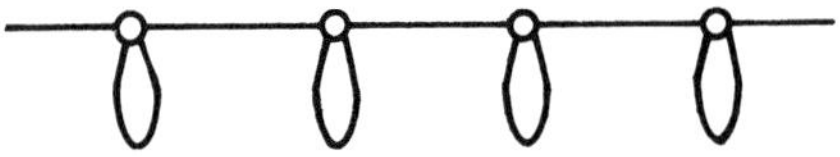

Fig. 1

The transition function (kernel) $Q$ of such a process is defined by the collection of positive numbers

$$q_k(y_k \mid x_{N(k)}) = \Pr(\xi_k^{t+1} = y_k \mid \xi_{N(k)}^t = x_{N(k)}) \tag{2}$$

where $k \in V$, $y_k \in Y$, $x_{N(k)} \in Y^{N(k)}$. We will also let the same letter $Q$ denote the corresponding linear operator $Q: \mathfrak{U} \to \mathfrak{U}$.

A Markov process $\xi^t$ $(t = 0, 1, 2, \ldots)$ with given $Q$ is defined by fixing the initial measure $\xi^0 = \eta \in \mathfrak{U}$; then $\xi^t = \eta Q^t$. We shall sometimes consider processes starting in the infinite past, that is, $\xi^t$ is defined for all integers $t$. If $\eta$ is an invariant measure for $Q$ $(\eta = \eta Q)$ then the generated process is stationary. We call a measure $\mu$ (or a kernel $Q$) strictly positive if $\mu(x_K) > 0$ for all $K \in K$, $x_K \in Y^K$ (or resp. if all of the $q_k$'s are positive).

Definition 2. A processes $\xi$ is called reversible if for any t and and $A_0, A_1 \in \mathcal{B}(X)$

$$\Pr(\xi^t \in A_0,\ \xi^{t+1} \in A_1) = \Pr(\xi^{t+1} \in A_0, \xi^t \in A_1)$$

Of course it is sufficient to check this condition for all cylinder sets $A_0$ and $A_1$. A reversible process is always stationary. To enable us to picture the common probability distribution at two successive time moments t and t + 1 for a process on $(\Gamma, Y)$, where $\Gamma$ is given, let us construct the doubled graph (bigraph) $\overline{\overline{\Gamma}}$:

Definition 3. The doubling of a given graph $\Gamma = (V, B)$ is the graph $\overline{\overline{\Gamma}} = (\overline{\overline{V}}, \overline{\overline{B}})$ where the site set $\overline{\overline{V}}$ is $\{(k, r), k \in V, r \in \{0, 1\}\}$ and the bonds connect sites of the form $(k, 0)$ and $(j, 1)$ where $(k, j) \in B$.

Figure 2 represents the doubling of the graph of Fig. 1. In general $\overline{\overline{V}}$ consists of two copies of V: $\overline{\overline{V}} = V_0 \cup V_1$ where

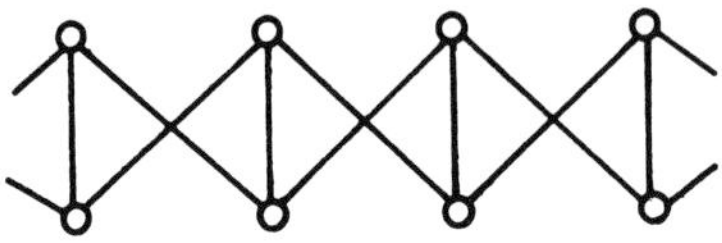

Fig. 2

$V_r = \{(k, r) : k \in V\}$, $r \in \{0, 1\}$. We interpret $X^{(0)} = Y^{V_0}$ and $X^{(1)} = Y^{V_1}$ as state spaces of the process at two successive time moments. All of the objects related to $\overline{\overline{\Gamma}}$ we mark with $=$ ($\overline{\overline{\mathcal{K}}}$, $\overline{\overline{N}}(k) = N_{\overline{\overline{\Gamma}}}(k)$ etc. ).

Each stationary process $\xi = (\xi^t)$ on $(\Gamma, Y)$ naturally generates a measure $\overline{\overline{\eta}} \in \overline{\overline{\mathcal{U}}}$ : for all $K, J \in \mathcal{K}$ such that if $K \subset V_1$, $\overline{\overline{N}}(K) \subset J \subset V_0$, $y_K \in Y^K$, $x_J \in Y^J$, then

$$\Pr(\xi_K^{t+1} = y_K | \xi_J^t = x_J) = \overline{\overline{\eta}}(y_K|x_J), \quad \Pr(\xi_J^t = x_J) = \overline{\overline{\eta}}(x_J) \tag{3}$$

(i. e., the projection of $\bar{\bar{\eta}}$ on $X^{(0)}$ is $\eta$). Of course the reversibility of $\xi$ is equivalent to $\bar{\bar{\eta}}$ being symmetrical, i. e. invariant under the reflection $(k, r) \rightarrow (k, 1-r)$ of $\bar{\bar{V}}$.

To formulate the main theorem describing reversible processes on $(\Gamma, Y)$ we use the following notion of Gibbsian kernel, analogous in some sense to the notion of Gibbsian conditional probabilities of random fields used in statistical physics.

Definition 4. A pair potential u on $(\Gamma, Y)$ is a collection $\{u_k : Y \rightarrow R;\ u_{kj} : Y \times Y \rightarrow R\}$ where $k \in V$ and $\{k, j\} \in B$, such that $u_{kj}(y, z) = u_{jk}(z, y)$ for all $\{k, j\} \in B$, $y \in Y$, $z \in Y$.

Definition 5. A symmetrical doubling of the pair potential u on $(\Gamma, Y)$ is the following potential $\bar{\bar{u}}$ on $(\bar{\bar{\Gamma}}, Y)$: $\bar{\bar{u}}_{(k,0)}(y) = \bar{\bar{u}}_{(k,1)}(y) = \bar{u}_k(y)$ and $\bar{\bar{u}}_{(k,0)(j,1)}(y, z) = \bar{\bar{u}}_{(k,1)(j,0)}(y, z) = u_{kj}(y, z)$.

Definition 6. A kernel Q on $(\Gamma, Y)$ is called Gibbsian with pair pair potential u if

$$q_k(y_k | x_{N(k)}) = \frac{\exp(-u_k(y_k) - \sum_{j \in N(k)} u_{kj}(y_k, x_j))}{\sum_{y \in Y} \exp(-u_k(y_k) - \sum_{j \in N(k)} u_{kj}(y_k, x_j))} \tag{4}$$

for $k \in V$, $y_k \in Y$, $x_{N(k)} \in Y^{N(k)}$.

Theorem 1. If Q is a strictly positive kernel on $(\Gamma, Y)$ a necessary and sufficient condition for the existence of at least one reversible process among the stationary process on $(\Gamma, Y)$ with Q, is that Q is Gibbsian with a pair potential u.

Before proving the theorem we recall the notions of Markov and Gibbs random fields.

Definition 7. Let $\tilde{\Gamma} = (\tilde{V}, \tilde{B})$ be a locally finite graph without loops. A random field $\eta$ on $(\tilde{\Gamma}, Y)$ is called Markovian if for all $K, J \in \mathcal{K}$,

$\tilde{N}(K) \subset J$ and $x_J \in Y^J$ such that $\eta(x_{J\setminus K}) > 0$ the equality $\eta(x_K|x_{J\setminus K}) = \eta(x_K|x_{\tilde{N}(K)\setminus K})$ holds.

Definition 8. A Gibbs potential on $(\tilde{\Gamma}, Y)$ is a collection u of the functions $u_C : Y^C \to R$ given for all cliques $C \subset \tilde{V}$ (full subgraphs) of $\tilde{\Gamma}$.

Given a Gibbs potential u and $J \in \mathcal{K}$, $x_J \in Y^J$ we denote the "energy" of the configuration $x_J$ by U:

$$U(x_J) = \sum_{C \subset J} u_C(x_C) \quad \text{(C's are cliques)} \tag{5}$$

and write for $K \subset J$

$$U(x_K|x_{J\setminus K}) = U(x_J) - U(x_K) \tag{6}$$

Definition 9. A random field $\eta$ on $(\Gamma, Y)$ is called Gibbsian with potential u if for $K, J \in \mathcal{K}$, $N(M) \subset J$, $x_J \in Y^J$

$$\eta(x_K|x_{J\ K}) = \frac{\exp(-U(x_K|x_{J\ K}))}{\sum_{z_K \in Y^K} \exp(-U(z_K|x_{J\ K}))}$$

Every strictly positive Morkov random field on $(\Gamma, Y)$ is Gibbsian and has a vacuum potential which is uniquely determined by "vacuum" $\bar{\bar{0}} \in Y$ (such that $u_0(x_0) = 0$ if $x_k = \bar{\bar{0}}$ for some $k \in C$), [10].

We fit these notions to $\bar{\bar{\Gamma}}$. Note that cliques of $\bar{\bar{\Gamma}}$ are only its sites and bonds, so that an arbitrary symmetrical (under reflection) Gibbs potential on $(\bar{\bar{\Gamma}}, Y)$ is the symmetrical doubling of a pair potential on $(\Gamma, Y)$. Now we will prove Theorem 1.

Necessity. If $\xi$ is reversible, one can easily check using (1) and (2) that the corresponding random field $\bar{\bar{\eta}}$ on $(\bar{\bar{\Gamma}}, Y)$ is

Markovian and therefore Gibbsian with vacuum potential $\bar{\bar{u}}$. Symmetry of $\bar{\bar{\eta}}$ and uniqueness of the potential $\bar{\bar{u}}$ imply that $\bar{\bar{u}}$ is symmetrical and so $\bar{\bar{u}}$ is the doubling of a pair potential $u$ on $(\Gamma, Y)$. It is easy to prove that $Q$ is Gibbsian with this potential $u$.

Sufficiency is almost obvious. If a symmetrical doubling $\bar{\bar{u}}$ of a given $u$ is considered, then there exists a symmetrical Gibbs random field $\bar{\bar{\eta}}$ on $(\bar{\bar{\Gamma}}, Y)$ with potential $\bar{\bar{u}}$. It is easy to verify that the reversible Markov process on $(\Gamma, Y)$ generated by $\bar{\bar{\eta}}$ satisfies condition (1) and corresponds to a Gibbs transition function on $(\Gamma, Y)$.

We need the following lemma to see that under the conditions of Theorem 1 the invariant measure $\eta$ of a reversible process is itself Gibbsian but "of the second order", i. e., Gibbsian on the graph $\hat{\Gamma}$ in which the neighborhood of any $k \in V$ is $\hat{N}(k) = N(N(k))$ (among the cliques $\hat{C}$ of this "second order" graph $\hat{\Gamma} = (V, \hat{B})$ are the sets $N(k)$, $k \in V$).

Lemma 1. Let $\bar{\bar{\eta}}$ be a strictly positive Gibbs measure on the bigraph $\bar{\bar{\Gamma}} = (V_0 \cup V_1, \bar{\bar{B}})$ with potential $\bar{\bar{u}}$; then

(A) for $K, J \in \mathcal{K}$, $K \subset V_1$, $\bar{\bar{N}}(k) \subset J \subset K_0$, $y_K \in Y^K$, $x_J \in Y^J$

$$\bar{\bar{\eta}}(y_K | x_J) = \prod_{k \in K} \frac{\exp(-U(y_k | x_{N(k)}))}{\sum_{y \in Y} \exp(-U(y | x_{N(k)}))}$$

(B) the projection $\eta$ of $\bar{\bar{\eta}}$ on $X^0 = Y^{V_0}$ is a Gibbs random field with the following potential $\hat{u}$ (the other $\hat{u}_{\hat{C}}$ being 0):

$$\hat{u}_k(x_k) = u_k(x_k); \quad k \in V \tag{7}$$

$$\hat{u}_{N(k)}(x_{N(k)}) = -\ln \sum_{y_k} \exp(-U(y_k | x_{N(k)}))$$

Proof: (A) is a direct consequence of the definition of a Gibbs random field. To check (B) one has to consider $I \subset V_0$ such that $K \supset \bar{\bar{N}}(I)$ and rewrite $\bar{\bar{\eta}}(x_I | x_{J\setminus I})$ as ratio of sums of some of the $\bar{\bar{\eta}}(x_K, x_J)$ using equalities

$$\sum_{y_K} \exp(-U(y_K, x_J)) = \prod_{j \in J} \exp(-u_J(x_j)) \prod_{k \in K} \sum_{y_k \in Y} (-U(y_k | x_{N(k)}))$$

Thus one gets the following:

Proposition 1. If a process $\xi$ on $(\Gamma, Y)$ is reversible and its kernel $Q$ is Gibbsian with potential $u$, then its invariant measure $\eta$ is Gibbsian with the potential $\hat{u}$ defined by (7).

A result related to Theorem 1 but dealing with finite homogeneous $V$'s (so called "periodic boxes" or "lattice tores") and $Y = \{0, 1\}$ is contained in [5]. The "second order" markov property of invariant measure of reversible process on $(\Gamma, Y)$ analogous to Proposition 1 was mentioned in [7] and other works of D. A. Dawson.

## III. CONVERGENCE THEOREMS AND EXAMPLES

In this section we study the behavior at infinity of the synchronous process $\xi^t$ with a Gibbs transition function $Q$. The space $\mathfrak{V} = \mathfrak{V}(X)$ of probability measures on $X = Y^V$ with the weak topology ($\mu_t \to \mu$ if $\mu_t(x_K) \to \mu(x_K)$ for all $x_K \in Y^K$, $K \in \mathfrak{K}$) is compact and so each continuous linear operator $Q : \mathfrak{V} \to \mathfrak{V}$ has a fixed point $\eta_Q$ in every convex closed $Q$ = invariant subset $N \subset \mathfrak{V}$.

Definition 10. A kernel $Q$ on $X$ is called ergodic on the set $N \subset \mathfrak{V}$ if the invariant measure $\eta_Q \in N$ of $Q$ is unique and $\eta Q^t \to \eta_Q$, when $t \to \infty$, uniformly in initial states $\eta \in N$.

We construct a process $\bar{\bar{\xi}}^t$ on $(\bar{\bar{\Gamma}}, Y)$ (see def. 3), which has its states in $V^{(0)}$ for even $t$ and in $V^{(1)}$ for odd $t$, as follows:

Definition 11. The symmetrical doubling of the process $\xi = (\xi^t)$ on $(\Gamma, Y)$ is the process $\bar{\bar{\xi}} = (\bar{\bar{\xi}}^t)$ on $(\bar{\bar{\Gamma}}, Y)$ such that for all $A_0 \in \mathcal{B}(X^{(0)})$, $A_1 \in \mathcal{B}(X^{(1)})$ we have

$$\Pr(\bar{\bar{\xi}}^t \in ((A_0 \times X^{(1)}) \cap (X^{(0)} \times A_1)) = \Pr(\xi^{t+s} \in A_0, \xi^{t+1+s} \in A_1)$$

where $s \in \{0, 1\}$, $s \equiv t \pmod 2$.

It is clear that the transition functions $Q_0, Q_1$ of $\bar{\bar{\xi}}^t$ ($Q_0$ operates at even $t$'s and $Q_1$ at the odd ones) are uniquely determined by the transition function $Q$ of $\xi^t$ and that $\xi$ and $\bar{\bar{\xi}}$ are uniquely determined by each other. If $\xi^t$ is stationary and reversible, then the measure $\bar{\bar{\xi}}^t = \bar{\bar{\eta}}$ (see (3)) does not depend on time; however, if $\bar{\bar{\eta}}$ does not depend on time, then $\xi^t$ may be either stationary (if $\bar{\bar{\eta}}$ is symmetrical) or periodic with period 2.

Definition 12. A sequence $\eta_n \in \mathcal{U}$, $n = 0, 1, 2, \ldots$ tends to a set $N \subset \mathcal{U}$ if for any open $\mathcal{O}$, $N \subset \mathcal{O} \subset \mathcal{U}$ there exists $n_0$ such that $\eta_n \in \mathcal{O}$ for $n > n_0$.

We shall prove that the process $\bar{\bar{\xi}}^t$ tends to a set of Gibbs random fields on $(\bar{\bar{\Gamma}}, Y)$ at least in the translation invariant case.

If $\Gamma = (\mathbb{Z}^d, B)$ is translation invariant (that is $(j,k) \in B \Rightarrow (j+i, k+i) \in B$ for all $i, j, k \in \mathbb{Z}^d$, where "+" denotes the usual sum of vectors) then one can define in a natural way the translation invariance of the measures $\eta$ processes, kernels etc. We denote the set of translation invariant $\eta$'s by $\mathcal{U}_T$.

Theorem 2. For any translation invariant $\xi^t$ on $(\Gamma, Y)$ related to the Gibbs kernel $Q$ (with pair potential $u$) the sequence $\bar{\bar{\xi}}^t \in \bar{\bar{\mathcal{U}}}_T$, $t = 0, 1, 2, \ldots$ tends to the set $\bar{\bar{\mathcal{G}}}_T$ of translation invariant Gibbs random fields (with potential $\bar{\bar{u}}$) on $(\bar{\bar{\Gamma}}, Y)$ and the convergence is uniform in initial states $\bar{\bar{\xi}}^0$ in $\mathcal{U}_T$.

A proof of the theorem using the decrease of free energy is contained in Section IV. If $\xi^t, \bar{\bar{\xi}}^t$ and Q are the processes and the kernel described in Theorem 2, then one can easily verify the following corollaries.

Corollary 1. $\xi^t$ $(t = 0, 1, 2, \ldots)$ tends to the set of translation invariant Gibbs random fields on $(\Gamma, Y)$ with potential u (defined in (7)).

Corollary 2. $\bar{\bar{\xi}}^t$ is stationary iff its initial state $\bar{\bar{\xi}}^0$ belongs to $\bar{\bar{\mathcal{G}}}_T$.

Corollary 3. Q is ergodic on $U_T$ iff $\bar{\bar{\mathcal{G}}}_T$ consists of only one random field.

Corollary 4. If $\Gamma = \mathbb{Z}^d$, then every translation invariant Gibbs kernel Q is ergodic.

Now we give some examples of Gibbs kernels with $Y = \{-1, +1\}$ (each element $k \in V$ has two states).

Example 1. Let $\Gamma$ be the graph shown in Fig. 1, i. e., $V = \mathbb{Z}$ and $N(k) = \{k-1, k, k+1\}$, and the probability $q_k(+1 | x_{N(k)})$ is equal to $\alpha^{3-r}/(\alpha^r + \alpha^{3-r}$ if r of the elements $j \in N(k)$ are in the states $x_j = +1$, $(r = 0, 1, 2$ or $3)$, $0 < \alpha < 1$; these $q_k$ determine the Gibbs kernel Q with pair potential $u_{jk}(x_j, y_k) = \beta x_j y_k$, $u_i \equiv 0$, where $\beta = \ln \alpha$. Then, according to Theorem 1 and Proposition 1, there exists a reversible process with having 2- Markov invariant measure $\eta_Q \in \mathcal{U}(\{-1, +1\}^{\mathbb{Z}})$; Corollaries 4 of Theorem 2, and 2 of Theorem 3 (Section V) show that Q is ergodic in $\mathcal{U}_T$ and even in $\mathcal{U}$.

Example 2. Suppose $V = \mathbb{Z}^2$ and each $N(k)$ contains the four nearest neighbors of k. Consider the Ising pair potential

$$u_{jk}(x_j, y_k) = \beta x_j y_k, \quad (j, k) \in B \tag{8}$$

Then for $\beta << 0$ (strong attraction case) there exist two regular translation invariant Gibbs random fields $\eta_{+1}$ and $\eta_{-1}$ with potential u (defined in (8)) such that $\eta_{+1}(x_k = +1) = \eta_{-1}(x_k = -1) > 1/2$.

The doubling graph $\overline{\overline{\Gamma}}$ consists of two parts $\Gamma^1$ and $\Gamma^2$ isomorphic to $\Gamma$ without connections among them: if $\mathbb{Z}^2$ is colored as a chessboard, the new states of the black (even) sites $k \in \mathbb{Z}^2$ depend only on the states of white (odd) ones and vice versa. Let $\overline{\overline{\eta}}_e$ $(e = +1)$ be the random field on $(\overline{\overline{\Gamma}}, Y)$ with projections on $\Gamma^1$ and $\Gamma^2$ which are independent and isomorphic to $\eta_e$. Then $\overline{\overline{\xi}}{}_e^0 = \overline{\overline{\eta}}_e$ generates a stationary reversible process $(\xi_e^t) = \xi^t$ and its kernel Q is Gibbsian (in the sense of Def. 6) with potential u. This kernel Q is the same for both of the processes $\xi_{+1}$ and $\xi_{-1}$; the probability $q_k(+1 | x_{N(k)})$ is equal to $\alpha^{4-r}/(\alpha^r + \alpha^{4-r})$ if r of the elements $j \in N(k)$ are in the states $x_j = +1$ $(r = 0, 1, 2, 3$ or $4)$, $0 < \alpha < 1$.

Thus for small $\alpha = \exp(2\beta)$ we get an example of a strictly stochastic translation invariant kernel Q on the lattice $\mathbb{Z}^2$, with two different invariant measures. A. Toom [6] has constructed a sophisticated example of such a kernel Q in which nonergodicity of Q is stable under slight distrubances of the parameters, while in our example 2, the loss of (+1) -- (-1) symmetry implies ergodicity.

Example 3. Consider the case $\beta >> 0$ in the previous example. Then there exists a regular, Gibbs random field $\xi$ on $(\Gamma, Y)$, invariant under translations by the vectors (1, 1) and (1, -1), such that $\zeta(x_k = +1) = \zeta(x_j = -1) > 1/2$ if k is any "even" site of $\mathbb{Z}^2$ and j is any odd one. Let $\overline{\overline{\zeta}}$ be the random field on $(\overline{\overline{\Gamma}}, Y)$ whose projections on $\Gamma'$ and $\Gamma''$ are independent and isomorphic (modulo translation) to $\zeta$. Then $\overline{\overline{\zeta}} = \overline{\overline{\xi}}{}^0$ generates a periodic (with period 2) process $\xi^t$ on $(\Gamma, Y)$ with Gibbs kernel corresponding to the potential u.

## IV. DECREASE OF FREE ENERGY

Let $\bar{\bar{u}}$ be the doubling of a pair potential $u$ on $(\Gamma, Y)$, where the graph $\Gamma = (\mathbb{Z}^d, B)$ and $u$ are translation invariants. Using the notation $U(\cdot)$ and $U(\cdot \mid \cdot)$ from (5) and (6) corresponding to $\bar{\bar{u}}$ we define the free energy $E_I$ of $\bar{\bar{\eta}} \in \mathcal{V}(\bar{\bar{X}})$ in a finite set $I \subset V$ as the difference between the mean value of its energy $U$ and its entropy: that is,

$$E_I(\eta) = \sum_{y_J} \bar{\bar{\eta}}(y_I) \cdot \bar{\bar{U}}(y_I) + \ln(\bar{\bar{\eta}}(y_I)$$

Denote by $K_1 \subset K_2 \subset \ldots$ a sequence of "lattice cubes" $K_n \in \mathcal{K}$, the union of which is $\mathbb{Z}^d$. One can show (using the method of [12], Theorem 7. 2. 3) that there is a limit

$$E(\bar{\bar{\eta}}) = \lim_{n \to \infty} \frac{1}{2|K_n|} E(\bar{\bar{K}}_n(\bar{\bar{\eta}}) \tag{9}$$

Here and later on, $\bar{\bar{K}} = \{(k, s) : k \in K, \; s \in \{0, 1\}\}$; $|K|$ is the number of elements in $K$, so $|\bar{\bar{K}}| = 2|K|$.

<u>Definition 13.</u> The value (9) is called specific free energy of $\bar{\bar{\eta}}$ (with respect to the potential $\bar{\bar{u}}$).

Let $\xi^t$ be a translation invariant process with Gibbsian kernel $Q$ corresponding to the pair potential $u$. We prove that $E(\bar{\bar{\xi}}^t)$ decreases when $t \to t + 1$ and preserves its value only if $\bar{\bar{\xi}}^t$ is a stationary process with a Gibbs (corresponding to $\bar{\bar{u}}$) initial state.

Let $Q_0$ and $Q_1$ be the transition functions of the doubled process $\bar{\bar{\xi}}^t$. Note that

$$\bar{\bar{\eta}}\, Q_s(y_I \mid x_J) = \frac{\exp(-U(y_I \mid x_J))}{\sum_{z_I} \exp(-U(z_I \mid x_J))} \tag{10}$$

for any $s \subset \{0,1\}$, $I, J \in \bar{\bar{K}}$, $I \subset V_s$, $\bar{\bar{N}}(I) \subset J \subset V_{1-s}$, $x_J \in Y^J$, $y_I \in Y^I$, $\bar{\bar{\eta}} \in \mathcal{V}(\bar{\bar{X}})$ and $\bar{\bar{\eta}}$ is Gibbsian with potential $\bar{\bar{u}}$ iff $\bar{\bar{\eta}} Q_0 = \bar{\bar{\eta}} Q_1 = \bar{\bar{\eta}}$.

<u>Lemma 2.</u> (A) <u>For any translation invariant Gibbs kernel</u> Q <u>and</u> $\bar{\bar{\eta}} \in \mathcal{V}_T(\bar{\bar{X}})$, $s \in \{0,1\}$, <u>we have the following inequality:</u>

$$\mathcal{D}_s(\bar{\bar{\eta}}) = E(\bar{\bar{\eta}}) - E(\bar{\bar{\eta}} Q_s) \geq 0$$

<u>which becomes an equality iff</u>

(B) <u>The function</u> $\mathcal{D}_s$ <u>is upper semicontinuous on</u> $\mathcal{V}_T(\bar{\bar{X}})$ <u>(i. e. for any</u> $\bar{\bar{\eta}} \in \mathcal{V}_T(\bar{\bar{X}})$ <u>and</u> $\epsilon > 0$ <u>there exists an open set</u> $\mathcal{O} \subset \mathcal{V}_T$ <u>such that</u> $\bar{\bar{\eta}} \in \mathcal{O}_k$, <u>and</u> $\mathcal{D}_{(s)}(\bar{\bar{\zeta}}) > \mathcal{D}_{(s)}(\bar{\bar{\eta}}) - \epsilon$ <u>if</u> $\zeta \in \mathcal{O}$).

The proof of the lemma follows [8] in essence. So we only sketch it briefly. Let $s = 1$. Introduce for all finite $I \subset V^1$, $J \subset V^0$, $J \supset \bar{\bar{N}}(I)$ and $x_J \in Y^J$ the conditional free energy of $\bar{\bar{\eta}}$ in I conditioned on $x_J$:

$$E_I(\bar{\bar{\eta}} \mid x_J) = \sum_{y_I} \bar{\bar{\eta}}(y_I \mid x_J)(U(y_I \mid x_J) + \ln \bar{\bar{\eta}}(y_I \mid x_J))$$

then

$$E_{I \cup J}(\bar{\bar{\eta}}) = E_J(\bar{\bar{\eta}}) + \sum_{x_J} \bar{\bar{\eta}}(x_J) E_I(\bar{\bar{\eta}} \mid x_J) \tag{11}$$

and

$$E_I(\bar{\bar{\eta}} \mid x_J) \geq E_I(\bar{\bar{\eta}} Q_1 \mid x_J) \tag{12}$$

which becomes an equality only if we have for all $y_I \in Y^J$

$$\bar{\bar{\eta}}(y_I \mid x_J) = \frac{\exp(-U(y_I \mid x_J))}{\sum_{z_I} \exp(-U(z_I) \mid x_J))} \tag{13}$$

To prove (12), (13) one may use the fact that $H(p_1, p_2, \dots, p_m) = \sum_m p_m(h_m - \ln p_m)$ (where the $h_m$ are fixed, $\sum_m p_m = 1$ and $p_m > 0$) reaches its minimum when $p_m$ are proportional to $\exp(-h_m)$.

The relations (11) and (12) imply

$$\mathcal{D}_{I \cup J}(\bar{\bar{\eta}}) = E_{I \cup J}(\bar{\bar{\eta}}) - E_{I \cup J}(\bar{\bar{\eta}}\, Q_I) \tag{14}$$

where the equality holds only if all of the $\bar{\bar{\eta}}(y_I | x_J)$ take the form (13).

For any lattice cube K denote by $T(K) = T_0(K) \cup T_1(K) \in \bar{\bar{\mathcal{K}}}$ the lattice trapesium with bases $T_1(K) = \{(k, 1) : k \in \mathcal{K}\}$ and $T_0(K) = \bar{\bar{N}}(T_1(K))$. One can easily check that $|E_{\bar{\bar{K}}}(\bar{\bar{\eta}}) - E_{T(K)}(\bar{\bar{\eta}})| < c_1 |K|$ where $c_1$ depends only on $\bar{\bar{u}}$ and $\bar{\bar{K}}$. So we can replace $\bar{\bar{K}}$ by $T(K)$ in (9) and get

$$\mathcal{D}_{(1)}(\bar{\bar{\eta}}) = \lim_{n \to \infty} \frac{1}{2|K_n|} \mathcal{D}_{T(K_n)}(\bar{\bar{\eta}}) \tag{15}$$

for any sequence $K_1 \subset K_2 \subset \dots$, $\bigcup_n K_n = \mathbb{Z}^d$ (the value $D_{T(K_n)}$ is defined as (14))

Now if the equal nonintersecting lattice cubes $K^{(1)}, K^{(2)}, \dots, K^{(2^d)}$ fill in a cube $K \subset \mathbb{Z}^d$, then

$$\mathcal{D}_{T(K)}(\bar{\bar{\eta}}) \geq \sum_{m=1}^{2^d} \mathcal{D}_{T(K^{(m)})}(\bar{\bar{\eta}}) \tag{16}$$

To prove (16) we represent D as the sum of the decrease in energy E and the increase in entropy H. Then for the first summand, inequality (16) turns out to be exact equality; to prove the inequality for the second summand one can express it in terms of the conditional entropies

$$H_{T(K)}(\bar{\bar{\eta}}) = H_{T_0(K)}(\bar{\bar{\eta}}) + \sum_{x_{T_0(K)}} \bar{\bar{\eta}}(x_{T_0(K)}) H_{T_0(K)}(\bar{\bar{\eta}} | x_{T_0(K)}).$$

and use the synchronous nature of the transition function $Q_1$.

If we fix a sequence $K_1 \subset K_2 \subset \ldots$, where $|K_n| = 2^d |K_{n-1}|$ and suppose $\bar{\bar{\eta}}$ translation invariant, then (16) implies $\mathcal{D}_{T(K_{n+1})}(\bar{\bar{\eta}}) \geq 2^d \mathcal{D}_{T(K_n)}$ so that (15) gives

$$\mathcal{D}_{(1)}(\bar{\bar{\eta}}) = \sup_n \frac{1}{2|K_n|} \mathcal{D}_{T(K_n)}(\bar{\bar{\eta}}) \tag{17}$$

If $\bar{\bar{\eta}} \neq \bar{\bar{\eta}} Q_1$ then (10) shows that (13) fails for some $n$, $x_{T_0(K_n)}, y_{T_1(K_n)}$ so $\mathcal{D}_{T(K_n)}(\bar{\bar{\eta}}) > 0$ and, according to (17), $\mathcal{D}_{(1)}(\eta) > 0$; this proves statement A of Lemma 2. Using (17) and the continuity of $\mathcal{D}_{T(K)}(\bar{\bar{\eta}})$ (as a function of $\bar{\bar{\eta}}$) we get statement B of the Lemma.

Proof of Theorem 2: If a translation invariant random field $\bar{\bar{\eta}}$ on $(\Gamma, Y)$ is not Gibbsian with potential $\bar{\bar{u}}$ then $\bar{\bar{\eta}} Q_0 \neq \bar{\bar{\eta}}$ or $\bar{\bar{\eta}} Q_1 \neq \bar{\bar{\eta}}$, hence A (Lemma 2) implies the continuity of $Q_s$, and B (Lemma 2) implies the upper semicontinuity of $\mathcal{D}$.

Take an open subset $\mathcal{O} \subset \bar{\bar{\mathcal{U}}}_T$ which contains the set $\bar{\bar{\mathcal{G}}}_T$ of the translation invariant Gibbs random field with potential $\bar{\bar{u}}$. The positive upper semicontinuous function $\eta \to D(\eta)$ takes its minimal value $\delta > 0$ on $\mathcal{U}_T \setminus \mathcal{O}$.

If $M = \max_{\bar{\bar{\eta}} \in \mathcal{U}_I} E(\bar{\bar{\eta}}) - \min_{\bar{\bar{\eta}} \in \mathcal{U}_I} E(\bar{\bar{\eta}})$, then for each $\bar{\bar{\xi}}^0 \in \mathcal{U}_T$, the process $\xi^t$ will be contained in $\mathcal{O}$ for all $t \geq 2U/\delta + 1$.

## V. MONOTONE TRANSITION FUNCTIONS

Let $Y$ be (partially) ordered. Then an ordering is naturally defined for $Y^V$ and $\mathcal{U} = \mathcal{U}(Y^V)$:

$$x, y \in Y^V,\ x \prec y \Leftrightarrow x_k \prec y_k \ (\text{or } x_k = y_k) \text{ for all } k \in K:$$

$$\mu, \nu \in U,\ \mu \prec \nu \Leftrightarrow \int_X f(x) d\mu(x) \leq \int_X f(x) d\nu(x)$$

for every nondecreasing function $f : X \to \mathbb{R}$ depending only on a finite set of coordinates $x_k$.

Of course it is sufficient to check the last inequality for all monotone (increasing) f's taking on only the two values 0 and 1.

We also need the following two properties of the relation $\prec$ in $\mathcal{U}$:

Lemma 3. If $\mu \prec \nu$ and for each $k \in V$ the projections $\mu_k$ and $\nu_k$ of $\mu$ and $\nu$ on the k-th coordinate are equal ($\mu_k = \nu_k$), then $\mu = \nu$.

Lemma 4. If $\mu$ and $\nu$ are independent (Bernoulli) measures on $Y^V$: $\mu = \prod_{k \in V} \mu_k$, $\nu = \prod_{k \in V} \nu_k$ and $\mu_k \prec \nu_k$ for every $k \in V$, then $\mu \prec \nu$.

One can check both of the statements first for $|V| = 2$ then for finite V by induction and for arbitrary V using the fact that the functions f considered depend only on a finite set of coordinates.

Definition 14. A pair potential u on $(\Gamma, Y)$ is convex if for all $y_1, y_2, z_1, z_2 \in Y$ such that $y_1 \prec y_2$, $z_1 \prec z_2$ and $i, j \in V$ we have

$$u_{ij}(y_1, z_1) + u_{ij}(y_2, z_2) \leq u_{ij}(y_1 z_2) + u_{ij}(y_2, z_1).$$

Definition 15. A transition function Q is called monotone iff $\mu \prec \nu$ implies $\mu Q \prec \nu Q$ or, equivalently, iff the operator Q' in the space of functions f on X generated by $Q(Q'f(x) = \int_X f \cdot d(\delta_x Q)$ maps the cone of monotone increasing functions into itself.

Lemma 5. A Gibbs kernel with convex potential u is monotone.

Proof: We must prove that $x \prec x'$ implies $\delta_x Q \prec \delta_{x'} Q$. Because of Lemma 4 one has to check only that the transition probabilities $q_k$ of Q satisfy the inequalities

$$\sum_{y \in Y} q_k(y_k | x_{N(k)}) \leq \sum_{y \in Y} q_k(y_k | x'_{N(k)})$$

for all $x < x'$ $(x, x' \in X)$, $y_k \in Y$, $k \in V$ and subsets $\overline{Y}$ of $Y$ with increasing indicators (i. e., $y \in \overline{Y}$, $y < z \Rightarrow z \in Y$).

We will also use the following notations later: $e$ is the maximal or minimal element of $Y$ $(e = \pm 1)$, $\delta_e$ (and $\bar{\bar{\delta}}_e$) is the measure in $\mathfrak{U}$ (and $\bar{\bar{\mathfrak{U}}}$) concentrated in the state "all $y_k = e$". A convex pair potential $u$ given, $\bar{\bar{\eta}}_e$ is a Gibbs random field (with potential $\bar{\bar{u}}$) which is the limit point of a sequence $\bar{\bar{\eta}}_e^{(n)}$, $n = 1, 2, \ldots$ where $\bar{\bar{\eta}}_e^{(n)}$ is Gibbsian with potential $\bar{\bar{u}}$ in $K_n \in \mathcal{K}$ $(K_1 \subset K_2 \subset K_3 \subset \ldots, \bigcup_n K_n = \bar{\bar{V}})$ and the projection of $\bar{\bar{\eta}}_e^{(n)}$ on $V \setminus K_n$ coincides with $\delta_e | K_n$. (As we shall see, $\bar{\bar{\eta}}_e$ is uniquely defined).

Note that the limits of the sequences $\eta_e^t$ (and $\bar{\bar{\xi}}_e^t$), $t = 0, 1, 2, \ldots$, corresponding to monotone $Q$ and initial states $\xi_e^0 = \delta_e$ exist. The equality of the limits for the initial states "all + 1" and "all -1" is equivalent to the ergodicity of $Q$:

Theorem 3. Let $(\bar{\bar{\xi}}_e^t)$ be the doubled process on $(\bar{\bar{\Gamma}}, Y)$ corresponding to the Gibbs kernel $Q$ with convex potential $u$ and $\bar{\bar{\xi}}_e^0 = \bar{\bar{\delta}}_e$, $e = \pm 1$. Then $\lim_{t\to\infty} \bar{\bar{\xi}}_e^t = \bar{\bar{\eta}}_e$ where $e = +1$ and $e = -1$ are maximal and minimal elements of the set $G$ of Gibbs random fields with the potential $\bar{\bar{u}}$. The condition $\bar{\bar{\eta}}_{+1} = \bar{\bar{\eta}}_{-1}$ is necessary and sufficient for ergodicity of $Q$.

Note the following corollaries of the theorem.

Corollary 1. Let $\lim_{t\to\infty} \delta_e Q^t = \eta_e$. If $\eta_{+1}(x_k) = \eta_{-1}(x_k)$ for each $k \in V$, $x_k \in Y$ then $Q$ is ergodic in $\mathfrak{U}$.

Corollary 2. If $V = \mathbb{Z}$ and the potential $u$ on $(\Gamma, Y)$ is convex and translation invariant, then $Q$ is ergodic in $\mathfrak{U}$.

The proof of Theorem 3 follows [9] in essence, so we sketch it briefly. Consider again the sequence $K_1 \subset K_2 \subset \ldots, \bigcup_n K_n = \bar{\bar{V}}$ and denote by $Q_e^{(n)}$ the superposition of $Q$ and the kernel which replaces all the $x_j$ by $j \in V \setminus K_n$, by $x_j = e$. There exists a doubling of the process related to $Q_{\bar{\bar{e}}}^{(n)}$, having invariant measure $\bar{\bar{\eta}}_e^t$ on $(\bar{\bar{\Gamma}}, Y)$ with initial measure $\bar{\bar{\delta}}_e$ and their restrictions to $K_n$ under arbitrary boundary conditions, one can show that if a random field $\bar{\bar{\eta}}^{(n)}$ is Gibbsian with potential $\bar{\bar{u}}$ in $K_n$ (and arbitrarily fixed in $V \setminus K_n$) then $\bar{\bar{\eta}}_{-1}^{(n)} \prec \eta^{(n)} \prec \eta_{+1}^{(n)}$ and for all $i < n$,

$$\eta_{-1}^{(i)} \prec \eta_{-1}^{(n)} \prec \lim_{t \to \infty} \bar{\bar{\xi}}_{-1}^{t} \prec \lim_{t \to \infty} \bar{\bar{\xi}}_{+1}^{t} \prec \eta_{+1}^{(n)} \prec \eta_{+1}^{(i)}$$

But $\lim_{n \to \infty} \bar{\bar{\eta}}_e^{(n)} = \lim_{t \to \infty} \xi_e^t$. All of the statements of Theorem 3 follow these relations.

## REFERENCES


1. L. N. Vasershtein, Markov processes on a countable product of spaces describing large systems of automata. Pr. Peredachi Informatzii, v. 5, No. 3, 64-71 (1969). (In Russian).
2. N. B. Vasilyev, M. B. Petrovskaya, I. I. Pyatetsky-Shaprio, A model of voting with random error. Automatika and Telemachanika, No. 10, 103-107 (1969). (In Russian).
3. N. B. Vasilyev, L. G. Mityushin, I. I. Pyatetsky-Shapiro, A. L. Toom, "Stavskaya operators", Moscow, Institute of Applied Mathematics, preprint, No. 12 (1973). (In Russian).
4. O. N. Stavskaya, I. I. Pyatetsky-Shapiro, Some properties of homogeneous nets of spontaneously active elements, Problemi Kibernetiki, v. 20, 91-106 (1968). (In Russian).
5. O. N Stavskaya, Gibbs invariant measures of Markov chains on finite lattices with local interactions, Math. sbornik, v. 92, No. 3, 402-419 (1973). (In Russian).
6. A. L. Toom, Nonergodic multidimensional systems of automata, Pr. Peredachi Informatzii, v. 10, No. 3, 70-79 (1974). (In Russian).

7. D. A. Dawson, Information flow in discrete Markov systems, J. Appl. Prob., No. 10, 63-83 (1973).

__________, Synchronous and asynchronous reversible Markov systems, Canad. Math. Bull., v. 75, No. 5, 633-649 (1975).

8. R. Holley, Free energy in a Markovian model of a lattice spin system, Comm. Math. Phys., v. 23, 87-99 (1971).

9. __________, An ergodic theorem for interacting systems with attractive interactions, Z. Wahrsheinlichkeitstheorie und vrw. Geb., v. 24, No. 4, 325-334 (1972).

10. M. B. Averintzev, Gibbsian description of random fields, the conditional probabilities of which may vanish. Pr. peredachi informatzii, v. 11, No. 4, 86-89 (1975).

11. Ali Nasr, Gibbs random fields for the Ising model. Trudi MMO (Proc. of Moscow Math. Soc.), v. 32, 187-209 (1975). (In Russian).

12. D. Ruelle, Statistical mechanics, New York, Amsterdam (1969).

chapter 15

# AN ALGORITHM–THEORETIC METHOD FOR THE STUDY OF UNIFORM RANDOM NETWORKS

G. I. Kurdumov

Abstract. Identical stochastic automata with finite state spaces are connected in an infinite chain. An automaton's state depends randomly on its own state and on the states of its two nearest neighbors at the previous moment. Section II contains a result concerning the algorithmic unsolvability of the ergodicity problem for chains of this type. In Section III the simultaneous dependence of automata on an input signal is introduced, and the chains are considered as devices for recognizing formal languages. It is shown that all languages are in some way recognizable by such networks. The method of proof in both sections is based on the simulation of Turing machines by networks.

## I. INTRODUCTION

Recall that a one-dimensional homogeneous random network S is an infinite automata system $\{s_i\}$ (i being an arbitrary integer) working in a discrete time $t = 0, 1, 2, \ldots$ . The state of an automaton $s_i$ at a moment t, denoted by $s_i^t$, is an element of a finite set X. It depends randomly on the states of the automata $s_{i-1}$, $s_i$, $s_{i+1}$

at the time $t-1$. In particular, if $s_{i-1}^{t-1} = x$, $s_i^{t-1} = x'$, $s_{i+1}^{t-1} = x''$, then $s_i^t = y$ with probability $\varphi_S^y(x, x', x'')$, where $\sum_{y \in X} \varphi_S^y(x, x', x'') = 1$. Furthermore, if the values of $s_i^j$ are given for all $i, j < t$ then all $s_i^t$ are mutually independent. The numbers $\varphi_S^y(x, x', x'')$ form a rectangular matrix of size $n^3 \times n$, $n$ being the cardinality of the set X. This matrix will be called the state transition matrix of an element of the network S.

The above definition coincides with that given in reference [1] for the case of a one-dimensional lattice with 3 neighbors (including the automaton itself). All the results in this paper can be easily generalized to the case of dependence on any finite number of neighbors and any finite dimension for the lattice.

A probability measure on the state space of a homogeneous random network is called invariant if it does not vary in time. A homogeneous random network is called ergodic if it has exactly one invariant measure. The problem concerning the ergodicity of homogeneous random networks has been investigated in a number of papers (see [1, 2] and some papers in this volume).

The result of Section II is as follows. In the general case the ergodicity problem for homogeneous random networks is algorithmically unsolvable. The problem is unsolvable even in the case in which all the transition probabilities belong to the set $\{0, 1, 1/2\}$.

In Section III homogeneous random networks are considered as devices for recognizing formal languages. We note that the concept of a homogeneous random network can be regarded as a generalization of the concept of an iterative automaton (see e. g. [3]). We give the definition of a recognizing network with parallel input. It is defined as a homogeneous random network in which there is also simultaneous dependence of all automata on an input signal. Initially, all the automata are in one and the same initial state. The input word is introduced into the network, symbol by symbol, during the first m moments, where m is the length of the input word. We define the output signal to be the proportion of those automata which are in states of a special type, called the acceptable ones. The input word is considered to be accepted if the corresponding proportion is at some time greater than a certain upper threshold $\beta$ (or equals $\beta$ in the case $\beta = 1$). The input word is considered to be rejected if the proportion is never greater than a lower threshold $\alpha$. The network $\alpha/\beta$-recognizes the language L if it accepts all the words of L and rejects all the other words. We present a number of theorems including Theorem 5 which is as follows. If $0 < \alpha \leq \beta < 1$, then any language can be recognized by a network with parallel input.

The main method used is the simulation of Turing machines. "Turing machine" means a deterministic or a probabilistic (see Section IV) one-tape machine which starts on the void tape (that is filled with symbols $\lambda$) and equipped with an input channel which gives a letter of the input alphabet A or the symbol $\lambda$ ($\lambda \notin A$). An input word $(a_0, a_1, \ldots, a_{n-1})$, where $a_i \in A$, is given one letter after the other during the first n moments. In all other moments the symbol $\lambda$ is given. An input word is said to be accepted if the machine stops sometime after having received it. A machine M recognizes a language L if it accepts a word w if and only if $w \in L$. The fact that the machine stops serves as the output signal. In the case of the simulation of a Turing machine by a homogeneous random network, the points corresponding to the initial positions of the tape are chosen randomly, and the simulation process is copied simultaneously at infinitely many places. In Section IV the corresponding construction is described and them used in the proofs of the theorems.

All non-trivial proofs (with the exception of Lemma 1 in Section IV) are given in Section V.

The work [10], published in English, was a preliminary version of the present paper. The difference is the addition of some new results on the recognition of languages and more concise wording of proofs.

The author would like to thank R. L. Dobrushin, A. L. Toom and V. L. Stefanyuk for their interest in the work and valuable discussions. He also expresses his gratitude to G. S. Plesnevich who put forward the hypothesis of the algorithmic unsolvability of the ergodicity problem for uniform networks of finite automata.

## II. THE ALGROITHMIC UNSOLVABILITY OF THE ERGODICITY PROBLEM

The main result of the section is the following theorem.

Theorem 1. There is no algorithm which determines for a given matrix $Q$ of size $n^3 \times n$ with elements from the set $\{0, 1, 1/2\}$ whether or not the uniform random network $S$ associated with the state transition matrix $Q$ is ergodic.

The main idea of proving this result is as follows. For any deterministic Turing machine, M, which starts its work on a blank tape and does not restore the blank symbol $\lambda$ on it, it is possible to construct efficiently a uniform random network, $S_M$ for which the following three statements are valid:

(1) If $s_i^0 = +$ for some $i$, then for any $j$ there is $t$, such that $\mathbb{P}[s_j^t = +] = 1$,

(2) If the machine M does not stop and for any $i$, $s_i^0 = \lambda$, then $\mathbb{P}[s_0^t = +] \equiv 0$,

(3) If the machine stope, then for any initial state of the network $\lim_{t \to \infty} \mathbb{P}[s_0^t = +] = 1$,

where $\lambda$ and $+$ are states of elements of the network $S_M$. Using these statements, we easily prove that the network $S_M$ is ergodic if and only if the machine M stops. Thus, the classical example of an algorithmically unsolvable problem, i. e., halting problem for

Turing machines, is reduced to the ergodicity problem for uniform random networks.

It is worth mentioning that the number 1/2 is inessential here. Moreover, it is possible to vary all the transition probabilities which are not equal to 0 or 1, within the bounds 0 and 1, the property of the network $S_M$ to be ergodic or not ergodic being invariant.

There is as yet a number of open questions: for example, those of the recursive enumerability of the set of ergodic networks, of the algorithmic unsolvability of ergodicity problems for deterministic networks and others.

## III. UNIFORM RANDOM NETWORKS AS DEVICES FOR RECOGNIZING LANGUAGES

Recall that an alphabet is a finite non-empty set of symbols; a word of the alphabet A is a finite (possibly empty) sequence of symbols belonging to A. The set of all the words of an alphabet A is denoted by $A^*$. A language L of the alphabet A is a subset of the set $A^*$.

In some papers, e. g. [4] and [5], the question of the recognizability of languages by ideal computing machines has been studied. In reference [4] deterministic and non-deterministic Turing machines are treated as recognizing devices. The following definition is equivalent to that of reference [4]. A Turing machine M recognizes a language $L \subseteq A^*$ if having begun its work with an input word $w \in A^*$, the machine M stops if and only if $w \in L$.

A finite automaton can be regarded as a special case of a Turing machine; namely, a Turing machine corresponding to the finite automaton is one which moves its tape only in one direction and has the automaton as its head. A language is called a finite state language if it can be recognized on a deterministic finite automaton. The concept of a finite state language corresponds to the concept of regular event, see [8].

In reference [3] iterative automata are considered as recognizing devices. These iterative automata are infinite uniform chains of deterministic finite automata. There are two modes of computing with iterative automata. According to the first one (called "on line") an input word is put, symbol by symbol, into a special network's element. In the second mode (called "off line") an input word is written simultaneously on a set of the network's elements, each element of the set getting exactly one symbol.

We suggest a new construction which is intermediate between the concepts "on line" and "off line".

A recognizing network with parallel input is defined as a uniform random network in which the state of each element at time $t$ depends also on the signal which the network has on its input. At time $t = 0$ all the network's elements are in an initial state $x_0$. We can put on the network's input any symbol of an alphabet $A$ (called the input alphabet) or the symbol $\lambda$ $(\lambda \notin A)$ which we put on the input at all moments $t \geq m$, $m$ being the length of the input word. We shall consider a special set $X_+ \subseteq X$ called the set of acceptable states.

Note that our network (unlike all the other computing systems) is completely space-uniform.

Let $\omega = {}_w c_i^t$ $(i = 0, \pm, 1, \pm 2, \ldots, t = 0, 1, 2, \ldots)$ be a realization of the evolution of the network $C$, consisting of automata $c_i$, when the input word $w$ is on the network input. Consider the following function

$$Q_\omega^\sim(t) = \lim_{n \to \infty} \sum_{-n \leq i \leq n} q({}_w c_i^t)$$

where $c_i = C$ and

$$q({}_w c_i^t) = \begin{cases} 1, & \text{if the automaton } c_i \text{ is in an acceptable state at time } t, \\ 0, & \text{otherwise} \end{cases}$$

This function will be called the acceptance function.

Remark 1. For any work realization of the recognizing network with $C$ on the input word $w$, the values of the function $Q(t)$ are well defined for any moment $t$ and coincide with the probabilities $\mathbb{P}[q({}_w c_0^t) = 1]$ of the event that the state of the automaton $c_0$ at time $t$ is acceptable.

In connection with the above, it is convenient to introduce the following notation:

$$_wQ_C(t) = \tilde{Q_\omega}(t) = \mathbb{P}[q({}_w C_0^t) = 1]$$

All the values of the function $_wQ_C(t)$ are non-random and can be computed provided we have a description of the recognizing network and of the input word.

Let $A$ be an alphabet, $L \subseteq A^*$, $C$ a recognizing network for which $A$ is the input alphabet, $\alpha$ and $\beta$ real numbers, $0 \leq \alpha \leq \beta \leq 1$.

We shall say that the network $C$ $\alpha/\beta$- recognizes the language $L$ if the following conditions are satisfied.

(1) $\forall(w \in L) \quad \exists t \quad ({}_wQ_C(t) > \beta) \vee ({}_wQ_C(t) = 1)$;

(2) $\forall(w \in A^* - L) \quad \forall t \quad ({}_wQ_C(t) \leq \alpha) \,\&\, ({}_wQ_C(t) < 1)$.

We shall say that a language $L$ is $\alpha/\beta$-recognizable if there is a network, which $\alpha/\beta$-recognizes this language.

Before we begin studying the $\alpha/\beta$-recognizability of various languages for various $\alpha$ and $\beta$, we cite Remark 2 which will be used in the proof of Theorem 2. The remark is a special case of the theorem.

Remark 2. A language L can be 0/1-recognized by a network with all the elements' state transition probabilities belonging to the set $\{0, 1\}$ if and only if L is a finite state language.

The proof of this fact is obvious. At any moment all the elements of the deterministic network will be in one and the same state. It is clear from the above why it makes no sense to consider in detail the deterministic recognizing networks with parallel input. It became known, however, that the condition of 0/1-recognizability is also too strong for probabilistic networks; the following theorem shows this.

Theorem 2. A language is 0/1-recognizable if and only if it is a finite state language.

In the following two theorems we use the concept of recursive enumerability of a language, which is well known in the theory of algorithms. A language is called recursively enumerable (r. e.) if it is enumerable as a set, i. e., it can be recognized by a deterministic Turing machine (in the above sense).

Theorem 3. Let $0 \leq \beta < 1$; then a language L is $0/\beta$-recognizable if and only if it is r. e. .

To prove the theorem we make essential use of a construction which will be outlined in Section IV. We construct a recognizing network that simulates the work of the corresponding Turing machine. The points corresponding to the initial positions of the tape heads are chosen at random at moment $t = 0$. The computing process is copied simultaneously in infinitely many places. The tape heads, which do not have sufficient space to complete the computing process, vanish (in the case when the machine simulated does not stop, this occurs sooner or later to every tape head). The work of the network is deterministic all the times, except $t = 0$. All probabilities which are not equal to 0 or 1, take on any value in the interval

(0,1). However, the fact that at the moment $t = 0$ the work of the network is not deterministic is very important, since this enables us to overcome the network's homogeneity and to organize a computing process on it.

The statement and the proof of Theorem 4 are very similar to those of Theorem 3.

Theorem 4. A language is 1/1-recognizable if and only if it is r. e..

The most unexpected result of this section is the following theorem. It answers the question, which was put by M. G. Shnirman.

Theorem 5. There exists a natural K, that for any $\alpha, \beta$, where $0 < \alpha \leq \beta < 1$, any alphabet A and any language $L \subset A^*$ there is a network, which $\alpha/\beta$-recognizes L, all elements of which have K states.

In the proof we shall use ideas, that have been developed in the paper [7]. The fact that there exists a one-one correspondence between the set of all the languages of a fixed alphabet and the set of all real numbers in the interval (0,1), is used in an essential way. This proof like the others, is based on the simulation of Turing machines; in this case they are probabilistic (see [7]). Since an alphabet A may have arbitrarily many symbols, while K is fixed, the input word w will be recorded in a statistical way in infinitely many places. To every letter which comes to the input of the network at a moment i ($i < m$, where m is the length of the input word) there corresponds a certain probability of the event that the symbol "1" is recorded in the i-th point of any position where w is recorded. All the letters of w having been given (after the moment m) the percentage of points marked with the sign "1" among i-th points of sufficiently many records of w is computed (the necessary number of records depends on $\alpha, \beta$, A and m). Hence the corresponding input word is restored to a sufficient degree of reliability. Thus all the input word is restored; thus the probability of an error must be much less than min $\{\alpha, 1-\beta\}$. The further action of the simulated Turing machine is, first, to perform sufficiently many statistical trials, which result in "1" with a probability of r, which is the code of the language L; second, to check (reliably enough) whether the input word belongs to the language, and third, depending on the result, to transfer or not to transfer the appropriate number of elements into the acceptable state.

Following the example of reference [7], we shall examine separately the case when all the transition probabilities are rational.

Remark 3. Assume that $\alpha < \beta < 1$. A language L can be $\alpha/\beta$-recognized on a network with rational transition probabilities if and only if L is r. e.

An analogous statement is valid for networks in which the transition probabilities are constructive real numbers (see [6]).

It should be noted that the network which is constructed in the proof of Theorem 5 can lose its property to recognize the corresponding language under any arbitrarily small variation of a non-zero transition probability. This is the main difference between the networks in Theorem 5 and those in Theorems 3 and 4. In order to clarify this difference, we cite the following definition and theorem.

Definition. We say that a newtork C $\alpha/\beta$-recognizes a language L in a stable way if it $\alpha/\beta$-recognizes L, and this property of the network C remains valid under any sufficiently small variations of all its transition probabilities which are not equal to 0 or 1.

Theorem 6. Assume that $\alpha < \beta < 1$. A language L is $\alpha/\beta$-recognizable in a stable way on some network if and only if L is r. e. .

In connection with the above facts it is interesting to mention that there are some other approaches to the definition of language recognizability on random networks with parallel input. For example, if we take as a principal consideration the existence of the acceptance function's limit or rather the value of this limit, then the fact that the function exceeds a given threshold, we shall have a concept which is similar to that of limit computability (see [9]). It is also possible to define in various ways the computability of functions on random recognizing networks. Also, it is interesting to study the question of the required time for the recognizing network's work, e.g., to find how the time for stable recognition of a language depends on various criteria of its complexity (see [4]).

Similarly, Shnirman (see [11]) suggests the consideration of networks in which there is stochastic dependence of element's states at time $t + 1$ on an acceptance function at time $t$. The idea to combine the local dependence on neighbors with the global dependence of all the elements on a common input has various analogies in engineering, namely, in computing networks. In chemistry and biology it is natural to associate an infinite chain of automata with a set of polymeric chains interacting with smaller molecules and ions in a solution. Some analogies can be found even in sociology.

## IV. THE BASIC CONSTRUCTION

In this section we construct two mappings $\Phi$ and $\Phi'$ of the set of all one-tape deterministic Turing machines into a subset of the set of our networks. These mappings will be used to prove some theorems.

Let us construct $\Phi$.

Let $M$ be a one-tape deterministic Turing machine, $Q$ the set of states of its tape head, $P$ the inner alphabet (e.g. the set of symbols written on the tape), $A$ the input alphabet. Let

$$M : P \times Q \times (A \cup \{\lambda\}) \rightarrow (P \times Q \times \{L, R\}) \cup \{\underline{\text{stop}}\}$$

be a program of the machine $M$, where the symbol "stop" corresponds to the machine's stop, the symbol $L$ means that the to the next cell to the left and the symbol $R$ means that the head goes to the next cell to the right. The symbol $\lambda \in P$ corresponds to an empty tape cell.

The network $\Phi(M)$ is deterministic with all the transition probabilities equal to 0 or 1. The state set $X$ of a network's element consists of the following parts: a set of passive states $\pi$, a set of active states $\alpha$ and the final state $+$.

The constructed network $\Phi(M)$ simulates the operation of $M$; an element of the network, which is in an active state, corresponds to a cell of the tape where a machine head is present; an element in a passive state corresponds to a cell where there is no head. The final state $+$ comes into existence when the machine stops, and then it expands in two ways along the network.

Let $\pi = P$, $\alpha = P \times Q$.

All the transitions of the states of elements of $\Phi(M)$ are determined by a function

$$f : (A \cup \{\lambda\}) \times X^3 \to X,$$

whcih we are going to define. If $c_{i-1}, c_i, c_{i+1}$ are successive elements of the network $\Phi(M)$, $c_{i-1}^t = x$, $c_i^t = x'$, $c_{i+1}^t = x''$, and $a \in A - \{\lambda\}$ is the letter given to the input at the moment $t$, then $c_i^{t+1} = f(a, x, x', x'')$. We set for any $a$: $f(a, x, x', x'') = x'$ in the following four cases:

(1) $\{x, x', x''\} \subseteq \pi$,

(2) $x' = \lambda$ and $\{x, x'\} \cap \{+, \lambda\} = \Phi$,

(3) $\{x, x'\} \subseteq \pi$ and $\mu(a, x'') = (p, q, R)$,

(4) $\{x', x''\} \subseteq \pi$ and $(a, x) = (p, q, L)$,

where the notation $\mu(a, x) = (p, q, d)$ $(d \in \{L, R\})$ means:

$\alpha \ni x = (p', q')$ and $\mu(a, p', q') = (p, q, d)$. In all the listed cases the state of the element remains unchanged; these cases correspond to: absence of a head; a single symbol $\lambda$; cells of the tape which are next to heads, the latter not moving to these cells according to the program.

The final state + is agressive, i. e., for all $a, x$ and $x'$,

$$f(a, +, x, x') = f(a, x, +, x') = f(a, x, x', +) = + .$$

If at least two elements of the set $\{x, x', x''\}$ belong to $\alpha$, and the third element is not +, then $f(\alpha, x, x', x'') = \lambda$. This assertion corresponds to the fact that a Turing machine, which works normally, has only one tape head.

If $\mu(a, x') = (p, q, d)$, and $\{x, x'\} \subseteq \pi$, then $f(a, x, x', x'') = p \in \pi$.

If $x \in \pi$, $x' \in \pi$ or $x = x' = \lambda$, and $\mu(a, x'') = (p, q, L)$, then $f(a, x, x', x'') = (x', q)$. Similarly, if $x' \in \pi - \{x\}$ $x'' \in \pi$ or $x' = x'' = \lambda$ and $\mu(a, x) = (p, q, R)$, then $f(x, x', x'') = (x', q)$.

The last two paragraphs characterize the Turing machine's action in the proper sense, that is, the action which is connected with changing records on the tape and states of the tape head.

If $\{x, x''\} \subseteq \pi$, $x' \subseteq \alpha$ and $\mu(a, x') =$ stop, then $f(a, x, x', x'') = +$. The appearance of the state + corresponds to the halting of the machine.

The network constructed simulates the work of the machine M. The network element, which is in an active state, corresponds to the head of the Turing machine, network elements in passive states correspond to tape cells. The final state + is formed at the moment the machine stops and begins to spread in both directions. The simulation of the action of the Turing machine is adequate only in the case in which the machine does not restore the symbol $\lambda$ on the tape, that is, if for all $p \in P$ and $q \in Q$ $\mu(a, p, q) = (\lambda, q', d)$. This can be explained by the fact that the state of an element of the network $\Phi(M)$ is stable if both neighbors are in states which are different from $+$ and $\lambda$ but the element itself is in the state $\lambda$. In other words, the head can meet the symbol $\lambda$ only if there is also $\alpha$ symbol $\lambda$ behind it.

Now we present two lemmas which illustrate some properties of the networks $\Phi(M)$ and will be used to prove the theorems.

Let $M$ be a Turing machine which does not restore the symbol $\lambda$, $q_0$ be the initial state of its head, $P$ be its inner alphabet, $A$ be the input alphabet, $w = (a_0, a_1, \ldots, a_{m-1}) \in A^*$, $\{s_i^t\}$ be the state of the network $\Phi(M)$ at any moment $t$.

Lemma 1. If the machine $M$ does not stop, the word $w$ is on the input, and $s_i^0 \in \{\lambda, (a_0, q_0)\}$ for all $i$, then for any $j$ and $t$, $s_i^t = +$.

Lemma 2. If the machine $M$ stops at a time $T_M(w)$, the word $w$ is on the input, and there is an $i$, for which

$$(s^0_{i-2Ti_M(w)}, \ldots, s^0_{i-1}, s^0_i, s^0_{i+1}, \ldots, s^0_{i+2Ti_M}$$

$$= (\underbrace{\lambda, \ldots, \lambda}_{2Ti_M(w)}, (a_0, q_0), \underbrace{\lambda, \ldots, \lambda}_{2Ti_M(w)}),$$

then at least one of the automata of the segment $[i - Ti_M(w), i + Ti_M(w)]$ will be in state $+$ by the moment $Ti_M(w)$.

Proof of Lemma 1: No automaton of the network is in the state $+$ at the moment $t = 0$. The state $+$ may appear in the network only as a simulation of the machines stopping. Since the machine $M$ does not stop, the word $w$ is on the input, and the network $\Phi(M)$ simulates the action of $M$, the state $+$ cannot appear. This is essentially based on the fact that for all $a$, if $\{x, x'\} \cap \{+, \lambda\} = \Phi$, then

$f(a, x, \lambda, x') = \lambda$, which guarantees that the information written on the tape by a head cannot be spoiled by other heads (in absence of active and final states, the passive states of the elements of $\Phi(M)$ remain unchanged).

Proof of Lemma 2: It is evident that all the changes of states in the segment $[i - Ti_M(w), i + Ti_M(w)]$ at all the moments $t < Ti_M(w)$ are caused only by the activity of the head which is simulated by the i-th automaton at the moment $t = 0$. Since the machine M does not restore the symbol and the number of cells visited by the head during the time $Ti_M(w)$ cannot exceed $Ti_M(w)$, and this time is simulated correctly; the machine stops by the moment $Ti_M(w)$ and this is simulated by the appearance state + at the corresponding point, q. e. d.

To prove Theorem 5, we need a notion of a probabilistic Turing machine which is rather different from that defined in [7], and a method of simulation of probabilistic Turing machines by uniform random networks. First let us define a mapping $\Phi'$ of some set of deterministic Turing machines into some set of networks.

Let M be a deterministic Turing machine, A be its input alphabet, P be its inner alphabet, Q be the set of the head states. We shall construct a network $\Phi'(M)$, which simulates the action of the machine M in a rather different way than the previously described network $\Phi(M)$ does. Let $p_+ \in P$, $q_\leftarrow \in Q$; this symbol of the inner alphabet and this state of the head of M will play a special role when constructing $\Phi'(M)$. Let

$$X_{\Phi'(M)} = X_{\Phi(M)} - \{+\} = \alpha \cup \pi$$

where $X_{\Phi'(M)}$ is the set of states of elements of the network $\Phi'(M)$. The state $p_+ \in \pi$ is the only acceptable state of elements of $\Phi'(M)$. The transitions of states of elements of $\Phi'(M)$ are determined by a function $f(a, x, x', x'')$, where $a \in (A \cup \{\lambda\})$, $\{x, x', x''\} \subseteq X_{\Phi'(M)}$, in the same way as transitions of states of elements of $\Phi(M)$, were determined by the function $f(a, x, x', x'')$. Let

$$f'(a,x,x',x'') = \begin{cases} (\lambda, q'), & \text{if } x' = \lambda, x'' = p, q_{\leftarrow} \text{ and } \mu(a,p,q) = (p'', q', L) \\ p, & \text{if } x' = (p, q) \text{ and } \mu(a, x') = \underline{\text{stop}} \\ f(a, x, x', x''), & \text{otherwise} \end{cases}$$

where $p, p' \in P$, $q', q'' \in Q$.

Speaking informally, this means that when a machine $M$ is simulated by $\Phi'(M)$, the head, being in the state $q_{\hookleftarrow}$, has "emergency powers": it may drive onto a symbol $\lambda$, coming from the left, even when this symbol $\lambda$ is single (unlike a head in any other state), thus penetrating into another head's territory. The other distinction of $\Phi'(M)$ from $\Phi(M)$ is that the former simulates the machines halting by the disappearance of the active state, without appearance of the sign $+$; on the other hand, the acceptable state of elements of $\Phi'(M)$ corresponds to a sign $p_+$ of the inner alphabet of $M$.

Now we define a probabilistic Turing machine. It differs from a deterministic Turing machine by the fact that its head's state (unlike the recorded sign and the direction of movement) at a moment $t$ depends (in addition to the head's state, the observed symbol and the input signal at the moment $t - 1$) also on some signals received at the moment $t - 1$ from a finite set of mutually independent random generators. To every i-th generator there corresponds a certain probability $\delta_i$ of the event that it generates 1 at any moment; otherwise it generates 0. This definition is similar to that given in [7]; the main difference is that in [7] only one generator is used.

Let $M_p$ be a probabilistic Turing machine, which has $h$ random generators, and $A, P$ and $Q$ are its input and inner alphabets and the set of head states respectively; $\delta_1, \dots, \delta_n$ are the probabilities that a 1 is produced by the random generators; and a mapping

$$\mu : \{0,1\}^h \times A \times P \times Q \to P \times Q \times \{L,R\}$$

is given, which satisfies the condition

$$\forall a \in A \; \forall p \in P \; \forall q \in Q \; \exists p' \in P \, \exists a \in \{L,R\} \; \forall (\gamma_1, \ldots, \gamma_h) \in \{0,1\}^h$$

$$\mu(\gamma_1, \ldots, \gamma_h, a, p, q) = (p', q', d)$$

where $q' \in Q$ is a program of the machine $M_p$. To every subset $\xi \subseteq \{1, \}, \ldots, h\}$ there corresponds in a natural way a deterministic Turing machine $M$, the program of which is of the form:

$$\mu_\xi(a,p,q) = \mu(g_1, g_2, \ldots, g_h, a, p, q),$$

where $a \in A$, $p \in P$, $q \in Q$, and $g_i$ equals 1 for $i \in \xi$ and 0 for $i \in \xi$. Let the matrix $M_p$ of transitions of elements of $\Phi'(M_p)$ be defined as follows:

$$M_p = \sum_\xi \left( \prod_{i \in \xi} \delta_i \right) \cdot \left( \prod_{i \notin \xi} (1 - \delta_i) \right) \cdot M_\xi$$

where $M_\xi$ is the matrix of transitions of elements of $\Phi'(M)$.

In other words, the network $\Phi'(M_p)$ simulates the operation of the probabilistic Turing machine $M_p$ just in the same way as the network $\Phi'(M)$ simulates the operation of the deterministic machine $M$. Various modes of operation of machine $M_p$ result in various transition rules for elements of the network $\Phi'(M)$, but the probabilities of corresponding variants are always equal. Correctness of the simulation (provided, of course, the machine $M_p$ does not restore the sign $\lambda$ with a probability of 1) is ensured by the fact, that a bifurcation of transitions in the network $\Phi'(M_p)$, which simulates $M_p$, is possible in only one element (namely that one in an active state). It is essential, that neither the sign recorded on the tape, nor the direction of the shift at a moment $t$ depend on the random signals, which came at the moment $t-1$.

Thus, we have constructed the mapping $\Phi'$ of the set of all stochastic Turing machines (because in fact our requirement for a state $q_{\leftarrow}$ of the head and a symbol $p_{+}$ of the tape causes no loss of generality) into the set of all recognizing networks, which provides (under some conditions) an adequate simulation.

The following note will be used to prove Theorem 5.

Note 4. The cardinality of the set $X_{\Phi'(M_p)}$ can be found by the formula

$$\|X_{\Phi'(M_p)}\| = \|P\| + \|P\| \cdot \|Q\|$$

where $P$ is the inner alphabet and $Q$ is the set of states of the head of the stochastic machine $M_p$. So $\|X_{\Phi'(M_p)}\|$ does not depend on other characteristics of $M_p$.

The proof is obvious, since $X_{\Phi'(M_p)} = P \cup (P \times Q)$.

## V. PROOFS

Proof of Theorem 1. Let us prove the algorithmic unsolvability of the ergodicity problem for any homogeneous random network with the transition probabilities $\{0, 1, 1/2\}$.

Let $M$ be a Turing machine. We shall assume, that the machine begins its work with an empty tape, does not restore the symbol $\lambda$, and on the right hand side and on the left hand side of the array of cells visited by the head, it writes two special symbols, restricting marks. We accept also that the machine works infinitely long if the head runs onto a symbol $\lambda$ before it arrives at the corresponding restricting mark (of course, the first two moments, when the restricting marks do not exist yet, are not considered here). It can be easily shown, that the halting problem for Turing machines is algorithmically unsolvable in this case as well. We suggest an algorithm, which constructs a transition matrix of a network $S_M$ such that the transition probabilities belong to the set $\{0, 1, 1/2\}$, the network being ergodic if and only if the machine M stops.

Let the state set X of the element of a network $S_M$ be the same as that of the network $\Phi(M)$ (see Section IV). The transition matrix of an element of $S_M$ is defined as the arithmetic mean of the transition matrix of $\Phi(M)$ and that of a network $\Psi_M$. The former is $\Phi(M)$ (and in the present case the probabilities of transitions of its elements do not depend on the input signal); the latter will now be described.

The network $\Psi_M$, as well as the network $\Phi(M)$, is deterministic; its transitions are described by a function $\Psi : X^3 \rightarrow X$.

Let $\{x, x', x''\} \subseteq X$. Then

$$(\{x, x', x''\} \subseteq \{\lambda\}) \ \& \ (+ \in \{x, x', x''\}) \rightarrow \Psi(x, x', x'') = \lambda ;$$

$$\Psi(+, x, x') = \Psi(x, +, x') = \Psi(x, x', +) = + ;$$

$$(\lambda, \lambda, \lambda) = (\lambda, q_0) \in \alpha ,$$

where $q_0$ is the initial state of the head of the machine M.

In other words, the network $S_M$, as well as the network $\Phi(M)$, simulates the work of the machine M. However, this simulation is not deterministic, and each cell (except those that are in the state "+" and their neighbors) at any moment can degrade into $\lambda$ with probability 1/2. The state "+" is agressive in the network $S_M$ too. If three successive elements of the network are in the state $\lambda$, then a new tape head in its initial state will be generated in the middle element with probability 1/2.

The following three assertions are valid.

(1) If $s_i^0 = +$ for some $i$, then for any $j$ there is $t$, such that $\mathbb{P}[s_j^t = +] = 1$.

(2) If the machine M does not stop and for any $i$, $s_i^0 = \lambda$,

then for any $t$, $\mathbb{P}[s_0^t = +] = 0$.

(3) If the machine M stops, then for any initial state of the network,

$$\lim_{t \to \infty} \mathbb{P}[s_0^t = +] = 1.$$

The first assertion is a trivial corollary of the fact that the state + is aggressive in the network $S_M$.

The proof of the second assertion follows from the fact that any tape head cannot get into the territory of any other tape head and the degradation into $\lambda$ of a set of automata storing information recorded by some head can lead to the disappearance of the given head, but not to the violation of its action.

Let us prove the third assertion. Let $Ti_M$ be the time which are needed for the machine M's stop. Let $\ldots s_1^0, s_0^0, s_1^0 \ldots$ be the initial state of the network. If there is $i$, such that $s_i^0 = +$, then the corresponding assertion is obvious since the state + is aggressive. If for any $i$, $s_i^0 \neq +$, then for any $j$ there is a positive probability of the fact that at the moment $t = 0$ all the automata in the segment $[j - 2Ti_M, j + 2Ti_M]$ move to the state $\lambda$; then, at the moment $t = 1$, a head appears at point $j$ and at no other point of the segment; and all later times up to the moment $t = Ti_M + 2$, all the transitions of the automata states in the above interval will correspond to the transitions of the network $\Phi(M)$. In this case, at time $t = Ti_M + 2$, "plus" will appear with probability 1 in some point of the network. Since the network is infinite, then, by the law of large numbers, at time $t = Ti_M + 2$, "plus" will appear at some point of the network with probability 1. This means that

$$\lim_{t \to \infty} \mathbb{P}[s_0^t = +] = 1$$

The third assertion implies the ergodicity of the network in the case when the machine M stops. Indeed, the only invariant measure of such a network is concentrated in the state $s_i \equiv +$.

We can prove using the first two assertions that the network is non-ergodic when the machine M does not stop. Indeed, some invariant measure will be concentrated (as in the previous case) on the state $s_i \equiv +$. Another invariant measure can be obtained by means of a construction with is analogous to that described in [2]. One has to take as the initial measure the one that is concentrated in the state $s_i \equiv \lambda$, then to consider all the subsequent measures and finally to take the limit of the mean-value over a subsequence.

The theorem is proved.

Proof of Remark 1. Let us show that for any t

$$Q^{\sim}_{\omega}(t) = \mathbb{P}[q({}_w c_0^t) = 1]$$

where $\omega = \{{}_w c_i^t\}$; $i = 0, \pm 1, \pm 2, \ldots$; $t = 0, 1, 2, \ldots$ .

Indeed, the recognizing network at time t is completely time-homogeneous. The output signal at each moment is also the same for all the elements of the network. Thus all the probability measures on the state space of the network which appear at the moments $t = 1$, $t = 2, \ldots$, are homogeneous. Consequently, for any t and i

$$\mathbb{P}[q({}_w c_i^t) = 1] = \mathbb{P}[q({}_w c_0^t) = 1]$$

Moreover, the recognizing network is infinite and the condition $i - j > 2t$ implies that $q({}_w c_i^t)$ and $q({}_w c_j^t)$ are independent random variables. Thus, by the law of large numbers, for any t,

$$Q^{\sim}_{\omega}(t) = 1/(2t+1) \sum_{0 \leq k < 2t+1} \lim_{n \to \infty} 1/2n \sum_{-n \leq i < n} q({}_w c^t_{(2t+1)i+k})$$

$$= \mathbb{P}[q({}_w c_0^t) = 1]$$

Proof of Theorem 2. It is obvious, that any finite state language is

0/1-recognizable. Let us prove the converse. Let C be a stochastic recognizing network, and C' be a deterministic recognizing network obtained from C so that $\varphi^y_{C'}(x, x', x'') = 1$ implies $\varphi^y_C(x, x', x'') > 0$.

Using the statement formulated in Remark 1, one can easily show that if the network C 0/1-recognizes a language, then the network C' also 0/1-recognizes it. Applying Remark 2, we obtain, that only a finite state language can be 0/1-recognizable.

The theorem is proved.

Proof of Theorem 3. First, let us prove that if $0 \leq \beta < 1$, A is an alphabet, $\lambda \notin A$, and L is an enumerable language for the alphabet A, then there exists a network C which $0/\beta$-recognizes the language L. Let M be a Turing machine, which does not restore the sign $\lambda$ and if the work w is on the input then M stops if and only if $w \in L$. The set of states of an element of the network C is $X_C = X_{\Phi(M)} \cup \{0\}$, where $X_{\Phi(M)}$ is the set of states of an element of $\Phi(M)$ (see Section IV). The state 0 is initial and is possible only at the moment $t = 0$. It makes a transition with probabilities 1/2 to each of the states $\lambda \in X_{\Phi(M)}$ and $(a_0, q_0) \in X_{\Phi(M)}$, where $a_0$ is the signal given to the input at the moment $t = 0$. All the subsequent transitions of elements of C are the same as those of $\Phi(M)$. Only the state $+ \in X_{(M)}$ is acceptable.

Assume $w \notin L$ so that the machine M does not stop and the word w is on its input; then, according to Lemma 1,

$$\mathbb{P}[c_0^t = +] \equiv 0$$

Now assume $w \in L$ is on the input; then M stops, say at a moment $Ti_M(w)$. We cut the network into segments, each of the length $l = 4Ti_M(w) + 1$. At the moment $t = 1$ each of these segments has a state resulting from Lemma 2, hence at the moment

$Ti_M(w) + 1$ there is a + in it with a positive probability P. Now, since + is aggressive, the following estimate holds

$$\mathbb{P}[c_0^t = +] \leq (1 - p)^{[t/l] - 1}$$

where $[\delta]$ is the largest integer not exceeding $\delta$. Therefore

$$\lim_{t \to \infty} \mathbb{P}[c_0^t = +] = 1$$

hence, in particular, there is a t such that

$$_wQ_C(t) = \mathbb{P}[c_0^t = +] > \beta$$

This means that the network C $0/\beta$-recognizes the language L.

It remains to prove, that if $0 \leq \beta < 1$ and a language L can be $0/\beta$-recognized on a network C, then L is r. e. .

Before completing the proof of Theorem 3, we establish the following lemma. Let S be a homogeneous random network with elements $\ldots, s_1, s_0, s_1, \ldots$. Let X be the set of all the possible states of an element of the network and t a positive integer.

Lemma 3. There exists an algorithm which makes it possible to obtain the list of states

$$\{x \in X \mid \mathbb{P}[s_0^t = x] > 0\}$$

for the automaton $s_0$ at time t provided the list

$$\{x \in X \mid \mathbb{P}[s_i^0 = x] > 0\}$$

is known for each i from the interval $-t \leq i \leq t$.

The proof of the lemma is obvious because of the fact that all the quadruples of states $x_1, x_2, x_3, y$ such that

$$\varphi_S^y(x_1, x_2, x_3) > 0$$

are known for the network S.

We return to the proof of Theorem 3. Assume that a network C $0/\beta$-recognizes a language L, where $0 \leq \beta < 1$.

We describe here in informal terms an algorithm which is applicable to the words of the alphabet A. Using this algorithm, we can finish our work if and only if the input word belongs to the language L. According to the Turing thesis this algorithm can be realized on a Turing machine, and this proves the recursive enumerability of the language L.

So, let w be an input word. Since the set of signals which can be introduced into the network C is finite and the states of all the automata at time $t = 0$ are known, we are able to find for each t the list

$$\{x \in X \mid \mathbb{P}[{}_w s_0^t = x] > 0\}$$

the input word w being used here as a datum. We shall obtain the corresponding lists step by step for all t, beginning with $t = 0$, until an acceptable state is found in one of the lists.

Taking into account Remark 1 and the fact, that the network C $0/\beta$-recognizes the language L, we can easily see that the work of the above algorithm will be completed if and only if $w \in L$.

The proof is now complete.

Proof of Theorem 4. The fact that $1/1$-recognizability of a language implies its enumerability can be proved in much the same way as has been done in Theorem 3. The same lemma is used. The only difference is as follows. In this case, when looking through the lists

$$\{x \in X \mid \mathbb{P}[{}_w s_0^t = x] > 0\}$$

we search for states which are not acceptable. This process goes on until we find a t such that the corresponding list consists of acceptable states only (this means exactly that ${}_{w}Q_{C}(t) = 1$). It is clear that such a $t$ exists if and only if $w \in L$.

To prove the 1/1-recognizability of a r. e. language, it is sufficient to alter the network $C$ constructed in the proof of Theorem 3 as follows. Passive states from $\pi \subseteq X_C$, as well as the state $+$ are to be made into acceptable ones.

Proof of Theorem 5. We shall prove that there is a natural number $K$ such that for any alphabet $A$, any language $L \subseteq A^*$ and any real $\alpha$ and $\beta$, where $0 < \alpha \leq \beta < 1$, there is a network $C$ with $K$ as the number of states of elements and which $\alpha/\beta$-recognizes the language $L$.

For any alphabet $A$ having $n$ symbols, any language $L \subseteq A^*$ and any $\alpha$ and $\beta$, where $0 < \alpha \leq \beta < 1$, we shall define a probabilistic Turing machine $M$ with $A$ as the input alphabet, an inner alphabet $P$ and a set of head states $Q$. We shall arrange that $P$ and $Q$ do not depend on the alphabet $A$, the language $L$, or the numbers $\alpha$ and $\beta$. The $A$ alphabet, $L$ language, and numbers $\alpha, \beta$ will affect the probabilities of the production of ones from tho random generators to $M$; in addition, the number of generators will depend on $A$. The only requirement of $M$ is that the network $\Phi'(M)$ will $\alpha/\beta$-recognize the language $L$; according to Note 4, this will prove the theorem.

We shall describe how the probabilistic machine $M$ operates, emphasizing the cases in which its head gets into the distinguished state $q$ or the symbol $p_+$, corresponds to the only acceptable state of the network $\Phi'(M)$. Note that the operation of the machine $M$ as such (one head starts to operate on a void tape) is not "meaningful"; only the operation of the network $\Phi'(M)$ is "meaningful". For the sake of concreteness, we may assume that $M$ starts operating on a tape which is void on the right side of the head, while on the left side there is a segment of any length, filled with signs "0", "1", "2", "3" (in any order) with two signs "D" at the end (see below).

Thus, the machine M operates, performing the following 15 stages one after the other.

(1) At the moment $t = 0$ a symbol 2 or 3 is recorded with probabilities $n^{-1}$ and $1 - n^{-1}$ respectively. The signal $a_{i_0}$ entered on the input at the moment $t = 0$, is remembered by the head itself. We assume $a_i \neq \lambda$, i. e., that the input word w is not the empty word $\Lambda$; otherwise the head moves right, recording in each cell $p_+$ if $\Lambda \in L$ and $p_-$ if $\Lambda \notin L$.

(2) All the time up to the moment $t = m$, where m is the length of the input word, the head moves right, recording on the tape signs 0 or 1 as follows: to every signal $a_i$ (where $A = \{a_1, \ldots, a_n\}$) which arrived at the input at the moment $t - 1$, there corresponds a probability of $i/n$ of the event that one is recorded at the moment t.

(3) The machine stops with probability $1 - \epsilon$ (the value of $\epsilon$ depends on $\alpha, \beta$ and n, in the way described below); it is simulated by the disappearance of the corresponding active state. In the other case the head, its state unchanged, moves one cell to the left. This process repeats over and over until the machine stops or the head drives into a symbol 2 or 3 (thus, the head "survives" with a probability of $\epsilon^m$).

(4) The head moves right and borders the right end of the string of ones and zeros recorded by it with two signs "D" (divider). The program of M is arranged to prevent the head in any state from crossing the divider; the latter needs to have two ones to prevent two heads from colliding with each other on it, which would result in their disappearance.

(5) The head makes a transition to the state $q_{\rightarrow}$ and moves left, leaving the signs 0, 1, 2, 3 unchanged, but replacing * for all signs $\lambda$. This continues until the head drives into a sign "D". (Of

course, operating on an empty tape, the machine M will never meet "D" and never stops; but, in simulation, any head which has survived will meet sooner or later a divider, recorded by another head). When driving into a sign "D" and being in the state $q_{\rightarrow}$, the head of M makes a transition into another (non-distinguished) state. The part of the tape (the part of the network in simulation) between two successive dividers, inclusive, will be called the zone of the head which is in it.

(6) Throughout its zone, the head replaces by the symbols * all the segments of symbols 0, 1, 2, 3 whose length is not equal to m + 1. These segments are relics of the operation of heads which have vanished before the moment m.

(7) Throughout its zone, the head counts and encodes on the tape the number $n_2$ of symbols "2" and the number N of symbols "2" and "3". The head performs all the computations required for that using only its own zone while retaining in each cell the symbol (0, 1, 2, 3) recorded into it at the two first stages and not moving (up to the 14-th stage) the boundaries of its zone.

(8) The natural number $n_?$ is computed to be the nearest integer to $N/\max\{n_2, 1\}$. Note here that the parameter $\epsilon$ (see the stage 3) will be chosen below to make

$$\mathbb{P}[n_? = n] << \min\{\alpha, 1 - \beta\}$$

(9) Throughout its zone, the head counts the number $b_1$ of symbols "1", which stand just next to symbols "2" or "3". This means, for simulation, that those symbols are counted which were recorded at the moment t = 1 as directed by the signal $a_i$ which arrived at the moment t = 0. Analogous countings are made of the symbols "1" standing 2 - d, 3 - d, . . . , m-th after symbols "2" or "3".

(10) The natural numbers $g_1, g_2, \ldots, g_m$ are computed to be those nearest to the values of $n_? \; b_i/N$ for $1 \leq i \leq m$. The sequence $g_1, g_2, \ldots, g_m$ is treated later as a code for the input word. The parameter $\epsilon$ (which will be chosen below) will also make

$$\mathbb{P}[(a_{g_1}, a_{g_2}, \ldots, a_{g_m}) = w] << \min\{\alpha, 1 - \beta\}$$

This is possible by virtue of stage 3.

(11) A natural number $V = 1 + \sum_{j=1}^{m} g_j n_?^j$ is encoded on the tape and is treated as a code for the input word w.

(12) A number of $G = 2^N$ of mutually independent trials are made, each one resulting in 1 with probability $\tau$ (we shall choose the value of $\tau$ below; its meaning is to encode the language L). The number F of ones thus obtained is encoded on the tape.

(13) The number F is divided by G; in fact only the 2V-th sign $\sigma_F$ of the binary expansion of F/G is computed and retained.

(14) If $\sigma_F = 1$, then all the cells of the zone (including both dividers bordering the zone) are changed to the state $p_+$; otherwise all of them are changed to the state $p_-$.

(15) The machine stops.

Note first of all that, according to well-known facts of the algorithmic theory, a machine M corresponding to this description, really exists. Some doubts are produced by the requirement, made in the descriptions of stages 4 and 7 of the machine's action, that all the computations must be made inside the zone. But any head's zone is no shorter than Nm. This is sufficient to perform all the work, the most voluminous of which are the stages 11, 12, and 13.

Now we choose the values of our parameters $\epsilon$ and $\tau$. As was mentioned, $\tau$ is a code for the language L. The function

$$V_n((a_{i_1}, a_{i_2}, \ldots, a_{i_k})) = 1 + \sum_{j=1}^{k} i_j \cdot n^j$$

maps any word in the n-letter alphabet into a natural number. The conjectured (but very probable, as we shall see) value of this function applied to our input word $w$ is computed at stage 11. We set

$$\tau = \sum_{w \in L} 2^{-2V_n(w)}$$

The parameter $\epsilon$, in its turn, determines the reliability, with which the language L is recognized by the segment of the network $\Phi'(M)$: the smaller $\epsilon$, the more reliable it is (but, correspondingly, the longer is the time required by it). We set the value of $\epsilon$ by the formula

$$\epsilon = \gamma^2/400\,n^4 \tag{1}$$

where $\gamma = \min\{\alpha, 1 - \beta\}$.

Now let us prove that our network $\Phi'(M)$ does $\alpha/\beta$-recognize the language L. For that we shall show

$$\left.\begin{array}{ll} & w \in L \text{ implies } \mathbb{P}[c_0^\infty = p_+] > 1 - \gamma \\ \text{but} & w \in L \text{ implies } \mathbb{P}[c_0^\infty = p_-] > 1 - \gamma \end{array}\right\} \tag{2}$$

where $c_0^\infty = x$ means, that the 0-th automaton of the network $\Phi'(M)$ remains forever from some moment of time in the state $x$. According to Note 1, the condition (2) implies that the network $\alpha/\beta$-recognizes L.

So, let us examine the process of simulation of M by the network $\Phi'(M)$. We term the head whose zone includes the 0-th automaton $c_0$, the zero-head. Let us imagine that all the denotations, used when describing the operation of M (that is $n_2$, N, $n_?$, etc.), refer to the zero-head. It is easy to see that the number N is equal to the number of complete records of the input word $w$. It implies that for any $K \geq 1$ $\mathbb{P}[N < K] < K \cdot \epsilon^m$; $\epsilon^m$ is the probability with which a head survives the third stage of its operation. Let $K = 40\, n^{4m}/\gamma$, where $m \geq 1$ is the length of the input word. According to (1),

$$\mathbb{P}[N < 40\, n^{4m}/\gamma] < (40\, n^{4m}/\gamma) \cdot (\gamma^{2m}/(400\, n^4)^m) \leq \gamma/10$$

In other words, the number of complete records of the input word, necessary to find the cardinality of the input alphabet (stages (7) and (8)) and the input word itself (stages (9) and (10)) is less than $40\, n^{4m}$ with probability less than $\gamma/10$. Let

$$N > 40\, n^{4m}/\gamma \tag{3}$$

Let us estimate the probability $\mathbb{P}[n_? \neq n]$ of an error in restoring the cardinality of the input alphabet. Examine the conditional mathematical expectation $(M(n_2))$ and the conditional dispersion $(D(n_2))$ of the value $n_2$ for the fixed $N$. Of course, $M(n_2) = N/n$ and $D(n_2) \leq N/4$. On the other hand, for any $\delta$

$$D(n_2) \geq \mathbb{P}[\,|n_2 - M(n_2)| \geq \delta] \cdot \delta^2$$

so, setting $\delta = N/4\, n^2$, we have

$$\mathbb{P}[\,|n_2 - N/n| > N/4n^2] \leq (1/4N)(4n^2/N^2 = 4n^4/N$$

Taking (3) into account, we obtain an estimate

$$\mathbb{P}[\,|n_2 - N/n| > N/4n^2] \leq \gamma/10$$

which does not require $N$ to be fixed. But the case of

$$|n_2 - N/n| \leq N/4n^2,$$

that is

$$N(4n - 1)/4n^2 \leq n_2 \leq N(4n + 1)/4n^2$$

that is

$$4n^2/(4n + 1) \leq N/n_2 \leq 4n^2/(4n - 1)$$

that is

$$|N/n_2 - n| \leq n/(4n - 1) < 1/3$$

corresponds to $n_? = n$. Thus (4) assumed, the cardinality of input alphabet is recognized incorrectly with a probability of at most $\gamma/10$.

It can be proved in a similar way, if $N \geq 40n^{4m}/\gamma$ and $n_? = n$ assumed, that the input word is recognized incorrectly with a probability of at most $\gamma/10$. It can be proved in a similar way that if $N \geq 40n^{4m}/\gamma$, $n_? = n$ and $(a_{g_1}, a_{g_2}, \ldots, a_{g_m}) = w$ are assumed, then the question whether the word w belongs to the language L is also answered incorrectly with a probability of at most $\gamma/10$. Thus, the total error probability is no more than $4\gamma/10$. But, the w word's belonging to L being recognized correctly, all the automata of the zone are transferred to the correct state $p_+$ (for $w \in L$) or $p_-$ (for $w \notin L$), whence the condition (2) is fulfilled for the network $\Phi'(M)$.

The theorem is proved.

Proof of Remark 3. The fact that in the case $\alpha < \beta < 1$, the recursive enumerability of a language L implies that it is possible to $\alpha/\beta$-recognize it on a network with rational transition probabilities has been actually established in the proof of Theorem 3.

On the other hand, the fact that any $\alpha/\beta$-recognizable language for $\alpha < \beta$ is enumerable, is inferred by Note 1 and the fact that if $c_0$ is an element of a network C, which $\alpha/\beta$-recognizes the L language, t is a moment of time, x is state of elements of C, and transition probabilities of C are rational, then the probability $\mathbb{P}[{}_w c_0^t = x]$ is a rational number, which can be computed exactly if the input word w is given. Let r be rational, $\alpha < r < \beta$. The operation of a Turing machine recognizing L should consist in computation of the values of the function $Q_C(t)$ (which are rational) for all t and comparing them with r. If there is a t, such that ${}_w Q_C(t) > r$, then $w \in L$.

Combining the two assertions we obtain: if $\alpha < \beta < 1$, then a language L can be recognized by a network with rational transition probabilities if and only if L is enumerable.

Proof of Theorem 6. Having proved Theorem 3, we have actually proved the possibility of $\alpha/\beta$-recognizing in a stable way any r. e. language in the case in which $\alpha < \beta < 1$. The converse follows from Remark 3 and the fact that the rationals are dense in (0, 1).

## REFERENCES

1. A. L. Toom, Non-ergodic multi-dimensional automata systems. Problemy Peredachi Informacii, 10, vol. 3, 70-79, 1974.
2. L. N. Vasershtein, Markov processes on a countable space product describing large automata systems, Problemy Peredachi Informacii 5, vol. 3, 64-72, 1969.
3. G. S. Plesnevich, Comparison of two concepts of computing on homogeneous iterative automata, The Conference on Automata Theory and Artificial Intellect, Tashkent, Abstracts of Communications, Computing Center of the USSR Academy of Sciences, Moscow, 1968.
4. L. Stockmeyer, The complexity of decision problems in automata theory and logic, Project MAC MIT, 1974.
5. N. Chomsky, G. A. Miller, Finite state languages, Information and Control, 1, 91-112, 1958.

6. P. Martin-Löf, Notes on constructive mathematics, Almquist & Weksell, Stockholm, 1970.
7. K. De Leeuw, E. F. Moore, C. E. Shannon, N. Shapiro, Computing on stochastic Turing machines, in: Automata Studies, C. E. Shannon and J. McCarthy eds. , Princeton, New Jersey, Princeton University Press, 1956.
8. S. C. Kleene, A representation of events in nerve networks and finite automata, in: Automata Studies, C. E. Shannon and J. McCarthy, eds. , Princeton, New Jersey, Princeton University Press, 1956.
9. R. V. Freivald, Limit computations on stochastic machines, Uchonye Zapiski Latviiskogo Gosudarstvennogo Universiteta, 210, No. 1, 1974, (in Russian).
10. G. L. Kurdumov, An Algorithm-Theoretic method in studying homogeneous random network. Lecture Notes in Mathematics, 1977.
11. M. G. Shnirman, On non-uniqueness in some homogeneous networks. Lecture Notes in Mathematics, 1977.

chapter 16

# COMPLETE CLUSTER EXPANSIONS FOR WEAKLY COUPLED GIBBS RANDOM FIELDS

V. A. Malyshev

It is well known [3] that cluster expansions used to derive uniqueness, analyticity and exponential decay of correlations for weakly coupled Gibbs random field are called vacuum cluster expansions. The reason is that they give uniqueness of the vacuum and the lower mass gap.

Our terminology "complete cluster expansion" follows closely one of Glimm, Jaffe, and Spencer [1]. They deal with N-particle cluster expansions where the cluster expansion provides some information about the spectrum of the Hamiltonian in the N-particle region. The methods of [1] allow such expansions for $N \leq N_0(\beta)$, where $\beta$ is a small coupling constant and $N_0(\beta) \to \infty$ when $\beta \to 0$.

Here, we treat the expansion which gives information about the N-particle region for all $N$ uniformly in $\beta$ such that $|\beta| < \beta_0$. In this article we consider only lattice Gibbs random fields with bounded interaction potential.

Our methods are quite different from [1]. We shall work in a special basis which was first used in [2].

The cluster expansion proofs are crucially based upon delicate estimates of semiinvariants together with rather complicated combinatorial lemmas. They seem to be of independent interest, and are essentially better than all known estimates.

This expansion seems to be the necessary step towards asymptotic completeness.

The main results are formulated in § I. The next paragraphs (§§ 2-4) contain detailed estimates for semiinvariants. The proof of the main result is given in § 5-7.

This article is a more general and refined version of [7]. It contains, in particular, the case of a gauge field.

## I. THE COMPLETE CLUSTER EXPANSION

We consider the set $T = T_o \times \mathbb{Z}$, where $\mathbb{Z}$ is the one-dimensional lattice and $T_o$ is an arbitrary denumerable set. Furthermore, we identify $T_o$ with $T_o \times \{0\} \subset T$. Assume that for each $t \in T$ we are given the probability triple $(\Omega_t, \Sigma_t, \mu_t^{(o)})$. We define cartesian products

$$\Omega = \underset{t\in T}{\times} \Omega_t \ ; \qquad \Sigma_A = \underset{t\in A}{\times} \Sigma_t \ , \qquad A \subset T,$$

$$\mu^{(o)} = \underset{t\in T}{\times} \mu_t^{(o)}, \qquad \Sigma = \Sigma_T \ , \qquad \Sigma_o = \Sigma_{T_o}$$

The function $F$ on $\Omega$ is said to be <u>local</u> iff it is measurable w. r. t. $\Sigma_A$ for some finite $A \subset T$. The minimal such $A$ is the <u>support</u> of $F$.

Also, assume that we are given a system $\mathfrak{A}$ of finite sets $A_i \subset T$ such that

$$|A_i| \le d_1 < \infty \tag{1. 1}$$

$$\sup_{t\in T} |\{i : t \in A_i\}| < d_2 < \infty \tag{1. 2}$$

for all i and all

$$t_1 = (t_1^{(o)}, \tau_1) \in A_i \tag{1.3}$$

$$t_2 = (t_2^{(o)}, \tau_2) \in A_i \qquad \text{with} \qquad t_i^{(o)} \in T_o$$

and

$$\tau_i \in \mathbb{Z} \qquad \text{such that} \quad |\tau_1 - \tau_2| \leq 1$$

For the proof of our main result we must also assume that $T_o$ is a lattice in $\mathbb{R}^\nu$, i. e. a denumerable set in $\mathbb{R}^\nu$ which is invariant w. r. t. translations by $\nu$ linearly independent vectors. We note that this assumption is used only in the proof of Lemmas 6. 1 and 1. 1. In this case we also assume that for all i

$$\text{diam } A_i \leq d_3 < \infty \tag{1.4}$$

For arbitrary $A_i \in \mathfrak{A}$, let $\Phi_{A_i}$ be a function with support $A_i$. Such a system $\Phi = (\Phi_{A_i})$ is called a potential. We will always assume that $\Phi_{A_i}$ is uniformly bounded, i. e.

$$|\Phi_{A_i}| \leq C_\Phi < \infty \tag{1.5}$$

For construction of a transfer matrix we will also need invariance w. r. t. translations

$$u_\tau(t^{(o)}, \tau^1) = u_\tau(t^{(o)}, \tau^1 + \tau)$$

and reflections

$$\theta t = \theta(t^{(o)}, \tau) = \theta(-t^{(o)}, \tau)$$

We assume that $u_\tau$ identifies $\Omega_t$ with $\Omega_{u_\tau t}$, etc.

In a standard way, one defines translations $U_\tau$ and reflections $\theta$ of random variables on $\Omega$. We assume that if $\Phi_{A_i} \in \Phi$ and $\theta(\Phi_{A_i}) \in \Phi$, then $U_\tau(\Phi_{A_i}) \in \Phi$ (1.6)

Let $\mu$ be the limit Gibbs measure which is a limit in the usual sense of measures $\mu_V$ with densities

$$\frac{d\mu_V}{d\mu^{(o)}} = Z_V^{-1} \exp\left(-\beta \sum_{\substack{A \subset V \\ A \in \mathfrak{A}}} \Phi_A\right) \tag{1.7}$$

The existence and properties of such limits for small $\beta$ easily follow from considerations given below. We shall not dwell any further on the matter. We denote $\mathcal{H} = L_2(\Omega, \Sigma_o, \mu)$, and let $P_{\mathcal{H}}$ be the orthogonal projection onto $\mathcal{H}$ in $L_2(\Omega, \Sigma, \mu)$. The transfer matrix $\mathfrak{T}$ is defined as the following operator in $\mathcal{H}$:

$$\mathfrak{T} = P_{\mathcal{H}} U_1 P_{\mathcal{H}} \,.$$

It has matrix elements

$$(\xi_1, \mathfrak{T}\xi_2)_{\mathcal{H}} = \langle \xi_1 (U_1 \xi_2) \rangle \tag{1.8}$$

where $\langle \cdot \rangle$ is the expectation w. r. t. Gibbs measure $\mu$.

We now define the special basis in $\mathcal{H}$. Let

$$g_x \in L_2(\Omega, \Sigma_x, \mu), \qquad x \in T_o$$

Let $T_o$ be well-ordered in some manner by the relation $<$ and let

$$T_{o,x} = \{y : y \in T_o,\ y < x\}$$

Furthermore, let $P_x$ be the orthogonal projection of $\mathcal{H}$ onto $L_2(\Omega, \Sigma_{T_{o,x}}, \mu)$, and put $\hat{g}_x = g_x - P_x g_x$.

For each finite $I \subset T_o$ and for an arbitrary array $G_I = \{g_x\}_{x \in I}$, we put $\hat{G}_I = \prod_{x \in I} \hat{g}_x$. It is clear that $\hat{G}_I$ and $\hat{G}_{I'}$ are orthogonal if $I \neq I'$.

<u>Lemma 1.1.</u> <u>Constants (i.e. $G_\phi$) and $\hat{G}_I$ (which belong to $\mathcal{H}$) span all $\mathcal{H}$.</u>

One can often see that a more detailed assumption about $\Phi_A$ is valid

$$\Phi_A = \sum_{\alpha=1}^{d} \prod_{t \in A} \varphi_t^{(\alpha)} \tag{1.9}$$

where $\varphi_t^{(\alpha)}$ are $\Sigma_t$-measurable and bounded uniformly in $\alpha$ and $t$. (We do not use it here.)

We note that in some cases one can prove that (see [4])

$$\mathcal{F} \geq 0 \tag{1.10}$$

We note that (1.9) takes place for gauge fields and (1.10) also but in radiation gauge [5].

We proceed now to our main result. Let us be given two arrays

$$G_I = \{g_x\}_{x \in I}, \quad G_{I'} = \{g'_x\}_{x \in I'}, \quad \text{with} \quad \|g_x\|, \|g'_x\| \leq 1 \tag{1.11}$$

From the definition of the semiinvariants $\omega(J)$, we know that

$$(\hat{G}_I, \mathcal{F}\hat{G}_{I'})_{\mathcal{H}} \equiv \langle \hat{G}_I(U_1\hat{G}_{I'}) \rangle = \Sigma\, \omega(J_1) \ldots \omega(J_k) \tag{1.12}$$

where the summation is through all partitions $J_1 \cup \ldots \cup J_k$ of the set $I \cup u_1 I' \subset T_o \cup T_1$ $T_1 = u_1 T_o$. $\omega(J_i)$ is the semiinvariant of $|J_i|$ random variables $\hat{g}_x$, $x \in I \cap J_i$ and $\hat{g}'_x$, $x \in (u_1 J_i) \cap I'$.

Due to the orthogonality of $\hat{g}_x$ for different $x$, the summation in (1.12) is in fact only made over partitions such that for all $i$,

$$J_i \cap I \neq \phi, \quad J_i \cap (u_1 I') \neq \phi \tag{1.13}$$

Furthermore, we denote by $C$ positive constants which differ from one proposition to another. All of them can be explicitly defined.

Theorem 1.2. There exist constants $C > 0$ and $\beta_0 > 0$ such that for all $\beta$, $|\beta| < \beta_0$, and all $G_I$, $G_{I'}$ which satisfy (1.11),

$$|\omega(J_i)| \leq (C\beta)^{d_{J_i}} \tag{1.14}$$

Here, $d_J$ is the minimal $d$ such that there exist $A_1, \ldots, A_d \in \mathfrak{A}$ such that the array $\Gamma(A_1, \ldots, A_d)$ is connected and $J_i \subset \bigcup_{j=1}^{d} A_j$. We emphasize that $C$ does not depend on $\beta$, $|J_i$, $G_I$, $G_{I'}$.

Definition of connectedness. For this we construct a graph $G_\Gamma$ with vertices $A_1, \ldots, A_d$. We connect $A_i$ and $A_j$ by a line iff $A_i \cap A_j \neq \phi$. $\Gamma$ is called connected iff $G_\Gamma$ is connected.

The proof of this theorem takes the larger part of the following text. We also obtain absolutely convergent series (in $\beta$) for $\omega(J)$ $\hat{g}_x$, etc. Let us consider the subspace $\mathcal{H}_N \subset \mathcal{H}$ spanned by all $\hat{G}_I$ with $|I| \leq N$; let $P_N$ be the projection onto $\mathcal{H}_N$.

Corollary 1.3. There exist constants $C > 0$ and $\delta > 0$ which do not depend on $\beta$ and $N$ and such that

$$\|(1 - P_N)\mathcal{F}\| \leq (C\beta)^{\delta(N+1)} \tag{1.15}$$

Proof of this proposition as a corollary of Theorem 1.2 for the case of the Ising model (where $\delta \equiv 1$) is given in [10]. One can in fact find much more detailed estimates there. The same estimates can be easily obtained in our case by using the method of [10].

Corollary 1.4. Let $E_a$ be the spectral family for $\mathcal{T}$. Then for $a > (C\beta)^{\delta(N+1)}$

$$(1 - E_a)\mathcal{H}_N = (1 - E_a)\mathcal{H} \tag{1.16}$$

This is evident from (1.15).

## II. BOUNDS FOR SEMIINVARIANTS OF PARTIALLY DEPENDENT RANDOM VARIABLES

Let $\sigma_1, \ldots, \sigma_N$ be random variables defined on $(\Omega, \Sigma, \mu)$ and assumed to have finite momenta. For each $A = \{i_1, \ldots, i_p\} \subset \{1, \ldots, N\}$, we define $\sigma_A = \Pi_{k=1}^{p} \sigma_{i_k}$ and $\langle \sigma_A \rangle^c = \langle \sigma_{i_1}, \ldots, \sigma_{i_p} \rangle$; the latter being the semiinvariant of random variables $\sigma_{i_1}, \ldots, \sigma_{i_p}$.

Let $\mathfrak{A}$ be the partially-ordered set of all partitions $\alpha$ of the set $\{1, \ldots, N\}$. We define $\alpha < \beta$ iff each block of partition $\alpha$ is contained in some block of partition $\beta$. $e_0$ is the minimal partition (each block consists of one point), $e_1$ is the maximal partition which consists of one block only. We define the following functions on $\mathfrak{A}$:

$$f(\alpha) = \prod_{i=1}^{k} \langle \sigma_{A_i} \rangle, \quad g(\alpha) = \prod_{i=1}^{k} \langle \sigma_{A_i} \rangle^c$$

where $\alpha = (A_1, \ldots A_k)$ is the partition with blocks $A_i$.

Let us denote

$$C_\sigma = \max_{\alpha \in \mathfrak{A}} |f(\alpha)| \tag{2.1}$$

That is, if $|\sigma_i| \leq C_i$ with probability 1, then

$$C_\sigma \leq \prod_{i=1}^{N} C_i \tag{2.2}$$

One can also bound $C_\sigma$ by using Hölder inequalities. The main result of this paragraph is

Theorem 2.1.

$$|\langle\sigma_1,\ldots,\sigma_N\rangle| \le C_\sigma \prod_{i=1}^{N} (3v_i) \tag{2.3}$$

where $v_i$ is the number of indices $j$, $j \neq i$ such that random variables $\sigma_i$ and $\sigma_j$ are connected with a line in the graph $\Gamma$ defined below.

Before proceeding to the proof of this theorem, we note that the definition of semiinvariants gives

$$f(\alpha) = \sum_{\beta \le \alpha} g(\beta) \tag{2.4}$$

By the Möbius inversion formula [8],

$$g(\alpha) = \sum_{\beta \le \alpha} f(\beta)\mu(\beta, \alpha) \tag{2.5}$$

That is, we have [8]

$$g(e_1) = \sum_{\beta} f(\beta)(-1)^{|\beta|-1}(|\beta| - 1)! \tag{2.6}$$

where $|\beta|$ is the number of blocks in $\beta$. Now we want to take into consideration the fact that some pairs of random variables $\sigma_i$ may be independent.

For this reason we shall define instead of $\mathfrak{A}$ the other partially ordered set $\mathfrak{A}_\Gamma$.

Assume that we are given a graph $\Gamma$ with vertices labeled by $\sigma_1, \ldots, \sigma_N$. Suppose that $\Gamma$ enjoys the following remarkable property: $\sigma_A = \sigma_{A_1}\sigma_{A_2}$ if $A_1 \cup A_2 = A$, $A_1 \cap A_2 = \phi$, and if there are no lines in $\Gamma$ between $A_1$ and $A_2$.

Let $\mathfrak{A}_\Gamma \subset \mathfrak{A}$ be the partially-ordered set of all partitions $\alpha = (A_1, \ldots, A_k)$ of the set of vertices of $\Gamma$ such that the subgraph $\Gamma_{A_i}$ of $\Gamma$ with vertices from $A_i$ is connected for all $i$. (Two vertices in $\Gamma_{A_i}$ are connected with a line iff the same condition is fulfilled in $\Gamma$.)

We denote by $\mu_\Gamma(\beta, \alpha)$ the Mobius function for $\mathfrak{A}_\Gamma$. Using

$$<\sigma_A>^c = 0$$

if $\Gamma_A$ is not connected, we obtain instead of (2. 4),

$$f(\alpha) = \sum_{\beta:\beta\leq\alpha,\ \beta\in\mathfrak{A}_\Gamma} g(\beta), \qquad \alpha \in \mathfrak{A}_\Gamma$$

The Modius inversion formula then gives

$$g(\alpha) = \sum_{\beta:\beta\leq\alpha,\beta\in\mathfrak{A}_\Gamma} f(\beta)\,\mu_\Gamma(\beta, \alpha) \tag{2. 7}$$

Our aim is to now calculate $\mu_\Gamma(\beta, \alpha)$. Let us define a circuit $T$ as a set of lines of $\Gamma$ such that every vertex of any line $\gamma \in T$ is a vertex of exactly one other line $\tilde{\gamma} \in T$.

Let us assume that all lines of $\Gamma$ are well-ordered in some way: $\gamma_1 < \gamma_2 < \dots < \gamma_n$. If $T = \{\gamma_{i_1}, \dots \gamma_{i_k}\}$, $\gamma_{i_1} < \dots < \gamma_{i_k}$, is a circuit then we call $T' = \{\gamma_{i_1}, \dots \gamma_{i_{k-1}}\}$ a broken circuit.

Lemma 2. 2.

$$\mu_\Gamma(e_o, e_1) = (-1)^{N-1}\, m_1 \tag{2. 8}$$

where $m_1$ is the number of subsets $G$ of the set of lines of $\Gamma$ such that $|G| = N - 1$ and such that $G$ does not contain any broken circuit.

Proof: We note that $\mathfrak{A}_\Gamma$ is a geometric lattice [8] or M-lattice (see [9], VII, §5, problem 9). Therefore, the lemma is a slightly revised formulation of proposition 1, p. 358, of [8].

Corollary 2. 3.

$$|\mu_\Gamma(e_o, e_1)| \leq \prod_{i=1}^{N} v_i \tag{2.9}$$

where $v_i$ is the number of lines incident with vertex $\sigma_i$.

Proof: We bound the number $\tilde{m}_1$ of subsets $\tilde{G}$ of lines of $\Gamma$ such that $|\tilde{G}| = N-1$ and $\tilde{G}$ does not contain any circuit. Clearly $m_1 \leq \tilde{m}_1$. For this reason, we define for each $\tilde{G}$ a one-to-one map $\varphi_{\tilde{G}}$ of $\tilde{G}$ into the set $V_\Gamma$ of vertices of $\Gamma$.

We first note that $\tilde{G}$ is a connected tree: it is a tree since $\tilde{G}$ does not contain circuits, and it is connected since $|\tilde{G}| = N-1$.

We consider an arbitrary line $\gamma_1 \in \tilde{G}$ and define $\varphi_{\tilde{G}}(\gamma_1)$ as an arbitrary vertex incident with $\gamma_1$. Let us assume inductively that $\gamma_1, \ldots, \gamma_m$ and $\varphi_{\tilde{G}}(\gamma_1), \ldots \varphi_{\tilde{G}}(\gamma_m)$ are already defined. Let $V_m$ be the set of vertices incident with at least one of $\gamma_1, \ldots \gamma_m$. We choose $\gamma_{m+1}$ in such a manner that one of its vertices belongs to $V_m$. Both vertices of $\gamma_{m+1}$ cannot belong to $V_m$, as we obtain a circuit in that case. Let us define $\varphi_{\tilde{G}}(\gamma_{m+1})$ as that of the two vertices of $\gamma_{m+1}$ which belongs to $V_m$ if it does not coincide with one of $\varphi_{\tilde{G}}(\gamma_1), \ldots, \varphi_{\tilde{G}}(\gamma_m)$, and the other if it does.

We then see that $\varphi_{\tilde{G}}^{-1}$ are different maps (since $\tilde{G}$ are different of $V_\Gamma$ into the set of lines of $\Gamma$, and that for each $\tilde{G}$ and each $v \in V_\Gamma$, $\varphi_{\tilde{G}}^{-1}(v)$ is incident with $v$. That is why the number of such $\varphi_{\tilde{G}}^{-1}$ (which is the number of $\tilde{G}$) does not exceed $\Pi v_i$.

Q. E. D.

Lemma 2. 4.

$$\sum_{\beta \in \mathfrak{A}_\Gamma} |\mu_\Gamma(\beta, e_1)| \leq \prod_{i=1}^{N} (3 v_i) \tag{2.10}$$

Proof: For arbitrary $\beta \in \mathfrak{A}_\Gamma$, we define the new graph $\Gamma_\beta$ which identifies vertices of $\Gamma$ which belong to the same block of $\beta$, i. e. vertices of $\Gamma_\beta$ are blocks of $\beta$. Between two new vertices $b_1$ and $b_2$ there is a line iff there is a line in $\Gamma$ between some vertices i $i_1 \in b_1$ and $i_2 \in b_2$. Note that

$$\mu_\Gamma(\beta_1 e_1) = \mu_{\Gamma_\beta}(e_o, e_1)$$

We shall label partitions $\beta \in \mathfrak{A}_\Gamma$ of $\Gamma$ by vertices of some tree $\mathcal{G}$.

We shall construct vertices of our tree $\mathcal{G}$ inductively by the following algorithm

1. Vertices of $\mathcal{G}$ have order $1, \ldots, N$ and may be red or blue.

2. To each vertex of $\mathcal{G}$ will correspond some $\Gamma_\beta$ and a subset $D_\beta$ of vertices of $\Gamma_\beta$.

3. The vertex of order 1 is unique. It is red and corresponds to $\Gamma_\beta \equiv \beta$, $D_\beta = \phi$.

4. Over an arbitrary vertex of order k, there is only one vertex of order k-1, and there is only one blue vertex of order k+1 under it.

5. Assume that a red vertex t or order k has already been constructed and that $\Gamma_\beta$ and $D_\beta$ correspond to it. Let $v_i$ be the number of lines of $\Gamma_\beta$ incident with vertex i of $\Gamma_\beta$, and let

$$v_1 \leq \cdots \leq v_m$$

We construct under t some number of red vertices not exceeding $v_1$ and 1 blue vertex. To this blue vertex will correspond the same $\Gamma_\beta$, but to $D_\beta$ we add vertex $v_1$.

From graph $\Gamma_\beta$, we can construct by not more than $v_1$ ways a new graph $\Gamma_{\beta_r}$ by identifying vertex 1 of $\Gamma_\beta$ with one of the $v_1$ other vertices of $\Gamma_\beta$ which are connected with 1 by a line, and do not belong to $D_\beta$. Each $\Gamma_{\beta_r}$ will correspond to a new vertex. All new new vertices have the same $D_\beta$. (This is correct since the vertices of $D_\beta$ are not identified in construction of $\Gamma_{\beta_r}$.)

We note that

A. for arbitrary $\beta \in \mathfrak{A}_\Gamma$, there is a red vertex which corresponds to $\Gamma_\beta$. This follows easily from construction.

B. if we put $\gamma_t = \Pi_{r=1}^m v_r$, where $v_r$ are assumed to correspond to $t$, then for any vertex $t$ of order $k$, and for any red vertex $t'$ of order $k+1$ under $t$,

$$\frac{\gamma_{t'}}{\gamma_t} \leq \frac{v_1 + v_r - 1}{v_1 v_r} \leq \begin{cases} \frac{2}{v_1} & v_1 \geq 2 \\ 1 , & v_1 = 1 \end{cases}$$

Furthermore,

$$\sum_{\beta \leq 1} |\mu(\beta, 1)| \leq \sum_{\text{red } t} \gamma_t$$

However, it follows from construction that

$$\sum_{\substack{\text{red } t \\ \text{of order } k+1}} \gamma_t \leq 2 \sum_{\substack{\text{red } t \\ \text{of order } k}} \gamma_t + \sum_{\substack{\text{blue } t \\ \text{of order } k}} \gamma_t$$

$$\leq 2 \sum_{\substack{\text{red } t \\ \text{or order } k}} \gamma_t + \sum_{\substack{\text{red } t \\ \text{of order } < k}} \gamma_t$$

The inequalities

$$a_k \leq 2 a_{k-1} + \sum_{i=1}^{k-2} a_i , \quad a_1 = 1$$

imply that $a_k \leq 3^k$. Thus Lemma 2. 4 is proved.

Theorem 2. 1 clearly follows from (2. 7) if we use (2. 10).

## III. BOUNDS ON SEMIINVARIANTS OF FUNCTIONALS OF INDEPENDENT RANDOM FIELDS

Let $T$ be a denumerable set, $\mathcal{X}$ an arbitrary measurable space with $\sigma$-algebra $\Sigma_t$ and probability measure $\mu_t$ defined for all $t \in T$. We put

$$\mathcal{X} = \underset{t\in T}{\times} \mathcal{X}_t, \quad \Sigma = \underset{t\in T}{\times} \Sigma_t, \quad \mu = \underset{t\in T}{\times} \mu_t, \quad \Sigma_T = \underset{t\in T}{\times} \Sigma_t$$

Lemma 3.1. Let $\sigma_1, \ldots, \sigma_N$ be local and have supports $B_1, \ldots, B_N$ We denote

$$I = \{1, \ldots n\}, \quad I' = \{n+1, \ldots, N\}$$

for $i \in I$ ($i \in I'$). We define $\tilde{V}_i$ to be equal to the number of $\sigma_j$ such that $j \in I(j \in I')$ and $B_i \cap B_j \neq \phi$. Then,*

$$|\langle\sigma_1, \ldots \sigma_N\rangle| \leq C_\sigma \prod_{i=1}^{N} (3\tilde{V}_i 4^{|B_i|}) \tag{3.1}$$

Proof: This lemma will clearly follow from Theorem 2.1 if we can prove that**

$$\prod_{j\in I'} v_j \leq \prod_{i\in I} 4^{|B_i|} \prod_{j\in I'} 3V_j \tag{3.2}$$

where $v_j$ is the number of $i \neq j$, $i = 1, \ldots N$ such that $B_i \cap B_j \neq \phi$.

For $j \in I'$, $v_j^{(1)} \equiv v_j^{(1)}, v_j^{(2)}$ is the number of $i \neq j$, $i = 2, 3, \ldots, N$, such that $B_i \cap B_j \neq \phi$. We have $v_j^{(1)} = v_j^{(2)} + \delta_j$, where $\delta_j = 1$ if $B_j \cap B_1 \neq \phi$, and is otherwise equal to 0.

---

* The factor $C^{|B_i|}$ was omitted in Lemma 3.1 and Corollary 3.3 of [7]. This does not affect the remaining results of [7].

** We define the graph $\Gamma$ with vertices $\sigma_1, \ldots, \sigma_N$ connecting $\sigma_i$ and $\sigma_j$ by a line iff $B_i \cap B_j = \phi$.

For each $t \in B_1$, we choose an arbitrary (perhaps empty) subset $\beta(t) \in I'$ satisfying the following assumptions:

1. $t \in \beta_i$ for each $i \in \beta(t)$,
2. $\beta(t) \cap \beta(t') = \phi$ if $t \neq t'$,
3. if $j \in I'$ is such that $B_j \cap B_1 \neq \phi$, then $j$ belongs to some $\beta(t)$.

It follows from these assumptions that

$$\prod_{j \in I'} v_j^{(2)} = \prod_{j \in I'} (v_j^{(1)} - \delta_j) \geq \prod_{j \in I'} - v_j^{(1)} [ \prod_{t \in B_1} \prod_{j \in \beta(t)} (1 - \frac{\delta_j}{v_j^{(1)}} )]$$

$$\geq \prod_{j \in I'} v_j^{(1)} \prod_{t \in B_1} (1 - \frac{1}{\alpha(t)})^{|\beta(t)|}$$

(where $\alpha(t) \geq 2$, $\alpha(t) \geq |\beta(t)|$)

$$\geq \prod_{j \in I'} v_j^{(2)} \prod_{t \in B_1} \frac{1}{4} = \prod_{j \in I'} v_j^{(1)} 4^{|B_1|}$$

From this we obtain

$$\prod_{j \in I'} v_j^{(1)} \leq 4^{|B_i|} \prod_{j \in I'} v_j^{(2)}$$

Inductively, we obtain the same inequalities for similarly defined $v_j^{(3)}, \ldots v_j^{(n)}$. From this (3.2), a symmetric inequality, the the next Lemma follow.

Let us now be given a system $\mathfrak{A}$ of finite subsets $A_i \in T$ satisfying the following assumptions:

1. $|A_i| \leq d_1 < \infty$,
2. $\sup_{t \in T} |\{i : t \in A_i\}| \leq d_2 < \infty$.

Let $\Phi_{A_i}$ be given with supports $A_i \in \mathfrak{A}$. We consider local $F_1, \ldots F_k$ with supports $B_1, \ldots, B_k$.

We denote by $\gamma_\sigma$ the constant defined in (2. 1) for the system $F_1, \ldots, F_k, \Phi_{A_1}, \ldots, \Phi_{A_n}$ of random variables.

We consider the partition of $\{A_1, \ldots, A_n\}$ into $m$ groups such that $A_i$ and $A_j$ are identical if they belong to the same group and are otherwise different. Let $n_1, \ldots, n_m$ denote the number of elements in these groups.

Lemma 3. 2.

$$|<F_1, \ldots, F_k, \Phi_{A_1}, \ldots \Phi_{A_n}>| \leq C_\sigma [\prod_{i=1}^{k} 3v_i 4^{|B_i|}] n_1! \ldots n_m! C^n \tag{3. 3}$$

where $v_i$ is the number of $j$, $j = 1, \ldots, k$ such that $B_i \cap B_j \neq \phi$ and

$$C = 3 \cdot 4^{d_1 [\;\;] \log_2 (d_2 + 1) [\;\;\;+1]} \tag{3. 4}$$

where $]a[ = a$ is a is an integer and $[a] + 1$ otherwise.

Proof: We use (3. 1) first in the case where

$$\sigma_1 = F_1, \ldots, \sigma_n = F_k, \sigma_{n+1} = \Phi_{A_1}, \ldots \sigma_N = \Phi_{A_n}$$

To bound

$$\prod_{j \in I'} V_j$$

we inductively apply (3. 2) with $I'$ being partitioned on the 2, 4, 8,... group by the dichotomic rule. For this we need the following

Lemma 3. 3. There exists a partition of $I'$ on $d_2 + 1$ subsets $I_1, \ldots, I_{d_2+1}$ such that $I_i \cap I_j = \phi$ for $i \neq j$, $\cup I_i = I'$ and if $i$ and $j$ belong to different subsets of this partition then $A_i \cap A_j = \phi$.

Proof: We first define the graph $G$ with $I'$ as the set of its vertices.

Vertices i and j are connected with a line iff $A_i \cap A_j \neq \phi$. We set the length of each line equal to 1. We set the distance $\rho(i,j)$ between two vertices equal to the length of the minimal path from i to j. (If there is no such path we put $\rho(i, j)$ equal to $\infty$.) We choose a subset I' of vertices of G such that:

1. the distance between two arbitrary vertices of I' is at least 2,

2. each vertex not belonging to I' is at distance 1 from I'.

We delete all vertices I' from G, and also all lines incident with them. In the remaining graph we choose $I_2$ in a similar way, etc. The procedure will be complete after $d_2 + 1$ steps. In fact, we can consider arbitrary vertex i. In the first step either i belongs to $I_1$ (and then everything is proved), or one of at most $d_2$ vertices j, with $\rho(i, j) = 1$, belongs to I'. We repeat this reasoning in the next step, and continue in the same manner. The Lemma is thus proved. The theorem follows immediately.

## IV. BOUNDS ON SEMIINVARIANTS OF THE GIBBS RANDOM FIELD IN A HIGH TEMPERATURE REGION

We will change the notation of the preceding paragraph slightly and will denote by $\mu_o$ the measure $\mu$ on $(\mathcal{X}, \Sigma)$. $\langle\cdot\rangle_o$ will denote the expectation of $\mu_o$.

We define the new measure $\mu_V$ with density

$$\frac{d\mu_V}{d\mu_o} = Z_V^{-1} \exp(-\beta \sum_{A \subset V} \Phi_A) \qquad (4.1)$$

where $\langle\cdot\rangle_V$ denotes expectation w. r. t $\mu_V$.

If $V \to \mathbb{Z}^\nu$ in the usual sense, we obtain the formal Taylor expansion

$$\langle F_1, \ldots, F_k \rangle^c \equiv \lim_V \langle F_1, \ldots, F_k \rangle_V^c = \sum_{n=0}^{\infty} b_n \beta^n \qquad (4.2)$$

where

$$b_n = \frac{(-1)^n}{n!} \sum_{A_1} \cdots \sum_{A_n} < F_1, \ldots, F_k, \Phi_{A_1}, \ldots, \Phi_{A_n} >_o \tag{4. 3}$$

One can easily see that for fixed n in the sum (4. 3) only a finite number of terms are different from zero.

We denote by $d = d(B_1, \ldots, B_k)$ the minimal integer $d \geq 0$ such that there exist $A_1, \ldots, A_d \in \mathfrak{A}$ for which the array $\Gamma = (B_1, \ldots, B_k, A_1, \ldots, A_d)$ is connected.

Theorem 4. 1. The series (4. 2) is absolutely convergent for $|\beta| < \beta_o$, $\beta_o > 0$ being sufficiently small, and

$$|<F_1, \ldots, F_k>^c| \leq C_\sigma \; 2[\prod_{i=1}^{k} 3 v_i \, 8^{|B_i|}] \, (2^{d_1 d_2} C \beta)^d \tag{4. 4}$$

(All notation is the same as in Lemma 3. 2. )

Proof: For a given ordered array $\Gamma = (A_1, \ldots, A_N)$, we denote by $\hat{\Gamma} = (A_{i_1}, \ldots, A_{i_p})$ the minimal unordered array such that:

1. $A_{i_1}, \ldots, A_{i_p}$ are all different,
2. Any $A_i \in \Gamma$ is equal to one of $A_{i_1}, \ldots, A_{i_p}$.

We note that all $b_n$, $n < d$, are equal to zero. Using (3. 3), we obtain for $b_n$, $n \geq d$, the following bound

$$|b_n| \leq 2C_\sigma \prod_{i=1}^{k} (3v_i 4^{|B_i|}) \sum_{\hat{\Gamma} : |\hat{\Gamma}| \geq d} (C\beta)^{|\hat{\Gamma}|} \tag{4. 5}$$

Let $R_N$ be the number of unordered arrays $\hat{\Gamma}$ such that $|\hat{\Gamma}| = N$, and the array $(B_1, \ldots, B_k) \cup \hat{\Gamma}$ is connected. We shall prove that

$$R^N \leq 2^{|B_1| + \ldots + |B_k|} R'_N \tag{4. 6}$$

where the number

$$R'_N \leq 2^{d_1 d_2 N} \tag{4.7}$$

is the number of arrays $\hat{\Gamma} = (A_1, \ldots, A_N)$, $|\hat{\Gamma}| = N$, such that some fixed point $t$ belongs to $\tilde{\Gamma} = \bigcup_{i=1}^{N} A_i$.

We shall prove (4.6) first. Let all points of $T$ be enumerated in some fixed manner. $\hat{\Gamma}$ is the union of $r$ maximal connected sub-arrays, $\hat{\Gamma}_1, \ldots, \hat{\Gamma}_r$. One can easily see that $r \leq |B_1| + \ldots + |B_k|$. In each $\hat{\Gamma}_i$, we take the first (in this fixed order) point

$$t_i \in \tilde{\Gamma}_i \cap (\bigcup_{j=1}^{k} B_j) , \quad \tilde{\Gamma}_i = \bigcup_{A_j \in \hat{\Gamma}_i} A_j .$$

The number of sequences of the form $(t_1, \ldots, t_T)$ does not exceed $2^{|B_1| + \ldots + |B_k|}$. This proves (4.6).

<u>Lemma 4.2.</u>

$$R'_N \leq 2^{d_1 d_2 N}$$

<u>Proof:</u> Let the set of all $A_i$ be well-ordered in some way. We shall give the algorithm for construction of an arbitrary array $\hat{\Gamma}$, with $t \in \tilde{\Gamma}$ ($t$ being fixed), with simultaneous ordering of each $\hat{\Gamma}$. We first order the elements of $\hat{\Gamma}$ containing $t$:

$$A_1 \leq \ldots \leq A_m , \quad 1 \leq m \leq d_2 .$$

This order is induced by the order in question of all systems of $A_i$. The number of such sequences $(A_1, \ldots, A_m)$ is not more than $2^{d_2} - 1$. We then take all elements of $\hat{\Gamma}$ intersecting $A_1$ (the number of such elements is not more than $2^{d_1 d_2}$) and order them in a similar manner after $A_m$. Repeating this procedure with $A_2$, etc., we demonstrate the Lemma.

## V. THE SERIES FOR CONDITIONAL EXPECTATION

Let us consider the Gibbs measure $\mu$ defined in §4, and let $P_\Lambda$ be the orthogonal projection onto $L_2(\Omega, \Sigma_\Lambda, \mu)$ in $L_2(\Omega, \Sigma, \mu)$. (We return now to the notations of §1.)

Our aim here is to explicitly find the absolutely convergent series for $P_\Lambda \ F_t$, $t \in \Lambda$, $F \in L_2(\Omega, \Sigma_t, \mu)$.

Let us consider finite $\Lambda$ first. Let $<\cdot>_o^{(\Lambda)}$ be the expectation of the measure $\mu_{o,\Lambda} = \underset{t \in I \setminus \Lambda}{\times} \mu_t^{(o)}$, and let $<\cdot>^{(\Lambda)}$ be the expectation of the Gibbs measure $\lim_V <\cdot>_V^{(\Lambda)}$, where for $\Lambda \subset V$, we put

$$<F>_V^{(\Lambda)} = \frac{<F \exp(-\beta U_{V \setminus \Lambda})>_o^{(\Lambda)}}{<\exp(-\beta U_{V \setminus \Lambda})>_o^{(\Lambda)}} ,$$

where

$$U_{V \setminus \Lambda} = \sum_{A \subset V \setminus \Lambda} \Phi_A$$

Then

$$P_\Lambda F_t = M(F_t | \Sigma_\Lambda) = \lim_V \frac{<F_t \exp(-\beta \sum_{A \in G_\Lambda} \Phi_A)>_V^{(\Lambda)}}{<\exp(-\beta \sum_{A \in G_\Lambda} \Phi_A)>_V^{(\Lambda)}}$$

$$= \frac{<F_t \exp(-\beta \sum_{A \in G_\Lambda} \Phi_A)>^{(\Lambda)}}{<\exp(-\beta \sum_{A \in G_\Lambda} \Phi_A)>^{(\Lambda)}}$$

where $G_\Lambda$ is the set of all $A \in \mathfrak{A}$ such that $A \cap \Lambda \neq \phi$, and

$A \cap (T \setminus \Lambda) \neq \phi$. From the last formula one can obtain

$$P_\Lambda F_t = \sum_{n=0}^{\infty} \frac{1}{n!} \sum_{|I'|=n} \eta_{I'} \tag{5.1}$$

where the summation is over all ordered arrays $I' = (A_1, \ldots, A_n)$, $A_i \in G_\Lambda$, and where

$$\eta_{I'} = (-\beta)^n < F_t, \Phi_{A_1}, \ldots, \Phi_{A_n} >$$

We can rewrite (5. 1) as the sum over all unordered arrays $I = (A_1, \ldots, A_n)$, $A_i \in G_\Lambda$ :

$$P_\Lambda F_t = \sum_{n=0}^{\infty} \sum_{|I|=n} \eta_I \tag{5.2}$$

where

$$\eta_I = (-\beta)^n \frac{1}{n_1! \ldots n_m!} < F_t, \Phi_{A_1}, \ldots, \Phi_{A_n} >^{(\Lambda)}$$

and where $n_1, \ldots, n_m$ are defined as in Lemma 3. 2. Using (4. 4), we have for some $C > 0$,

$$\| \eta_I \| \leq \| F_t \| (C\beta)^{n+d_{I,t}} \tag{5.3}$$

where $d_{I,t}$ is the minimal $p$ such that there exist $\tilde{A}_1, \ldots, \tilde{A}_p$, $\tilde{A}_i \subset T \setminus \Lambda$, $\tilde{A}_i \in \mathfrak{A}$, such that the array $(t, A_1 \Lambda, \ldots, A_n \Lambda, \tilde{A}_1, \ldots, \tilde{A}_p)$ is connected. We note that $C$ does not depend on $t$, $n$, $I$.

Theorem 5. 1. If $\Lambda$ is infinite, then (5. 2) also holds, $\eta_I$ satisfy the bound (5. 3), and the series

$$\sum_{n=0}^{\infty} \sum_{|I|=n} \| \eta_I \|_n \tag{5.4}$$

is convergent for sufficiently small $\beta$.

We omit the rather straightforward proof of this statement.

## VI. PROOF OF THE MAIN THEOREM

We want to prove the estimate (see §1)

$$|\omega(J)| \leq (C\beta)^{d_J} \tag{1.14}$$

For $I$ as in (5.2), we write

$$I = (A_1 \setminus \Lambda) \cup \ldots \cup (A_n \setminus \Lambda)$$

and

$$\alpha_B = \sum_{I:\tilde{I} = B} \eta_I \tag{6.1}$$

From (5.3), it easily follows that

$$\|\alpha_B\| \leq \|F_t\| (C\beta)^{n_B + d_{B,t}} \tag{6.2}$$

with a different constant $c > 0$. Here, $d_{B,t} \equiv d_{I,t}$ for $\tilde{I} = B$, and $n_B$ is the minimal $n$ such that there exists $I$ with $\tilde{I} = B$ and $|I| = n$.

To prove (1.14), we insert into $\omega(J)$

$$\hat{g}_x = g_x - P_x g_x \equiv \alpha_x - \Sigma_B \alpha_B$$

where $\alpha_x \equiv g_x$. (We put $F_t = g_x$.) Then, $\omega(J)$ will be expanded into the sum of terms

$$\langle \alpha_{B_1}, \ldots, \alpha_{B_m} \rangle$$

where $m = |J|$, $J = (x_1, \ldots, x_m)$. From this we obtain

$$|\omega(J)| \leq \sum |\langle \alpha_{B_1}, \ldots, \alpha_{B_m} \rangle|$$

$$\leq \sum_{B_1} \cdots \sum_{B_m} [\prod_{i=1}^{m} \|\alpha_{B_i}\| v_i C^{|B_i|}] \cdot (C\beta)^{d(\hat{B}_1, \ldots, \hat{B}_m)} \tag{6. 3}$$

where $v_i$ is the number of $j$ such that $B_j \cap B_i \neq \phi$, $j = 1, \ldots, m$. Here, we have used (4. 4). $C$ is some constant which does not depend on $\beta$ and $B_i$ and $\hat{B}_i = B_i \cup \{x_i\}$.

Lemma 6. 1. Assume that $m$ (pairwise different) subsets $\hat{B}_1, \ldots, \hat{B}_m \subset T$ are given. Let $v_i$ be the number of $j$ such that $\hat{B}_i \cap \hat{B}_j \neq \phi$, $j = 1, \ldots, m$. Then, there exists a $C > 0$ such that for all $m$ and all $\hat{B}_1, \ldots, \hat{B}_m$,

$$\sum_{i=1}^{m} d_{\hat{B}_i} \geq C \sum_{i=1}^{m} \ln v_i \tag{6. 4}$$

We note that $C$ depends only on $T$ and $\mathfrak{A}$. The rather complicated proof of this Lemma for the case $T = \mathbb{Z}^{\nu}$ is given in [7]. (See Lemma A. 4.) In the general case, the proof is quite similar. We note that

$$\sum d_{B_i, x_i} \geq \sum d_{\hat{B}_i} \tag{6. 5}$$

Using (6. 2), (6. 4), and (6. 5), we now bound $|\omega(J)|$. We insert (6. 2) and (6. 4) into (6. 3), and obtain

$$|\omega(J)| \leq \sum_{B_1} \cdots \sum_{B_m} \prod_{i=1}^{m} C^{n_{B_i} + d_{B_i, x_i}} \cdot (C\beta)^{n_{B_i} + d_{B_i, x_i} + d(\hat{B}_1, \ldots, \hat{B}_m)}$$

with some (new) constant $C > 0$. Changing the constant and notation, we rewrite the last inequality as

$$|\omega(J)| < \sum_{R} (C\beta)^{\delta_R} \tag{6. 6}$$

where the summation is over all $R = (\hat{B}_1, \ldots, \hat{B}_m)$, and where

$$\delta_R = d(\hat{B}_1, \ldots, \hat{B}_m) + \sum_{i=1}^{m} d_{B_i, x_i}$$

It remains to prove that

$$\sum_R (C\beta)^{\delta_R} \leq (C_1\beta)^{d_J} \tag{6.7}$$

for some $C_1 > 0$. But,

$$\sum_R (C\beta)^{\delta_R} = \sum_{D=1} (C\beta)^D \sum_R{}' (C\beta)^{d_{B_1, x_1} + \ldots + d_{B_m, x_m}} \tag{6.8}$$

where in $\sum'$ the summation is over all $R$ satisfying

$$d(\hat{B}_1, \ldots, \hat{B}_m) = D \tag{6.9}$$

We shall prove the auxiliary

<u>Lemma 6.2.</u> <u>Let</u> $\mathcal{R}$ <u>be the set of all arrays</u> $R = (\hat{B}_1, \ldots, \hat{B}_m)$. <u>Then for each subset</u> $\mathcal{R}' \subset \mathcal{R}$,

$$\sum_{R \in \mathcal{R}'} (C\beta)^{d_{B_1, x_1} + \ldots + d_{B_m, x_m}} \leq (C_1\beta)^d$$

<u>where</u>

$$d = \min_{R \in \mathcal{R}'} (d_{B_1, x_1} + \ldots + d_{B_m, x_m})$$

<u>Proof:</u> The problem evidently reduces to the case when $\mathcal{R}'$ is the set of all arrays $R$ such that

$$d_{B_1, x_1} + \ldots + d_{B_m, x_m} \geq d$$

Let us fix an ordered array $(r_1, \ldots, r_m)$ of integers with

$$0 \le r_i \le d, \quad \sum r_i = d$$

The number of such arrays is at most $4^d$. If we can prove that

$$\sum_{B_1 : d_{B_1, x_1} \ge r} (C\beta)^{d_{B_1, x_1}} \ge (C_2 \beta)^r \tag{6. 10}$$

then our result with $C_1 = 4C_2$ will follow. But the proof of (6. 10) is quite standard.

To prove (6. 7), we take $\mathcal{R}' = \mathcal{R}_D$ as the set of all R satisfying (6. 9). Then

$$|\omega(J)| \le \sum_{D=1}^{\infty} (C\beta)^{D + d(D)} \tag{6. 11}$$

where

$$d(D) = \min_{R \in \mathcal{R}_D} (d_{B_1, x_1} + \dots + d_{B_m, x_m})$$

But

$$D + d(D) \ge d_J \tag{6. 12}$$

which follows from the definitions.

From (6. 12) and (6. 11), the proof of the main result (1. 14) is straightforward.

## VII. PROOF OF LEMMA 1. 1

We choose some $x_o \in T_o$ and choose finite subsets $R = \{x_1, \dots x_k\} \subset T_o$ satisfying the following conditions:

1. $x_o > x_1 > \dots > x_k$, where $>$ is defined by lexicographic order on $T_o$,
2. $R = \{x : x < x_o, \rho(x, x_o) \le d\}$, where $\rho$ is the euclidean distance in $T_o \subset R^\nu$.

Let us first assume that $\{x : x_i < x < x_{i+1}\}$ is empty for each $i = 0, 1, \dots, k-1$. We want to approximate bounded $g_{x_o}$ within arbitrary degree of accuracy by products of $\hat{g}_{x_o}, \hat{g}_{x_1}, \dots, \hat{g}_{x_k}$. It is easy to prove (we omit it) that $g_{x_o} - P_{x_k} g_{x_o}$ can be approximated with arbitrary accuracy by such products. Using the exponential bound

$$\| P_{x_k} g_{x_o} \| \le \| g_{x_o} \| (C\beta)^{\delta d} \tag{7. 1}$$

where $\delta > 0$ and $C > 0$ do not depend on $\beta$ and $d$, we obtain the proof. The bound (7. 1) can be proved in the following way. We apply

$$\| P_{x_k} g_{x_o} \|^2 = \langle M(g_{x_o}^2 | \Sigma_{\{x < x_k\}}) \rangle \cdot M(g_{x_o}^2 | \Sigma)$$

which can be calculated for any given "boundary conditions" on $\{x : x < x_k\}$ by using e. g. (4. 2), (4. 3) for $F_1 \equiv g_{x_o}$ and $k = 1$.

One easily observes from the estimates of §3 that the "boundary terms give a negligible contribution, and we obtain (7. 1).

In the general case (i. e. when $\{x : x_i < x < x_{i+1}\}$ are not empty), we use the following decomposition

$$x_o = x^{(o)} + x^{(1)} + \dots + x^{(k+1)} \tag{7. 2}$$

where

$$x^{(o)} = x_o - P_{x_o} x_o, \quad x^{(1)} = P_{x_o} x_o - \hat{P}_{x_1} x_o,$$

$$x^{(2)} = \hat{P}_{x_1} x_o - \hat{P}_{x_2} x_o, \dots,$$

$$x^{(k)} = \hat{P}_{x_{k-1}} x_o - \hat{P}_{x_k} x_o, \quad x^{(k+1)} = \hat{P}_{x_k} x_o$$

and $\hat{P}_x$ is the projection onto $L_2(\Omega, \Sigma_{\{x', x\}}, \mu)$. Furthermore, we proceed as earlier, using bounds (7.1) for each $P\{x_i < x < x_{i+1}\}x_0$. Since K is bounded by $d^\nu$, the exponential bound (7.1) dominates. The proof is complete.

## REFERENCES


1. J. Glimm, A. Jaffe, T. Spencer, The Wightman Axioms and Particle Structure in the $P(\varphi)_2$ quantum field model. Annals of Math., 100, 5θ5-632 (1974).
2. R. Minlos, Ya. Sinai, Investigation of the spectra of some stochastic operators arising in the lattice gas models. Theoret. i Mathemat. Phys., 2, 230-243 (1970).
3. J. Glimm, A. Jafee, T. Spencer, The Particle structure of the weakly coupled $P(\varphi)_2$ Model and other applications of high temperature expansions. I, II. Lect. Notes Phys., 25, 132-242 (1973).
4. R. S. Schor, The Particle Structure of $\nu$-Dimensional Ising Models at Low Temperatures. Preprint. Rockefeller Univ., N.Y., 1977.
5. K. Osterwalder, E. Seiler, Gauge field Theories on the Lattice. Preprint. Harvard Univ., 1977.
6. F. Abdulla-Zade, R. Minlos, S. Pogosian, Cluster properties of Gibbs random fields and their applications (this volume).
7. V. A. Malyshev, One-particle states and scattering for Markov processes. Lect. Notes Math., 1977.
8. G.-C. Rota, On the Foundations of Combinatorial Theory. I. Theory of Möbius Functions. Zeit. Varsch. Theorie 2, 340-368 (1964).
9. G. Birkhoff, Lattice Theory, N. Y., 1948.
10. V. A. Malyshev, R. A. Minlos, N-particle spectrum of clustering operator. Commun. Math. Phys. (to appear).

chapter 17

# THE CENTRAL LIMIT THEOREM FOR RANDOM FIELDS WITH MIXING PROPERTY

B. S. Nahapetian

Let $\xi_t$, $t \in Z^\nu$, be a random field which takes values in a standard measurable space $(X, \sigma)$.* For $J \subset Z^\nu$, let $(X_J, \sigma_J)$ denote the product of the spaces $(X_t, \sigma_t)$, $t \in J$, where $(X_t, \sigma_t)$ is a copy of $(X, \sigma)$ for every $t \in Z^\nu$, and let $P_J$, $J \subset Z^\nu$, $|J| < \infty$,** denote the finite-dimensional distributions of the random field $\xi_t$, $t \in Z^\nu$. We say that the random field $\xi_t$, $t \in Z^\nu$, satisfies the uniform strong mixing condition (u. s. m. condition) if for any finite $I, V \subset Z^\nu$, $I \cap V = \phi$, a real-valued function $\varphi_I(d)$, $d \in R^{1)}$, exists such that

$$\gamma(I, V) = \sup_{A\in\sigma_I,\ B\in\sigma_V,\ P_V(B)>0} |P_{I/V}(A/B) - P_I(A)| \le \varphi_I(d(I, V)),$$

where $d(I, V)$ denotes the distance between $I, V$, $I, V \subset Z^\nu$, and for fixed $I \subset Z^\nu$, $|I| < \infty$,

$$\varphi_I(d) \to 0, \quad d \to \infty$$

* We shall call the measurable space $(X,\sigma)$ standard if there is a separable complete metric space $\tilde{X} \supset X$ such that $X$ is an element of the Borelian $\sigma$-field $\tilde{\sigma}$ of the subsets of $\tilde{X}$ and if $\sigma = \{B \cap X : B \in \tilde{\sigma}\}$.

** We denote by $|\cdot|$ the number of elements of any finite set.

Let $m(x)$, $x \in X$, be a real-valued measurable function, and for any finite $I \subset Z^{\nu}$, let

$$U_I^m(x) = \sum_{t \in I} m(x_t), \quad x = (x_t, t \in I) \in X_I$$

We write $\overline{n} = (n_i, i \in \nu) \to \infty$, $\overline{\nu} = (1, 2, \ldots, \nu)$, $n_i \in Z^1$, $i \in \nu$, if $\min_{i \in \nu} \{n_i\} \to \infty$ and $|\frac{n_i}{n_j}| < C$, $0 < C < \infty$, $i, j \in \overline{\nu}$. We now formulate the main results of this publication, which was announced in [4]

Theorem 1. *Let the stationary random field* $\xi_t$, $t \in Z^{\nu}$, *satisfy the u. s. m. condition be such that*

1. $\varphi_I(d) \leq |I| \varphi(d)$, $\quad \sum_{s \in Z^{\nu} \setminus \{0\}} \varphi^{\frac{1}{2}}(\|s\|) < \infty$, *

2. *the second moment* $\| m(\xi_t) |^2 = C^{(1)} < \infty$,

3. *for any* $\nu$-*dimensional finite cube* $I \subset Z^{\nu}$,

$$\sigma_{I,m}^2 \geq C^{(2)} |I|, \quad 0 < C^{(2)} < \infty,$$

*where* $\sigma_{I,m}^2$ *is a variance of* $U_I^m(\xi_t, t \in I)$.

*Then for any sequence of increasing cubes* $I_{\overline{n}}$ *with center at the origin and with length of sides* $2n_i$, $i \in \overline{\nu}$, $n_i \in Z^1$,

$$\lim_{\overline{n} \to \infty} \Pr\left(\frac{U_{I_{\overline{n}}}^m(\xi_t, t \in I_{\overline{n}}) - MU_{I_{\overline{n}}}^m(\xi_t, t \in I_{\overline{n}})}{\sigma_{I_{\overline{n}}, m}} < 2\right) = \frac{1}{\sqrt{2\pi}} \int_{\infty}^{\alpha} e^{-\frac{s^2}{2}} ds$$

*for* $\alpha \in R^1$.

The proof of this theorem is based on the Bernstein method of

---

* For $s \in Z^{\nu}$, $\|s\| = \max_{i \in \nu} |s_i|$.

reduction of independent random variables. For $\nu = 1$, we have an analogue of Ibragimov's well-known result. The difficulty of generalization for $\nu > 1$ is that the convergence of $\varphi_I(d)$ to zero is not uniform with respect to I. It is necessary to note that the case $\gamma(I, V) \leq \varphi(d(I, V))$, i. e. the case of uniform convergence to zero with respect to I, does not take place even in very simple cases (see [9], p. 209); furthermore, there are some publications in which the central limit theorem has been proven under condition $\gamma(I, V) \leq \varphi(d(I, V))$. (See for example [8].) The central limit theorem for the random fields under different mixing conditions is proved in [5, 6]. We also note article [7], in which the central limit theorem is proved for the Gibbs random field without mixing conditions. Now let $\xi_t$, $t \in Z^\nu$, be a Gibbs random field with potential $\Phi$ and phase space $(X, \sigma, \mu)$, $0 < \mu(X) < \infty$. (For details, see [2].) By the potential $\Phi$ we mean an arbitrary measurable function on $\bigcup_{J: J \subset Z^\nu,\ |J| < \infty} (X_J, \sigma_J, \mu_J)$. We say that the potential $\Phi$ is a v-potential if a fixed element $\theta \in X$ ("vacuum") exists, such that $\mu(\varphi) = 1$ and $\Phi(x) = 0$ for any $x = (x_t, t \subset J)$ such that $x_t = \theta$ for at least one $t \in J$.

Let $X^* = X \setminus \{\theta\}$, $\sigma^* = \{B \cap X^* : B \in \sigma\}$, and $\mu^*$ be the restriction of the measure $\mu$ on the $\sigma$-field $\sigma^*$. We consider the following classes of potentials: The class of v-potentials

$$N_\mu^\theta = \Phi : \begin{cases} 1. & \Phi(x) = \Phi(\overline{x}),\ x \in X_I,\ \overline{x} \in X_{I+t},\ x_s = \overline{x}_{s+t},\ s \in I \\ & \text{for any } t \in Z^\nu \\ 2. & C_{\Phi,\mu}^{(1)} (1 + C_{\Phi,\mu}^{(2)}) < 1 \end{cases}$$

$$C_{\Phi,\mu}^{(1)} = e^{\|\Phi\|} \mu^*(X^*) (1 + e^{\|\Phi\|} \mu^*(X^*))^{-1}$$

$$C_{\Phi,\mu}^{(2)} = 2(1 + 2e^{\|\Phi\|} \mu^*(X^*)) (\exp\{e^{\|\Phi\|} - 1\} - 1)$$

$$\|\Phi\| = \sum_{J \in H_0} \sup_{x \in X_J} |\Phi(x)| < \infty$$

$$H_0 = \{J \subset Z^\nu : 0 \in J \subset Z^\nu,\ |J| < \infty\},$$

and the class $N_\mu$ of general potentials

$$N_\mu = \left\{ \Phi: \begin{array}{l} 1.\quad \Phi(x) = \Phi(\overline{x}),\ x \in X_I,\ \overline{x} \in X_{I+t},\ x_s = \overline{x}_{s+t},\ s \in J \\ \qquad \text{for any } t \in Z^\nu \\ 2.\quad (\mu(X) e^{\|\Phi\|})^2 (1 + \mu(X) e^{-2\|\Phi\|}) \sum_{\substack{J \in H_0 \\ |J| > 2}} |J| \sup_{x \in X_J} |\Phi(x)| < 1 \end{array} \right.$$

$$\|\Phi\| = \sum_{J \in H_0} |J| \sup_{x \in X_J} |\Phi(x)| < \infty$$

It is easy to see from [2] that the Gibbs random field with potential from classes $N_\mu$ or $N_\mu^\theta$ satisfies the u. s. m. condition and $\varphi_I(d) \leq |I| \varphi(d)$, where $\varphi(d)$ is decreasing exponentially to zero when $d \to \infty$ for finite potentials, and is decreasing as $\frac{\beta(d)}{d^\gamma}$, $\gamma > 0$, $\beta(d) = 0(d)$, $d \to \infty$ for potentials such that

$$\sum_{J \in H_0} |J| (D(J))^\gamma \sup_{x \in X_J} |\Phi(x)| < \infty, \quad \Phi \in N_\mu, \quad \gamma > 0,$$

$$\sum_{J \in H_0} (D(J))^\gamma \sup_{x \in X_J} |\Phi(x)| < \infty, \quad \Phi \in N_\mu^\theta, \quad \gamma > 0$$

$D(J)$ is the diameter of the set $J \subset Z^\nu$. We say that the potential $\Phi$ is degenerate if there is a function $F(\overline{x})$, $\overline{x} \in X_{Z^\nu \ \{0\}}$, such that

$$\sum_{J \in H_0} \Phi(x_0, \overline{x}_{J \setminus \{0\}}) = F(\overline{x}),\ x_0 \in X_0 \quad *$$

* If $x = (x_t, t \in I) \in X_I$ and $J \subset I$, then by $x_J$ we denote $(x_t, t \in J)$.

We also note (see [3]) that for potentials $\Phi \in N_\mu^\theta$ and non-degenerate potentials from $N_\mu$, the condition 3 of Theorem 1 holds. We will assume that $m(x) = \Phi(x)$, $x \in X$.

So, as a corollary of Theorem 1, we have

Theorem 2. For any Gibbs random field with v-potential $\Phi \in N_\mu^\theta$ satisfying the condition

$$\sum_{J \in H_0} (D(J))^{2\nu} \sup_{x \in X_J} |\Phi(x)| < \infty$$

the central limit theorem holds.

Theorem 3. For any Gibbs random field with non-degenerate potential $\Phi \in N_\mu$ satisfying the condition

$$\sum_{J \in H_0} |J| (D(J))^{2\nu} \sup_{x \in X_J} |\Phi(x)| < \infty$$

the central limit theorem holds.

Proof of Theorem 1: Note that the random field $\eta_t = m(\xi_t)$, $t \in Z^\nu$ satisfies the u. s. m. condition with the same function $\varphi_I(d)$ as the random field $\xi_t$, $t \in Z^\nu$. The proof of Theorem 1 is based on the following Lemmas.

Let $\mathfrak{M}_J$, $J \subset Z^\nu$ be the $\sigma$-algebra generated by the random variables $\eta_t$, $t \in J$, and $M\eta_t = 0$, $t \in Z^\nu$.

Lemma 1. Let $X_i$, $i = 1, 2$, be a $\mathfrak{M}_{I_i}$-measurable function,

$$I_1 \cap I_2 = \phi, \quad I_i \subset Z^\nu, \quad i = 1, 2,$$

and let

$$M|X_1|^p < \infty, \quad M|X_2|^q < \infty, \quad p, q > 1$$

Then,

$$|MX_1X_2 - MX_1MX_2| \le 2(\varphi_I(d(I,V)))^{\frac{1}{p}} M^{\frac{1}{p}}|X_1|^p M^{\frac{1}{q}}|X_2|^q$$

The proof of Lemma 1 does not depend on the dimension of $Z^\nu$. (For the case $\nu = 1$, see [1], p. 392.)

Lemma 2. For any $J \subset Z^\nu$, $|J| < \infty$,

$$\sigma_J^2 = M(\sum_{t\in J} \eta_t)^2 \le C^{(3)}|J|, \quad 0 < C^{(3)} < \infty$$

Proof: Using Lemma 1 for $p = q = \frac{1}{2}$ and conditions (1) of Theorem 1, we have

$$\sigma_J^2 \le C^{(1)}|J| + \sum_{t\in J}(\sum_{s \in Z^\nu\setminus\{t\}} M\eta_t\eta_s) \le C^{(1)}|J|(1+2\sum_{s\in Z^\nu\setminus\{0\}} \varphi^{\frac{1}{2}}(\|s\|))$$

$$= C^{(3)}|J|$$

$$C^{(3)} = C^{(1)}(1 + 2\sum_{s\in Z^\nu\setminus\{0\}} \varphi^{\frac{1}{2}}(\|s\|))$$

Consider the following system of integer-valued functions:

$$q = q(n) = \left[ n^{1-\frac{1}{(1+\nu)^4}} \right], \quad p = p(n) = \left[ n^{1-\frac{1}{(1+\nu)^2}} \right], \quad k = k(n) = \left[\frac{2n}{p(n)+q(n)}\right],$$

$n = 1, 2, \ldots$ . It is obvious that

$$q(n),\ p(n),\ k(n) \to \infty,\quad n \to \infty,\quad p = 0(n),\quad q = 0(p),\quad n \to \infty$$

Let

$$I_n(i) = [-n+ip+iq,\ -n+(i+1)p+iq] \subset Z^1, \quad i = 0, 1, \ldots, k-1,$$

$$I_{\bar{n}}(i_s,\ s \in \bar{\nu}) = \{t \in Z^\nu : t^{(s)} \in I_{n_s}(i_s),\ s \in \bar{\nu}\}$$

$$i_s = 0, \ldots, k_s - 1,\ k_s = k(n_s),\ n_s \in Z^1,\ s \in \bar{\nu},$$

$$I'_{\bar{n}} = \bigcup_{(i_s,\ s\in\bar{\nu})} I_{\bar{n}}(i_s,\ s\in\bar{\nu})$$

$$I''_{\bar{n}} = I_{\bar{n}} \setminus I'_{\bar{n}}$$

We have

$$\eta_{I_{\bar{n}}} = \eta_{I'_{\bar{n}}} + \eta_{I''_{\bar{n}}} \quad {}^{*}$$

Lemma 3.

$$M(\eta_{I''_{\bar{n}}})^2 \cdot \sigma^{-2}_{I_{\bar{n}}} \to 0 \quad \underline{\text{as}} \quad \bar{n} \to \infty$$

Proof:

$$M(\eta_{I''_{\bar{n}}})^2 \le C^{(3)} |I''_{\bar{n}}| = C^{(3)}\Big(\prod_{i=1}^{\nu} (2n_i) - \prod_{i=1}^{\nu} k_i p(n_i)\Big)$$

and therefore

$$M(\eta_{I''_{\bar{n}}})^2 \sigma^{-2}_{I_{\bar{n}}} \le \frac{C^{(3)} \prod_{i\in\bar{\nu}} (2n_i) - \prod_{i\in\bar{\nu}} k_i p(n_i))}{C^{(2)} \prod_{i\in\bar{\nu}} (2n_i)} \to 0$$

as $\bar{n} \to \infty$. Also,

$$1 - \prod_{i\in\bar{\nu}} \frac{p(n_i)}{p(n_i) + q(n_i)} \to 0, \quad n \to \infty$$

Therefore, the sequence of the random variables $\eta_{I''_{\bar{n}}} \sigma^{-1}_{I_{\bar{n}}}$ converges in probability to zero, and hence the limiting distribution of the random variables $\eta_{I_{\bar{n}}} \sigma^{-1}_{I_{\bar{n}}}$ coincides with the limiting distribution of

---

* Here and in the following, $\eta_J = \sum_{t\in J} \eta_t$, $J \subset Z^{\nu}$, $|J| < \infty$.

$\eta_{I'_{\overline{n}}} \cdot \sigma^{-1}_{I_{\overline{n}}}$.

<u>Lemma 4.</u>

$$\left| M e^{it\eta_{I_{\overline{n}}}(i_s,\, s\in\overline{\nu})\cdot\sigma^{-1}_{I_{\overline{n}}}} - \prod_{(i_s,\, s\in\overline{\nu})} \psi^{(\overline{n})}_{(i_s,\, s\in\overline{\nu})}(t) \right| \to 0, \quad \underline{\text{as}}\ \overline{n} \to \infty$$

$\psi^{(\overline{n})}_{(i_s,\, s\in\overline{\nu})}(t)$ <u>is the characteristic function of</u> $\eta_{I_{\overline{n}}}(i_s,\, s\in\overline{\nu})\cdot\sigma^{-1}_{I_{\overline{n}}}$.

<u>Proof:</u> Without loss of generality, we may assume that $\varphi(d)$ is a monotonically decreasing function. Therefore, we can write

$$\sum_{s\in Z^{\nu}\setminus\{0\}} \varphi^{\frac{1}{2}}(\|s\|) \geq \sum_{n=1}^{\infty} \sum_{s\in V_n\setminus V_{n-1}} \varphi^{\frac{1}{2}}(\|s\|) \geq \sum_{n=1}^{\infty} (2n)^{\nu-1} \varphi^{\frac{1}{2}}(n)$$

where

$$V_n = \{t\in Z^{\nu} : t^{(s)} \in (-n, n),\ s\in\overline{\nu}\}, \qquad n = 0, 1, 2, \ldots$$

$$\sum_{r\geq\frac{n}{2}} r^{\nu-1}\varphi^{\frac{1}{2}}(r) \geq \varphi^{\frac{1}{2}}(n)\left(\frac{n}{2}\right)^{\nu-1}\left(\frac{n}{2}\right) = \varphi^{\frac{1}{2}}(n)\left(\frac{n}{2}\right)^{\nu}$$

and

$$\varphi(n) \leq \frac{\beta^2(n)}{n^{2\nu}}$$

where

$$\beta(n) = 2^{\nu} \sum_{r\geq\frac{n}{2}}^{\infty} r^{\nu-1}\varphi^{\frac{1}{2}}(r) \to 0 \qquad \text{as} \qquad n\to\infty$$

Let $j = j(i_s,\ s\in\overline{\nu}0 = 1, \ldots, \prod_{s\in\overline{\nu}} k_s$ be any integer-valued enumeration of the sets $(i_s,\ s\in\overline{\nu})$, $i_s = 0, \ldots, k_{s-1}$, $s\in\overline{\nu}$. We have, that

$$|Me^{it\eta_{I'_{\bar{n}}}\cdot\sigma^{-1}_{I_{\bar{n}}}} - \prod_j \psi_j^{(\bar{n})}(t)| \le C^{(4)} \prod_{s\in\bar{\nu}} k_s \psi_{I_{\bar{n}(j)}} \cdot q_0$$

$$\le C^{(4)} \prod_{s\in\bar{\nu}} k_s \cdot p_s \cdot \frac{\beta(q_0)}{q_0^{2\nu}} \to 0, \quad \text{as} \quad \bar{n}\to\infty$$

where

$$q_0 = \min_{i\in\nu} \{q_i\}, \quad p_s = p(n_s)$$

and
$$0 < C^{(4)} < \infty$$

since

$$\prod_{s=1}^{\nu} k_s \cdot p_s \cdot \frac{1}{q_0^{2\nu}} \le C^{(5)} \prod_{s=1}^{\nu} \left(\frac{2n_s}{q_0^2}\right) \to 0, \quad \text{as} \quad \bar{n}\to\infty, \quad \text{for} \quad 0 < C^{(5)} < \infty$$

Now consider the sequence of the series of the independent random variables

$$\eta_{\bar{n}1}, \eta_{\bar{n}2}, \ldots, \eta_{\bar{n}\tau(\bar{n})}\ldots, \quad \text{where} \quad \tau(\bar{n}) = \prod_{s\in\bar{\nu}} (k_s - 1)$$

and assume that the distribution of the random variable $\eta_{\bar{n}j}$, $j = 1, \ldots, \tau(\bar{n})$ coincides with the distribution of the random variable $\sigma^{-1}_{I_{\bar{n}}} \cdot \eta_{I_{\bar{n}}(j)}$, $j = 1, \ldots \tau(\bar{n})$. Then it follows from Lemma 4, that the limit distribution of the $\sigma^{-1}_{I_{\bar{n}}} \cdot \eta_{I_{\bar{n}(j)}}$ coincides with the limit distribution of the sums $\eta_{\bar{n}1} + \ldots + \eta_{\bar{n}\tau(\bar{n})}$. It is known that the last limit distribution is $N(0,1)$ if the Lindeberg condition

$$\sum_{j=1}^{\tau(\bar{n})} \int_{|\ell|>\epsilon} \ell^2 \, dPr(\eta_{\bar{n}j} < \ell) \underset{\bar{n}\to\infty}{\longrightarrow} 0, \quad \text{for any } \epsilon > 0, \ \epsilon \in R^1$$

is satisfied.

Lemma 5. Let $M|\eta_t|^{2+\delta} < \infty$ for some $\delta > 0$, $\delta \in R^1$. Then,

$$M|\eta_{I_{\bar{n}}}|^{2+\delta} \le C_\nu |I_{\bar{n}}|^{\frac{2+\delta}{2}} \quad \text{where} \quad 0 < C_\nu < \infty,$$

$$I_{\bar{n}} = \{t \in Z^\nu : -n_i \le t^{(i)} \le n_i,\ i \in \bar{\nu}\}, \text{ and}$$

$$\bar{n} = (n_i,\ i \in \bar{\nu}),\quad n_i \in Z^1.$$

Suppose that Lemma 5 is true. Then we can write

$$\sum_{j=1}^{\tau(\bar{n})} \int_{|\ell|>\epsilon} \ell^2 d\Pr(\eta_{\bar{n}j} < \ell) = \sigma_{I_{\bar{n}}}^{-2} \sum_{j=1}^{\tau(\bar{n})} \ell^2 d\Pr(\eta_{I_{\bar{n}}(j)} < \ell)$$

$$\le \sigma_{I_{\bar{n}}}^{-(2+\delta)} \cdot \epsilon^{-\delta} \sum_{j=1}^{\tau(\bar{n})} \int_{-\infty}^{\infty} |\ell|^{2+\delta} d\Pr(\eta_{I_{\bar{n}}(j)} < \ell)$$

$$\le \frac{C^{(s)} \tau(\bar{n}) \sigma_{I_{\bar{n}(j)}}^{2+\delta}}{\sigma_{I_{\bar{n}}}^{2+\delta} \cdot \epsilon^{\delta}} \sim \frac{C^{(5)}}{\epsilon^{\delta}} \frac{\prod_{s=1}^{\nu} k_s \left(\prod_{s=1}^{\nu} p_s\right)^{\frac{2+\delta}{2}}}{\left(\prod_{s=1}^{\nu} 2n_s\right)^{\frac{2+\delta}{2}}} \to 0 \text{ as } \bar{n} \to \infty$$

So if $M|\eta_t|^{2+\delta} < \infty$, then Theorem 1 is true. Note that we can remove this condition in analogy with the case $\nu = 1$ (see [1], p. 438) by means of truncation.

Proof of Lemma 5: Consider the function $m = m(n) = [n^{\frac{1}{2}} \beta^{\frac{1}{2\nu}}(n)]$. We may assume that $m(n) \to \infty$, as $n \to \infty$. Let

$$m_s = m(n_s),\ s \in \bar{\nu},\ m_0 = [n_0^{\frac{1}{2}} \beta^{\frac{1}{2\nu}}(n_0)],\ n_0 = \min_{i \in \nu} \{n_i\}$$

$$\alpha_{n_s} = [-2(n_s + m_0),\ 2m_0] \subset Z^1,\quad s \in \bar{\nu}$$

$$\gamma_{m_0} = [-2m_0,\ 2m_0] \subset Z^1$$

$$\hat{\alpha}_{n_s} = [2m_0, 2(n_s + m_0)] \subset Z^1, \quad s \in \overline{\nu}$$

are intervals on $Z^1$. For any $T, \hat{T} \subset \overline{\nu}$ such that $T \cap \hat{T} = \phi$, denote

$$B_n^{T,\hat{T}} = \{t \in Z^\nu : t^{(i)} \in \alpha_{n_i}, \ i \in T, \ t^{(j)} \in \hat{\alpha}_{n_j}, j \in \hat{T}, \ t^{(s)} \in \gamma_{m_0}, \ s \in \overline{\nu} \ (T \cup \hat{T})\}$$

$$B_n^{T,\hat{T}} = B_n^T \quad \text{if} \quad T \cup \hat{T} = \overline{\nu}$$

Let $T_i$, $i = 1, 2, \ldots, 2^\nu$, be the sequences of the different subsets of $\overline{\nu}$. Let

$$a_{\overline{n}} = M|\eta_{T_i}|^{2+\delta}, \quad \sigma_{\overline{n}} = \sigma_{B_{\overline{n}}^{T_i}}, \quad i = 1, 2, \ldots, 2^\nu,$$

$$\eta_{T_i} = \eta_{B_{\overline{n}}^{T_i}}, \quad i = 1, 2, \ldots, 2^\nu$$

Lemma 6. The inequality

$$M|\sum_{i=1}^{2^\nu} \eta_{T_i}|^{2+\delta} \leq (2^\nu + \epsilon_{\overline{n},\nu}^{(1)}) a_{\overline{n}} + C_\nu^{(1)} \sigma_{\overline{n}}^{2+\delta} \tag{1}$$

holds, where $0 < C_\nu^{(1)} < \infty$, $\epsilon_{\overline{n},\nu}^{(1)} \to 0$ as $\overline{n} \to \infty$

Proof: We have

$$M|\sum_{i=1}^{2^\nu} \eta_{T_i}|^{2+\delta} \leq M(\sum_{i=1}^{2^\nu} |\eta_{T_i}|)^2 (\sum_{k=1}^{2^\nu} |\eta_{T_k}|^\delta) \leq 2^\xi a_{\overline{n}}$$

$$+ \sum_{i \neq j = k} M|\eta_{T_i}||\eta_{T_j}|^{1+\delta} + \sum_{k \neq i = j} M|\eta_{T_k}|^\delta |\eta_{T_j}|^2$$

$$+ \sum_{j \neq i = k} M|\eta_{T_i}| M|\eta_{T_k}|^{1+\delta} + \sum_{i \neq j \neq k} M|\eta_{T_i}||\eta_{T_k}||\eta_{T_j}|^\delta \tag{2}$$

Using Lemma 1 with $p = 2 + \delta$, $q = \frac{2+\delta}{1+\delta}$, Liapunov's moment inequality and Condition I of Theorem 1 yield

$$M|\eta_{T_i}||\eta_{T_k}|^{1+\delta} \leq 2|B_{\overline{n}}^{T_i}|^{\frac{1}{2+\delta}} (\varphi(2m_0))^{\frac{1}{2+\delta}} a_{\overline{n}} + \sigma_{\overline{n}}^{2+\delta} \tag{3}$$

Applying Lemma 1, with $p = \frac{2+\delta}{2}$, $q = \frac{2+\delta}{\delta}$, and the Schwartz inequality, we have

$$M|\eta_{T_i}|^2|\eta_{T_k}|^{\delta} \leq 2|B_{\overline{n}}^{T_i}|^{\frac{2}{2+\delta}} (\varphi(2m_0))^{\frac{2}{2+\delta}} a_{\overline{n}} + \sigma_{\overline{n}}^{2+\delta} \tag{4}$$

$$M|\eta_{T_i}||\eta_{T_j}||\eta_{T_k}|^{\delta} \leq 2|B_{\overline{n}}^{T_i}|^{\frac{2}{2+\delta}} (\varphi^{\frac{2}{2+\delta}}(2m_0)$$

$$\times M^{\frac{2}{2+\delta}} |\eta_{T_i}|^{\frac{2+\delta}{2}} |\eta_{T_j}|^{\frac{2+\delta}{2}} \cdot M^{\frac{\delta}{2+\delta}} |\eta_{T_k}|^{2+\delta}$$

$$+ M|\eta_{T_i}||\eta_{T_j}||\eta_{T_k}|^{\delta} \leq 2|B_{\overline{n}}^{T_i}|^{\frac{2}{2+\delta}} (\varphi(2m_0))^{\frac{2}{2+\delta}} a_{\overline{n}} + \sigma_{\overline{n}}^{2+\delta} \tag{5}$$

Putting (3), (4) and (5) into (2), we obtain (1) with

$$C_{\nu}^{(1)} = 2^2 + 2^{1+\frac{2\nu}{2+\delta}}$$

$$\epsilon_{\overline{n},\nu}^{(1)} = 2^{1+\frac{2\nu}{2+\delta}} \left(2^{\nu} \prod_{i\in\overline{\nu}} (n_i\varphi(2m_0))^{\frac{1}{2+\delta}} + 2\left(\prod_{i=1}^{\nu} n_i\varphi(2m_0)\right)^{\frac{2}{2+\delta}}\right)$$

$$\epsilon_{\overline{n},\nu}^{(1)} \to 0 \quad \text{as} \quad \overline{n} \to \infty.$$

Since

$$\prod_{i=1}^{\nu} n_i\varphi(2m_0) \leq m_0^{2\nu} \prod_{i\in\overline{\nu}} n_i\,\beta^2(2m_0) = \beta(2m_0) \to 0 \quad \text{as} \quad \overline{n} \to \infty$$

Now we finish the proof of Lemma 5. Consider the intervals

$$\pi_{n_s} = [-2(n_s + m_0),\ -2n_s] \subset Z^1$$

and

$$\pi_{n_s} = [2n_s, 2(n_s + m_0)] \subset Z^1$$

and let

$$C_{\bar{n}}^{T,\hat{T}} = \{t \in Z^\nu \cdot t^{(i)} \in \pi_{n_i}, i \in T, t^{(j)} \in \hat{\pi}_{n_j}, j \in \hat{T}, t^{(s)} \in (-2n_s, 2n_s), s \in \bar{\nu} \setminus (T \cup \hat{T})\}$$

where $T, \hat{T} \subset \bar{\nu}$, $T \cap \hat{T} = \phi$. We have

$$I_{2(\bar{n}+\bar{m}_0)} = \bigcup_{T,\hat{T} \subset \nu} B_{\bar{n}}^{T_1 \hat{T}} = \bigcup_{i=1}^{2^\nu} B_{\bar{n}}^{T_i} + \bigcup_{T,\hat{T} \subset \nu,\ T \cup \hat{T} \neq \bar{\nu}} B_n^{T,\hat{T}}$$

where $\bar{m}_0 = \underbrace{(m_0, \ldots, m_0)}_{\nu}$,

$$I_{2(\bar{n}+\bar{m}_0)} = I_{2\bar{n}} + \bigcup_{T,\hat{T} \subset \bar{\nu},\ T \cup \hat{T} \neq \phi} C_{\bar{n}}^{T,\hat{T}}$$

and

$$I_{2\bar{n}} = \bigcup_{i=1}^{2^\nu} B_{\bar{n}}^{T_i} + \bigcup_{\substack{T,\hat{T} \subset \bar{\nu} \\ T \cup \hat{T} \neq \bar{\nu}}} B_{\bar{n}}^{T,\hat{T}} \quad \bigcup_{\substack{T,\hat{T} \subset \bar{\nu} \\ T \cup \hat{T} \neq \phi}} C_{\bar{n}}^{T,\hat{T}}$$

Applying the Minkowski inequality to (6), we obtain

$$a_{2\bar{n}} \le \Big(M^{\frac{1}{2+\delta}} \Big|\sum_{i=1}^{2^\nu} \eta_{T_i}\Big|^{2+\delta} + \sum_{T,\hat{T} \subset \bar{\nu}, T \cup \hat{T} \neq \bar{\nu}} M^{\frac{1}{2+\delta}} |\eta_{B_{\bar{n}}^{T,\hat{T}}}|^{2+\delta}$$

$$+ \sum_{T,\hat{T} \subset \bar{\nu}, T \cup \hat{T} \neq \phi} M^{\frac{1}{2+\delta}} |\eta_{C_{\bar{n}}^{T,\hat{T}}}|^{2+\delta}\Big)^{2+\delta}$$

Furthermore, we note the following relations:

$$B_{\bar{n}}^{T,\hat{T}} = \bigcup_{t^{(i_0)}=-2m_0}^{2m_0} B_{\bar{n},t}^{T,\hat{T}}(i_0), \quad T \cup \hat{T} \neq \bar{\nu}, \quad i_0 \in \bar{\nu} \setminus (T \cup \hat{T})$$

and

$$B^{T,\hat{T}}_{\overline{n},t^{(i_0)}} = \{t \in Z^{\nu}: t^{(k)} \in \alpha_{n_k}, k \in T, t^{(j)} \in \hat{\alpha}_{n_j}, j \in \hat{T}, t^{(s)} \in \gamma_{m_0}, s \in \overline{\nu}\ [T \cup \hat{T} \cup \{i_0\}], t_{i_0}\}$$

For $t_0 \in T$, let

$$C^{T,\hat{T}}_{\overline{n},t^{i_0}} = \{t \in Z^{\nu}: t^{(k)} \in (-2n_k, 2n_k), k \in \nu\ (T \cup \hat{T}), t^{(s)} \in \hat{\pi}_{n_j}, j \in \hat{T}, t^{(s)} \in \pi_{n_s}, s \in T\}$$

Then

$$C^{T,\hat{T}}_{\overline{n}} = \bigcup_{t^{(i_0)} = 2m_{i_0}}^{2(n_{i_0}+m_0)} C^{T,\hat{T}}_{\overline{n},t^{(i_0)}}$$

Applying the Minkowski inequality, we have

$$a_{2\overline{n}} \leq (M^{\frac{1}{2+\delta}} |\sum_{i=0}^{2^{\nu}} \eta_{T_i}|^{2+\delta} + \sum_{\substack{T,\hat{T}\subset\overline{\nu} \\ T\cup\hat{T}\neq\overline{\nu}}} \sum_{t^{(i_0)}=-2m_0}^{2m_0} M^{\frac{1}{2+\delta}} |\eta_{B^{T,\hat{T}}_{\overline{n},t^{(i_0)}}}|^{2+\delta}$$

$$+ \sum_{T,\hat{T}\subset\overline{\nu},\, T\cup\hat{T}\neq\phi} \sum_{t^{(j_0)}=-2n_{j_0}}^{2(n_{j_0}+m_0)} M^{\frac{1}{2+\delta}} \quad C^{T,\hat{T}}_{\overline{n},t^{(j_0)}} \quad |^{2+\delta})^{2+\delta}$$

Using Lemma 6, we can write

$$a_{2\overline{n}} \leq ((2^{\nu} + \epsilon^{(1)}_{\overline{n},\nu})a_{\overline{n}} + C^{(1)}_{\nu}\sigma^{2+\delta}_{\overline{n}})(1 + \omega_{\overline{n},\nu})^{2+\delta}$$

and

$$\omega_{\overline{n},\nu} = ((2^{\nu}+\epsilon^{(1)}_{\overline{n},\nu})a_{\overline{n}} + C^{(1)}_{\nu}\sigma^{2+\delta}_{\overline{n}})^{-\frac{1}{2+\delta}}$$

$$\times \sum_{T,\hat{T}\subset\overline{\nu},\, T\cup\hat{T}\neq\overline{\nu}} \; \sum_{t^{(i_0)}=-2m_0}^{2m_0} M^{\frac{1}{2+\delta}} |\eta_{B^{T,\hat{T}}_{\overline{n},t^{(i_0)}}}|^{2+\delta}$$

$$+ \sum_{T,\hat{T}\subset\overline{\nu},\, T\cup\hat{T}\neq\phi} \; \sum_{t^{(j_0)}=-2m_0}^{2m_0} M^{\frac{1}{2+\delta}} |\eta_{C^{T,\hat{T}}_{\overline{n},t^{(j_0)}}}|^{2+\delta} \qquad (6)$$

It is easy to see that for $\nu = 1$, Lemma 7 follows from (6) and so (see [1], p. 434) Lemma 5 is valid. Now let $\nu > 2$ and assume Lemma 5 has been demonstrated for $\nu - 1$. Then

$$M|\eta_{B^{T,\hat{T}}_{\overline{n},t^{(i_0)}}}|^{2+\delta} \leq C^{(1)}_{\nu-1}|I_{\overline{n},\nu-1}|^{\frac{2+\delta}{2}}, \text{ where } 0 < C^{(1)}_{\nu-1} < \infty$$

and

$$M|\eta_{C^{T,\hat{T}}_{\overline{n},t^{(j_0)}}}|^{2+\delta} \leq C^{(2)}_{\nu-1}|I_{\overline{n},\nu-1}|^{\frac{2+\delta}{2}}, \text{ where } 0 < C^{(2)}_{\nu-1} < \infty$$

and

$$L_{\overline{n},\nu-1} = \{t \in Z^{\nu-1} : -n_i \leq t^{(i)} \leq n_i,\quad i \in \overline{\nu-1}\}$$

In a manner analogous to the case $\nu = 1$, (6) implies that for any integers $r_1, r_2, \ldots, r_\nu \in Z^1$,

$$a_{\overline{n}} \le C^{(6)} \sigma^{2+\delta}_{I(2^{r_1}, 2^{r_2}, \ldots, 2^{r_\nu})}, \quad 0 < C^{(6)} < \infty$$

Let

$$n_i = \sum_{k_i=0}^{r_i} \nu_{r_i-k_i} 2^{r_i-k_i}, \quad \nu_{r_i-k_i} = \begin{cases} 1 \\ 0 \end{cases}, \quad i = 1, 2, \ldots, \nu$$

We have

$$I_{\overline{n}} = \bigcup_{k_1=0, k_2=0, \ldots, k_\nu=0}^{r_1 r_2 \ldots r_\nu} I_{(2^{r_1-k_1}, 2^{r_2-k_2}, \ldots, 2^{r_\nu-k_\nu})}$$

Using the Minkowski inequality, we obtain

$$a_{\overline{n}} \le \Big( \sum_{k_1=0, k_2=0, \ldots, k_\nu=0}^{r_1 r_2 \ldots r_\nu} M^{\frac{1}{2+\delta}} |\eta_{I(2^{r_1-k_1}, 2^{r_1-k_2}, \ldots, 2^{r_\nu-k_\nu})}|^{2+\delta}$$

and then

$$\sum_{k_1=0, k_2=0, \ldots, k_\nu=0}^{r_1 r_2 \ldots r_\nu} M^{\frac{1}{2+\delta}} |\eta_{I(2^{r_1-k_1}, 2^{r_2-k_2}, \ldots, 2^{r_\nu-k_\nu})}|^{2+\delta}$$

$$\le C_\nu^{(4)} \sum_{k_1=0}^{r_1} 2^{\frac{r_1-k_1}{2}} \sum_{k_2=0}^{r_2} 2^{\frac{r_2-k_2}{2}} \cdots \sum_{k_\nu=0}^{r_\nu} 2^{\frac{r_\nu-k_\nu}{2}}$$

$$\le C_\nu^{(5)} 2^{\frac{r_1+\ldots+r_\nu}{2}} \le C_\nu^{(5)} \prod_{s=1}^{\nu} n_s^{\frac{1}{2}} \le C_\nu^{(5)} \Big( \prod_{s=1}^{\nu} n_s \Big)^{\frac{1}{2}}$$

where $0 < C_\nu^{(4)} < \infty$, and $0 < C_\nu^{(5)} < \infty$.

## REFERENCES

1. I. A. Ibragimov, Ju. V. Linnik, Independent and stationary sequences of random variables. Moscow: Nauka, 1965. English transl. Groningen: Noordhoff, 1971.

2. B. S. Nahapetian, Strong mixing property of Gibbs random field with discrete parameter and some of its applications. Isvestia A. N. Arm. S. S. R. "Matematika", X. 3. 242-254. 1975.

3. R. L. Dobrushin, B. S. Nahapetian, Strong convexity of pressure for a lattice system of classical statistical physics. Teor. Mat. Fis. 20, 2, 223-234, 1974.

4. B. S. Nahapetian, Central limit theorem for the random field with strong mixing property. Dokl. A. N. Arm. S. S. R. 61. 4. , 210-213, 1975.

5. V. A. Malishev, Central limit theorem for the Gibbs random fields, Dokl. A. N.,U. S. S. R. , 224, 1975.

6. A. V. Bulinsky, I. G. Jurbenko, Central limit theorem for the random fields. Dokl. A. N. U. R. S. S. , 226, 1, 1976.

7. R. A. Minlos, A. M. Halfina, Central limit theorem for the energy and number of particles in lattice models of gas. Isvestia A. N., U. S. S. R., "Matematika" 34. 5, 1173-1191, 1970.

8. C. Deo, Central limit theorem for random field, An. Probl, 3, 708, 1975.

9. R. L. Dobrushin, The description of the random field by its conditional distribution and its regularity conditions. Teor. ver i ee prim. 13. 2, 201-229, 1968.

chapter 18

# STABLE AND ATTRACTIVE TRAJECTORIES IN MULTICOMPONENT SYSTEMS

A. L. Toom

This paper contains a group of interrelated results on local interacting multicomponent systems with a small random noise. Though many notions are introduced here in a new form, in essence this paper follows the direction presented in [1, 2, 3] in particular. Besides definitions, our Section I formulates our main Theorem 1, which gives a sufficient condition of stability. The Section II proves it. Sections III and IV are based on Theorem 1. In Section III, for any finite number, we construct a system with no fewer stable trajectories. In Section IV we give a criterion for stable and attractive trajectories in a special case. Two counter-examples given in Section V show directions in which the results of Section IV cannot be generalized.

## I. BASIC DEFINITIONS AND PRIME FORMULATIONS

Throughout this paper, V is a countable set, the elements of which are called <u>points.</u> There is a finite set $U(a) \subset V$ for every point $a \in V$. For any $A \subset V$ we denote $U(A) = \bigcup_{a \in A} U(a)$ and $U^{k+1}(A) = U(U^k(A))$, where $U^0(A) = A$, and $U^\infty(A) = \bigcup_{k=0}^{\infty} U^k(A)$. We assume

$$\forall a \in V \quad \exists k : U^k(a) = \phi \qquad (1)$$

We call the maximal $k$, for which $U^k(a)$ is non-empty, the depth of point $a$. We term a point $a \in V$ boundary if $U(a)$ is empty and inner if $U(a)$ is non-empty. The set of boundary points is called boundary and denoted by $W$.

Now there corresponds to any point $a$, a finite set $X_a$. We denote $X_A = \prod_{a \in A} X_a$. The elements of sets $X_a$, $X_A$ are called states of the point $a$ and of the set $A$, respectively, and are denoted by $x_a, x_A$. So $x_A$ is a tuple, components of which are $x_a$, $a \in A$. We write $X$ for $X_V$ and simply call its elements states. We term the states $x_W$ of the boundary bases. Of course, a base $x_W$ is a restriction of a state $y$ if $x_a = y_a$ in all $a \in W$.

A map $\varphi_a : X_{U(a)} \to X_a$ is given for any inner point $a$. We term a state $x \in X$ a trajectory if $x_a = \varphi_a(x_{U(a)})$ in all inner $a$. By virtue of the condition (1), for any $x_W$ base there is just one trajectory, the restriction of which to $W$ is $x_W$. We denote this trajectory by $\mathrm{tr}(x_W)$. In fact, if $x_W$ is given, all the components of $\mathrm{tr}(x_W)$ can be found successively, their depth increasing. We term the whole construction described a combine. Thus, a combine is determined by an $X$ set and systems of $U(a)$ sets and $\varphi_a(x)$ maps, (1) being assumed.

Now we introduce our main definition of a stable trajectory. Let $M$ stand for the set of normed measures on the $\sigma$-algebra generated by all cylinder subsets of $X$. Let $\epsilon$ be a parameter, $0 \leq \epsilon \leq 1$. To every value of $\epsilon$ there corresponds a subset $M_\epsilon \subset M$. A measure $\mu \in M$ belongs to $M_\epsilon$ if for any finite $A \subset V$

$$\mu(x_a \neq \varphi(x_{U(a)}) \quad \text{for all} \quad a \in A) \leq \epsilon^{|A|}$$

Here and further, $|A|$ denotes the cardinality of any finite set $A$.

To every base $x_W$ there corresponds a subset $M(x_W) \subset M$. A measure $\mu \in M$ belongs to $M(x_W)$, if the projection of $\mu$ to $X_W$ is a $\delta$-measure concentrated in the state $x_W$. Finally, $M_\epsilon(x_W)$ denotes the intersection $M_\epsilon(x_W) = M_\epsilon \cap M(x_W)$.

<u>Definition 1.</u> We term a trajectory $y = tr(y_W)$ stable if

$$\lim_{\epsilon \to 0} \sup_{\substack{\mu \in M_\epsilon(y_W) \\ a \in V}} \mu(x_a \neq y_a) = 0 \tag{2}$$

<u>Theorem 1.</u> <u>A combine and a state</u> $y$ <u>in it are given.</u> <u>Let there be</u> $n$ <u>real functions</u> $L_1(\cdot), \ldots, L_n(\cdot)$ <u>on</u> $V$ <u>and two numbers</u> $r > 0$, $R > 0$, <u>so that the following four conditions hold for all points</u> $a, b$, <u>all</u> $k$ <u>from</u> $1$ <u>to</u> $n$, <u>and all</u> $x_{U(a)} \in X_{U(a)}$:

(1) $|U(a)| \leq R$; $|\{c : a \in U(c)\}| \leq R$.

(2) $b \in U(a) \Rightarrow |L_k(b) - L_k(a)| \leq 1$.

(3) $\sum_{i=1}^{n} L_i(a) = 0$.

(4) $\varphi_a(x_{U(a)}) \neq y_a \Rightarrow \exists\, c \in U(a)$ <u>such that</u> $x_c \neq y_c$ <u>and</u> $L_k(c) - L_k(a) \geq r$.

<u>Then</u> $y$ <u>is a stable trajectory.</u>

We shall prove this in Section II. Note here that $y$ is obviously a trajectory, and we assume it be proved. This theorem generalizes the result and method of proof in [1]. The author was moved to obtain Theorem 1 by a special case (see [2]), not subject to former proofs; we shall describe it in Section III as Example 1. Now we define an attractive trajectory and illustrate it by a simple theorem.

<u>Definition 2.</u> A trajectory $y$ is termed <u>attractive</u> if the following holds for any trajectory $x$: if the set $\{a \in W : x_a \neq y_a\}$ is finite, then the set $\{a \in V : x_a \neq y_a\}$ is also finite.

Theorem 2. Given the conditions of Theorem 1, the trajectory y is attractive.

Proof: Let the trajectory x be such that the set $A_0 = \{a \in W : x_a \neq y_a\}$ is finite. We denote $A_m = \{a \in V : U^m(a) \cap A_0 \neq \phi\}$, $m = 1, 2, \ldots$. All $A_m$ are finite since $|A_m| \le |A_0| \cdot R^m$. Of course $\{a \in V : x_a \neq y_a\} \subset \bigcup_{m=0}^{\infty} A_m$. We shall prove our theorem by finding an M for which $\{a \in V : x_a \neq y_a\} \subset \bigcup_{m=0}^{M} A_m$. In fact, we set

$$M = [\,(nr)^{-1} \sum_{i=1}^{n} \max_{a \in A_0} L_i(a) + 1]$$

Suppose there is a point $a \notin \bigcup_{m=0}^{M} A_m$, in which $x_a \neq y_a$. The condition (4) of Theorem 1 implies that for any k from 1 to n there is a point sequence $a_0 = a, a_1, a_2, \ldots, a_Q \in W$ such that $L_k(a_q) - L_k(a_{q-1}) \ge r$ and $x_{a_q} \neq y_{a_q}$ for all q from 1 to Q. Hence, $a_Q \in A_0$ and $L_k(a_Q) \ge L_k(a_0) + Qr$. On the other hand, $a \notin \bigcup_{m=0}^{M} A_m$ infers $Q > M$. Therefore, $\max_{b \in A_0} L_k(b) > L_k(a) + Mr$, whence $\sum_{k=1}^{n} \max_{b \in A_0} L_k(b) > Mrn$. But this contradicts our formula for M, which proves the theorem.

## II. PROOF OF THEOREM 2

The proof consists of four parts, the first three of which are auxiliary.

Part 1. We denote $N = \{1; 2; \ldots; n\}$, where n is the number of functions $L_1(\cdot), \ldots, L_n(\cdot)$ in the formulation of the theorem. We term any map $\pi : N \to A$ a polar $\pi$ on a set A. The image $\pi(k)$ of a number $k \in N$ is called the k-th pole of the $\pi$ polar. Similarly, $\pi(N)$ is the domain of values of $\pi$ and $\pi(N') = \{\pi(k), k \in N'\}$ for any $N' \subset N$. For any polar $\pi$ on V we term the value of $\sum_{k=1}^{n} L_k(\pi(k))$ the extent of $\pi$.

A polar $\pi$ on the set of vertices of a graph $g$ is termed an edger on $g$ if $\pi(N) = \{a; b\}$ and there is an edge in $g$ which connects $a$ and $b$. When treating graphs, we assume that the ends of any edge differ. Everywhere, except in emphasized cases, we assume that any two vertices $a, b$ are connected by no more than one edge, denoted by $(a, b)$ in this case.

The following two special kinds of polars are the most important for us: A polar $\pi$ on $V$ is called an arrow if $\pi(N) = \{a; b\}$, where $a \in U(b)$ and there is such a $k \in N$ that $\pi(k) = a$, $\pi(N \setminus k) = b$ and $L_k(a) - L_k(b) \geq r$. A polar $\pi$ on $V$ is called a fork if $\pi(N) = \{a;b\} \subset U(c)$, where $a \nmid b$, $c \in V$. We can introduce a graph $\gamma$ with $V$ as the set of vertices; $a, b \in V$ are connected with an edge in $\gamma$ is $a \in U(b)$ or $\exists c : \{a; b\} \subset U(c)$. Both arrows and forks are edgers on $\gamma$. It is clear that no more than $R^2 + 2R$ edges start from any vertex in $\gamma$.

Suppose we have two sequences of polars on a set $A$. We say that these sequences are equivalent if one can be turned into the other by a permutation of members. So all sequences of polars on $A$ partition into equivalence classes, which we term trusses. We designate a truss $\Pi$ by any sequence in its class: $\Pi = (\pi_1, \ldots, \pi_m)$. There is an empty truss, which contains no polar. For any truss $\Pi = (\pi_1, \ldots, \pi_m)$, we denote $\Pi(N) = \bigcup_{k=1}^{m} \pi_k(N)$. If two trusses $\Pi^1 = (\pi^1_1, \ldots, \pi^1_k)$ and $\Pi^2 = (\pi^2_1, \ldots, \pi^2_m)$ are given, then $\Pi^1 * \Pi^2 = (\pi^1_1, \ldots, \pi^1_k, \pi^2_1, \ldots, \pi^2_m)$ is the truss obtained by the juxtaposition of their sequences.

We term a truss $\Pi = (\pi_1, \ldots, \pi_m)$ on a set $A$, p-even or even on a subset $B \subset A$, if for any $k \in N$ the sequence $\pi_1(k), \ldots, \pi_m(k)$ has one and the same number $p$ of members belonging to $B$. We term a truss $\Pi$ on $A$ overall even on $B \subset A$ if it is even on all subsets of $B$.

Let a truss $\Pi = (\pi_1, \ldots, \pi_m)$ on $V$ consist only of edgers on the graph $\gamma$ introduced above. We form a graph $\gamma(\Pi)$, which has $m$ edgers and $\Pi(N)$ as the set of vertices. To every edger $\pi_k$, $1 \leq k \leq m$, there corresponds biuniquely an edge in $\gamma(\Pi)$, which connects the two points that are elements of $\pi_k(N)$. (Several edges here may connect two given points.) We call the truss $\Pi$ connected if the corresponding graph $\gamma(\Pi)$ is connected.

Lemma 1. If a truss on $V$ is overall even on $V$, then the sum of the extents of its polars equals zero.

The proof is quite easy.

Lemma 2. Suppose that a truss on $V$ is overall even on $V$ and consists of $k$ arrows and $m$ forks. Then $rk \leq 2nm$.

Proof: The extent of any arrow is no less than $r$. The extent of any fork is no more than $2n$ in absolute value. Therefore, the sum of extents in our truss is no less than $rk - 2nm$. But from Lemma 1, this sum is zero, priving the lemma.

Lemma 3. Let a point $v \in V$ be given. The number of different connected trusses $\Pi$ on $V$, consisting of $m$ edgers on $\gamma$, and with $v \in \Pi(N)$, does not exceed $[2^n(R^2 + 2R)]^{2m}$.

Proof: It is known that in any connected graph there is a closed loop, which passes every edge twice. For any truss $\Pi$ in the lemma, we choose such a closed loop in $\gamma(\Pi)$ starting from $v$. Now we shall code each of our trusses with a sequence of $2m$ symbols. Each symbol codes for a specific polar, and the order of the symbols concurs with the order in which our loop passes through the edges corresponding to symbols.

Now estimate how many variants every sumbol must have in order to recover the truss uniquely from the sequence. Suppose we have recovered the polars corresponding to the edges of our loop up to a point a. To determine the next polar, it is sufficient to know the following: (1) which of the edges starting from a in the $\gamma$ graph continues our loop; this allows no more than $R^2 + 2R$ variants; (2) how the poles of the next polar are distributed among the ends of this edge; this allows less than $2^n$ variants. Thus, $2^n(R^2 + 2R)$ variants per symbol will do. Then, there are $[2^n(R^2 + 2R)]^{2m}$ sequences and no more differing trusses than that.

Lemma 4. For any polar $\pi_0$ on a set A of vertices of a connected graph g, there is a truss $\Pi$ of edgers on g, the truss $(\pi_0) * \Pi$ being 0-even or 1-even on every vertex of g.

Proof: Let $g^1$ be a minimal connected subgraph of g, containing $\pi_0(N)$. Of course, $g^1$ is a tree. For any edge (a, b) of the tree $g^1$ we form an edger $\pi_{(a,b)}$ on g, by the following rule: each of its poles $\pi_{(a,b)}(k)$ coincides with vertex a or b; namely with that one of the two, which needs the edge (a, b) to be connected with $\pi_0(k)$. The truss which consists of the edgers $\pi_{(a,b)}$ for all edgers (a, b) of $g^1$, taken once each, is the one sought.

Part 2. Now we choose a point $v \in V$ and a state $x \in X$, in which $x_v \neq y_v$. All the constructions in Parts 2 and 3 are made for the chosen v and x, and we shall not mention this any more. Suppose that a point $a \in V$ is such that $\varphi_a(x_{U(a)}) = x_a \neq y_a$. In this case, for every $k \in N$ we choose a point $u_k(a) \in U(a)$, in which $x_{\bar{u}_k}(a) \neq y_{\bar{u}_k}(a)$ and $L_k(\bar{u}_k(a)) - L_k(a) \geq r$. The existence of such a $\bar{u}_k(a)$ is ensured by condition (4) in the formulation of the theorem. We denote $\bar{U}(a) = \{\bar{u}_k(a),\ k \in N\}$. If $\varphi_a(x_{U(a)}) \neq x_a$ or $x_a = y_a$, then let $\bar{U}(a)$ be empty by definition. For any $A \subset V$ we denote $\bar{U}(A) = \bigcup_{a \in A} \bar{U}(a)$; then $\bar{U}^{k+1}(A) = \bar{U}(\bar{U}^k(A))$, where $\bar{U}^0(A) = A$, and

$\overline{U}^{\infty}(A) = \bigcup_{k=0}^{\infty} \overline{U}^{k}(A)$. We also denote $\hat{U} = \{a \in \overline{U}^{\infty}(v) : \overline{U}(a) = \phi\}$.

Lemma 5. All points $a \in \hat{U}$ are inner and $x_a = \varphi_a(x_{U(a)})$ in all of them.

Proof: It is easy to prove by induction, that $x_a \neq y_a$ in all $a \in \overline{U}^{\infty}(v)$. In particular, this is true for all $a \in \hat{U}$. This easily proves the lemma.

Now we define an oriented graph G as follows: Its set of vertices is $\overline{U}^{\infty}(v)$. An edge leads from vertex a to vertex b in G, if $a \in \overline{U}(b)$. It is clear that G is finite and contains no oriented loops. Therefore, an integer function $t : \overline{U}^{\infty}(v) \to Z$ can be defined such that $a \in \overline{U}(b) \Rightarrow t(a) < t(b)$. We choose a function $t(\cdot)$ for this.

Now we reconstruct G into another oriented graph $\tilde{G}$. This is done independently for all edges of the graph G. Let us choose two vertices a, b in G such that $a \in \overline{U}(b)$. If $t(b) > t(a) + 1$, we interpose between a and b new vertices $\alpha_{a,b}^{k}$, where $1 \leq k \leq t(b) - t(a) - 1$. The new vertices are not termed points since they do not belong to V. Now we reject the old edge (a, b) and add $t(b) - t(a)$ new edges, which lead successively: from a to $\alpha_{a,b}^{1}$, from $\alpha_{a,b}^{k}$ to $\alpha_{a,b}^{k+1}$ for $1 \leq k \leq t(b) - t(a) - 2$ and, finally, from $\alpha_{a,b}^{t(b)-t(a)-1}$ to b. Having done this with all the edges (a, b) of the graph G, we obtain the new graph $\tilde{G}$. We define our function $t(\cdot)$ for the new vertices by the formula $t(\alpha_{a,b}^{k}) = t(a) + k$. For any vertex a of the graph $\tilde{G}$ let $\tilde{U}(a)$ stand for the set of those vertices in $\tilde{G}$, from which edges lead straight to a. Of course, $b \in \tilde{U}(a) \Rightarrow t(a) - t(b) = 1$. We introduce these notations for any set A of vertices in $\tilde{G}$: $\tilde{U}(A) = \bigcup_{a \in A} \tilde{U}(a)$, then $\tilde{U}^{k+1}(A) = \tilde{U}(\tilde{U}^{k}(A))$, where $\tilde{U}^{0}(A) = A$, and $\tilde{U}^{\infty}(A) = \bigcup_{k=0}^{\infty} \tilde{U}^{k}(A)$.

Now we reconstruct $\tilde{G}$ into another oriented graph T (which will appear to be a tree). In order to do it, we separate all the vertices of G into equivalence classes by the following three rules, where $a \sim b$ means "a is equivalent to b" : (1) if $t(a) = t(b)$ and $\tilde{U}^{\infty}(a) \cap \tilde{U}^{\infty}(b)$ is nonempty, then $a \sim b$; (2) if $a \sim b$ and $b \sim c$, then $a \sim c$; (3) only those pairs a, b are equivalent, for which equivalence is implied by the rules (1), (2). The equivalence classes thus obtained will simply be called classes. These classes serve as vertices of the new graph T, defined as follows. An edge leads from a class A to a class B in the graph T if there are such $a \in A$ and $b \in B$ that $a \in \tilde{U}(b)$. Thus T is defined. It is natural to define a function $t(\cdot)$ on the classes as follows: $t(A) = t(a)$, where $a \in A$. Let $U_T(A)$ be the set of classes, from which edges of the T graph lead to the class A. We denote for any S set of classes: $U_T(S) = \bigcup_{A \in S} U_T(A)$, then $U_T^{k+1}(S) = U_T(U_T^k(S))$, where $U_T^0(S)$, and $U_T^{\infty}(S) = \bigcup_{k=0}^{\infty} U_T^k(S)$.

The following is easy to prove. The vertex v of the graph $\tilde{G}$ is equivalent to no other and forms a class $\{v\}$. The set of those vertices $a \in \tilde{U}^{\infty}(v)$, for which $\tilde{U}(a)$ is empty, coincides with $\hat{U}$. Every vertex $a \in \hat{U}$ is equivalent to no other and forms a class $\{a\}$. If $\{A; B\} \subset U_T(D)$ and $A \neq B$, then $U_T^{\infty}(A) \cap U_T^{\infty}(B)$ is empty. The graph T is a tree, in which all edges are oriented towards the vertex $\{v\}$.

Now for any A class, for which $U_T(A)$ is nonempty, we introduce a (non-oriented) graph $g(A)$ with $U_T(A)$ as the set of vertices. Classes $B, B^1 \in U_T(A)$ are connected with an edge in the graph $g(A)$, if $B \neq B^1$ and there are such points $b, b^1, a \in V$ that $b \in \tilde{U}^{\infty}(B) \cap \bar{U}(a)$ and $b^1 \in \tilde{U}^{\infty}(B^1) \cap \bar{U}(a)$.

<u>Lemma 6.</u> <u>Any such graph</u> $g(A)$ <u>is connected.</u>

<u>Proof:</u> Let B and $B^1$ be different vertices of $g(A)$. By definitions

of $g(A)$ and $U_T(A)$ there are such $b \in B$, $b^1 \in B^1$, $a$, $a^1 \in A$ that $b \in \tilde{U}(a)$ and $b^1 \in U(a^1)$. First assume $a = a^1$. Then $a \in V$ and there are $d \in \tilde{U}^\infty(b) \cap \bar{U}(a)$ and $d^1 \in \tilde{U}^\infty(b^1) \cap \bar{U}(a)$, whence $B$ and $B^1$ are connected with an edge in $g(A)$. Now assume $a \neq a^1$. But, since $a \sim a^1$, there is such a sequence $a_0 = a, a_1, a_2, \ldots, a_{k-1}$, $a_k = a^1$, consisting of different elements of $A$, that intersections $\tilde{U}^\infty(a_{m-1}) \cap \tilde{U}^\infty(a_m)$ are nonempty for all $m$ from 1 to $k$. Then for every $m$ there are such $d_{m-1} \in \tilde{U}(a_{m-1})$ and $d_m^1 \in \tilde{U}(a_m)$ that $\tilde{U}^\infty(d_{m-1}) \cap \tilde{U}^\infty(d_m^1)$ is nonempty. Besides, $t(d_{m-1}) = t(d_m^1) = t(A) - 1$. Therefore $d_{m-1} \sim d_m^1$ and get into one class $D_m$. Now it is easy to prove that in the sequence $B, D_1, D_2, \ldots, D_k, B^1$ every two next classes either coincide or are connected with an edge in $g(A)$, q.e.d.

Part 3. Now $q$ is an integer parameter, which runs from 0 to $Q$. The value of $Q$ will be chosen later. For every value of $q$ we shall form two trusses $\Pi_q^1$ and $\Pi_q^2$ on $\bar{U}^\infty(v)$, a set $S_q$ of classes and a (non-oriented) graph $G_q$, which is defined as follows. The set $(\Pi_q^1 * \Pi_q^2)(N)$ is the set of vertices of $G_q$. Two vertices $a, b$ are connected with an edge in the graph $G_q$ if $a \neq b$ and at least one of the two following conditions holds: (1) there is such a polar $\pi$ in the truss $\Pi_q^1 * \Pi_q^2$, that $\pi(N) = \{a; b\}$; (2) there is such a class $A$ in the set $S_q$, that $\{a; b\} \subset \tilde{U}^\infty(A)$. We shall construct $\Pi_q^1$, $\Pi_q^2$, $S_q$ (and also $G_q$ by that inductively when $q$ increases from 0 to $Q$. Based on this construction, which we shall describe below, one can prove by induction that $\Pi_q^1$, $\Pi_q^2$, $S_q$ and $G_q$ possess the following seven properties for all values of $q$. We denote $\Pi_q = \Pi_q^1 * \Pi_q^2 * \Pi^0$, where $\Pi^0$ consists of one polar, all the poles of which coincide with $v$.

1. The truss $\Pi_q^1$ contains arrows only, the truss $\Pi_q^2$ contains forks only.

2. If classes $A$, $B$ are different and both of them belong to $S_q$,

then A does not belong to $U_T^\infty(B)$.

3. For any point $a \in \Pi_q(N)$ there is such a class $A \in S_q$ that the set $\tilde{U}^\infty(a) \cap \tilde{U}^\infty(A)$ is nonempty.

4. The truss $\Pi_q$ is overall even on the set $\overline{U}^\infty(v) \setminus \tilde{U}^\infty(\bigcup_{A \in S_q} A)$.

5. The truss $\Pi_q$ is 1-even on $\tilde{U}^\infty(A)$ for every $A \in S_q$.

6. The number $|S_q|$ of elements in $S_q$ equals the number of forks in $\Pi_q^2$ plus one.

7. The graph $G_q$ is connected.

Now we describe the inductive construction by three items.

1. The initial case $q = 0$: We set the trusses $\Pi_0^1$, $\Pi_0^2$ empty. The set $S_0$ consists only of the class $\{v\}$. All the seven properties obviously take place in this case.

2. When the induction ends: Suppose we have constructed $\Pi_q^1$, $\Pi_q^2$, $S_q$. If $U_T(A)$ is empty for all $A \in S_q$, then we set Q equal to this value of q and stop the process of constructing.

3. The induction step: Let $\Pi_q^1$, $\Pi_q^2$, $S_q$ be constructed already and possess the seven properties. Let $A \in S_q$ be a class such that $U_T(A)$ is nonempty. We choose such a class A and construct $\Pi_{q+1}^1$, $\Pi_{q+1}^2$, $S_{q+1}$. The new trusses have the form $\Pi_{q+1}^1 = \Pi_q^1 * \Pi^1$ and $\Pi_{q+1}^2 = \Pi_q^2 * \Pi^2$, where $\Pi^1$ and $\Pi^2$ are to be defined.

If $A \cap \Pi_q(N)$ is empty, then the truss $\Pi^1$ is empty also. Now let $A \cap \Pi_q(N)$ be nonempty. The 5-th property for $\Pi_q$ implies for every k that no more than one polar in $\Pi_q$ may have its k-th pole in A. Let $N^1 \subset N$ be the set of those values of k for which A contains the k-th pole of some polar in $\Pi_q$. We denote this pole by $\alpha_k$ and form an arrow $\pi_k$ by the rule: the k-th pole of $\pi_k$ is $\bar{u}_k(\alpha_k)$; all its other poles coincide with $\alpha_k$. The truss $\Pi^1$ consists of all these arrows $\pi_k$ for all $k \in N^1$, each taken once. Note here that the truss $\Pi_q * \Pi^1$ is overall even on A and is 1-even on $U_T^\infty(A) \setminus (A)$. The last assertion allows us to form a polar $\pi_0$ on

$U_T(A)$ by the following rule: $\pi_0(k) = B$, where $B$ is that element of $U_T(A)$, for which $\tilde{U}^\infty(B)$ contains the k-th pole of some polar in $\Pi_q * \Pi^1$. Further, using Lemmas 4 and 6, we choose such a truss $(\pi_1, \ldots, \pi_m)$ of edgers on $g(A)$ that the truss $(\pi_0, \pi_1, \ldots, \pi_m)$ is 0-even or 1-even on every vertex of $g(A)$. For every edger $\pi_i$, $1 \leq i \leq m$ on $g(A)$ we form a fork $\bar{\pi}_i$ by the following rule. Let $\pi_i(N) = \{B, B^1\} \subset U_T(A)$. Since $B$ and $B^1$ are connected with an edge in $g(A)$, there are such points $b, b^1, a \in V$ that $b \in \tilde{U}^\infty(B) \cap \bar{U}(a)$ and $b^1 \in \tilde{U}^\infty(B^1) \cap \bar{U}(a)$. We choose such $b, b^1$ and form the fork $\bar{\pi}_i$ as follows: if $\pi_i(k) = B$, then $\bar{\pi}_i(k) = b$, and if $\pi_i(k) = B^1$, then $\bar{\pi}_i(k) = b^1$. The truss $\Pi^2$ consists of the $m$ forks $\bar{\pi}_1, \ldots, \bar{\pi}_m$ thus constructed.

Now we define the set $S_{q+1}$. It is obtained from $S_q$ by rejecting the class $A$ and inserting those classes $B \in U_T(A)$ for which $\tilde{U}^\infty(B) \cap \Pi_{q+1}(N)$ is nonempty.

The induction step is described. It is easy to infer our seven properties for $\Pi^1_{q+1}, \Pi^2_{q+1}, S_{q+1}, G_{q+1}$ from those of $\Pi^1_q, \Pi^2_q, S_q, G_q$.

The amount of classes in $U_T^\infty(S_q)$ strictly decreases at every induction step, whence the construction is sure to have an end. When it is over, we obtain the trusses $\Pi^1_Q$ and $\Pi^2_Q$ and the set $S_Q$. We denote $\Pi = \Pi^1_Q * \Pi^2_Q$. Obviously, every class in $S_Q$ consists of one point. This allows us to introduce a set $S = \{a : \{a\} \in S_Q\}$. Of course $S \subset \hat{U}$. Applying our seven properties to the case $q = Q$ we get the following. The truss $\Pi$ consists of arrows and forks; the number of forks in $\Pi$ equals the number of points in $S$ minus one. The truss $\Pi$ is connected and overall even.

Part 4. Now only the point $v$ remains fixed. Let us consider all $x \in X$ such that $x_v \neq y_v$. For each of them we have constructed a truss $\Pi$ and a set $S$, which we denote by $\Pi(x)$ and $S(x)$ now.

Lemma 7. $\Pi(x) = \Pi(x^1)$ implies $S(x) = S(x^1)$.

The lemma is proved by the following representation of the set $S$, which can be checked directly. The set $S$ consists of all those points $a \in \Pi(N)$ for which there is no such arrow $\pi$ in $\Pi$ that $\pi(N) = \{a; b\}$ where $b \in U(a)$.

Now let $M_k$ stand for the number of different sets $S(x)$, which contain $k+1$ points. Let us estimate $M_k$. For these $x$ the truss $\Pi_Q^2$ contains $k$ forks, whence $\Pi(x)$ contains no less than $k$ and no more than $k(1+2nr^{-1})$ polars (see Lemma 2). Hence and from Lemma 3, the number of the corresponding trusses $\Pi(x)$, and from Lemma 7, the value of $M_k$ also is no more than

$$M_k \leq \sum_{m=k}^{[k(1+2nr^{-1})]} [2^n(R^2+2R)]^{2m}$$

Now we estimate the value of $\mu(x_V \neq y_V)$ for $\mu \in M_\epsilon(y_W)$. For all $x$ such that $x_V \neq y_V$, we introduce a cylinder set $C_x \subset X$:

$$C_x = \{x^1 : x_a^1 \neq \varphi_a(x^1_{U(a)}) \quad \text{for all} \quad a \in S(x)\}.$$

Every $x$ belongs to the corresponding $C_x$ from Lemma 5. Therefore $\{x : x_V \neq y_V\} \subset \bigcup_{x : x_V \neq y_V} C_x$, where it is sufficient to retain only different sets $C_x$ on the right side. On the other hand, $\mu \in M_\epsilon(y_W)$ implies $\mu(C_x) \leq \epsilon^{|S(x)|}$. All this implies that

$$\mu(x_V \neq y_V) \leq \sum_{k=0}^{\infty} M_k \epsilon^{k+1}$$

Substituting our estimation for $M_k$ here, we obtain that the series $\sum_{k=0}^{\infty} M_k \epsilon^{k+1}$ converges for sufficiently small $\epsilon > 0$ and its sum tends to $0$, when $\epsilon \to 0$. Whence follows our theorem.

## III. TESSELLATIONS WITH MANY STABLE TRAJECTORIES AND UNIFORM RANDOM NETWORKS WITH MANY STATIONARY MEASURES

Based on Theorem 1, this section examines one example and proves two theorems. First we define tessellations in suitable terms; these are the kind of combines, which we shall consider. We term a combine a tessellation if it is constructed with the following three items.

1. A $(d+1)$-dimensional integer lattice $Z^{d+1}$ is given. We denote its elements by $v = (s, t)$, $s \in Z^d$, $t \in Z$. Our set $V$ is a half of this lattice:

$$V = Z^d \cdot Z_+ = \{(s, t) : t \geq 0\}$$

2. A finite set $U \subset Z^{d+1} \setminus V$ is given; such that $(s, t) \in U \Rightarrow t < 0$. We denote $R = |U|$ and $t_W = \max_{(s,t) \in U} |t|$. A point $(s, t) \in V$ is inner if $t \geq t_W$; in other words $W = \{(s, t) : 0 \leq t < t_W\}$. For any inner $v$ point $U(v) = v + U$. An enumeration of $U$ is used: $U = \{(u_1, t_1), \ldots, (u_R, t_R)\}$, where $u_1, \ldots, u_R \in Z^d$, $t_1, \ldots, t_R < 0$. We call a tessellation without memory if $t_W = 1$, that is $t_1 = \ldots = t_R = -1$.

3. All the sets $X_v$ for $v \in V$ and $v \in U$ coincide with a certain finite set $X_0$. A map $\varphi : X_0^R \to X_0$ is given and all the maps $\varphi_v : X_{U(v)} \to X_v$, $v \in V$ W coincide with $\varphi(\cdot)$ in the following sense:

$$\varphi_v(x_{U(v)}) = \varphi(x_{v+(u_1, t_1)}, \ldots, x_{v+(u_R, t_R)})$$

It may be convenient to define first a map $\varphi_0 : X_U \to X_0$, which coincides with $\varphi(\cdot)$ in a similar sense:

$$\varphi_0(x_U) = \varphi(x_{(u_1, t_1)}, \ldots, x_{(u_R, t_R)})$$

Example 1. A tessellation without memory has $d = 2$, $X_0 = \{0;1\}$, $R = 3$, $U = \{(0,0,-1); (0,1,-1); (1,0,-1)\}$. The value of $\varphi(x_1, x_2, x_3)$ equals that one of 0 and 1, which the majority of the arguments $x_1, x_2, x_3$ equal.

The states "all zeros" and "all ones" are obviously trajectories; these are analogous because zeros and ones are symmetric in this example. Behavior of this tessellation (among others) in a random noise was first examined in [2] by computer simulation. The two trajectories were found to be stable (using our term). Now the trajectories "all zeros" and "all ones" are proved to be stable since our Theorem 1 can be applied to both of them with $r = \frac{1}{2}$ and $L_1(s,t) = -\frac{1}{2}(3s_1 + t)$, $L_2(s,t) = -\frac{1}{2}(3s_2 + t)$, $L_3(s,t) = \frac{1}{2}(3s_1 + 3s_2 + 2t)$, where $s_1, s_2$ are components of $s$.

The following Tneorems 3 and 4 generalize this example. We call a state $y \in X_0^{Z^{d+1}}$ of our lattice $Z^{d+1}$ a translate of another state $x \in X_0^{Z^{d+1}}$ if there is such $w$ that $y_v = x_{v+w}$ for all $v \in Z^{d+1}$. We call a state $x \in X_0^{Z^{d+1}}$ periodic if it has only a finite number of different translates. We call $x \in X_0^{Z^{d+1}}$ stationary if $x_{(s,t)}$ does not depend on $t$.

Theorem 3. Let $y^1_V, \ldots, y^n_V$ be restrictions to $V$ of any periodic states $y^1, \ldots, y^n \in X_0^{Z^{d+1}}$, where $V = Z^d \cdot Z_+$ and $d \geq 2$. Then there is a tessellation with this $V$ and $X_0$, in which all $y^1_V, \ldots, y^n_V$ are stable trajectories. If in addition all $y^1, \ldots, y^n$ are stationary the tessellation can be chosen without memory.

The restriction $d \geq 2$ is quite necessary for our method of construction. The paper [4] described a one-dimensional (that is, with $V = Z \cdot Z_+$) but non-uniform combine with two stable trajectories.

Proof of Theorem 3: Let $y^1, \ldots, y^N$ be all the different translates of the states $y^1, \ldots, y^n$. We denote $C_{\sigma,\tau}(s^0, t^0) = \{(s,t) \in Z^{d+1} : |s-s^0| \leq \sigma,\ t^0 - \tau \leq t > t^0\}$, where $|\cdot|$ is Euclidean norm and

$\sigma > 0$, $\tau > 0$, $s^0 \in Z^d$, $t^0 \in Z$ are parameters. We can choose $\sigma, \tau$ so large that all the restrictions of $y^1, \ldots, y^N$ to $C_{\sigma,\tau}(s^0, t^0)$ will differ for any $s^0, t^0$. In particular, if all $y^1, \ldots, y^n$ are stationary, one can take $\tau = 1$. We fix $\sigma, \tau$ for that. Now choose any three homogeneous linear functions $L_1, L_2, L_3 : Z^d \to R$ provided

$$\forall\, s \in Z^d : L_1(s) + L_2(s) + L_3(s) = 0$$

and every two of the functions $L_1(\cdot)$, $L_2(\cdot)$, $L_3(\cdot)$ are linearly independent. We introduce three subsets of $Z^d$:

$$Q_1 = \{ s : L_1(s) \geq 1;\ L_2(s) \geq 1 \};$$

$$Q_2 = \{ s : L_2(s) \geq 1;\ L_3(s) \geq 1 \};$$

$$Q_3 = \{ s : L_3(s) \geq 1;\ L_1(s) \geq 1 \}$$

Each of $Q_1, Q_2, Q_3$ contains arbitrary large spheres in $Z^d$. Using this, we choose such $s^1, s^2, s^3 \in Z^d$ that $\{s : |s - s^i| \leq \sigma\} \subset Q_i$ for $i = 1, 2, 3$. Now we define our tessellation. We set $U = C_1 \cup C_2 \cup C_3$, where $C_i = \{ (s, t) \in Z^{d+1} : |s - s^i| \leq \sigma,\ -\tau \leq t < 0 \}$, $i = 1, 2, 3$. We define the map $\varphi_0 : X_U \to X_0$ as follows:

$$\varphi_0(x_U) = \begin{cases} y^k_{(0,0)}, & \text{if at least two of the three following conditions hold:} \\ & (1)\ x_{C_1} = y^k_{C_1};\ (2)\ x_{C_2} = y^k_{C_2}; \\ & (3)\ x_{C_3} = y^k_{C_3}, \text{ where } k \in \{1, \ldots, N\} \\ \text{is arbitrary if the previous condition holds for no } k \in \{1, \ldots, N\}. & \end{cases}$$

The consistency of this definition is easy to check. So we have a tessellation. It is clear that all the conditions of Theorem 1 hold for all $y_v^1, \ldots, y_v^N$ with suitably scaled $L_1(\cdot)$, $L_2(\cdot)$, $L_3(\cdot)$ as introduced above and $r > 1$, q. e. d.

Theorem 4 repeats Theorem 3 in essence, but it is formulated in more common terms of uniform random networks (called briefly operators here) investigated in a number of papers (see [1, 2, 3] in particular). There are a finite list of vectors $u_1, \ldots, u_R \in Z^d$, a finite set $X_0$ and nonnegative numbers --probabilities $\theta(b \mid a_1, \ldots, a_R)$, where $b$, $a_1, \ldots, a_R \in X_0$; such that

$$\forall a_1, \ldots, a_R : \sum_{b \in X_0} \varphi(b \mid a_1, \ldots, a_R) = 1$$

Let $M$ stand for the set of normed measures on the $\sigma$-algebra generated by all cylinder subsets of $X = X_0^{Z^d}$. We term any linear continuous (in the product-topology) map $P : M \to M$ an <u>operator</u>, the following being assumed: For any $\delta$-measure $\delta_y$ concentrated in a state $y \in X$ the image $P\delta_y$ is Bernoullean and

$$(P\delta_y)(x_s = b) = \theta(b \mid y_{s+u_1}, \ldots, y_{s+u_R})$$

We call a measure $\mu$ stationary for $P$ if $\mu = P\mu$. If two operators $P$ and $P^0$ have the same $d$, $u_1, \ldots, u_R$, $X_0$, only their probabilities $\theta(b \mid a_1, \ldots, a_R)$, $\theta^0(b \mid a_1, \ldots, a_R)$ varying, we denote

$$|P, P^0| = \max_{b, a_1, \ldots, a_R} |\theta(b \mid a_1, \ldots, a_R) - \theta^0(b \mid a_1, \ldots, a_R)|$$

We call an operator $P$ deterministic if all its $\theta(b \mid a_1, \ldots, a_R)$ equal zero or one.

<u>Theorem 4.</u> <u>Let</u> $y^1, \ldots, y^n$ <u>be any periodic states in</u> $X = X_0^{Z^d}$. <u>Then there is a deterministic operator</u> $P^0$ <u>acting on the set</u> $M$ <u>of measures on</u> $X$, <u>for which the measures</u> $\delta_{y^1}, \ldots, \delta_{y^n}$ <u>are stationary; and</u>

any operator $P$ has an invariant measure for any $k \in \{1, \ldots, n\}$, which tends to $\delta_{y^k}$ when $|P, P^0| \to 0$. Therefore, there is an $\epsilon > 0$, such that any $P$, for which $|P, P^0| < \epsilon$, has no less than $n$ different stationary measures.

Proof: Let $z^k$, $1 \le k \le n$, be an element of $X_0^{Z^d \cdot Z_+}$ defined as follows: $z^k(s,t) = y^k(s)$ in all $s \in Z^d$, $t \in Z_+$. Using Theorem 3, we construct a tessellation without memory, for which all $z^k$ are stable trajectories. This tessellation (as any tessellation without memory) may be considered as a temporal development of a deterministic operator $P^0$ with the same $d, u_1, \ldots, u_R$ and

$$\theta(y \mid x_1, \ldots, x_R) = \begin{cases} 1, & \text{if} \quad y = \varphi(x_1, \ldots, x_R) \\ 0, & \text{otherwise} \end{cases}$$

This operator $P^0$ is the one we sought. In fact, it is easy to deduce from the stability of $Z^k$, that

$$\lim_{\substack{\epsilon \to 0 \\ |P, P^0| \le \epsilon}} \sup_{s,t} (P^t \delta_{y^k})(x_s \ne y_s^k) = 0$$

Now one can use the fixed-point theorem, as was done in [3], to prove Theorem 4.

## IV. MONOTONIC BINARY TESSELLATIONS

This section treats tessellations (defined in general in Section III) in which $X_0 = \{0; 1\}$ and the function $\varphi(\cdot)$ is monotonic, that is

$$(\forall v \in U : x_v \le x_v^1) \Rightarrow \varphi(x_U) \le \varphi(x_U^1)$$

Then we may assume, without loss of generality, that $\varphi(0, 0, \ldots, 0) = 0$

and $\varphi(1, 1, \ldots, 1) = 1$, since the contrary implies $\varphi(x_U) = \text{const}$, which is trivial. These assumptions made, we obtain a class of monotonic binary tessellations or MBT. Of oucrse, the state "all zeros" is a trajectory in all MBT. We shall find out here when it is attractive and stable. Attractivity of "all zeros" was examined in [5] for MBT without memory and in [6] for one-dimensional monotonic tessellations with memory and with $X_0 = \{0;1;\ldots;n\}$.

Theorem 5. In any MBT the trajectory "all zeros" is stable if and only if it is attractive.

We shall prove this at the end of this section; first we find out when "all zeros" is attractive. We term a set $A \subset U$ a zero-set if the condition $\forall\ v \in A : x_v = 0$ ensures $\varphi(x_U) = 0$. We term a zero-set minimal if it contains no other zero-set. Throughout all of this section $Z_1, \ldots, Z_Q$ is the list of all the minimal zero-sets of a given MBT. Now we immerse our lattice $Z^{d+1}$ into a real space $R^{d+1}$ with the same coordinate axes. Thus any set in $Z^{d+1}$ becomes a set in $R^{d+1}$. Conv (A) is the convex hull of any $A \subset R^{d+1}$.

Theorem 6. The trajectory "all zeros" is attractive if and only if no ray in $R^{d+1}$ starts from the origin $0$ and meets all $\text{conv}(Z_q)$, $1 \le q \le Q$.

We shall use the following notations, where $c \in R^{d+1}$, $A, B \subset R^{d+1}$, $k \in R$:

$$kA = \{ka,\ a \in A\},\ -A = (-1)\cdot A,\ A + c = \{a + c,\ a \in A\},$$

$$A + B = \{a + b;\ a \in A,\ b \in B\},\quad A - B = A + (-B)$$

We denote also $\text{ray}(A) = \bigcup_{k \ge 0} kA$ and introduce a set of importance

$$\sigma = \sigma(\text{MBT}) = -\bigcap_{q=1}^{Q} \text{ray}(\text{conv}(Z_q))$$

<u>Another wording of Theorem 6. The trajectory "all zeros" is attractive if and only if $\sigma = \{0\}$, that is $\sigma$ consists of only the origin</u> 0.

We shall prove this in parts, bringing by the way other theorems describing properties of trajectories generated by finite perturbations of the base "all zeros". For any $x_W$ base and any $x$ state we denote

$$I(x_W) = \{a \in W : x_a = 1\} \quad \text{and} \quad I(x) = \{a \in V : x_a = 1\}$$

<u>Lemma 8.</u> <u>For any</u> $x_W$ <u>base</u>

$$I(\mathrm{tr}\,(x_W)) \subset \bigcap_{q=1}^{Q} (I(x_W) - \mathrm{ray}\,(\mathrm{conv}\,(Z_q))$$

<u>Proof:</u> We first prove the containment

$$I(\mathrm{tr}(x_W)) \subset I(x_W) \cup (I(x_W) - Z_q) \cup (I(x_W) - Z_q - Z_q) \cup \ldots$$

$$= I(x_W) \cup \bigcup_{i=1}^{\infty} (I(x_W) - \sum_{j=1}^{i} Z_q) = E_q$$

where $q$ is any integer from 1 to $Q$. The last equality defines $E_q$. Evidently, $(v + Z_q) \cap E_q$ is empty if $v \notin E_q$. On the other hand, if $v$ is inner and $v \in I(\mathrm{tr}(x_W))$, then $(v + Z_q) \cap I(\mathrm{tr}(x_W))$ is nonempty. Thus, if the set $I(\mathrm{tr}(x_W)) \setminus E_q$ contains an inner point $v$, it contains a point in $U(v)$ also. Hence we obtain, by inductive reasoning, that if $I(\mathrm{tr}(x_W)) \setminus E_q$ is nonempty, it contains a boundary point; which is clearly wrong. So $I(\mathrm{tr}(x_W)) \setminus E_q$ is empty, that is the containment formula is right. It remains right when substituting $\mathrm{conv}\,(Z_q)$ for $Z_q$. But $\sum_{j=1}^{i} A = iA$ for any convex $A$. Using this, we obtain

$$I(\mathrm{tr}(x_W)) \subset I(x_W) - \bigcup_{i=0}^{\infty} i\ \mathrm{conv}\ (Z_q) \subset I(x_W) - \mathrm{ray}(\mathrm{conv}\,(Z_q))$$

Intersecting this over all $q$ from 1 to $Q$ proves the lemma.

Now let $Sp(\rho)$ stand for the ball in $R^{d+1}$ with a center 0 and a radius of $\rho$. It is clear that $A + Sp(\rho)$ is the commonly termed $\rho$-neighborhood of A. We call any set $\{v \in R^{d+1} : L(v) \leq 0\}$ a halfspace, where the function $L : R^{d+1} \to R$ is linear, homogeneous and nonconstant. In other words, a halfspace is a closed full-dimensional set in $R^{d+1}$ bordered by a hyperplane passing through the origin.

Lemma 9. For any finite ensemble $P_1, \ldots, P_M$ of sets, all of which are intersections of several halfspaces, there is a number $\lambda \geq 1$ such that

$$\forall \rho > 0 : \bigcap_{m=1}^{M} (P_m + Sp(\rho)) \subset \bigcap_{m=1}^{M} P_m + sp(\lambda\rho)$$

The proof is left to the reader.

Theorem 7. For any MBT there is a number $\lambda \geq 1$ such that

$$\forall \rho > 0 : I(x_W) \subset Sp(\rho) \Rightarrow I(tr(x_W)) \subset \sigma + Sp(\lambda\rho)$$

This follows from Lemmas 8 and 9 and from the fact that every ray (conv $(Z_q)$) is an intersection of several halfspaces.

Theorem 7 implies a half of Theorem 6. When $\sigma = \{0\}$, then from what we have proved, if $I(x_W)$ is finite, $I(tr(x_W))$ is too. Now we pass to the case $\sigma \neq \{0\}$. We term a set $A \subset R^{d+1}$ obtuse for $B \subset R^{d+1}$ if

$$\forall c \in R^{d+1}((A + c) \cap B = \phi \Longrightarrow (A + c) \cap conv(B) = \phi)$$

Lemma 10. If A is obtuse for B, then any $A + A^1$ is obtuse for B.

Proof is obvious.

Theorem 8. Let A set be obtuse for all $Z_1, \ldots, Z_Q$ and $(s, t) \in A \Rightarrow t < 0$. Then the set $P = (A + \sigma) \cap V$ possesses the following property: $(P \cap W) \subset I(x_W) \Longrightarrow P \subset I(tr(x_W))$.

Proof: Suppose the set $P \setminus I(tr(x_W))$ is nonempty. Choose a point

$v$ in it. If $v$ is boundary, then $(P \cap W) \setminus I(x_W)$ is nonempty, q. e. d. Now let $v$ be inner. Since $v \in A + \sigma$, we can represent it as $v = a + (s, t)$, where $a \in A$, $(s, t) \in \sigma$, $t \geq t_W$. Then $(s, t) + \text{conv}(Z_q)$ meets $\sigma$ for all $q$, whence $v + \text{conv}(Z_q)$ meets $A + \sigma$. This, and the fact that $A + \sigma$ is obtuse for $v + Z_q$, imply that $v + Z_q$ meets $A + \sigma$. Therefore $v + Z_q$ meets $P$ for all $q$ from 1 to $Q$. On the other hand, $v \notin I(\text{tr}(x_W))$ implies that there is a $q$ for which $v + Z_q$ does not meet $I(\text{tr}(x_W))$. Thus, if the set $P \setminus I(\text{tr}(x_W))$ contains an inner point $v$, it meets $U(v)$; whence, by induction reasoning, it contains a boundary point, q. e. d.

<u>Lemma 11.</u> <u>For any</u> $V \subset R^{d+1}$ <u>the set</u> $-(d + 1)\,\text{conv}(B)$ <u>is obtuse for</u> $B$.

<u>Proof:</u> We leave to the reader to prove it, when $B$ is the set of vertices of a simplex. This being assumed, we prove the lemma. Let some $A = c - (d + 1)\,\text{conv}(B)$ meet $\text{conv}(B)$. There is a homothetic transformation $H : R^{d+1} \to R^{d+1}$ with a center $h$ and a coefficient $-(d + 1)$ such that $H(\text{conv}(B)) = A$. As $A$ meets $\text{conv}(B)$, so $h \in \text{conv}(B)$. From Caratheodory's theorem, $h$ belongs to a simplex $S$ whose vertices lay in $B$. As $S$ meets $H(S)$, some vertex of $S$ belongs to $H(S)$, since $H(S)$ is obtuse for $S$. Hence, $B$ meets $A$, q. e. d.

Theorem 8 and Lemmas 10 and 11 imply the rest of Theorem 6. In fact, let $\sigma \neq \{0\}$. Then we define $A$ by the formula

$$A = -(d + 1) \sum_{q=1}^{Q} \text{conv}(Z_q) + C + e$$

Here $C$ is a unit cube in $R^{d+1}$. It is added to make $A + \sigma$ contain points $(s, t) \in V$ for all $t \in Z_+$. A vector $e \in R^{d+1}$ is added to make $(s, t) \in A \Rightarrow t < 0$. From Lemmas 10 and 11, this $A$ is obtuse for all $Z_1, \ldots, Z_Q$. Now define a base $x_W$ by the formula

$I(x_W) = (A + \sigma) \cap W$. Theorem 8 implies in this case that $\forall t \in Z_+ \; \exists s \in Z^d : (tr(x_W))_{(s,t)} = 1$, that is the trajectory "all zeros" is not attractive. Thus Theorem 6 is completely proved. Proof of Theorem 5 requires yet another lemma.

Lemma 12. $\sigma(MBT) = \{0\}$ if and only if there are homogeneous linear functions $L_1, \ldots, L_n : R^{d+1} \to R$, where $n \leq d + 2$, and such an $r > 0$, such that $\forall v : \sum_{k=1}^{n} L_k(v) = 0$ and all sets $\{v \in U : L_k(v) \geq r\}$ are zero sets, $1 \leq k \leq n$.

Proof: Suppose there are $L_1(\cdot), \ldots, L_n(\cdot)$ and $r > 0$ possessing the mentioned properties and nevertheless $\sigma$ contains a point $a \neq 0$. Then some $k$ makes $L_k(a) \leq 0$ and $\forall m \geq 0 : L_k(ma) \leq 0$, whence

$$\forall q \in \{1, \ldots, Q\} \; \exists b \in Z_q : L_k(b) \leq 0$$

But, since $\{v \in U : L_k(v) \geq r\}$ is a zero set, there is such a $q$ that $b \in Z_q \Rightarrow L_k(b) \geq r$. This contradiction proves a half of our lemma.

Now we assume $\sigma = \{0\}$ and seek $L_1(\cdot), \ldots, L_n(\cdot)$, $r > 0$ with the mentioned properties. We represent all ray(conv$(Z_q)$) as intersections of several halfspaces. Let $H_1, \ldots, H_M$ be all these halfspaces. Let $f_m : R^{d+1} \to R$, where $1 \leq m \leq M$, stand for a linear function such that $H_m = \{v \in R^{d+1} : f_m(v) \leq 0\}$. We know, that there is no $a \neq 0$ such that $\forall m \in \{1, \ldots, M\} : f_m(a) \leq 0$. This allows us to apply Theorem 21.3 in [7] (a variant of Helly's theorem) to our functions $f_1(\cdot), \ldots, f_M(\cdot)$ and the set $C = \{(s,t) \in R^{d+1} : t \leq -1\}$. Thus we obtain $n$ functions $f_1^1(\cdot), \ldots, f_n^1(\cdot)$ from among the $f_1(\cdot), \ldots, f_M(\cdot)$, where $n \leq d + 2$, and positive numbers $\lambda_1, \ldots, \lambda_n, \epsilon$ such that $\forall v \in C : \sum_{k=1}^{n} \lambda_k f_k^1(v) \geq \epsilon$. Hence, $\sum_{k=1}^{n} \lambda_k f_k^1(v) = -\delta t$, where $\delta > 0$. Then the functions $L_k = -(\lambda_k f_k^1 + \delta t n^{-1})$, $1 \leq k \leq n$ and the number $r = \delta n^{-1}$ are the ones we sought.

Evidently $\forall v : \sum_{k=1}^{n} L_k(v) = 0$. Further, all sets $\{v \in U : f_k^1(v) \leq 0\}$, $1 \leq k \leq n$ are zero sets, since they contain some $Z_q$. But $v \in U$ and $f_k^1(v) \leq 0$ imply that $L_k(v) \geq \delta n^{-1}$, which proves the lemma.

Now we shall prove Theorem 5. First assume that the trajectory "all zeros" is attractive. Then we apply Theorem 1 to it with the $L_1(\cdot), \ldots, L_n(\cdot)$ and the $r > 0$ found by Lemma 12, and prove one half of Theorem 5.

Now we assume that "all zeros" is not attractive and prove that it is unstable. (This is like Proposition 1 in [8]). We use the following construction here and in Section V. We introduce an auxiliary space $\Omega = \{0;1\}^{V \setminus W}$. Elements of $\Omega$ are of the form $\omega = (\omega_v)$, where $\omega_v \in \{0;1\}$, $v \in V \setminus W$. We define a map $F : \Omega \to X$ by the formula

$$x_a = \begin{cases} 0, & \text{if } a \in W \\ \max\{\varphi_a(x_{U(a)}), \omega_a\}, & \text{if } a \in V\ W \end{cases}$$

Let $\mu_\epsilon$ be a Bernoullean measure on $\Omega$ defined as follows: $\mu_\epsilon(\omega_a = 1 \text{ in all } a \in A) = \epsilon^{|A|}$ for all finite $A \subset V \setminus W$. $\mu_\epsilon$ designates also the measure on $X$ induced by the measure $\mu_\epsilon$ on $\Omega$ with the map F. It is clear, that $\mu_\epsilon \in M_\epsilon(0)$, where $0$ stands for the base "all zeros". The trajectory "all zeros" can be attractive only if $\lim_{\epsilon \to 0} \sup_{v \in V} \mu_\epsilon(x_v = 1) = 0$. We shall prove, on the contrary, that $\lim \mu_\epsilon(x_{(s,t)} = 1) = 1$ for any $\epsilon > 0$. We choose such a base $x_W$ that $I(x_W)$ is finite, but $\forall t\, \exists s : (\mathrm{tr}(x_W))_{(s,t)} = 1$. For any $t$ we choose some $s$, say $s_t$, corresponding to $t$ in this formula; that is $\forall t : (\mathrm{tr}(x_W))_{(s_t, t)} = 1$. Now, if $\omega_v = 1$ in all $v \in (s^0, t^0) + I(x_W)$, then $\forall t \geq 0 : (F(\omega))_{(s^0 + s_t, t^0 + t)} = 1$. Thus a condition $(F(\omega))_{(s,t)} = 1$

is ensured if there is some $\tau \in \{0;\ldots;t - t_W\}$ such that $\omega_v = 1$ in all $v \in I(x_W) + (s - s_\tau, t - \tau)$. We confine ourselves to values of $\tau$ which are multiples of $t_W : \tau \in \{0; t_W; 2t_W;\ldots;[t t_W^{-1} - 1] t_W\}$, where $t_W = \max_{(s,t) \in W} t + 1$. For these values of $\tau$ all the events "$\omega_v = 1$ in all $v \in I(x_W) + (s - s_\tau, t - \tau)$" are independent of each other. The probability of every quoted event equals $\epsilon^{|I(x_W)|}$, whence at least one of them takes place with the probability of

$$1 - (1 - \epsilon^{|I(x_W)|, [t\, t_W^{-1} - 1]}$$

which tends to 1 when $t \to \infty$. All the more

$$\lim_{t \to \infty} \mu_\epsilon (x_{(s,t)} = 1) = 1. \quad \text{q. e. d.}$$

## V. TWO COUNTEREXAMPLES

Alas, simple examples beyond the class of MBT breach the equivalence of "All Zeroes" being stable and attractive as stated by Theorem 5. Let us describe two combines in which the trajectory "all zeros" is attractive, but unstable. The corresponding proofs are published in [8] .

Example 2. A tessellation without memory, in which $d = 1$, $X_0 = \{0;1;2\}$, $R = 3$, $u_1 = -1$, $u_2 = 0$, $u_3 = 1$, and

$$\varphi(x_1, x_2, x_3) = \begin{cases} 0, & \text{if } x_1 = 0,\ x_2 = x_3 = 1; \\ 1, & \text{if } x_1 = x_2 = 2,\ x_3 = 1; \\ 2, & \text{if } x_1 = x_2 = x_3 = 2; \\ \text{the nearest integer to } (x_1 + x_2 + x_3)/3 & \\ \text{in all the other cases.} & \end{cases}$$

It is easy to check, that "all zeros" is an attractive trajectory. Its being unstable is based on the following. Let the space $\Omega$ and the measure $\mu_\epsilon$ on it be the same as in the proof of Theorem 5. Let

Let $F:\Omega \to X$ be defined by the formula

$$x_v = \begin{cases} 0, & \text{if } v \in W \\ \min\{2; \varphi_v(x_{U(v)}) + \omega_v\}, & \text{if } v \in V\setminus W \end{cases}$$

Then $\lim_{t\to\infty} \mu_\epsilon (x_{(s,t)} = 2) = 1$ for any $\epsilon > 0$.

Example 3. A combine in which $V = \{-1;1\} \cdot Z \cdot Z_+$. We denote elements of $V$ by $v = (r, s, t)$, where $r \in \{-1;1\}$, $s \in Z$, $t \in \{0;1;2;\dots\}$. A point $(r, s, t)$ is inner if $t \geq 1$; in this case

$$\varphi(r, s, t)(x_{U(r, s, t)}) = \min\{x_{(r,s-r,t-1)}; \max\{x_{(r,s,t-1)}, x_{(-r,s,t-1)}\}\}$$

and the set $U(r, s, t)$ consists of the three points which are indices on the right-hand side. The combine is defined. It is easy to check that "all zeros" is an attractive trajectory. Its being unstable can be proved by taking a space $\Omega$, a measure $\mu_\epsilon$ and a map $F:\Omega \to X$ the same as in the proof of Theorem 5. Then $\lim_{t\to\infty} \mu_\epsilon (x_{r,s,t)} = 1) = 1$ for any $\epsilon > 0$.

## REFERENCES


1. A. L. Toom, Non-ergodic multidimensional automata systems. Problems of Information Transmission 10 (3), 70-79, 1974.
2. N. B. Vasilyev, M. B. Petrovskaya, I. I. Pyatetski-Shapiro, Simulation of voting with random errors. Automatika i Telemechanika, 10, 103-107, 1969 (In Russian).
3. L. N. Vasershtein, Markov processes on a countable space product, describing large automata systems. Problems of Information Transmission 5 (3), 64-71, 1969.
4. B. S. Zirelson, Non-uniform local interaction can produce "far-range order" in a one-dimensional system. Theor. Probability Appl., 21 (3), 681-683, 1976.

5. A. L. Toom, Monotonic binary tessellation automata. Problems of information transmission 12 (1), 48-54, 1976.

6. G. A. Galperin, One-dimensional local monotonic operators with memory. Soviet Math. Dokl. 228 (2), 277-280, 1976.

7. R. T. Rockafellar, Convex Analysis. Princeton, New Jersey, Princeton Univ. Press, 1970.

8. A. L. Toom, Non-stable multicomponent systems. Problems of information transmission 12 (3), 78-84, 1976.

chapter 19

# ONE SYSTEM OF AUTOMATA WITH LOCAL INTERACTIONS

S. S. Vallander

ABSTRACT. We study the invariant measures of one transform the space of probability measures on the space $\{0,1\}^{\mathbb{Z}}$ and t havior of iterations of this transformation.

## I. INTRODUCTION

Let $X = \{0,1\}^{\mathbb{Z}}$ be the set of all doubly infinite seque symbols 0 and 1 with usual (Tichonov) topology. Let M(X note the set of all Borel probability measures on X and $M_0$ subset of all stationary (translation invariant) probability me We define convergence in M(X) as the convergence on all f dimensional cylinder sets.

Let us define a transformation $T: X \to X$ and its indu formation $P: M(X) \to M(X)$ by the formulas

$$(Tx)_n = |x_n - x_{n+1}|, \qquad x = (x_n) \in X,$$

$$(P\mu)(A) = \mu(T^{-1}A), \qquad \mu \in M(X),$$

$$A \subset X \text{ - Borel set.}$$

It is evident that $P(M_0(X)) \subset M_0(X)$.

We will study the asymptotic properties of the powers $P^t$ of the transformation $P$ in the space $M_0(X)$ as $t \to \infty$. A part of our results was announced in [1].

The informal interpretation of our problem is the following. We have automata (lamps) with two states 1 and 0 (on and off) at all integer points of the line. At every moment $t = 0, 1, 2, \ldots$ the state $\xi_n(t)$ of automaton $n$ is random, and the sequence of random variables $\xi_n(t)$ $(-\infty < n < \infty)$ is stationary. The evolution of the system is described by our formulas. So the lamp $n$ is on at the next step if and only if she and her right neighbor are in different states at the current step.

## II. INVARIANT MEASURES OF THE TRANSFORMATION P

It is natural to begin the study of the asymptotics of powers $P^t$ with the study of P-invariant measures.

Theorem 1. The set of stationary P-invariant measures is infinite.

Proof: Let us consider sequences $x^{(k)} \in X$ $(k \geq 1)$, which are defined by

$$x_n^{(k)} = \begin{cases} 0, & n = (2^k - 1)\ell, \quad \ell \in \mathbb{Z}, \\ 1, & \text{in other cases} \end{cases}$$

Let us prove that $x^{(k)}$ is a fixed point of the transformation $T^{2^k-1}$. It is convenient to use the "squared paper". We consider the horizontal infinite strip of squared paper of squares of edge $2^k$, choose the initial square and write down the sequence $x^{(k)}$ in the top line of the strip, so that in every square will be one symbol 0 or 1. In the second line we write down the sequence $Tx^{(k)}$ etc. We need to

prove that there will be the sequence $x^{(k)}$ in the last line. By reasons of periodicity we can consider only the columns with numbers $0, \ldots, 2^k-1$ (so the first and the last of them coincide).

Now let us describe constructively some process of filling the squares of our strip. It will be almost evident that every next line in this filling can be obtained as the T-image of the previous. The last line will coincide with the first one by construction. So we have a big square of $2^k$ lines (with numbers $1, \ldots, 2^k$) and $2^k$ columns (with numbers $0, \ldots, 2^{k-1}$). The first line is already filled. We fill the last line in the same manner as the first one and write down 1 on all the remaining places in columns numbered 0 and $2^k-1$. Next we use a sequence $K_i$ of "checked" squares filled by symbols 0 and 1. This sequence will be described later. Note only, that the side of square $K_i$ is equal to $2^i$. Now let us continue the filling of the strip. We have an unfilled square of size $2^k-2$. We being the process of filling from its "right lower" corner, that is from the corner of intersection of line and column with maximal numbers. In this corner we mark the square of size $2^{k-1}$ and add squares of size $2^{k-2}$ on the left below and above on the right to it. We have "stairs" with steps of size $2^{k-2}$. Next add 4 squares of size $2^{k-3}$ and obtain stairs with steps of size $2^{k-3}$ etc., up to the moment we obtain stairs with steps of size 2. The squares of size $2^i$, which we already added, we fill as $K_i$ $(i = 1, \ldots k-1)$. After this we will the rest of the squares with zeros.

Now we describe by recursion the sequence $K_i \cdot K_0$ is the square of size 1 with 0 in it. If the square $K_i$ (and all previous squares) is already defined, we need to define $K_{i+1}$. So we have a square of size $2^{i+1}$. Let us construct "stairs" from its right lower corner: mark the square of size $2^i$, add to it 2 squares of size $2^{i-1}$ etc. up to the moment we obtain stairs with steps of size 1. All already added squares of size $2^j$ we fill as $K_j$ $(0 \leq j \leq i)$. Next write down 0 in the left upper square and 1 in all other squares of line 1 and of column 0. The rest of the squares we will with zeros. This is the end of the recursive construction.

It is not difficult to prove by induction that the every next line of the strip can be obtained by T from the previous line. We omit the details.

So it is already proved that $x^{(k)}$ is a fixed point of the transformation $T^{2^k-1}$. Now let us consider the measure $\mu^{(k)}$, concentrated on the points $T^i x^{(k)}$ $(i = 0,\dots 2^k-2)$ with equal weights $(2^k-1)^{-1}$. It is evident that $\mu^{(k)}$ is P-invariant. Its averaging by shifts gives us the stationary P-invariant measure $\overline{\mu}^{(k)}$. So we have an infinite set of stationary P-invariant measures $\overline{\mu}^{(k)}$ $(k \geq 0)$ and the theorem is proved.

We remark now that there are also other P-invariant stationary measures. The next theorem gives us an invariant measure which is important for the rest of the paper.

We introduce now the symbol $\beta_p$ for the measure corresponding to the sequence of independent identically distributed random variables, which take the value 1 with probability p and the value 0 with probability 1-p.

<u>Theorem 2.</u> <u>The measure</u> $\beta_{1/2}$ <u>is</u> P-<u>invariant</u>.

<u>Proof:</u> It is sufficient to consider the common distributions of random variables with successive numbers. For simplicity of notation we consider the three-dimensional distribution. We have

$$P\{\xi_0(1) = i_0,\ \xi_1(1) = i_1,\ \xi_2(1) = i_2\}$$
$$= P\{|\xi_0(0) - \xi_1(0)| = i_0,\ |\xi_1(0) - \xi_2(0)| = i_1,\ |\xi_2(0)-\xi_3(0)| = i_2\}$$
$$= P\{\xi_0(0) = 0, \xi_1(0) = i_0, \xi_2(0) = |i_1 - i_0|, \xi_3(0) = |i_2 - |i_1 - i_0||\}$$
$$+ P\{\xi_0(0) = 1, \xi_1(0) = 1-i_0, \xi_2(0) = |i_1 - 1 + i_0|, \xi_3(0) = |i_2 - |i_1 - 1 + i_0||\}$$
$$= 1/16 + 1/16 = 1/8 = P\{\xi_0(0) = i_0,\ \xi_1(0) = i_1,\ \xi_2(0) = i_2\}$$

It is evident that the general case can be considered in the same way, so the theorem is proved.

## III. THE EVOLUTION OF PRODUCT-MEASURES

In this paragraph we consider the behavior of the sequence $P^t\beta_p$ as $t \to \infty$. Our results show that this sequence is not a convergent one (except of the trivial cases $p = 1/2$, $p = 0$, $p = 1$). But apparently the sequence $P^t\beta_p$ is convergent in the sense of Cesaro. We are only able to prove that the corresponding two-dimensional distributions converge in the sense of Cesaro.

Theorem 3. Let $p \in (0, 1)$. Then the sequence of one-dimensional distributions of measures $P^t\beta_p$ converges in the sense of Cesaro to the one-dimensional distribution of the measure $\beta_{1/2}$. If $p \neq 1/2$, this sequence is not convergent in the usual sense.

Proof: We need a simple result about the auxiliary transformation $P_0$. The transformation $P_0$ acts in the space of probability measures on $\{0,1\}$ and transforms the distribution of random variable $\xi$ with values 0 and 1 into the distribution of random variable $|\xi - \eta|$, where $\eta$ has the same distribution as $\xi$, and is independent of $\xi$.

Lemma 1. Let $\mu$ be a probability measure on $\{0, 1\}$, which is not concentrated in one of the points 0, 1. Then the sequence of *measures* $P_0^t\mu$ *converges to the uniform measure on* $\{0,1\}$.

The proof of Lemma 1 is very simple (see [2], where a considerably more general result is proved) and we omit it.

Now we formulate the next lemma.

**Lemma 2.** *Let* a, b, c *be numbers equal to* 0 *or* 1. *Then:*

(a) $|a - b| = a + b \pmod 2$.

(b) $\|a - b| - |b - c\| = |a - c|$.

The proof is trivial.

Now we pass to the proof of Theorem 3. We will compute the one-dimensional distribution of the measure $P^t\beta_p$ in terms of the transformation $P_0$ Namely we will prove that the one-dimensional distribution of the measure $P^t\beta_p$ is equal to

$$P_0^{f(t)}\mu_p$$

where $\mu_p$ is the one-dimensional distribution of the measure $\beta_p$ and $f(t)$ is the number of ones in the binary representation of $t$. Let us now show how to deduce the statement of Theorem 3 from this description. First of all we see that the one-dimensional distribution of the measure $P^{2^k}\beta_p$ $(k = 1, 2, \ldots)$ is $P_0 3_p$, and this measure is not a uniform measure on $\{0,1\}$ (except of the case $p = 1/2$). So the second statement of the theorem is the consequence of the first one. On the other hand "almost all" numbers have many units in binary representation so the values of $t$ with no more than fixed number of units in binary representation give a neglected contribution in the Cesaro averaging and all the other values of $t$ correspond to the measures close to the uniform one (by Lemma 1). This proves the Cesaro convergence of the one-dimensional distributions of the measures $P^t\beta_p$ to the uniform measure on $\{0,1\}$.

Now we pass to the proof of previously mentioned description of the one-dimensional distributions of the measures $P^t\beta_p$. With Lemma 2 we have

$$\xi_0(t) = \sum_{i=0}^{t} \binom{t}{i}\xi_i(0)\,(\mathrm{mod}\ 2) = \sum_{i\,:\,\binom{t}{i}\ \mathrm{odd}} \xi_i(0)\ (\mathrm{mod}\ 2)$$

We will see that the number of summands in the last sum is equal to $2^{f(t)}$. It then follows from this fact that the distribution of $\xi_0(t)$ is $P_0^{f(t)}\mu_p$. So we need to prove that the number of odd coefficients in $(1+x)^t$ is equal to $2^{f(t)}$. Firstly let $t$ be $2^k$. Then all the coefficients except the first one and the last one are even. The coefficient of $x^m$ $(0 < m < t)$ has the form

$$\frac{2^k(2^k-1)\ldots(2^k-m+1)}{m!}$$

It suffices to verify that number of factors 2 in the numerator is greater than the same number in the denominator. We note now that the numbers $i$ and $2^k - i$ are divisible on the same power of $2 (i = 1, \ldots m-1)$. In addition we have in the numerator factors of 2 from $2^k$ and in the denominator we have no more than $k-1$ factors of 2 (from $m$). So the numerator is divisible by a greater power of 2 than the denominator and there are in the decomposition of $(1+x)^{2^k}$ $2 (2 = 2^{f(2^k)})$ odd coefficients. Now we pass to the general case $t = 2^{k_1} + 2^{k_2} + \ldots 2^{k_r}$ $(k_1 > k_2 > \ldots k_r)$. We have

$$(1+x)^t = (1+x)^{2^{k_1}}(1+x)^{2^{k_2}}\ldots(1+x)^{2^{k_r}}$$

There are 2 terms with odd coefficients in every factor so after the multiplication there will be $2^r$ terms with odd coefficients. Note that all of them have different powers of $x$, so they will not disappear after the reduction of similar terms.

We have finished the proof of the description of the one-dimensional distributions of the measures $P^t\beta_p$ and at the same time the proof of Theorem 3.

Theorem 4. Let $p \in (0,1)$. Then the sequence of the two-dimensional distributions of measures $P^t\beta_p$ converges in the sense of Cesaro to the two-dimensional distribution of the measure $\beta_{1/2}$.

Proof: As in the proof of Theorem 3 we will describe the two-dimensional distributions of the measures $P^t\beta_p$ with help of the transformation $P_0$. The subsequent part of the proof is analogous to the corresponding part of the proof of Theorem 3. For definiteness we consider the common distribution of two next variables $\xi_0(t)$ and $\xi_1(t)$. In general case there are some unessential changes in this description. We omit also some bulky technical details.

The two-dimensional distributions of the measures $P^t\beta_p$ will be described with help of a certain two-parameter family of distributions $\mu_{rs}$. Firstly we describe how to find the pair $(r,s)$ for given t. If $t = 2^k-1$ we take $(r,s) = (k,0)$. If $2^k \leq t < 2^{k+1} - 1$ we find first of all the pair $(r_0, s_0)$ for $t - 2^k$ and then take $r = r_0$, $s = s_0+1$. Now we describe the structure of distributions $\mu_{rs}$. $\mu_{0s}$ is the product of two distributions $P_0^s\mu_p$, where $\mu_p$ as in the Theorem 3 is the one-dimensional distribution of the measure $\beta_p$. In the general case

$$\mu_{rs}(0,1) = \mu_{rs}(1,0) = (P_0^s\mu_p)(1)(P_0^s\mu_p)(0)$$

$$\mu_{rs}(1,1) = (P_0^{r+s}\mu_p)(1) - \mu_{rs}(0,1)$$

It is evident that $\mu_{rs} \to \mu_{1/2} \times \mu_{1/2}$ as $s \to \infty$.

Let us prove that for $t = 2^k-1$ the two-dimensional distribution of $\xi_0(t)$ and $\xi_1(t)$ is $\mu_{k,0}$. As we already know from Theorem 3 each of the variables $\xi_0(t)$, $\xi_1(t)$ has the distribution $P_0^k\mu_p$, so it is sufficient to prove that

$$P\{\xi_0(t) = 1,\ \xi_1(t) = 0\} = \mu_{k,0}(1, 0) = p(1 - p)$$

We know from the proof of Theorem 3 that

$$\xi_0(2^k-1) = \sum_{i=0}^{2^k-1} \xi_i(0) \pmod 2$$

$$\xi_1(2^k-1) = \sum_{i=1}^{2^k-1} \xi_i(0) \pmod 2$$

So

$$\begin{aligned}
&P\{\xi_0(2^k-1) = 1,\ \xi_1(2^k-1) = 0\} \\
&= P\{\sum_{i=1}^{2^k-1} \xi_i(0) = 0 \pmod 2,\ \xi_0(0) = 1,\ \xi_0(2^k) = 0\} \\
&+ P\{\sum_{i=1}^{2^k-1} \xi_i(0) = 1 \pmod 2,\ \xi_0(0) = 0,\ \xi_0(2^k) = 1\} \\
&= p(1 - p)P\{\sum_{i=1}^{2^k-1} \xi_i(0) = 0 \pmod 2\} \\
&+ p(1 - p)P\{\sum_{i=1}^{2^k-1} \xi_i(0) = 1 \pmod 2\} = p(1 - p)
\end{aligned}$$

q. e. d. In the general case there are no additional difficulties and we omit the corresponding proof.

## IV. THE EVOLUTION FOR THE LARGER CLASS OF INITIAL MEASURES

Theorem 5. Let the measure $\mu$ correspond to a sequence of symmetrically dependent random variables. Then the two-dimensional distributions of the measures $P^t\mu$ converge in the sense of Ceasro to the two-dimensional distribution of the invariant measure

$\lambda\bar{\mu}^{(1)} + (1-\lambda)\beta_{1/2}$, where $\bar{\mu}^{(1)}$ is the measure from the proof of Theorem 1 and $\lambda$ a $\mu$-dependent number from $[0,1]$ ($\lambda$ is computed in the proof).

Proof: Let us use the so-called de Finetti representation of any symmetric measure as a mixture of the measures $\beta_p$ ([3], Ch. 7, § 4):

$$\mu = \int_{[0,1]} \beta_p \sigma(dp)$$

where $\sigma(dp)$ is the probability measure on $[0,1]$. It is clear that

$$P^t\mu = \int_{[0,1]} P^t\beta_p \sigma(dp)$$

and we obtain the statement of Theorem 5 with

$$1-\lambda = \int_{(0,1)} \sigma(dp)$$

from the statement of Theorem 4.

Before the formulation of Theorem 6 we describe a certain representation of the measures on X, which is like the de Finetti representation. It has the form

$$\mu = \int_{[0,1]} \mu_\theta \sigma(d\theta)$$

where $\sigma(d\theta)$ is the probability measure on $[0,1]^{\mathbb{Z}}$ and $\mu_\theta = \prod_{i\in\mathbb{Z}} \mu_{\theta_i}$ is the product-measure (the designation $\mu_p$, $p \in [0,1]$ was introduced in Theorem 3). We will call this representation an integral one. It can be shown that every probability measure on X has an integral representation.

Theorem 6. Let measure $\mu$ (in general non-stationary) have such integral representation that the support of measure $\sigma$ lies in the set $[a, b]^{\mathbb{Z}}$ for certain $0 < a \leq b < 1$. Then the two-dimensional distributions of the measures $P^t\mu$ converge in the sense of Cesaro to the two-dimensional distribution of the measure $\beta_{1/2}$.

Proof: As in the proof of Theorem 5 we reduce the proof to the case of product-measure (in general non-stationary). The condition on the support of $\sigma$ will give us the uniformity of the convergence for these product-measures. The proof of the convergence itself is based on the spacely inhomogeneous generalization of Lemma 1 and on the same ideas as the proof of Theorems 3 and 4. We only formulate the generalization of Lemma 1 and omit the subsequent part.

Lemma 3. Let $\xi_1$ and $\xi_2$ be independent random variables with values 0 and 1 and let for some $\lambda < 1$

$$|1 - 2P\{\xi_i = 1\}| \leq \lambda \qquad (i = 1, 2)$$

Then

$$|1 - 2P\{|\xi_1 - \xi_2| = 1\}| \leq \lambda^2 .$$

## REFERENCES

1. S. S. Vallander, On the System of Automata with Local Interaction. In: Theses of reports on the IV International Symposium on Information Theory, part I, pp. 26-27. Leningrad-Moscow, 1976.

2. S. S. Vallander, Weak convergence of certain sequences of distributions of probabilities. Vestnik Leningradskogo Universiteta 13, 21-31 (1972). (In Russian).

3. W. Feller, An Introduction to Probability Theory and Its Applications, v. II. New York, London, Sydney: Wiley, 1966.

chapter 20

# SOME RESULTS OF NUMERICAL EXPERIMENTS RELEVANT TO THE SPECTRUM OF ONE–DIMENSIONAL SCHRÖDINGER EQUATION WITH RANDOM POTENTIAL

E. B. Vul

The question of the nature of eigenfunctions of the Schrödinger equation with a random potential arises in plenty of problems, e. g. , in connection with problems of conductivity in disordered media, sound propagation in random media etc. Extensive physical literature covers this subject. The mathematical aspect of this problem was described in the well-known survey of L. A. Pastur [1]. Recently new important results have been obtained by I. Ya. Goldsheid, S. A. Moldhanov and L. A. Pastur [2, 3]. Goldsheid and Molchanov have shown in [2] that under natural assumptions on the distribution of probabilities of the potential, with probability one the spectrum of the Schrödinger equation contains a discrete component; in the work of Goldsheid, Molchanov and Pastur [3] it has been shown that the whole spectrum is purely discrete.

The present paper presents and discusses some results of numerical experiments directly related to [2, 3]. We shall study the difference analogue of the one-dimensional Schrödinger equation which appears in the problem of the frequency spectrum of a chain of linear oscillators with random masses ([4]):

$$u_{n+1} - 2u_n + u_{n-1} + q_n u_n = Eu_n, \quad -\infty < n < \infty \tag{1}$$

Here $q_n$ is the sequence of random values which we assume as independent and equally distributed on the segment $[0,1]$, $E$ is the spectrum parameter. For this case, the results of [2, 3] show that with probability one on the segment $[-3, 2]$ there exists a numerical set of values which are related to eigenfunctions forming an orthogonal basis in the Hilbert space $\ell_2(-\infty, \infty)$. The investigation of the spectrum of (1) can be reduced to the analysis of the product of random unimodular matrices of the second order. Let

$$A_n(E) = \begin{Vmatrix} (2 + E - q_n) & -1 \\ 1 & 0 \end{Vmatrix}$$

It is evident that if the two-dimensional vector $e_n = (u_n, u_{n-1})$, then $e_n = A_n(E) \cdot A_{n-1}(E) \cdot \ldots \cdot A_1(E) e_1$. Similarly, we can construct matrices of the second order to obtain $e_{-n}$ from $e_1$ at $n > 0$. Eigenvalues are determined by the fact that for them there can be found a vector $e_1$ for which the vectors $u_n$, $-\infty < n < \infty$, form a sequence belonging to $\ell_2(-\infty, \infty)$. Since $q_n$ are independent random values, the matrices $A_n(E)$ are independent and identically distributed. According to the known results of G. Furstenberg [5] for almost all $E$ on any measure we can find a vector $e^+_{-1}$ for which $u_n$ tends to 0 as $n \to \infty$ exponentially fast. The vector having this property as $n \to -\infty$ will be denoted as $e^-_1$. I. Ya. Goldsheid has shown ([6]) that the vector-functions $e^+_1(E)$, $e^-_1(E)$, defined on the spectrum of the operator on the left-hand side of (1) is discontinuous at each point with probability one. Those $E$ at which the vectors $e^+_1(E)$, $e^-_1(E)$ exist will be called the Furstenberg points.

The present work studies the values of $E$ which are not Furstenberg points and in particular the question of how these values of

E are related to the eigenvalues of (1). Another facet of the same problem is whether the eigenfunctions decrease exponentially, as has not been assumed by Mott et al ([7]) or whether they decrease more slowly. For this purpose, with the aid of random numbers, we constructed a realization of independent random values $q_n$, $-\infty < n < \infty$, equally distributed on the segment $[0,1]$. For this fixed realization we studied at various E products of matrices $A_n(E) \dots A_1(E)$ when E changed with a step of 0.0005 on the segment from $-3$ up to 2. The calculation was performed either up to $n = 800$ or until the sum of squares of the matrix elements exceeded 1000. As a result we had the values E at which the product was retained 800 times. We obtained the following groups of values:

1) -0.1500 - (-0.1400); 2) 0.2545 - 0.2570;

3) 0.2905; 4) 0.5885 0 0.6580; 5) 1.0635

When the step size was shortened to 0.0001 there appeared two new values: 0.3410 and 0.3520. Evidently, the points which are not Furstenberg points, are positioned densely on the operator spectrum but the boundary for the sum of squares of matrix elements enables us to find only a small portion of them.

In accordance with the methods of [2] at the obtained values we considered various initial conditions of the form $u_1 = \cos\alpha$, $u_0 = \sin\alpha$ and chose the values $\alpha$ for which the sum of squares $\sum_n u_n^2$ was the least. Then for the value $\frac{1}{\max |u_n|}\sqrt{\sum_n u_n^2}$ we obtained the following values:

Table I

| E | -0.150 | 0.255 | 0.290 | 0.589 | 0.601 | 0.608 | 0.620 | 0.630 |
|---|---|---|---|---|---|---|---|---|
| $\frac{1}{\max \lvert u_n \rvert}\sqrt{\sum_n u_n^2}$ | $\frac{3640}{124}$ | $\frac{3599}{43}$ | $\frac{12520}{137}$ | $\frac{4568}{56}$ | $\frac{24083}{1485}$ | $\frac{5106}{162}$ | $\frac{51232}{1665}$ | $\frac{42477}{3086}$ |

It follows from these data that for the majority of the obtained values $(\max u_n^2)^{-1} \sum_n u_n^2$ constitutes a few percent of the length of the interval equal to 800. This shows that the found solutions do not increase exponentially and therefore may be close to the eigenfunctions of the operator.

Furthermore, let us for each E of Table I construct distributions of probabilities $p_n(E) = u_n^2 / \sum_k u_k^2$. An important feature of this distribution related to the localization properties is its dispersion equal to $\sigma(E) = \sum_n n^2 p_n(E) - (\sum_n n p_n(E))^2$. For this quantity we obtained the following values:

Table II

| E | -0.150 | 0.255 | 0.290 | 0.589 | 0.601 | 0.608 | 0.620 | 0.630 |
|---|---|---|---|---|---|---|---|---|
| $\sigma(E)$ | 6483 | 9272 | 10501 | 14327 | 19897 | 11237 | 18713 | 10807 |

Finally, to find out to what extent the obtained functions $u_n(E)$ are close to the eigenfunctions of the discrete spectrum we found eigenvalues $\lambda_{min}(E)$ for the product of matrices $A_{800}(E) \cdot \ldots \cdot A_1(E)$ for 8 values of E in Table I. Here are the values:

| E | -0.150 | 0.255 | 0.290 | 0.589 | 0.601 | 0.608 | 0.620 | 0.630 |
|---|---|---|---|---|---|---|---|---|
| $\lambda_{min}(E)$ | 0.007 | 0.009 | 0.006 | 0.008 | 0.006 | 0.006 | 0.013 | 0.008 |

The data show that the norm of the product of matrices $A_n(E)$ lies within the limits 100 - 150, while in case of the Furstenberg behavior it should be of the order of $e^{\text{const.}\ 800}$. At the same time the above set of values E shows that the found functions imitate well the operator eigenfunctions $u_n(E)$ on the segment of order 100 but are far from the eigenfunctions of the operator (1) with a periodic boundary condition on the whole segment from 0 to 800.

## REFERENCES

1. L. A. Pastur, Spectra of Random Operators, Uspekhi Mat. Nauk, v, XXVIII, No. 1, 3-63, 1973.
2. I. Ja. Goldsheid, S. A. Molchanov, On a Problem by Mott, Doklady Akademii Nauk SSSR, v. 230, No. 4, 761-764, 1976.
3. I. Ja. Goldsheid, S. A. Molchanov, L. A. Pastur, Random One-dimensional Schrödinger Operator has a Purely Point Spectrum, Functional Analysis and Applications, v. XI, No. 1, 1-10, 1977.
4. F. Dyson, The Dynamics of a Disordered Linear Chain, Physical Review, v. 92:6, 1331-1338, 1953.
5. H. Furstenberg, Noncommuting Random Products, Transactions Amer. Math. Soc., v. 108, No. 3, 377-428, 1963.
6. I. Ja. Goldsheid, Asymptotics of Products of Random Matrices Depending on a Parameter, Doklady Akademii Nauk SSSR, v. 224, No. 6, 1248-1251.
7. H. F. Mott, Electrons in Disordered Media, Advances in Physics, v. 16, No. 61, 49, 1967.

chapter 21

# A STRUCTURALLY STABLE MECHANISM OF APPEARANCE OF INVARIANT HYPERBOLIC SETS

E. B. Vul and Ya. G. Sinai

Recently, both in the physical and mathematical literature, there appeared a number of papers devoted to dynamical systems of an extremely simple form which exhibit a stochastic behavior of trajectories. By stochasticity we understand the existence in the phase space of trajectories with "random" behavior, i e. spending in different parts of the phase space random times not described by any simple law. In this connection one of the systems most discussed is the so called Lorenz model (re. [1], which is the result of cutting an infinite system of equations of the Galerkin type for the Bernard heat convention problem). A thorough study of this model in relation to the turbulence problem is contained in the survey by McLauglin and Martin [2], and the study of the mathematical properties of this model is contained in the paper by Ruelle [3]. Systmes of analogous behavior were numerically investigated in the paper by Gertzenstein and Schmidt [4].

From the mathematical point of view we deal with the phenomenon in which in very simple systems of nonlinear ordinary differential equations appear closed invariant sets which are not manifolds and have the property of the exponential divergence of trajectories called the hyperbolic property (re [5]). If such an invariant set is attractive, it is called a "strange attractor" (re. [6]). We know three mechanisms for the appearance of strange invariant sets: (1) destruction of separatrix of an integrable system under small perturbations of the right-hand sides and the formation of homoclinic and heteroclinic trajectories; (2) Smale's "horseshoe"; (3) availability of solenoids (re. the survey by S. Smale [5] and the book by Z. Nitecki [7]). The first mechanism does not lead to the appearance of strange attractors but is common to hamiltonian systems of general form. The third mechanism is discussed in the well-known work by Ruelle and Takens [6], but the possibility of its appearance in simple systems of differential equations is still open to discussion.

The present paper is devoted to one more mechanism leading to the appearance of strange invariant sets similar to the mechanism resulting in the appearance of Smale's "horseshoe". It is based on the assumption that there exists a fixed point with a stable separatrix of codimension I. If one constructs a small cell of codimension I transversally intersecting the stable separatrix, then the Poincarè mapping of this cell is not defined on the intersection with the separatrix and is discontinuous in its neighborhood. We write out the conditions of the inequality type for this mapping which lead to the appearance of a closed invariant hyperbolic set forming a set of the Cantor perfect set type in a certain section.

Definition of a mixer. As numerous examples show, the stochastic behavior of dynamical systems appears in cases where there are special cells of the phase space having the following property: when a trajectory comes to this cell it behaves as if it casts lots and chooses to what part of the phase space it should go next. We call such cells mixers. Their properties and positioning are determined by the behavior of separatrix of the fixed points. Now we shall dwell upon the simplest possibility.

Let a system of three differential equations be given:

$$\frac{dx_i}{dt} = f_i(x_1, x_2, x_3), \quad i = 1, 2, 3 \tag{1}$$

where the right-hand sides are of the class $C^\infty$, and $f_i(0,0,0) = 0$, i. e., $0 = (0,0,0)$ if a fixed point of the system (1). Let us assume that the stable separatrix of the point $(0,0,0)$ is two-dimensional and therefore the unstable separatrix of the same point is one-dimensional. Let us study the region of the phase space (re. Fig. 1) which can be represented as a direct product of a closed two-dimensional area W and a time interval $[0,t]$. Let us further assume that W is represented in the form of a union of the two parts $W_1, W_2$, the intersection of which $W_1 \cap W_2$ is a curve, the intersection of W with the stable separatrix (re. Fig. 2). The unstable separatrix $\Gamma^{(u)}(0)$ is a curve which if continued from different sides from the point $(0,0,0)$ can intersect W. Let us assume that this intersection actually occurs, in the points $A_1 \in W$ and $A_2 \in W$. Recall that the Poincarè mapping T transforms each point of W to the point of intersection of its positive semi-trajectory with W.

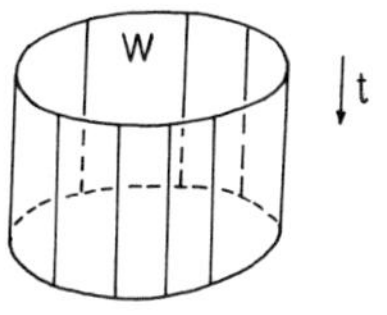


Fig. 1

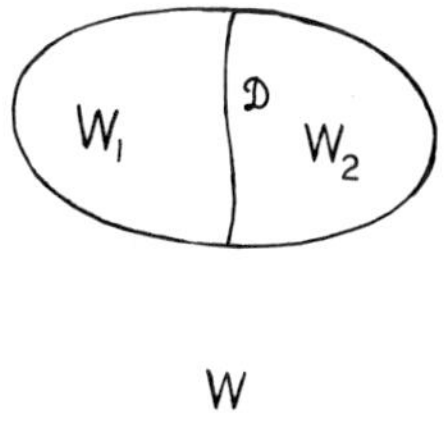


Fig. 2

<u>An important remark.</u> The Poincarè mapping can be defined only on $W - D$. If it is extended continuously from $W_1 (W_2)$ we obtain $TD = A_1 (A_2)$. In principle the discontinuous nature of the mapping leads to the appearance of strange hyperbolic invariant sets.

It is convenient to formulate the precise conditions by assuming that coordinates $(x, y)$ transforming $W$ into a unit square have been introduced in such a way that $D$ is the y-axis and $W_2 (W_1)$ is the right- (left) half of the square. Usually the choice of this system of coordinates does not cause any difficulties. The mapping will be written down in the form: $T(x, y) = (x', y')$, $x' = h_1(x, y)$, $y' = h_2(x, y)$.

Let us assume that $W_1 (W_2)$ can be divided into three subsets:

$$W_{11} = \{(x, y) : -\tfrac{1}{2} \leq x < f_1^{(1)}(y)\} \qquad (W_{21} = \{(x,y) : f_2^{(1)}(y) < x < \tfrac{1}{2}\})$$

$$W_{12} = \{(x, y) : f_1^{(1)}(y) \leq x < f_1^{(2)}(y)\} \qquad (W_{22} = \{(x, y) : f_2^{(2)}(y) < y \leq f_2^{1}(y)\})$$

$$W_{13} = \{(x, y) : f_1^{(2)}(y) \leq x < 0\} \qquad (W_{23} = \{(x, y) : 0 < x \leq f_2^{(2)}(y)\})$$

(re. Fig. 3) such that

($S_1$) $TW_{11} \subset W_{21}$, $\quad TW_{21} \subset W_{11}$;

($S_2$) $TW_{13} \subset W_{11}$, $\quad TW_{23} \subset W_{11}$;

($S_3$) $TW_{12} \subset W_{12} \cup W_{13} \cup W_2$; $\quad TW_{22} \subset W_{21} \cup W_{23} \cup W_1$;

moreover, $T(\{(x, y) : x_1 = f_1^{(2)}(y)\}) \subset \{(x, y) : x = f_1^{(1)}(y)\}$

and similarly, $T(\{(x, y) : x = f_2^{(2)}(y)\}) \subset \{(x,y) : x = f_2^{(1)}(y)\}$.

($S_4$) The hyperbolicity condition on $W_{11}, W_{12}$ and $W_{21}, W_{22}$. More precisely let the Poincarè mapping, written in terms of the coordinates, have the form:

$$T(x, y) = (h_1(x, y), h_2(x, y))$$

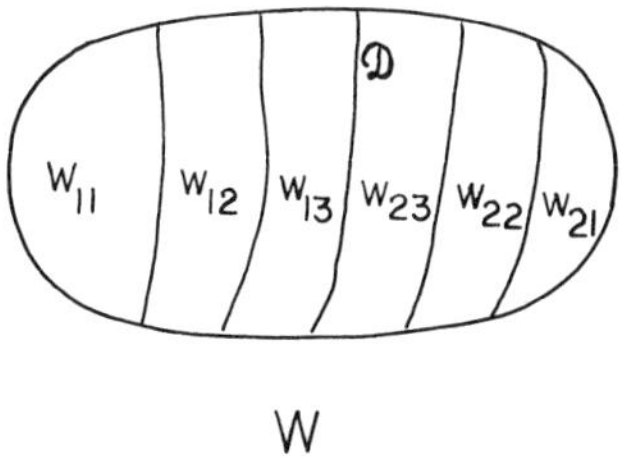


Fig. 3

We shall construct the Jacobi matrix

$$\partial T = \begin{Vmatrix} \dfrac{\partial h_1}{\partial x} & \dfrac{\partial h_1}{\partial y} \\ \dfrac{\partial h_2}{\partial x} & \dfrac{\partial h_2}{\partial y} \end{Vmatrix}$$

Then we can find constants $C$, $0 < C < \infty$ and $\lambda, \lambda < 1$, such that if $K^{(u)}(K^{(s)})$ is the cone of the two-dimensional vectors $e$ for which $|e_2/e_1| < C$, $(\frac{e_1}{e_2} < C)$, then for any point $(x, y) \in W_{11} \cup W_{12}) \cup W_{21} \cup W_{22})$ and any $e \in K^{(u)}$

$$\partial T(x, y)\, e \in K^{(u)}, \quad \| \partial T(x, y) e \| \geq \lambda^{-1} \| e \|$$

and for any point $(x, y) \in T(W_{11} \cup W_{12}) \cup T(W_{21} \cup W_{22})$ and any $e \in K^{(s)}$

$$\| \partial T^{-1}(x, y) e \in K^{(s)}, \quad \| \partial T^{-1}(x,y) e \| \geq \lambda^{-1} \| e \|$$

In principle, conditions $(S_1)$--$(S_4)$ show how the cells $W_{ij}$ are transformed under the action of $T$. Condition $(S_2)$ means that the cell $W_{13}$ is transformed into a domain lying in $W_{11}$ with $A_1$ within the boundary of this domain. The shape of this domain is determined by the character of the singular point of the vector field induced by system (1) of the stable separatrix $\Gamma^{(s)}(O)$. In case of a focus this domain will have the shape of a spiral mounted on $A_1$. In case of a tie it will be a wedge-shaped domain with $A_1$ serving as a vertex. The same is valid for $W_{21}$. Condition $(S_4)$ characterizes the stretching of the cells $W_{11}, W_{12}$ and $W_{21}, W_{22}$. A condition of this kind was introduced for the definition of Smale's "horseshoe" (re. [7]). It was also used in the well-known papers by V. M. Alexeev [8].

Theorem 1. Let properties $(S_1)$--$(S_4)$ be fulfilled. Then $\lim_{n\to\infty} T^n W = A$ is an attractor, i. e. , $TA = A$ and for each point $z \in W$ $\lim_{n\to\infty} \text{dist}(T^n z, A) = 0$. The attractor $A$ contains an invariant hyperbolic set for the points of which the dimension of the unstable subspace is positive.

From the topological point of view the invariant hyperbolic set which appears in Theorem 1 is a countable union of direct products of the Cantor perfect set by a segment. Fig. 4 represents the form $TW$, $T^2W$, $T^3W$ for the case when a tie is induced on the stable separatrix.

Statistical properties of attractors. The question of the statistical properties of invariant hyperbolic sets and of the relationship of these properties with the results of computer experiments is not simple. One of the most successful is the approach of Bowen and Ruelle [9], who consider any smooth probability measure $\mu_0$ on $W$ and then study its evolution $(T^*)^n\mu_0$. If the sequence of measures $(T^*)^n\mu_0$ has a weak limit $\bar{\mu}$ as $n \to \infty$ which is an invariant measure for $T$ with good mixing properties, then from the statistical viewpoint this corresponds to the fact that a random choice of initial data with the distribution $\mu_0$ averaged in time and various time correlation functions will be determined by the limit measure. Particularly, if with respect to the measure $\bar{\mu}$, $T$ is K-automorphism

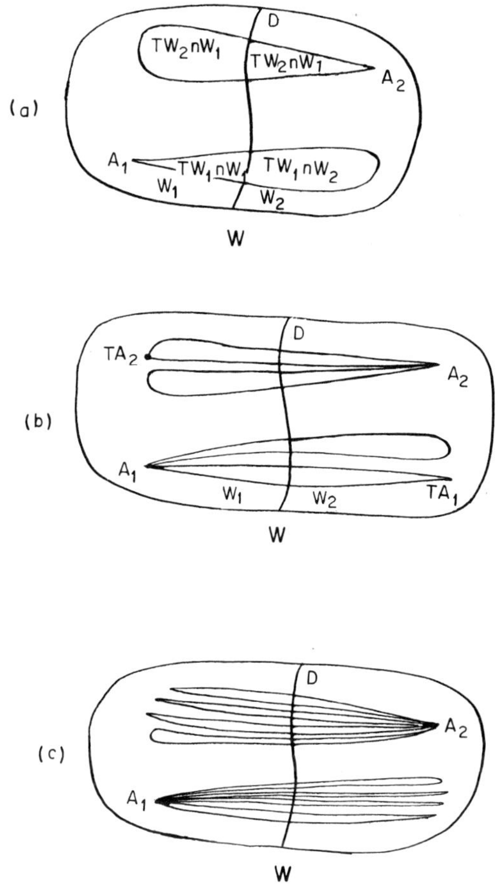


Fig. 4

(a) Form of TW, (b) Form of $T^2W$, (c) Form of $T^3W$

or B-automorphism, this will lead to the availability of continuous spectrum in T, which is often related to stochasticity. In the case under consideration condition $(S_1)$ entails that $T^2W_{11} \subset W_{11}$, $T^2W_{21} \subset W_{21}$, implying the existence of periodic points in $W_{11}, W_{21}$, which may be stable. Therefore it may happen that the image of the smooth measure is contracted into these points. At the same time, just as in the case of Smale's "horseshoe" (re. [5]), we shall consider the set of points $x \in W$ for which $T^k x \in W_{12} \cup W_{22}$ for any $k$. It is easy to show that this set can be represented as a continuous image of the space of binary sequences, and under this mapping the shift transforms into the Poincaré mapping. Hence it follows that $T$ is characterized by invariant measures with good properties of mixing, and by positive topological entropy.

Some remarks: 1. The definition of a mixer and the analogue of conditions $(S_1)$--$(S_4)$ can also be formulated for the multi-dimensional case. The necessary requirement is the existence of a fixed point with a stable separatrix of codimension I. Then we can study the Poincaré mapping of a plane of codimension I transversal to the separatrix. The stable separatrix divides this plane into 2 subsets and is their common boundary. The Poincare mapping is not defined on this boundary, and, moreover, is discontinuous in its vicinity, which may result in the appearance of the stochastic regime described above.

2. The similarity with Smale's "horseshoe" lies in the existence of a hyperbolic set of the Cantor perfect set type. One of the differences is that the set of "edges" is infinite in our case while in case of Smale's "horseshoe" it consists of two points.

Preliminary results of numerical experiments. We have performed a numerical experiment to determine the applicability of the above representation to the stochastic behavior of the Lorenz model observed in [1]. The Lorenz model is a system of three ordinary equations:

$$\begin{aligned} \frac{dx_1}{dt} &= -\sigma x_1 + \sigma x_2 \\ \frac{dx_2}{dt} &= rx_1 - x_2 - x_1 x_3 \\ \frac{dx_3}{dt} &= -bx_3 + x_1 x_2 \end{aligned} \tag{2}$$

Like Lorenz we shall consider the following values of the parameters: $\sigma = 10$, $r = 28$, $b = \frac{8}{3}$. System (2) has three fixed points: $(0,0,0)$, $(\sqrt{b(r-1)}, \sqrt{b(r-1)}, r-1)$, $(-\sqrt{b(r-1)}, -\sqrt{b(r-1)}, r-1)$, which are unstable at the chosen values of the parameters. The point $(0,0,0)$ has a two-dimensional stable separatrix on which a tie is induced and a one-dimensional unstable separatrix. We have investigated the Poincaré mapping which arises on the plane $z = 28$. It is more convenient to pass over to new coordinates $\overline{x}_1, \overline{x}_2, \overline{x}_3$ where $x_1 = \overline{x}_1 + \overline{x}_2$, $x_2 = (\lambda_1\sigma^{-1} + 1)\overline{x}_1 + (\lambda_2\sigma^{-1} + 1)\overline{x}_2$, $x_3 = \overline{x}_3$. In these variables the linear part of system (2) is diagonized. The further results will be represented in the variables $\overline{x}_1, \overline{y}_2, \overline{x}_3$. The one-dimensional unstable separatrix was constructed as a trajectory outgoing from the point $\overline{x}_1 = 0$, $\overline{x}_2 = 10^{-5}$, $\overline{x}_3 = 0$ in the negative $t$ direction. The trajectory was constructed in accordance with a certain modification of the Euler method with the step repeated in time $\Delta t = 10^{-3}$. It turned out that the first intersection of the unstable separatrix with the plane $\overline{x}_3 = 28$ occurred near the fixed point $(\sqrt{b(r-1)}, \sqrt{b(r-1)}, r-1)$, and the projection of this separatrix on the neighborhood of the intersection point has the shape of a spiral unwound from the fixed point (re. Fig. 5).

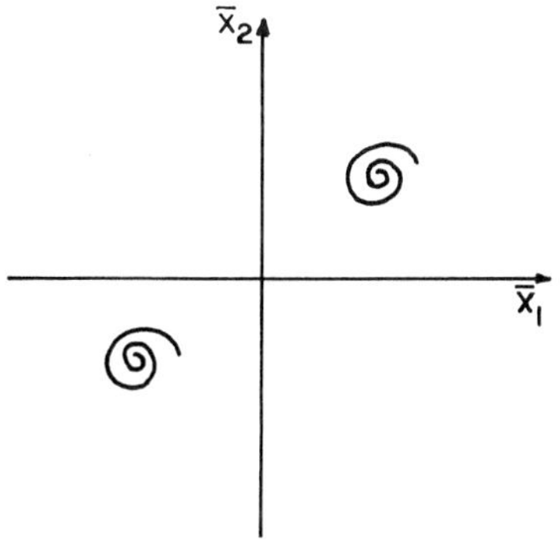


Fig. 5

The stable separatrix and its intersection with the plane $\overline{x}_3 = 28$ were found by two methods. Using the first method we considered the segment $\overline{x}_3 = 10$, $\overline{x}_2 = 0$, $|\overline{x}_1| \leq 1$ and for each $\overline{x}_1$ with the step $\Delta\overline{x}_2 = 10^{-4}$ we constructed trajectories until the first exit out of a certain previously chosen neighborhood of 0. The separatrix was defined by the fact that a change $\Delta\overline{x}_2$ across the separatrix gives rise to a change in the sign of the coordinate $\overline{x}_2$ at the moment of the first exit out of the above neighborhood. For the points thus found we constructed trajectories in the negative t direction until the first intersection with the plane $\overline{x}_3 = 28$. Using the second method, on the plane $\overline{x}_3 = 28$ we construct the Poincaré mapping of the cell $|\overline{x}_1| \leq 7$, $|\overline{x}_2| \leq 15$ with the step $\Delta\overline{x}_1 = \Delta\overline{x}_2 = 0,5$ and take only the points where the Poincaré mapping is disrupted. The first method is fairly traditional whereas the second method in all probability is much more efficient. The approximate intersection of the separatrix with the plane $\overline{x}_3 = 28$ is given in Fig. 6. A characteristic feature is the presence of two branches of the separatrix. Some other results show that the separatrix is a smooth curve which gradually reaches $\overline{x}_1 \sim 30$, $\overline{x}_2 \sim 5$ and then comes back. In Fig. 6 we see a branch of the separatrix after it comes back. The most probable candidates for the role of a mixer are the cells $2 \leq \overline{x}_1 \leq 4$, $4 \leq \overline{x}_2 \leq 13$; $-4 \leq \overline{x}_1 \leq 2$, $-13 \leq \overline{x}_2 \leq -4$. The results of the study of the Poincaré mapping of these cells will be published later. Recently very interesting results relevant to the Lorenz model have been obtained in the work of V. S. Afraimovitch, V. V. Bykov and L. P. Shilnikov [9]. We have also recently learned of the works [10, 11], which contain many statements relevant to the results of this paper (re. also [12, 13]).

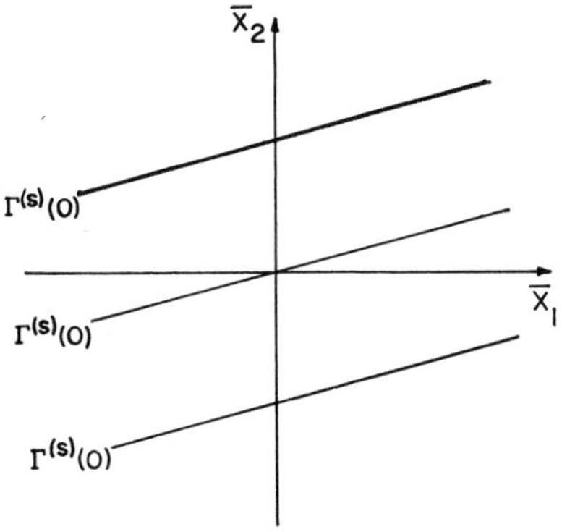


Fig. 6

## REFERENCES


1. E. Lorenz, Deterministic Nonperiodic Flow, Jour. of Atmos. Sci., v. 20, 130-141, 1963.
2. J. B. McLauglin, P. C. Martin, Transition to Turbulence of a Statistically Stressed Fluid System, Phys. Rev. A 12, 186-203, 1975.
3. D. Ruelle, The Lorenz Attractor. Preprint, 1976.
4. S. Ja. Gerzenstein and W. M. Schmidt, Concerning Interactions of Waves of Finite Amplitude in case of convection Instability, Doklady Adademii Nauk SSSR, V. 225, No. 1, 59-62, 1975.
5. S. Smale, Differentiable Dynamical Systems, Bull. Amer. Math. Soc., v. 73, 747-817, 1967.
6. D. Ruelle and F. Takens, On the Nature of Turbulence, Comm. in Math. Physics, v. 20, 167-192, 1971.
7. Z. Nitecki, Differential Dynamics, The MIT Press, Cambridge, Mass. and Lond, Eng., 1971.
8. B. M. Alekseev, Quasi-Random Dynamical Systems, Mat. Sbornik, v. 76, No. 1, 72-134, 1968; v. 77, No. 1, 545-601, 1968; v. 78, No. 1, 3-50, 1969.

9. V. S. Afraimovitch, V. V. Bykov, L. P. Shilnikov, Concerning the Appearance and the Structure of Lorenz Attractor, Doklady Akademii Nauk SSSR, v. 234, No. 2, 336-339, 1977.

10. R. Williams, The Structure of Lorenz Attractor. Preprint, 1976.

11. D. Guckenheimer, Strange Strange Attractor, Appl. Math. Sci., v. 19, 368-381, 1976.

12. Ya. G. Sinai, Stochasticity of Dynamical Systems, Proc. of Winter School "Nonlinear Waves", Gorki, 1977 (in print).

13. L. A. Bunimovitch and Ya. G. Sinai, Stochasticity of Lorenz Attractor, Proc. of Winter School "Nonlinear Waves", Gorki, 1977 (in print).